BIOLOGY AS SOCIETY, SOCIETY AS BIOLOGY: METAPHORS

Sociology of the Sciences

A YEARBOOK – VOLUME XVIII – 1994

Managing Editor:

R.D. Whitley, *Manchester Business School, University of Manchester*

Editorial Board:

Y. Ezrahi, *The Hebrew University of Jerusalem*
B. Joerges, *WZB, Berlin*
E. Mendelsohn, *Harvard University*
Y. F. Murakami, *University of Tokyo*
H. Nowotny, *Institut für Wissenschaftstheorie und Wissenschaftsforschung, Vienna*
T. Shinn, *Groupe d'Etude des Méthodes de l'Analyse Sociologique, Paris*
P. Weingart, *University of Bielefeld*
B. Wittrock, *SCASSS, Uppsala*

The titles published in this series are listed at the end of this volume.

BIOLOGY AS SOCIETY, SOCIETY AS BIOLOGY: METAPHORS

Edited by

SABINE MAASEN
University of Bielefeld

EVERETT MENDELSOHN
Harvard University

and

PETER WEINGART
University of Bielefeld

KLUWER ACADEMIC PUBLISHERS
DORDRECHT / BOSTON / LONDON

A.C.I.P. Catalogue record for this book is available from the Library of Congress

```
Biology as society, society as biology : metaphors / edited by Sabine
Maasen, Everett Mendelsohn, Peter Weingart.
       p.   cm.
    ISBN 0-7923-3174-5 (hardback : acid-free paper)
    1. Sociobiology.  2. Human biology--Philosophy.  3. Human
evolution--Philosophy.  I. Maasen, Sabine, 1960-      .
II. Mendelsohn, Everett.  III. Weingart, Peter.
GN365.9.B544   1995
304.5--dc20                                              94-36251
```

ISBN 0-7923-3174-5

Published by Kluwer Academic Publishers,
P.O. Box 17, 3300 AA Dordrecht, The Netherlands.

Kluwer Academic Publishers incorporates
the publishing programmes of
D. Reidel, Martinus Nijhoff, Dr W. Junk and MTP Press.

Sold and distributed in the U.S.A. and Canada
by Kluwer Academic Publishers,
101 Philip Drive, Norwell, MA 02061, U.S.A.

In all other countries, sold and distributed
by Kluwer Academic Publishers Group,
P.O. Box 322, 3300 AH Dordrecht, The Netherlands.

Printed on acid-free paper

All Rights Reserved
© 1995 Kluwer Academic Publishers
No part of the material protected by this copyright notice may be reproduced or
utilized in any form or by any means, electronic or mechanical,
including photocopying, recording or by any information storage and
retrieval system, without written permission from the copyright owner.

Printed in the Netherlands

TABLE OF CONTENTS

Introduction: SABINE MAASEN, EVERETT MENDELSOHN, PETER WEINGART / Metaphors: Is There a Bridge over Troubled Waters?   1

PART I: METAPHORS REVALUED

SABINE MAASEN / Who is Afraid of Metaphors?   11

LORRAINE DASTON / How Nature Became the Other: Anthropomorphism and Anthropocentrism in Early Modern Natural Philosophy   37

RAPHAEL FALK / The Manifest and the Scientific   57

ROBERT WOKLER / The Nexus of Animal and Rational: Sociobiology, Language, and the Enlightenment Study of Apes   81

PART II: "STRUGGLE"

PETER J. BOWLER / Social Metaphors in Evolutionary Biology, 1870–1930: The Wider Dimension of Social Darwinism   107

PETER WEINGART / "Struggle for Existence": Selection and Retention of a Metaphor   127

PART III: "EVOLUTION" AND "ORGANISM"

PETER M. HEJL / The Importance of the Concepts of "Organism" and "Evolution" in Emile Durkheim's Division of Social Labor and the Influence of Herbert Spencer   155

ANTONELLO LA VERGATA / Herbert Spencer: Biology, Sociology, and Cosmic Evolution   193

SANDRA D. MITCHELL / The Superorganism Metaphor: Then and Now   231

GREGG MITMAN / Defining the Organism in the Welfare State: The Politics of Individuality in American Culture, 1890–1950   249

PART IV: ECONOMICS

TIMOTHY L. ALBORN / A Plague Upon Your House: Commercial Crisis and Epidemic Disease in Victorian England   281

MARY S. MORGAN / Evolutionary Metaphors in Explanations of American Industrial Competition   311

GEOFFREY M. HODGSON / Biological and Physical Metaphors in Economics   339

METAPHORS: IS THERE A BRIDGE OVER TROUBLED WATERS?

SABINE MAASEN
University of Bielefeld

EVERETT MENDELSOHN
Harvard University

PETER WEINGART
University of Bielefeld

It would seem to be a simple and unexciting fact that metaphors, analogies, and models, are taken from one area of human thought and practice to another. They very term "analogy" denotes this process. And yet, at least with respect to science, this fact has attracted the attention of numerous authors in more than 6,000 books and articles throughout the last decades.[1] And the attention is by no means limited to the use of metaphors in science, i.e., between disciplines, but can also be found when they are transferred from science to the social and political arena and vice versa. Hence, although this transfer of concepts and ideas apparently is a customary feature of both scientific and nonscientific discourses, there must be something about this feature which elicits irritation. What are the reasons for this irritation? What accounts for the often passionate debates about the pros and cons of the use of metaphors?

First of all, our intent is not to duplicate the plethora of works on metaphors, and expound on fine-grained terminological distinctions between metaphors, images, analogies, models, rhetoric, and systems of thought. Whatever their significance may be for the linguist or the philosopher of science, in the context that is interesting to us here they are, by and large, the same phenomenon: the transfer of ideas and concepts, thus "pieces of meaning" from one delineable discourse to another. Thus, the term "metaphor" is used to represent all the other variants as well. The answer to the questions mentioned above does not lie in the conceptual distinctions but in the perceived *functions* of metaphors and whether in the concrete case they are judged positive or negative. The ongoing debates reflect these concerns quite clearly, namely that metaphors are judged on the basis of supposed *dangers* they pose and *opportunities* they offer. These are the criteria of evaluation that are obviously dependent on the context in which the transfer of meaning occurs. Our fundamental concern is indeed the transfer itself, its prospects and its limits.

Looking at possible functions of metaphors is one approach to understanding and elucidating sentiments about them. The papers in this volume illustrate, by quite different examples, three basic functions of metaphors: illustrative, heuristic, and constitutive. These functions represent different degrees of transfer of meaning. Metaphors are illustrative when they are used primarily as a literary device, to increase the power of conviction of an argument, for example. Although the difference between the illustrative and the heuristic function of metaphors is not great, it does exist: metaphors are used for heuristic purposes whenever "differences" of meaning are employed to open new perspectives and to gain new insights. In the case of "constitutive" metaphors they function to actually replace previous meanings by new ones. Sabine Maasen in her paper introduces the distinction between transfer and transformation. The most frequent example is the transfer of models from one discipline to another, where obviously the transfer of meaning is no longer illustrative or even decorous but can imply the redefinition of the subject matter. By becoming integrated, in the case of interdisciplinary transfer, into the body of theories, methods, and basic definitions of subject matter, the transfer may involve their transformation. If the transfer takes place from science into nonscientific discourses it may involve, at least in extreme cases, the replacement of worldviews.

These functions of metaphors do not only occur within science, i.e., between disciplines and specialties, but also between everyday discourse and specialized discourses. In fact, the whole issue of metaphors only arises between demarcated areas of meaning and language. The answer to the question of why metaphors cause irritation and excitement, may thus be summarized simply: as long as the perceived difference in meaning is slight, or serves the purpose to create additional insights, metaphors are judged to be good. As soon as the use of metaphors implies change of meaning, even change of the disciplinary identity, they are judged to be bad.

One area of scientific inquiry and political discourse where the dangers of metaphors are particularly conspicuous is in the relation between nature and society. Here the battle thunder rolls about *reductionism* and *reification*. "Biologization," i.e., the interpretation of social phenomena in terms of biological concepts and metaphors, has assumed pejorative connotations at least since Social Darwinism has become identified, rightly or wrongly, with the Nazis' race-hygiene programs and their catastrophic consequences. Although well entrenched in the academy's collective memory, a closer look at the history of biological metaphors in political ideology reveals the recency of the judgment that we are familiar with and hold to be self-evident. Before this traumatic experience occurred, biological concepts were used indiscriminately on all sides of the political spectrum with contradicting ideological attributions or none at all (see the articles by Peter Bowler and Peter Weingart in this volume). Indeed, as Lorraine Daston shows in her paper, the moral evaluation of anthropomorphisms, i.e., equating humans and animals (perhaps the most fundamental use of metaphor) has changed many times throughout the history of science, depending on the time-bound notions of what is considered the proper access to knowledge. Thus political, moral, and epistemological contexts have to be considered in order to evaluate the impact of metaphors which is the aim of the first section of this volume:

Maasen's paper gives an overview of competing theories of metaphors – pragmatic, semantic and constructivist – and attempts to integrate them into a model of innovation in science, thereby avoiding the futile alternatives that characterize the metaphor debate. By elaborating on the various functions of metaphors both within science and between science and other types of discourses this chapter can be read as a guide to the case studies that follow.

Daston, in a way, does the same from an historical perspective by linking the history of evaluations of anthropomorphism to that of epistemology in general. Again, by revealing the relativity of that connection in terms of the historical changes it sustains, one is forced to concede that any simple equation of metaphor and ideology is unwarranted.

Raphael Falk exemplifies a similar point with a particular case: the history of the gene concept is taken to demonstrate that metaphors or models have a dual function insofar as they open new perspectives and provide new insights, but at the same time constrain the scope of investigation.

Robert Wokler addresses continuities and discontinuities between Enlightenment scientists and modern sociobiologists in their attempts to show the uniqueness of our species and its links with all other creatures. Speculative anthropologists, particularly, were concerned with the articulation of images and metaphors as the divide between man and ape. In doing so, they showed caution in the use of interspecific transfers of metaphors, whereas sociobiologists, according to Wokler, all too commonly tend to reify social phenomena so as to explain them by innate, genetically coded, properties. Taken altogether the essays of the first chapter support a reevaluation of metaphors.

The following sections contain detailed historical studies that deal with the different ways in which central concepts of biological theory have been applied to explain social phenomena. However, the introduction of biological categories into the realm of social analysis has not been the only or even the primary direction of transfer. As is well known, Darwin drew on the social philosophy of his day when formulating his theory of "natural selection." Thus, the most influential biological theory in social thought was forged from social metaphors that were transferred into biology before they were reified as laws of nature and reapplied to the interpretation of society. The early eugenicists designated their field of research as a "social biology," a proximity of term of which modern sociobiologists are surely not aware. This was at a time, around the turn of the century, when no particular value judgment, least of all an ideological evaluation, was applied to the transfer of biological metaphors to the social sciences and vice versa. It was also a time when the boundary between the two disciplines, biology and sociology, was just being drawn and institutionalized. Once the distinction between the "natural" and the "social" had become available as a mode of self-observation of society, it developed its own dynamics, including the perception of the inappropriateness of describing one in terms of the other. Today, to describe nature as society and society as nature is at least inappropriate in categorical terms, and it is considered an ideological sin where the respective historical memory exists.

There may be some good reasons to discard the import of concepts, metaphors or models from science into political discourse as nothing but "ideological." Usually this is done in retrospect, by imputing, for example, that Darwin's selectionist theory was used to legitimize colonialism or that race theories served to support the feelings of identity of the emerging nation states. The same may be said for more recent developments: after race theories have been renounced and the perspective has shifted from blood ties to the individual, the discoveries of modern genetics are used to elucidate phenomena of social deviance. Even more fundamentally, genetics is combined with the paleontologists' find of "Lucy" to support the principle of human equality. This may be more palatable to the liberal academic community, but it is ideological nonetheless.

The conviction that the use of biological metaphors or models is ideological is seemingly corroborated by bouts of disciplinary imperialism such as E.O. Wilson's propagandistic declaration that "sociology and the other social sciences, as well as the humanities, are the last branches of biology waiting to be included in the Modern Synthesis."[2] But even though the battle over disciplinary demarcations drew remarkable public attention during the "sociobiology debate," the excitement in a few intellectual periodicals and literary reviews is hardly representative of developments and prominent concerns in science and society at large. As compelling as some cases may be, the ideology label does not get anywhere near providing and understanding of the fundamental importance of metaphors. Historical contingency and the empirical proof of a nondeterminist relationship between metaphors and their political use are reason enough to abandon this simplistic persuasion.

The debate over the ideological functions of "biologization" is only an indicator and a very special case of the fascination and irritation about the transfer of metaphors both between scientific disciplines and between scientific and nonscientific discourses, especially political ones. Why is there such a fascination with biological metaphors, with biological explanations and narratives in general? The facts that this particular theme is so prominent, so varied and stirs more excitement in the current discussion about metaphors than any other, and because it encompasses the interchange between the realms of science and politics, were the reasons to make it the focus of this book. Detailed historical case studies aim to show the tortuous paths by which metaphors enter and transgress a discipline, sometimes leaving the latter unimpressed, sometimes changing it markedly. The following studies focus on such concepts as "struggle" and "organism" as central concepts of biological theory that have an intricate record as metaphors in social theory and in societal discourses as well.

STRUGGLE: Peter Bowler tracks the Spencerian impact on Darwin's theory and the political origins as well as labels attached to them. Peter Weingart's paper gives a detailed study of the political impact of Darwin's "struggle" metaphor in late nineteenth and early twentieth – century Germany. Both essays show that contrary to common belief (though in accordance with more recent scholarship) the political label attached to Darwinism (and to Spencer) is empirically untenable. In fact, it may be said, as Weingart shows for the history of the "struggle" metaphor that its role as the Social Darwinist credo par excellence is a myth constructed by historians. More importantly, the point emerging from this is that any presumed determinism between biological metaphors and their ideological use does not hold water. Such a notion simply does not take into account the complexity of contextual variance.

EVOLUTION: Peter Hejl and Antonello La Vergata both explore the intricacies of the concept of "organism," which has had a complex career both within social theory and biology as well as between them. Hejl shows that concepts like "organism" and "evolution" played and still play a constitutive role in sociology shaped by historically, socially, and theoretically contingent experiences and contexts. Durkheim, too, in reaction to his reading of Spencer, elaborated on the "organismic model." According to Hejl, the usage of this metaphor should be conceived as part of tendency, common to social philosophy, economics, biology and sociology at the turn of the century, to refer to general pictures in order to describe composite entities, e.g., societies.

La Vergata inquires into the constitutive role of the metaphors "social organism" and "struggle for existence" in Spencer's thought. Both helped to construe reality in terms of spontaneous development and beneficient self-adjustment. They underscored Spencer's idea of a nature-culture continuum: biological, physical, and social phenomena were all governed by universal physical causation.

ORGANISM: Susan Mitchell is concerned with the "superorganism" metaphor which takes the individual organism as a model of functional integration of parts, i.e., to explore social groups of individuals. Yet, with respect to Wilson and Sober's attempt to revive this metaphor by applying it to social insects, it turns out that this is a case where a metaphor actually fails to reveal crucial features of social insect organization which this model was supposed to highlight.

Gregg Mitman studies the strong resonance between American political thinking from the early twentieth century onward and biological discussions of organisms. He considers this to be an important heuristic tool used by scientists in order to gain support of diverse audiences.

By placing biological discussions of individuals within a changing political discourse (from positive to negative freedom, from common good to individual right), Mitman reveals that these changes helped to create an environment conducive to a modern, genetic definition of the self.

ECONOMICS: Although biological metaphors, especially those of Darwinism, are commonly associated with economic life in particular, the actual import is much smaller than it would appear. Studies on Darwinism by Bannister (1979) and Kelly (1981) have already moderated the legend of the Darwinism-capitalism connection. Timothy Alborn explores explanations of and institutional responses to economical crisis throughout the Victorian period. Using medical concepts of cause and effect, framed into the language of epidemiology, Victorian economists called attention to a higher moral responsibility in the economic realm. At a period when biological and medical explanations were much less differentiated than today, it must have been obvious to conceive of an economic crisis as one of a society's ills.

Mary Morgan's study corroborates the assessment that Darwin's theory and metaphors did not have much impact on economic life but then goes on to identify a particular case where it did: American economists were concerned with the dynamics of monopoly and olygopoly. More or less restricted to the realm of large-firm competition, economists like Veblen took evolutionary ideas into account. With their help, contemporary economists analyzed both the behavior of firms in the market place and that of between-firm competition.

Geoffrey Hodgson looks at the history of the discipline of economics in order to trace the role played by evolutionary theory. Here, he focuses on the disciplinary fate of evolutionary metaphors that still have to compete against the prevalence of mechanistic metaphors derived from classical mechanics and physics before 1860. According to Hodgson, mechanistic metaphors are bound to entrap economics in equilibrium schemata: neither systematic nor cumulative errors can be perceived. Only recently has literature on economics evolution and technical change inquired into biological concepts, if mostly limited to the narrow focus on evolutionary selection mechanisms.

There is a consensus that can be extracted from this collection of historical and sociological studies. Metaphors are, first of all, unavoidable just as much as the use of language is generally. Thus, principled discussions about the use or avoidance of metaphors are moot. From this it follows that metaphors transport and determine meaning and, likewise, reshape or even cancel other contents. In short, they structure perception. So far this is neither original nor surprising. But it should also be taken seriously by anyone engaging in a dogmatic debate over positive and negative effects of metaphors. Their challenging impact is propagated by those in favor of pushing new models and concepts, and only time itself can distinguish between lasting innovation and shortlived fads. The dangers of reification, which are particularly ominous in science, and of its analogue in everyday and political discourses are foreboded by purists, rationalists and moralists who, in a way, fight a Don Quixotian cause against disciplinary opportunists, ideological cynics, and the nature of human communication.

One lesson to be drawn from these studies remains implicit, however. The flow of metaphors among science, politics, literature, and everyday discourses is continuous, recursive, and selective. At the same time, metaphors or parts of thought systems cannot be controlled by the context from which they emerge but, once they have left it, they begin to lead a life of their own, by way of use and interpretation in new contexts. Although the precise rules of this process (if there are any at all) are unknown, one can assume an underlying mechanism that accounts for the ongoing translation of meanings between different discourses and, thus, for the sociopolitical significance of (scientific) "systems of metaphors."[3] Thus, although the water is troubled, i.e., scientists still feel uneasy about metaphors, a bridge could well be built: the mechanism of "metaphor circulation," if decoded, could provide access to an explanation of the co-evolution of scientific theories and political "worldviews" – an explanation that reflects the complexity of the process as it is presented in the different chapters of this book.

Notes

1. Jean-Pierre van Noppen et al., *Metaphor. Bibliography of Post-1970 Publications* (Amsterdam, Philadelphia: John Benjamins Publishing Company, 1985); Noppen and Edith Hols, *Metaphor II. A Classified Bibliography of Publications, 1985–1990* (Amsterdam, Philadelphia: John Benjamins Publishing Company, 1990).
2. E. O. Wilson, *Sociobiology: The New Synthesis* (Cambridge, Mass: Harvard University Press, 1975).
3. George Lakoff and Mark Johnson, *Metaphors We Live By* (Chicago: University of Chicago Press, 1980).

I: Metaphors Revalued

WHO IS AFRAID OF METAPHORS?

SABINE MAASEN
University of Bielefeld

"Every one knows what is meant and implied by . . . metaphorical expressions, and they are almost necessary for brevity."[1]

Judging from this quote, Charles Darwin certainly was not afraid of metaphors and seemed to be convinced that no one should be. Against the background of such an innocent approach the recent flood of publications on the notion of metaphor appears to be rather mysterious. Indeed, a look at the bibliographies of metaphors compiled by Noppen and others in 1985 and 1990 provides a first insight into what Noppen calls the "metaphormania" of the intellectuals since the 1970s.[2] Although both bibliographies are not considered exhaustive, they already contain more than 6,000 entries; the recommendations for beginners alone amount to more than 200 entries. Not only is the sheer amount of publications impressive but so is the variety of disciplines and research areas covered by them: pertinent studies are to be found in linguistics, semiotics, rhetoric, literature, as well as in philosophy, psychology, sociology, history, political sciences, medicine, or artificial intelligence.

The moral of this enormous scientific effort in the analysis of metaphor is twofold. First, these bibliographies could be a reason for reconsidering the idea of contributing yet another analysis. Second, with respect to Darwin's remark, the bibliographies highlight the fact that Darwin referred to an everyday notion of metaphor. Meanwhile, however, this very notion has come under intensive investigation by various disciplines: "metaphor" has become established as a subject matter of scientific inquiry by itself. Because of this remarkable scientific interest one should not expect a shared understanding of "what is meant and implied by metaphorical expressions" as Darwin assumed. Would this then be a reason to be afraid of metaphors, particularly in science?

Especially with regard to evolutionary metaphors, Darwin would perhaps be surprised to see that some scholars in the social sciences – above all, historians and sociologists – still feel uneasy toward the use of evolutionary metaphors. In the light of German history any attempt toward a "biologization" of the social sciences became almost synonymous with racism: apart from these historical and political reasons sociologists also have general reservations about what they perceive as a biological reduction of their discipline. On the other hand, Darwin would see other scholars in the social sciences, namely psychologists and anthropologists, whose work is remarkably inspired by evolutionary metaphors: approaches such as "evolutionary psychology," "Darwinian anthropology," or "social physiology" all make use of biological concepts.

This highly ambiguous reaction to the transfer of concepts and images between biology and the social sciences coincides with the impression that the use of (evolutionary) metaphors in science is at the same time influential and inevitable.

Increasingly, it has become recognized that almost no scientific theory is a pure logical construct. It both takes root within a particular sociopolitical context and feeds back into it. Evolutionary metaphors have a particularly powerful influence on our humanistic perspectives which underpin all social and political action.[3]

Neither a rigid rejection nor an enthusiastic embrace of the use of metaphor in science is appropriate. Rather, in my view, the diversity of the above observations call for a change in perspective. The extent to which evolutionary metaphors have entered a variety of disciplines, where they have been both welcomed and strongly criticized, underscores the increasing urgency of an inquiry into which functions metaphors do in fact have for scientific theorizing, namely with respect to disciplinary innovations. Whereas almost nobody would deny that they are of *heuristic* value, it is a hotly debated issue whether they are of *constitutive* value for scientific reasoning.

Thus, the following inquiry will pursue two basic questions:
1) Are there any theoretical conceptions supporting the assumption that metaphors can indeed have a constitutive function for scientific theories?
2) How can such conceptions be theoretically mapped onto empirical cases of the transfer of metaphors between disciplines (as well as non-scientific discourse)?

My claim is that the use of metaphors *is* constitutive for scientific theorizing. To support this, I will briefly review three different, but connectable accounts of metaphors on a philosophy and sociology of science level. I also claim that we cannot avoid scientific discourses becoming "infected" by concepts of other disciplines and of nonscientific discourses, such as everyday communication, literary and political discourses. But, as I will argue, at least on a scientific level there are ways to control this kind of influence by metaphors. To this end I will propose that, as far as the process of scientific innovation is concerned, which is initiated by a transfer of metaphors, a distinction should be made between *transfer* and *transformation*. Whereas the transfer of a (metaphorically formulated) concept has a heuristic function and inspires the scientist to generate hypotheses, the subsequent transformation refers to the discipline-specific processing of that metaphor with the help of the existing corpus of theories, methods, and objects.

This proposal is intended to provide a preliminary answer to the question of how to make use of metaphors in science without being afraid of inaccuracy or unwarranted reductionism, to mention only the most frequent objections.

On the one hand, such a model of distinguishing between transfer and transformation would adequately reflect the observation that, rather than scientific revolutions, it is successive modifications that guide disciplinary innovation. From this perspective, the fruitful processing of the stimuli provided by metaphors might be conceived of as one of the pertinent procedures. On the other hand, only from this angle should one discuss the prevalence of certain metaphors in science and their interdependence with social and political phenomena. The transfer/transformation model could be applied to the movement of metaphors between different types of discourses (e.g., literature and everyday life, in addition to the sciences) and their specific metaphoric applications (poetic, phrasal). This aspect is especially important because only through the continuous diffusion of scientific metaphors in literature[4] and everyday communication, and its feedback on scientific discourse, does a political and social significance of metaphors – or, as Lakoff and Johnson called them, "system of metaphors"[5] – become established. My ultimate conclusion will be that there is really no reason to be afraid of metaphors.

On the Scientific Use of Metaphors: Semantic, Pragmatic, Constructivist Theories

Semantic Theories of Metaphors

The writings of Max Black (1962)[6] and Mary Hesse (1964)[7] run counter to the widely held view (obviously influenced by a certain literary usage) that a metaphor is something purely decorative – "the propose is to entertain and divert."[8] Black's and Hesse's works reflect an early interest in the scientific relevance of metaphors and the relationship between models and metaphors. Hesse is in favor of supplementing the deductive model of a scientific explanation with a view that decribes "theoretical explanation as a metaphoric redescription of the domain of the explanandum."[9] In this respect she joins Black's interactionist view of the metaphor: it states that a metaphor's impact could be attributed to the fact that the phenomena of a primary system are described in terms of a secondary system. Such a transfer might presuppose a certain similarity or comparability of phenomena and their associate meanings. According to Black, however, it is the metaphor that *evokes* this comparability and subsequently contributes to the domains of meaning that reciprocally influence each other. For example, to say "man is a wolf" would make man more wolf-like, but would also make the wolf more human. This effect is a result of the selectivity of the domains of meaning that works reciprocally for the latter:

Any human traits that can without undue strain be talked about in 'wolf-language' will be rendered prominent, and any that cannot will be pushed into the background. The wolf-metaphor suppresses some details, emphasizes others – in short organizes our view of man[10]

– and vice versa.

The continued use of a metaphor leads to a further dynamic development: ". . . an expression initially metaphoric may become literal (a 'dead' metaphor), and what is at one time literal may become metaphoric."[11] However, Black does not answer the question as to what makes some metaphors "successful," i.e., those that change the meaning of their objects and eventually acquire a literal meaning, and others fail.[12]

Generally speaking, one could say that, according to Black's interaction view, a metaphor functions almost like a pair of glasses through which the metaphoric object is observed, i.e., reorganized. Those metaphors which turn out to be successful eventually establish a privileged perspective on an object or constitute "the" object and by doing so, disappear *as* metaphors.

Black's theory of metaphors operates mainly on the level of semantics – where meanings are shifting or individual aspects take on a different weight. The critical test would be a "translation": Black concedes that, yes, the "meaning" of a metaphor could be translated literally, but the translation would always remain deficient.

> . . . the implications, previously left for a suitable reader to educe for himself with a nice feeling for the relative priorities and degrees of importance, are now presented as explicitly as though having equal weight. The literal paraphrase inevitably says too much – and with the wrong emphasis. One of the points I most wish to stress is that the loss in such cases is a loss in cognitive content.[13]

Some authors, such as Andrew Ortony, do not associate Black with a predominantly semantic theory of metaphors. In his volume *Metaphor and Thought*,[14] he lists Black as a representative of the pragmatic theory of metaphors: "Black's conclusion that metaphors are recognized as such by users of the language in particular contexts, places metaphors in the domain not of semantics – the study of meaning – but of pragmatics – the study of the speech acts and the contexts in which they occur."[15] But even if Black assumes there exists a "tacit knowledge" of the difference between the literal and the metaphoric use of a metaphor, he still considers the meaning of the metaphor to be the crucial phenomenon that needs to be explained.

So the writer is employing conventional means to product a non-standard effect, while using only the standard syntactic and semantic resources of his speech community. *Yet the meaning of an interesting metaphor is typically new or "creative," not inferable from the standard lexicon.*[16]

Pragmatic Theories of Metaphors

A claim diametrically opposed to Black's is made by Donald Davidson. His "thesis is that metaphors mean what the words, in their most literal interpretation, mean, and nothing more."[17] Davidson's theory of metaphors is not primarily based on semantics but on pragmatics; a "metaphor belongs to the domain of use. It is something brought off by the imaginative employment of words and sentences and depends entirely on the ordinary meanings of those words and hence on the ordinary meanings of the sentences they comprise."[18]

Samule R. Levin, too, is of the opinion that the words of a metaphor should be taken literally, but that these would not constitute the entire meaning of a metaphor,

> It is in this taking of the meaning, this processing of the linguistic expression, that the metaphoric work consists – in the relations, resemblances, and analogies that the reader, in the course of processing the metaphor, imposes upon or draws out of the expression's literal meaning. The interpretation of the metaphor can thus be said to grow out of the literal meaning of the expression, but it is not, as it is on my approach, an interpretation of that meaning.[19]

But what is the goal of "processing" a metaphor? To answer this question, Levin introduces a distinction between "actual" (=ordinary) and "deviant" (=e.g., metaphoric) usage of language as well as between the "actual" world and "deviant" (=possible) worlds According to Levin, deviant usage starts from the literal meaning of the metaphor, but since this is incompatible with the actual world, it would refer to the construction of a deviant, possible world: "On this approach, in other words, what is metaphoric is not the language, but the world.[20] In terms of this approach the metaphor is the key to an alternative worldview; in Levin's terminology we are dealing with a "conception" he regards as the result of mental states or precesses.[21]

Analogously, the following could be said with respect to science:

> If conceptions are expressed in a semantically deviant language, it follows that the state of affairs thereby conceived of will in some sense lie beyond conventional notions of how the world is constituted. In this respect metaphoric conceptions resemble conceptions that arise in the progress of science, where such conceptions also entail states of affairs (aspects of nature) hitherto unthought of.[22]

But to talk of scientific metaphors here would be out of place. The author points to some important differences between scientific and literary conceptions, using the notion "conservation of energy" as an example.

> We pointed out that the new meaning gained for the notion of conservation was established only with the thinking out and formulation of a comprehensive theory of thermodynamics. Once the theory is formed the extended or modified meanings of the words assume fixed, well-defined positions in the semantic field. But behind this process there typically lies a series of experiments

and the testing of various hypotheses. During the course of these procedures the new meanings are inchoate and tentative. It is only with the formation of this theory that the shifts of meaning actually are accomplished. And once this state is achieved, the words in their modified meanings take standing positions in the semantic, and correlatively, in the conceptual field. As such, they represent new concepts. The theory then defines relations between these concepts. All this is unlike what occurs in metaphor. In the first place, on my analysis of metaphor there is no extension of meaning. The novelty of the expression consists in the hitherto unconceived of relation that it projects between preexisting concepts. The novelty is thus the conception, not the concepts. . . . in science the terms, that is, the concepts, are new, the novelty of the relation being a concomitant and automatic consequence. In metaphor the relation is new, but since the concepts related belong to different conceptual subspaces, there is nothing automatic about the relation.[23]

Starting from examples taken from natural science Levin reconstructs innovative effects as a discovery of new concepts which correct or supplement existent knowledge. In this Levin is guided by a correspondence theory of truth: a phenomenon and its contextual linkages once discovered can then be addressed directly by questions concerning scientific truth. In Levin's theory the scientific use of concepts apparently figures as a counterconcept force of the metaphor in literature: compared to the generation of an unlimited number of alternative worlds, the discovery of the First Law of thermodynamics and its application to different natural phenomena can only be a disciplined variant that always relates to the one reality. The reality of the actual world and the discipline-specific truth that has been discovered about it constitute criteria which, by definition, forbid the use of metaphors for scientific purposes: alternative worldviews as well as linguistic and social construction of conditions of truth are not envisaged by Levin. According to a pragmatically-oriented theory, metaphors are suited to literature rather than science.[24]

Constructivist Theories of Metaphors

At this point, a stocktaking of the accounts previously discussed seems to be appropriate: which functions do they ascribe to metaphors with regard to innovative scientific theorizing?

The semantic accounts suggest that the interaction of a metaphoric construct with the existing corpus of disciplinary objects, theories, and methods may bring a cognitive gain because a metaphor reorganizes an object and, in this introduces new questions. It remains unclear, though, what the introduction of a metaphor stimulates and what makes some metaphors successful in terms of an innovative impact on a discipline. The pragmatic accounts suggest that such an effect is brought about by the use and "processing" of constructs that have so far not been part of a discipline. It remains unclear, though, as to whether such a transfer would be legitimate in the scientific domain whose aim is the search for truth.

In this connection, toward the end of his essay "What Metaphors Mean," Davidson insists on distinguishing between "seeing as" and "seeing that,"

> Metaphors make us see one thing as another by making some literal statement that inspires or prompts the insight. Since in most cases what the metaphor prompts or inspires is not entirely, or not even at all, recognition of some truth or fact, the attempt to give literal expression to the content of the metaphor is simply misguided.[25]

This statement differentiates sharply between pragmatism and constructivism, as represented by Richard Rorty and "Laboratory Studies," a school of thought in the sociology of science. The aim here is the (social) construction of scientific facts, i.e., of criteria that determine whether facts are to be considered true.

> We must accept that there are no adequate grounds for establishing criteria of truth except the grounds that are employed to grant or concede it – truth is conceivable only as a socially organized upshot of contingent courses of linguistic, conceptual, and social courses of behavior.[26]

On this basis Rorty has reformulated Davidson's argument: the difference between the literal ("dead" metaphor) and the metaphoric corresponds to a distinction between familiar and unfamiliar noises and marks. Unfamiliarity, states Rorty, is always correlated with a need for theory. "The literal uses of noises and marks are the uses we can handle by our old theories about what people will say under various conditions. The metaphoric use is the sort which makes us get busy developing a new theory."[27]

However, this new theory would have to be more than a formal definition of the new meanings of old words (i.e., a metaphor), "for that would only be a new combination of old ideas. It is given by "relational univocity," that is, by the reaction of the context with the old sense. A strange context forces the word to take a new sense."[28]

In terms of this argument the tranfer of concepts and models of one discipline to another becomes conceivable. If we follow Rorty, this would be a process of replacing an "old vocabulary" with a "new" one. This process

works holistically and pragmatically. It says things like "try to think of it this way" – or more specifically, "try to ignore the apparently futile traditional questions by substituting the following new and possibly interesting questions." It does not pretend to have a better candidate for doing the same old things which we did when we spoke in the old way. Rather, it suggests that we might want to stop doing those things and do something else. But it does not argue for this suggestion on the basis of antecedent criteria common to the old and the new language games. For just insofar as the new language is really new, there will be no such criteria.[29]

However – and this is my working hypothesis – such a radical view would require a modification that is oriented to the *process of scientific activity*. Among analyses advocating this perspective are the microsociological studies conducted by Karin Knorr-Cetina. One such example, which I will outline below, is her work "The Scientist as Analogical Reasoner: A Critique of the Metaphor Theory of Innovation."[30]

On the basis of laboratory observations in natural science research and the "ethnotheories" of scientific innovation proposed by the scientists involved, Knorr readjusts the role of metaphor from a process-oriented perspective. To begin with, she considerably reduces its role: in the context of the regularly occurring analogical reasoning, says Knorr, the metaphor only represents a special case of a figurative relation of similarity. In actual fact, however, the imaginative stimulus contains a persuasive force that should not be underrated,[31] because scientists

link innovation to the making of analogies in a much more general sense than postulated by the metaphor-account of innovation. When scientists were asked to tell the story of the origin of an idea or of a research effort which they considered to be innovative, they regularly displayed themselves as analogical reasoners who built their "innovative" research upon a perceived similarity between hitherto unrelated problem contexts.[32]

Presenting quotations from various interviews, Knorr shows that such relations of similarity lead to a transfer of methods and procedures as well as to a transfer of solutions already obtained in other fields. "In the process of extension, this knowledge regularly becomes modified according to the particularities of the new situation, and the result of this process will transform the interpretation of the original situation.[33]

Knorr thus follows the model of interaction between metaphor and "metaphorized" object as developed by Black and Hesse; she misses, however, an answer to the question as to when an innovative idea normally occurs and what connection it has to previous research. On these aspects, which are crucial for a constructivist sociology of science, semantic and pragmatic metaphor-accounts of scientific innovation remain silent.

By contrast, the perspective of the constructivist sociology of science, which is directed to the process of scientific production, considers the everyday aspects of practical scientific (here: natural science) research which is oriented to "feasibility" and "success." From this angle we note that analogies have already figured as "solutions" in the transferring field (and have indeed been present on an ethnotheoretical basis[34]) and, as "hypotheses" for a new field, promise a successful transfer to the latter. The goal, as Knorr puts it, is "to make a solution work." This goal refers us to scientific work as "a process of production whose products are specified by what can be done."[35] Therefore, the character of scientific innovation inspired by analogical reasoning is called "conservative" by Knorr: scientific work starts from a "solution" so far not realized, translates the latter into a promising "hypothesis," and in doing so relies on available knowledge.

The social-constructivist conception of the scientifically innovative importance of metaphors accounts for the aspects emphasized both by semantics and by pragmatics. The interaction between metaphor and "metaphorized" object (Black, Hesse) emphasized by semantic, in this sense presents itself as a reciprocal reconstruction of concepts and methods between different fields or disciplines. The theory-inducing effect stressed by pragmatics, which is the result of an unconventional use of noises and sign (Davidson), presents itself as a translation of successful constructs into promising hypotheses.[36]

Both procedures have become constituents of the routine "displacement of concepts."[37] They have become procedures of the process of scientific research aimed at success and feasibility. These procedures generate relations that are subject to constant change: in this sense, the metaphor, whose role had first been limited to the special case of a figurative relation of similarity, now assumes a paradigmatic significance for a constructivist theory of science. The similarities suddenly recognized by a scientist

> include an element of decision and persuasion, and consequently also of change. In that sense the similarities which underlie a metaphor or an analogy are complex rather than primitive, fragile and temporary rather than basic and stable relationships. Because of their figurative character, metaphors show this perhaps more clearly than literal interpretation.[38]

In the light of recent studies of scientific innovations (e.g., Rheinberger, Latour, and Lenoir),[39] this perspective on metaphors appears to be part of a broader project which might be entitled "the role of basic ideas in the epistemology of innovative scientific practice." Rheinberger is convinced that the construction of basic terms is intertwined with experimental practices in such a way as to produce something new. He agrees with Lenoir: "The very construction of the concepts is intertwined with the practices which operationalize them, give them empirical reference, and make them function as tools for the production of knowledge. In order to develop new conceptual structures, it is necessary to expand, alter and reshape the body of empirical practices."[40] The role of metaphors, as of basic ideas in general, consists in creating conditions for the emergence of something which Rheinberger calls a *not anticipatable event* ("unvorwegnehmbares Ereignis").[41] Unlike Knorr, Rheinberger follows Latour in that the eventual result of scientific practice guided by basic ideas, or metaphors in particular, is neither entirely determined by natural laws *nor by social constructions*. Rather, innovative scientific research takes place within the quadrature of theory and practice, nature and society: scientific innovations are the result of a series of displacement, *partly independent* of the procedures that brought them about in the first place. Hence, it turns out that the major objection against metaphors ("they are bound to be replaced in the course of scientific research") is really a statement about the fate of scientific tools in general, notwithstanding the fact that it applies to metaphors more obviously.

Transfer and Transformation of Metaphors: A Conservative Model of Innovation

In my view, the semantic, pragmatic, and constructivist accounts of metaphor each highlight a crucial aspect of its role for scientific research, namely scientific innovation. Constructivist accounts contribute the idea that metaphors are introduced in order to gain access to a problem when disciplinary tools are conceived of as insufficient. In other words, constructivist accounts stress the point that metaphors are concepts from other disciplines or research areas which promise to be successful solutions for disciplinary research problems. Pragmatic conceptions contribute the idea that metaphors as unfamiliar elements of a disciplinary language game may eventually evoke the discovery of a new aspect of the phenomenon in question. In other words, pragmatic accounts stress the point that metaphors have to be 'processed' with disciplinary tools. Semantic accounts of metaphors contribute the idea that this processing takes place in an interactive way: namely, the metaphorically introduced concepts or methods interact with the concepts or methods already being part of the discipline in question. In other words, semantic accounts stress the point that metaphors and disciplinary tools reorganize each other.

These different accounts of metaphor thus connected permit a more sophisticated notion of my thesis that metaphors play a constitutive role in scientific theorizing. A further elaboration of this thesis should consist in a systematic account for the *variety of ways* in which this constitutive role occurs.

To that end I suggest a thesis according to which the scientific use of metaphors should be regarded as a process characterized by two forms or stages: *transfer* and *transformation*. The impossibility of accounting for an object with the aid of available disciplinary tools may "prompt" the researcher to resort to a metaphor, or in view of a metaphor the scientist or a school of thought might notice a challenge to their disciplinary object or tools. Thus, the metaphor may offer information on a possible approach to the problem concerned and its possible solution. To begin with, the *transfer* of a metaphor leads to a reorganization of the phenomena and thus to a novel way of problematizing their study. In this form, or at this stage, metaphors occur as heuristics. This kind of usage is considered primarily in terms of their pedagogical value: are metaphors capable of illustrating unfamiliar connections through familiar ones? If so, this novel way of problematizing extant phenomena may give rise to the expectation that this might by done more successfully with the procedures, concepts, or questions suggested by the metaphoric construct. A further, though not a necessary, step might be the *transformation* of the metaphor in question. This would proceed along disciplinary tracts that are characterized by specific methods and/or a corpus of theories and objects. A transformation of a metaphor would thus aim at a full-fledged investigation of the imported concept.

For example, one would not only say "oh, this protein really looks just like sand" (an example by Knorr), but also try to explore further similarities between the protein and sand, using discipline-specific procedures to test these similarities, i.e., "hypotheses." Another example: if as a social scientist one is intrigued by the idea that "Social Systems Evolve as Species Do" (William H. Durham), one might want to explore the notions associated with the "social-evolution" metaphor in greater detail. This metaphor might generate new kinds of questions about one's anthropological material and thus reorganize it.[42]

In terms of the cognitive significance of a metaphor, this proposal follows Susan Haack; according to her, metaphors, initially rich but vague in content, are "ladders that science must aim to kick away."

> It is precisely because metaphorical statements are unspecific or open-textured that they are apt for representing novel conjectures in their initial and undeveloped stages, and for prompting investigation of what might be significant respects of resemblance. But eventually, of course, the goal is that new conjectures be made precise and rigorous, and hence are capable of test and appraisal.[43]

As the metaphor fades (via scientific transformations), its initial polysemic character is rendered fixed, i.e., it becomes less ambiguous in terms of the similarity to the existing disciplinary concepts. Hence, in the course of disciplinary transformation the cognitive significance of a metaphor is bound to diminish.[44]

The ensemble of possible procedures of transformation could be visualized along a *continuum* whose poles are marked by a low, or high, degree of integration. Procedures that require a relatively small integrative effort are those in which metaphors (models) are fitted into disciplinary "gaps." Procedures requiring a comparatively large measure of integrative effort are those that process metaphorically inspired innovations to make them into disciplinary constructs, or that engage in the differentiating of interdisciplinary fields. But all procedures along this continuum of possible integration are characterized by the "filtering" of a metaphor's innovating impact: this "filtering" is done by the importing discipline with the aid of its existent set of objects, theories, and procedures. Even in a case where a metaphorically introduced construct eventually leads to the reconceptualization of a disciplinary object, to the application of a new approach, or to the development of a new theory, what is usually required first, is that one refers to the discipline-specific vocabulary. At the stage of transformation, in other words, a scientist or a school of thought looks for ways to make use of a metaphor and to control the kind of influence it has by assigning to it a circumscribed role in the importing discipline. At this point, different schools of thought may have different views on whether and how a metaphor should be processed. Thus, the stage of transformation is likely to be a stage of intensive debate which eventually may result in a paradigm shift but the process of transformation may also reveal that the metaphorically introduced construct is misleading or inadequate. One possible reason for "failure" might be that the metaphor in question cannot be processed with disciplinary tools (i.e., the given data, methods, etc.), or that the disciplinary environment does not lend itself to the required modification.

These considerations suggest a view that, in terms of the innovative impact of metaphors, is at the same time more radical and more conservative than the one proposed by Rorty. Such a view is more radical because it claims that even the heuristic use of a metaphor will inevitably lead to perspectival reorganizations, if applied to the discipline in which this metaphor is supposed to illustrate something; and it is more conservative because it grants to the discipline and its corpus of objects, methods, and theories sufficient stabilizing power to ensure that metaphorically stimulated innovations predominantly remain within the framework of disciplinary differentiation and modification. All these forms of discipline-specific transformations accomplish what, following Levin, may be called the "processing of metaphors."

Thomas Kuhn seems to support the idea that the distinction between transfer and transformation is not only applicable to individual concepts but also to models. He distinguishes metaphors from so-called "metaphor-like processes" on whose effectiveness models are based. According to him, models, in addition to possessing a pedagogical-heuristic value, are also based on an interactive process that produces similarities with the "primary system." He gives the following, illustrative example:

Bohr and his contemporaries supplied a model in which electrons and nuclei were represented in tiny bits of charged matter interacting under laws of mechanics and electromagnetic theory. That model replaced the solar system but not, by doing so, a metaphor-like process. Bohr's atom model was intended to be taken only more-or-less literally; electrons and nuclei were not thought to be exactly like small billiard or Ping-Pong balls; only some of the laws of mechanics and electromagnetic theory were thought to apply to them; finding out which ones did apply and where the similarities to billiard balls lay was a central task in the development of the quantum theory. Furthermore, even when that process of exploring potential similarities has gone as far as it could (it has never been completed), the model remained essential to the theory. Without its aid, one cannot even today write down the Schrödinger equation for a complex atom or molecule, *for it is to the model, not directly to nature, that the various terms in that equation refer.*[45]

The constitutive role of models is demonstrated in this extreme example by the fact that they remain irreplaceable: they do not successively turn into discipline-specific constructs or methods, but the latter remain oriented to the model. The "primary system" almost leaves a blank spot in this case (although we should not refrain from disciplinary procedures to process this model). With respect to this variant, "transformation" means an addition to the discipline-specific tools.[46]

The transformation of individual metaphors may thus occur in the form of different variants: as translation into the discipline-specific vocabulary, as filler of disciplinary gaps, even as differentiation of new research fields. Following Darden and Maull, the last-mentioned variant of the transformation of metaphorically stimulated reconceptualizations can be made plausible with the help of the so-called "interfield theories."[47] These theories connect research fields that cannot solve their problems using their own instruments and, through this connection, introduce new research methods, objects, and results. Such theories become established in conceptual interfields, for "fields," contrary to "disciplines," emphasize the conceptual rather than the institutional aspects of research. It is thus quite possible that conceptual differentiations proceed within the institutional framework of one discipline, but also, that they lead to the establishment of an "inter-discipline."

However, for all these variants, "transformation" does not, as a rule, imply a replacement of disciplinary vocabulary, as assumed by Rorty, but rather a successive fruitful application within the framework of disciplinary procedures. This, too, is a complex process that may actually have far-reaching consequences, but need not necessarily blur disciplinary boundaries and perspectives, though it may cause them to shift.

The distinction between transfer and transformation gives rise to a conception according to which a metaphor challenges the concepts and perspectives of a "target discipline" to interact with it. This also has an impact on the imported construct: the discipline-specific transformation of the construct extinguishes the specifics of the "exporting" discipline previously attached to the metaphor: the metaphor is integrated into the disciplinary language games of the "target discipline."[48] If this thesis is correct the following result, although counter-intuitive, would most likely be that the exchange of concepts and models does not abolish disciplinary boundaries, but stabilizes them. In the case of a "mere" transfer, metaphors remain visible as disciplinary "intruders;"[49] in the case of transformation, the construct is gradually absorbed into the disciplinary corpus, modifying the latter in the course of this happening.[50]

On the Transfer of Metaphors Between Science and Other Types of Discourse

The discussion so far may have given the impression that metaphors are limited to the scientific and literary domains and that, what at a certain time is conceived of as an "adequate" metaphor, somehow comes out of the blue. But in fact metaphors are constituent parts of all discourses or of "inter-discursive" exchange. Some recent studies claim that even everyday language is interspersed with metaphors, or systems of metaphors. Lakoff and Johnson, for instance, hold that "our ordinary conceptual system, in terms of which we both think and act, is fundamentally metaphoric in nature."[51] A similar stance is defended by Hesse whose recent contributions to this issue include the idea that "all language is metaphorical."[52] In this, she relies on Wittgenstein's concept of "family resemblances": especially with regard to shifts of meaning undergone by predicates applied in family resemblances classes, she draws attention to the similarity with respect to metaphoric shifts of meaning. Both are context-dependent with respect to perceived similarities and differences. Metaphoric shifts of meanings are only "the more striking examples of something that is going on all the time in the changing and holistic network that constitutes language."[53]

Everyday discourse is not only itself largely metaphorically structured, but also provides a privileged source for a great many metaphors that are passed on to other types of discourse. These include scientific discourses as well as nonscientific, albeit stylized, discourses, such as literary or political ones. The linkage of all these different types of discourses is made possible, inter alia, through the transfer of metaphors; within the general societal discourse, the scientific discourse constitutes only one aspect of the cultural enterprise.

Appreciating the role of models and metaphorical thinking in science will make it possible to see that science really is a cultural enterprise, continuous with other aspects of our culture. . . . Models are often drawn from technology – craftsmen, machines, clocks, telephone exchanges, computers – and as technology develops the models will change. Such models play a role not only in science, but also in the humanities, in art and religion. As the models come and go, theories change – in science, art, religion, management, politics.[54]

Particularly "successful" metaphors recur in all types of discourse; what is more: they actually stimulate a permanent feedback between them. Gillian Beer gives an illustrative example of the influence of Darwinist metaphors on everyday discourse.

We now live in a post-Freudian age: it is impossible, in our culture, to live a life which is not charged with Freudian assumptions, patterns for apprehending experience, ways of perceiving relationships, even if we had not read a word of Freud, even – to take the case to its extreme – if we have no Freudian terms in either our active or passive vocabulary. . . . This was the nature also of Darwin's influence on the generations which succeeded him. Everyone found themselves living in a Darwinian world in which old assumptions had ceased to be assumptions, could be at best beliefs, or myths, or, at worst, detritus of the past.[55]

The effect of a "Darwinization of our life world" can be taken as an example of the recent permeability of all types of discourse for biological metaphors. An uncritical permeability, however, may also entail negative consequences, according to Hubert Rottleuthner. Referring to the use of biological metaphors in legal thought he makes the following distinction, using as examples the terms "organism" and "development": on the one hand, it was possible that the metaphors referred to some biological aspect; but it might as easily be just a figure of speech.[56] The author concedes, though, that a distinction of this kind between the intentional or merely phrasal use of biological metaphors is difficult to make. But he rightly points out that metaphors do not "belong" to any particular discipline. Studies in the history of science and the history of concepts would show that individual constructs came to be *assigned to* a particular discipline and how they were subsequently *made into* metaphors.

But the distinction between the intentional and the merely phrasal use of metaphors is interesting also from a discourse-analytical perspective, not because it refers to the orthodox or even the merely heretical usage of metaphors. On the contrary, the above outline has already shown that the emergence, use, and significance of metaphors is effectuated by an ensemble of different discourses and their various, specific (always meaning deviant) applications. Empirically, it can be shown, though, that different discourses and types of discourse use metaphors in a multitude of ways, for example phrasally, illustratively, or as "theses."

To illustrate this conception, one might visualize it in terms of a coordinate graph: everyday life, literature, science, politics, economics, and law would be examples forming the axis of discourses involving a lively exchange of metaphors; the other axis would represent possible forms of use. Such a graph would enable an observation of the transfer of concepts, methods, and models, i.e., the construction of (scientific) metaphors and their fate. As Ho and Fox have stated, metaphors can actually be found in several places in the coordinate system at the same time. Constructs such as "struggle for survival" do not only have an hypothesis-generating effect that operates between disciplines, but can also function as slogans in political jargon. In such cases metaphors function as interfaces of different, discourse-specific forms of usage.

This aspect necessitates an expansion of the "conservative model of innovation" proposed above. This still rested on the assumption of the transfer of a construct from one discipline to another. The image of the coordinate system, however, according to which a metaphor can operate in different discourses and in various ways, suggests a *nonlinear model* of the transfer of metaphors. According to this notion, the continuous transfer of particular metaphors, or systems of metaphors, generates what Foucault has called a "dispositif": a network of social, political, and scientific discourses, which – in Mitman's words – generate "a general field of meaning" (see Mitman, this volume). Within this field, metaphors can be defined as the smallest vehicles of concepts, which are characterized by two functions: a) their spontaneous transferability (as "ideas") enables their rapid distribution and modification of their conceptual content; b) in the course of their distribution they link the discourses they have entered, turning them into a "dispositif" or "general field of meaning" (Mitman), something that makes metaphors significant in the first place.

Both functions – their transferability and their linkage function – contribute to the political-social, but also to the scientific significance of metaphors. In a recursive process, systems of metaphors which come to be accepted into different fields (e.g., "organism"-metaphor), are considered "successful," generate a "general field of meaning" (e.g., assignment of parts to a functional whole), and thus, as "adequate" metaphors, successfully enter other fields.[57]

This conception, I believe, meets the one outlined by James J. Bono.[58] He looks at metaphors as *sites and media of exchange* both in the intrascientific and extrascientific domains. Such exchanges, which "trade on" the capacity of metaphoric language to shift meaning, "create an 'ecological' network driven by the tension-fraught need or desire *both* to 'fix' meanings *and* to disrupt, generate, and transform them."[59] Among other things, his analysis points to the role of metaphors as valuable sources for historiographical models of scientific development: according to Bono, scientific change, rather than being a result of "unexplicable gestaltlike change as with Kuhn and Foucault,"[60] can be rooted in the destabilizing tendencies inherent in any scientific language. Metaphors play a crucial role in Bono's scenario.

> Metaphors and tropes may be transmitted over time, but their meaning must always be reconstituted synchronically . . . this very process of reconstituting the meaning of metaphors subjects them to the interference of other discourses – and, I might add, other metaphors – which, indeed allows them to speak reasonantly to communities of individuals . . . : the metaphors and tropological features of extrascientific discourses – whether religious, political, social, economic, or "literary" – through individual acts of interference and interaction work to "fix" the meanings of inherited terms, and metaphors, within a newly constituted scientific language. By fixing meanings in highly specific, local, though still plastic, ways, the diachronic dimensions of scientific discourse come to constitute a synchronically coherent, if now metaphorically reordered and situated, language. Such a language constitutes a particular discourse and makes possible its production of theories.[61]

From this perspective scientific languages are revealed as "hybrid"; even the most coherent of them will prove inherently unstable, and, when "exacerbated by the interference of social, cultural, or ideological factors, such tendencies" may prove disruptive and produce change. Hence, along with Roy Porter, Bono deems *negotiation* rather than *revolution* to be a more fitting model of scientific change – given the complexity of metaphoric exchange underlying it.[62]

This consideration also leads to another aspect regarding the processing of metaphors – control. From a macroperspective of scientific change it becomes apparent that once a metaphor is part of a discourse and its mechanics, the capacity of scientists or even scientific communities to control them is limited. "Rather than exhibiting unerring conscious design and authorial control, such scientific metaphors adapt themselves to a larger ecology of contesting social and cultural values, interests, and ideologies."[63] If I understand Bono correctly, this assessment should be taken as a word of caution but not as counterargument against the conscious scientific use of metaphors. A metaphor's embedding ecology of (non)scientific discourses, cultural values and the like, is not only "contesting" but also "confirming." Although the stabilizing or destabilizing effects of discursive networks cannot be settled theoretically but have to be assessed on a case-by-case basis, I plead, as Bono implicitly does, in favor of the scientific use of metaphors. Scientists should not be afraid of metaphors since the innovative – which always means destabilizing – effect of metaphors is counterbalanced by a number of stabilizing factors. On the one hand, a linkage of metaphors into a "dispositif" (or ecological network) ensures that not just any metaphor is used at any historical moment of the general and scientific discourse. On the other hand, a discourse-specific processing of metaphors ensures that "connectable" knowledge is produced that (to a certain degree) connects with the tools of the importing discipline, the body of knowledge of other disciplines and discourses, and the societal discourse.

Acknowledgments

I would like to thank Terry Shinn and Peter Weingart for thoughtful criticism and stimulating suggestions. Also, I am indebted to Peter Sloep whose comments on earlier versions of this paper were both challenging and encouraging.

Notes

1. Charles Darwin, *The Origins of Species* (New York: Collier ed., 1962), p. 216.
2. Jean-Pierre van Noppen, et al., *Metaphor. Bibliography of Post-1970 Publications* (Amsterdam, Philadelphia: John Benjamins Publishing Company, 1985); Noppen, and Edith Hols, *Metaphor II. A Classified Bibliography of Publications, 1985 to 1990* (Amsterdam, Philadelphia: John Benjamins Publishing Company, 1990).
3. Mae-Wan Ho, and Sydney W. Fox, *Evolutionary Process and Metaphor* (Chichester: New York, 1988), p. 1.
4. Generally, the relationship between literature and science manifests itself mainly in two ways: 1) as an exchange of concepts across the boundaries of the different types of discourse; and 2) through its literary usage the metaphor provides, implicitly or explicitly, a model also for scientific use.

    As far as the extent and recognition of the transfer of individual concepts between science and literature is concerned, both are subject to continuous change. Until the mid-19th century, no rigid boundary existed between men of letters and scientists. Beer, for instance, discusses Darwins's influences on the literature of Kingsley, G. Eliot, and Hardy (Gillian Beer, *Darwin's Plots. Evolutionary Narrative in Darwin, George Eliot and Nineteenth-Century Fiction* (London: Routledge & Paul, 1983) p. 5. In turn, Darwin based his insights, inter alia, on Malthus's *On Population* as well as on *Beagle: The Poetical Works of John Milton*; Beer, *op. cit.*, 1983, p. 9). A countercurrent is associated with the name of T. H. Huxley: even if science was a powerful and convincing aspect of everyday life to which writers and poets should devote their attention, they would find nothing but "raw material." "For Huxley language is a great chasm that separates science from literature; on one side parsimony and discipline, on the other effulgence and abandon" (Stephen J. Weininger, Introduction: The Evolution of Literature and Science as a Discipline, in F. Amrine (ed.), *Literature and Science as Modes of Expression*, Dordrecht (Boston, London: Kluwer, 1989), XV). In Weininger's view, Huxley at least insists on a discourse-typical processing of scientific materials.

    The divisional perspective shared by a number of authors has again been countered since the 1960s by the research undertaken in literary criticism, philosophy, and history of science: it addresses the influence of scientific constructs on literature, but also the linguistic foundations of science (e.g., Beer, *op. cit.*, 1983; Weininger, *op. cit.*, 1989; Stephen Mulkay, "Action and Belief or Scientific Discourse? A Possible Way of Ending Intellectual Vassalage in Societal Studies of Science," *Philosophical Studies of Science* **11** (1981), 163–171; S. Shapin, "Talking History: Reflections on Discourse Analysis, *Isis* **75** (1984), 125–130; Stephen Wooglar, "On the Alleged Distinction Between Discourse and Praxis," *Social Studies of Science* **16** (1986), pp. 309–317. Apparently, the metaphor is of special relevance for both directions of research.
5. George Lakoff, and Mark Johnson, *Metaphors We Live By* (Chicago: University of Chicago Press, 1980).
6. Max Black, *Models and Metaphors. Studies in Language and Philosophy* (Ithaca: Cornell University Press, 1962).

7. Mary Hesse, "The Explanatory Function of Metaphor" in Y. Bar-Hillel (ed.), *Logic, Methodology and Philosophy of Science* (Amsterdam: North-Holland, 1972).
8. Black, *op. cit.*, 1962, p. 33.
9. Hesse, *op. cit.*, 1972, p. 249.
10. Black, *op. cit.*, 1962, p. 41.
11. Hesse, *op. cit.*, 1972, p. 253.
12. Cf. Black, *op. cit.*, 1962, p. 45.
13. Black, *op. cit.*, 1962, p. 46.
14. Andrew Ortony, "Metaphor: A Multidimensional Problem" in Andrew Ortony (ed.), *Metaphor and Thought* (Cambridge, London, New York, Melbourne: Cambridge University Press, 1979).
15. *Ibid.*, p. 5.
16. Black, *op. cit.*, 1962, p. 23, emphasis added. Such ambiguities in the classification of metaphors into semantic and pragmatic accounts have a systematic reason, i.e., this distinction always describes an analytic division of the phenomenon. As a rule, the authors usually prefer one of the two aspects without entirely denying the other. In the next two sections I will propose a solution for this perspectival dilemma.
17. Donald Davidson, "What Metaphors Mean," in Mark Johnson (ed.), *Philosophical Perspectives on Metaphor* (Minneapolis: University of Minnesota Press, 1981), p. 201.
18. *Ibid.*, p. 202.
19. Samuel R. Levin, *Metaphoric Worlds* (New Haven and London: Yale University Press, 1988), p. 17.
20. *Ibid.*, p. 2.
21. *Ibid.*, p. 25.
22. *Ibid.*, p. xii.
23. *Ibid.*, p 103.
24. Leatherdale has formulated a relativistic counterposition: "Obviously there are great differences between the literary metaphorical use and the scientific use, and the scientific use has a much more rigorously circumscribed context of purpose and intention, particularly in that it employs a stock diction and is under limiting control of a number of phenomenal considerations which are regarded as relevant; moreover the scientific use is required at least to harmonize with fairly complex logical structures and systems, including mathematical ones. However, in both cases, the poetic and scientific, we are dealing with *a* truth, not *the* truth," in W. H. Leatherdale, *The Role of Analogy, Model, and Metaphor in Science* (Amsterdam, Oxford: North-Holland Publishing Company, New York: American Elsevier Publishing Co., Inc., 1974), p. 197. For practitioners of research the latter aspect is not merely of theoretical importance: in spite of a more rigorously circumscribed context in science, the processing of a certain metaphor, rather than being strictly determined is influenced by biases brought to bear upon it by individual scientists or schools of thought.
25. Davidson, *op. cit.*, 1985, p. 218.
26. P. McHugh, "On the Failure of Positivism," in J. D. Douglas (ed.), *Understanding Everyday Life* (London: Routledge & Paul, 1971), pp. 337–354, here: p. 329.
27. Richard Rorty, *Contingency, Irony, and Solidarity* (Cambridge: Cambridge University Press, 1989), p. 17.

28. Robinson, quoted in Leatherdale, *op. cit.*, 1974, p. 132. It should be noted that the dimension of the metaphor's interaction within the scientific domain has greatly increased. Whereas the man/wolf example describes the interaction of a concept with another concept, Robinson goes further by referring to the entire context of existent concepts and theories – something in which he is joined by Leatherdale (*cf.* note 24) and Rorty: having been "alienated" by the import of a metaphor, this context has to be capable of integrating the metaphor into its language game or dismiss it as inappropriate or misleading. In both cases a whole corpus of theories, methods, and classic objects is affected or: has to react to the "intruder."
29. Rorty, *op. cit.*, 1989, p. 9.
30. Karin Knorr, "The Scientist As An Analogical Reasoner: A Critique Of The Metaphor-Theory Of Innovation," *Communication & Cognition* **13**, 2/3 (1980), 183–208.
31. *Ibid.*, p. 111.
32. *Ibid.*, p. 187.
33. *Ibid.*, p. 186.
34. *Ibid.*, p. 195.
35. *Ibid.*, p. 194.
36. Ortony, too, places Black in a constructivist context (Ortony, *op. cit.*, 1979, p. 7).
37. D. A. Schön, *Displacement of Concepts* (London: Tavistock, 1963); Mulkay, *op. cit.*, 1981.
38. Knorr, *op. cit.*, 1980, p. 186.
39. Hans-Jörg Rheinberger, *Experiment, Differenz, Schrift. Zur Geschichte epistemischer Dinge* (Marburg/Lahn: Basilisken Presse, 1992); Bruno Latour, "Postmodern? No, Simply Amodern! Steps Towards an Anthropology of Science," *Studies in History and Philosophy of Science* **21** (1990), 145–171; Timothy Lenoir, "Practice Reason, Context: the Dialogue Between Theory and Experiment" *Science in Context* **2** (1988), 1, 3–22.
40. Lenoir, *op. cit.*, 1988, p. 11.
41. Rheinberger, *op. cit.*, 1992, p. 14.
42. As a reformulation of this process one might say that the transfer of a concept to a new discipline turns this concept into a metaphor. In the course of transformation the goal of processing this metaphor is to turn it into a disciplinary tool. Thus, the distinction between transfer and transformation, from this perspective, refers to the process of "metaphorizing" and "literalization" of concepts.
43. Susan Haack, "Surprising Noises: Rorty and Hesse on Metaphor," *Proceedings of Aristotelian Society* (1987–1988), 293–301, here, 299.
44. Deidre Gentner regards this course of development as the most likely one and points to journals of scientists like Kepler, Maxwell, Poincaré and Feynman revealing that they entertained initially unruly metaphors. In addition, he cites Polya's advice to those who wish to develop mathematical insight: "And, remember, do not neglect analogies. Yet, if you wish them respectable, try to clarify them" (Polya in Deidre Gentner, "Are Scientific Analogies Metaphors" in David S. Miall (ed.), *Metaphor: Problems and Perspectives* (Sussex: Harvester Press, 1982), pp. 106–132, here, 128).
45. Thomas Kuhn, "Metaphor in Science" in Andrew Ortony (ed.), *Metaphor and Thought*

(Cambridge, London, New York, Melbourne: Cambridge University Press, 1979), p. 414, emphasis added.
46. Recently, Hesse claimed that with regard to the question whether scientific models are "true" we should distinguish the construction of an imaginative ontology with its internal system of meaning relations from the claim that successful empirical test justifies pragmatic acceptance of a model" (Michael Arbib, and Mary Hesse, *The Construction of Reality* (Cambridge: Cambridge University Press, 1986), p. 161). This distinction, in my view, maps onto the one between transfer and transformation insofar as it addresses (and thus, adds) an ontological perspective: "Scientific models are a prototype, philosophically speaking, for imaginative creations or schemas based on natural language and experience, but they go beyond it by metaphorical extension to construct symbolic worlds that may or may not adequately represent certain aspects of the empirical world" (Arbib, Hesse, *op. cit.*, 1986, p. 161). The criteria for models being "adequate" or, in my view, for being successfully transformed, following Hesse, are to be found in an answer to the question whether or not scientific models meet the purposes of prediction and control, i.e., refer to pragmatic reasoning.
47. Lindley Darden, and Nancy Maull, "Interfield Theories," *Philosophy of Science* **44** (1977), 43–64, here, p. 43.
48. As of this moment the concept concerned develops differently according to the use that both the importing and the exporting discipline make of it. I will enlarge on this aspect in the following section.
49. This thesis is supported by Schön's conception who believes it is possible that such metaphors can be "translated back": "The actual language of ta theory contains metaphors which are vestiges and signs of the old theories whose displacement helped to form the new one. These metaphors are clues to the identity of the old theories" (Schön, *op. cit.*, 1963, p. 113). Generally speaking, at this stage metaphors are clues to the identity of the exporting discipline as "the other" and, hence, expose both, the importing and the exporting disciplines with the aid of concepts which are made into "their" respective concepts.
50. On this Knorr gives an example: "Naturally, what we called "transfer" involved modification and adaptation, for example working with different enzymes. It thus required an actual transformation of the procedure involved" (Knorr *op. cit.*, 1980, p. 191). Although Knorr uses the terms "transfer" and "transformation" she does not integrate them to become a systematic account of scientific processing of metaphors.
51. Lakoff, Johnson, *op. cit.*, 1980, p. 454.
52. Mary Hesse, "The Cognitive Claims of Metaphor," *The Journal of Speculative Philosophy* **2** (1988a); Mary Hesse, "Theories, Family Resemblances, and Analogy" David Helman (ed.), *Analogical Reasoning* (Dordrecht: Kluwer, 1988b), pp. 317–340.
53. Hesse, *op. cit.*, 1988a, p. 3.
54. Bo Dahlbohm, "The Role of Metaphors in Science" Mats Furberg (ed.), *Logic and Abstraction* (Göteborg: Acta Universitatis, 1986), pp. 95–118, here, p. 98. In this quotation Dahlbohm refers especially to technological concepts as primary source for scientific metaphors. This cannot be but part of an answer to the question on what makes which kind of metaphors conceived of as appropriate at a certain socio-

historical time. In what follows I will address this issue on a purely formal level although I am convinced that one might be able to trace back the prevalence of certain metaphors in science. Such a project might be called "Evolutionary Metaphorology": presumably one would find several (perhaps interdependent) branches, like technological, physical, biological, and medical metaphors in science which develop parallel to each other over time. Questions such as: are there epistemological or ontological reasons for major changes in privileged (systems of) metaphors? might be answered within the framework of such a project.

55. Beer, *op. cit.*, 1983, p. 5.
56. Hubert Rottleuthner, "Biological Metaphors in Legal Thought: in G. Teubner (ed.), *A New Approach to Law and Society* (Berlin, New York: de Gruyter, 1988), p. 101.
57. The transfer of metaphors which was specifically assigned the role of heuristics from the microperspective of disciplinary innovations, thus assumes a further significance from the macroperspective in which scientific discourses form an element of societal discourse: the permanent transfer of successful metaphors contributes to the constitution of a general field of meaning. This system, which comprises politically, socially, and scientifically relevant concepts, will continue to exist even if the discipline-specific processing of metaphors in some cases succeeds only partly or not at all: for some organisms, for instance, "society" metaphors may not work, except perhaps for certain aspects. Even so, against the background of a "dispositif," which for instance recommends the assignment of parts to a functional whole, such a transfer may suggest itself and may thus confirm the dispositif – even if the transformation proves unsuitable in some cases.
58 James J. Bono, "Science, Discourse, and Literature. The Role/Rule of Metaphor in Science" in St. Peterfreund (ed.), *Literature and Science. Theory & Practice* (Boston: Northeastern University Press, 1990), pp. 59–89.
59. *Ibid.*, p. 73, emphasis added.
60. *Ibid.*, p. 78.
61. *Ibid.*, p. 77.
62. *Cf, ibid.*, p. 81.
63. *Ibid.*, p. 81.

HOW NATURE BECAME THE OTHER: ANTHROPOMORPHISM AND ANTHROPOCENTRISM IN EARLY MODERN NATURAL PHILOSOPHY

LORRAINE DASTON
University of Chicago

Metaphors demand a chasm to bridge: connections joined between neighboring categories or kindred concepts hardly qualify as figurative. It is because we are persuaded that such a chasm yawns between the natural and the human that we so often dignify (or revile) the concourse between the biological and social sciences as "metaphorical." Conversely, those who insist upon analogies between these domains also typically insist upon the proximity of the terms compared, narrowing the gap between them until the figurative collapses into the literal: society *is* an organism; physiological specialization of body organs *is* a division of labor. My concern here is not with any particular metaphor liking the biological and social realms, but rather with the historical preconditions for metaphor-making between them, with the emergence of a chasm wide enough to create a potential for figurative sparks. How did the natural come to be so other than the human? More pointedly, how and why did anthropomorphism – the conceptual habit of seeing the non-human in human terms – come to be suspect in science?

Despite the several upheavals in the historiography of the Scientific Revolution since the early nineteenth century, one part of the story has survived more or less intact. It was sometime in the middle decades of the seventeenth century, we are told, that natural philosophers began explicitly and emphatically to abandon anthropomorphic modes of description as applied to nature. Whether it is Michel Foucault relating the decline of sympathies and antipathies,[1] or Alexandre Koyré marking the passing of formal and final causes,[2] or Max Weber sighing over the "Entzauberung der Welt,"[3] or Auguste Comte applauding the transition of physics and astronomy from the theological to the metaphysical phase,[4] the story is more or less the same. For good or ill (and there are voices on both sides of even this short list), after circa 1700 it was no longer possible, so the argument goes, for reputable natural philosophers to apply the language of human beliefs and desires to nature.[5] Nature had become irretrievably "the other." Views on whether this was a good or bad thing vary widely, from Comte's enthusiasm to Schiller's regret: "Unbewußt der Freuden, die sie schenket,/Nie entzückt von ihrer Herrlichkeit, . . . Gleich dem toten Schlag der Pendeluhr,/Dient sie knechtisch dem Gesetz der Schwere,/Die entgötterte Natur."[6] However, there is surprising unanimity not only that seventeenth-century natural philosophy banned anthropomorphism, but also that this anthropomorphism taboo is partly constitutive of what is modern about "modern science." In this paper I want to challenge both claims: I shall argue first, that the seventeenth-century anthropomorphism taboo resembles our own in name only, and was moreover purchased at the price of other anthropomorphisms; and second, that it was opposed to, not part of, the secularization of science.

Anthropomorphism means to describe the nonhuman in human terms, and it was a cardinal religious sin long before it became a cardinal scientific sin.[7] For my purposes, it is important to be clear about why it counts as a sin, as opposed to a mere error. After all, to impute "anger, terror, jealousy, and love" to insects, as Darwin did, or to ascribe a sense of humor of nature, as Renaissance naturalists such as Pierre Belon and Girolamo Cardano did, is in our view first and foremost a category mistake, not a moral failing. We no longer believe that nature is the kind of entity that could relish jokes such as "expressing the figures of fish and other animals in stones" – Belon's explanation of fossils.[8] Our own categories, and the sharp boundaries we draw between them, force us to construe such claims charitably as metaphors, or uncharitably as mistakes. But why should either metaphors or mistakes be grounds for moral censure? The reason why we do not just reject but also abhor anthropomorphism in the study of nature must go beyond our belief that humans and nature are different kinds of entities.

By our late twentieth-century account, the reason why anthropomorphism is to be abhorred as well as rejected is because it is the expression of anthropocentrism. Anthropomorphism means the projection of human traits onto a nonhuman object, including nature at large; anthropocentrism means privileging human viewpoints and desires above all others. We regard anthropocentrism as the childish vanity that makes man the measure of all things, and as a violation of the ideal of aperspectival objectivity, which insists that science abandon all merely particular viewpoints, be they of class, nationality, gender, epoch, or even species, in favor of what philosopher Thomas Nagel has called "the view from nowhere."[9] It is because we believe that anthropomorphism is *caused* by anthropocentrism that we find it reprehensible as well as erroneous, arrogant as well as confused. We judge anthropomorphism and anthropocentrism both to be scientific felonies, and we further consider them to be as closely allied as "breaking" and entering" are in the police ledger.

However, distaste for anthropocentrism cannot explain the ban on anthropomorphism introduced so forcefully and enduringly in seventeenth-century natural philosophy. The grounds on which thinkers such as Blaise Pascal and Robert Boyle attacked scientific anthropomorphism were not only different from our current grounds for the same position, the positions are arguably incompatible with one another. We are antianthropomorphism because we are antianthropocentrism; they were antianthropomorphism because they were *pro*anthropocentrism. Leading seventeenth-century proponents of the new mechanical philosophy reacted vigorously against anthropomorphism both local (i.e., as applied to animals) and global (i.e., as applied to nature as a whole). Their newly won Cartesian convictions told them that it was an error to credit soulless animals even with pain, much less with human emotions and thoughts; those same convictions robbed nature of her whimsy, plastic powers, and autonomy. Cartesian dualism privileged the human and (a *fortiori* the divine) in an even more dramatic fashion than the medieval hybrid of Christianity and Aristotelianism had, albeit for different reasons. By granting God and man a monopoly on *res cogitans*, Descartes also secured for them a monopoly on all activity in the world. Without the quickening power of human or divine initiative, nature and all its contents remain immobile and inert – "brute, passive, stupid matter," in seventeenth-century parlance. For all the epistemological and ontological radicalism of the Cartesian system, it actually fortified the special status of humans in creation.

Like us, seventeenth-century natural philosophers condemned anthropomorphism not only as an intellectual but also as a moral error: failure to recognize that nature was the other was not simply wrong, but dangerously wrong. Wherein lay the dangers, and what were their consequences for the new natural philosophy, of the latter half of the seventeenth century? What were the conceptual preconditions for this first emergence of the anthropomorphism taboo in the study of nature, and how did it transform not only the category of the natural, but also that of the human and the divine? In short, how and why did nature become the other, and what kind of other?

In the following I shall address these questions by paying close attention to three closely intertwined themes in the seventeenth-century debate over anthropomorphism in natural philosophy: the relationship between nature and art, and between nature's art, man's art, and God's art; the meaning of intelligence and also of stupidity, as well as the degrees in between; and the problematic notion of natural law. I shall argue, *pace* Schiller, Comte, Weber et al., that the seventeenth-century ban on anthropomorphism represents a vigorous imposition of Judeo-Christian theology upon natural philosophy, rather than a step towards secularization.

Nature's Art and Nature as Art

When Sir Thomas Browne wrote in *Religio Medici* (1642) that "All things are artificial, for Nature is the Art of God,"[10] he deliberately affronted the expectations of his educated readers, who would have habitually opposed art to nature. The locus classicus of this conventional distinction was in Aristotle's *Physics* (192b12–34): according to Aristotle, natural objects have an "innate impulse to change" whereas artificial ones do not. If we plant a bed, and the wood from the bed sprouts leaves, it does so *qua* wood, not *qua* bed. However, final causes apply to bother the natural and the artificial.

> Now action is for the sake of an end; therefore the nature of things also is so. Thus if a house, for example, had been a thing made by nature, it would have been made in the same way as it now is by art; and if things made by nature were made not only by nature, but also by art, they would come to be in the same way as by nature. The one, then, is for the sake of the other; and generally art in some cases completes what nature cannot bring to a finish, and in others imitates nature. If, therefore, artificial products are for the sake of an end, so clearly also are natural products. (*Physics*, 199a10–19)[11]

Teleology does not, however, commit Aristotle to anthropomorphism: he is clear and consistent on the point that an organism (or nature as a whole) may have ends, without being able to deliberate upon them. Citing the examples of the spider's web and the bird's nest – examples that were still touchstones in seventeenth-century debates over anthropomorphism – Aristotle insists that "animals other than man . . . make things neither by art nor after inquiry or deliberation. . . . It is absurd to suppose that purpose is not present because we do not observe the agent deliberating" (*Physics*, 199a10; 199b16). Human art is only one special case of purposeful change, anomalous in its reliance on inquiry an deliberation. Both art and nature stand opposed to chance and to necessity – these latter being defined by the absence of purpose – as the sources of regularity in the universe.

The changes in seventeenth-century natural philosophy did not so much transform these conceptual categories as to shuffle their references. Where major innovations did occur, as when natural laws replaced final causes as the direct guarantors of regularity in nature, they posed almost insoluble problems just because their most ardent proponents still clung to so much of the Aristotelian framework. An innate principle of change, or "self-activity" as seventeenth-century natural philosophers often put it, remained a special property, but its application was now restricted to God, humans, and perhaps also angels and demons. Final causes or purposes underwent a similar contraction of domain, but ultimately still bore the bulk of the burden for explaining regularities in nature, especially organic nature. For example, the English chemist and mechanical philosopher Robert Boyle embraced the Aristotelian notion of chance as absence of purpose, and was as adamant as Aristotle that only purpose – in this case, divine purpose – could satisfactorily account for "things in nature so curiously contrived, and so exquisitely fitted for certain operations and uses."[12] Where the many-cornered battle between Aristotelian, mechanical, neo-Platonist and magical natural philosophies was really joined was over the possibility of such internal principles and final causes *without* deliberation. Whereas Aristotle (and many late seventeenth-century natural philosophers such as Ralph Cudworth and Gottfried Wilhelm Leibniz) could conceive of an art that did not deliberate, of (to use Aristotle's own oft-cited example) the shipwright's art being somehow in the timber, the stricter mechanical philosophers such as Boyle could only regard such usages as intolerable anthropomorphisms. It was this discomfort with the notion of nature's art that placed the Aristotelian opposition of art and nature under such conceptual strain in the middle decades of the seventeenth century. In order to understand why, we must first reconstruct what was meant by nature's art in the decades directly preceding these debates.

It is a platitude among historians of the Scientific Revolution that the seminal thinkers of the seventeenth century, most notably Bacon and Descartes, abolished the nature/art distinction by subsuming the artificial under the natural.[13] However, it would be just as, if not more, accurate to claim that the distinction was collapsed in the manner of Sir Thomas Browne, by subsuming the natural under the artificial. What characterizes late sixteenth– and early seventeenth-century discussions about art and nature is not so much a reduction of one to another, although examples can be found in both directions, but rather a melting of the categories into one another, which made such reductions thinkable. This melting took two forms, which must be distinguished both by chronology and import: the earlier form likened nature to active artist or artisan; the later form, to passive work of art. Moreover, the kind of artistry in question, whether active or passive, also gradually changed its aesthetic character in the second half of the seventeenth century, from intricately wrought *objet de luxe* to well-designed engine.

Although the topos of nature as artist has an antique lineage – Pliny spoke of *natura pictrix* in connection with fossils and landscape marble – it was only among the Renaissance naturalists such as Conrad Gesner, Girolamo Cardano, Pierre Belon, Ambroise Paré, Levinus Lemnius, Fortunio Liceti, and others that it was elevated to an explanatory principle not only for images in stones and plants, but also for resemblances between the creatures of land and sea, for heteroclite organisms like the zoophyte Scythian lamb (half-plant, half-animal) or the bat, for Ovidian metamorphoses from caterpillar to butterfly or insect to reptile, and more generally for any extraordinary object or phenomenon, from a fly frozen in amber to gigantic turnip to two-headed cat. These oddities were evidence not just of nature's creativity, but also of her whimsy, as the term *lusus naturae* (sport or joke of nature) suggest.[14] For example, when the French surgeon Ambroise Paré could not figure out how a hyper-symmetric African monster with feet and eyes pointing in all four compass directions could ever decide on which direction to go, he concluded that here "nature was at play."[15] Apparent defiance of function was often the departure point for reflections on nature's craftsmanship or arch sense of humor – as when Cardano noted that certain sea creatures like crayfish have feet, although they are useless for swimming. Authors of paeans to nature's cornucopia-like variety and fecundity underscored her wit as well as skill, even if they sometimes despaired, as the sixteenth-century French ichthyologist Guillaume rondelet did, of a perfected natural philosophy because nature's infinite variety so far outstripped human understanding.[16] Nature as portrayed in much sixteenth-century natural history and natural philosophy bears a strong resemblance to the goldsmith Benvenuto Cellini as portrayed in his autobiography: an artisan of prodigious dexterity and ingenuity, possessed of a penchant for the whimsical and for the practical joke, as well as of a refined aesthetic sense.

By identifying nature with the virtuoso artisan, Renaissance naturalists deliberately subverted the distinction between the works of nature and art. The *Wunderkammern* or cabinets of curiosities of the late sixteenth and early seventeenth centuries bore material witness to this studied conflation. Objects that blurred the boundary between art and nature featured prominently in these collections: scenes of coral arranged in scenes of mountains and forests, fragments of marble incorporated into paintings as clouds or water, natural objects such as minerals assembled into artificial tableaux as in the *lapides manuales*, or coconuts and nautilus shells embellished with the goldsmith's art.[17] Among the painting most coveted by collectors were those that were all but indistinguishable from nature, still life and *trompe d'oeil*; the natural objects most prized were those that seemed to approximate the fine craftsmanship of the luxury trade – seashells, crystals, fossils, and the like.[18] It was an aesthetic that fully exploited the ambiguities of the traditional oppositions of art and nature – art imitating nature, art assisting and improving nature, nature excelling art – by singling out hybrid objects and underscoring resemblances of detail as well as overall organization between natural and artificial products.

In light of this hybrid aesthetic of the extraordinary, it is no longer so clear that when Francis Bacon called for a "history of nature wrought," i.e., for a history of the artificial, as the third part of a refurbished natural history that he was thereby simply subsuming the artificial under the broader rubric of the natural. In order to connect the history of "nature wrought" with that of "nature in course" and, especially, "nature erring," i.e., the history of nature's extraordinary products, Bacon tacitly appealed to the image of nature as artisan. Art and nature approached each other at the extremes, in the most curious and ingenious products of each: as Bacon wrote, "Now it is an easy passage from miracles of nature to miracles of art."[19] It is by the close study of nature's most highly wrought, most "artificial" objects that we may improve the human arts and trades. The mechanical philosophers, starting with Descartes, were to conversely claim that the products of human art, in particular machines, were the best model for understanding the inner workings of nature.

Although the extraordinary works of nature might provoke the prototypically religious emotions of wonder and admiration, sometimes even tinged with awe, they could not be directly imputed to God in all instances without a measure of irreverence. Exquisite artistry, the delicate devetailing of organ to function, even beauty and variety could all be praised as divine workmanship – but it verged on impiety to suggest that God had a fondness for a good joke. Thus Nature's distinct identity as God's "minister," "viceroy," or "chambermaid," as she was variously apostrophized, served to protect God's dignity: God may benevolently provide for the welfare of His creatures or even ornament the world for their pleasure, but it was not seemly that He should relish surprises and practical jokes. Granting Nature a certain limited autonomy safeguarded divine dignity in yet another way: many sixteenth- and seventeenth-century authors echoed the Aristotelian opinion that it was indecorous for kings, and *a fortiori* for the deity, to

> appear himself to administer all things and to carry out his own wishes and superintend the government of his kingdom. . . . and just as puppet-showmen by pulling a single string make the hand and the neck and shoulder and eye and sometimes all the parts of the figure move with certain harmony, so too the divine nature, by simple movement of that which is nearest to it, imparts its power to that which next succeeds, and thence further and further until it extends over all things (*On the Heavens*, 398b1–24).

As Issac Newton put it with characteristic severity at the conclusion of the *Principia*, "God is a relative word, and has respect to servants; and Deity is the dominion of God not over his own body, as those imagine who fancy God to be the soul of the world, but over servants."[20] The nature of dignified labor, divine and otherwise, is a point to which I shall return in the context of the late seventeenth-century controversy over natural law.

Degrees of Soul

The Renaissance naturalists had thus granted nature a measure of autonomy and discretion through their peculiar vision of nature's art that exceeded Aristotle's indwelling final causes but that stopped short of outright personification. Almost no one, including the neo-Platonists, went so far as to assert that nature was capable of conscious deliberation. However, the image of nature as genial artisan and God's servant, along with the hybrid aesthetic of the extraordinary works of art and nature, made it plausible to endow nature with a "lowest level of soul," as the Cambridge neo-Platonist Ralph Cudworth put it in his treatise *The True Intellectual System of the Universe* (1678).[21] This intermediate soul, possessed of "vital energy" and "plastic powers" but neither of understanding nor sensation – Cudworth likened it to the mental state of a baby or of a person deep in sleep – straddled the gulf dividing brute, passive, inert matter from intelligent, active mind in the mechanical philosophy. Although the seventeenth-century mechanists attacked Aristotle and his followers for anthropomorphizing nature, their real target was the ensouled but not fully human nature of the Renaissance naturalists. Only when read in the light cast by Cardano, Paré, Rondelet, and others could the scholastic slogans ridiculed by the mechanists – nature does nothing in vain, nature always seeks the shortest path, nature abhors a vacuum – be construed as anthropomorphisms rather than as shorthand. Here is a typically literal-minded reading by Pascal, writing to his brother-in-law Florin Perrin concerning his experiments in air pressure.

> I am now working ... on finding experiments that will reveal whether the effects attributed to the horror of the vacuum must really be attributed to the horror of the vacuum, or whether they should be [attributed] to the weight and pressure of the air. For, to tell you my ideas frankly, I have difficulty believing that nature, which is neither animated nor sensitive, is susceptible to horror, since the passions presuppose a soul capable of feeling them. . . .[22]

Pascal is here refuting a caricature, for it would have been difficult to find a seventeenth-century natural philosopher, even a hylozoist, who would have granted nature a sensitive soul, susceptible to passions. However, it was a necessary caricature for the mechanical philosophy, which could not concede *degrees* of soul in its two-valued ontology of passive matter and active mind. Once wedged into this procrustean schema, nature ensouled to any degree whatsoever must aspire to a rational soul that perceives, understands, judges, and reason – and would therefore appear to be presumptuously anthropomorphic. Note that the mechanists preserved Aristotle's fundamental categories of activity and passivity, as well as his asymmetric valuation of them: active soul set passive matter in motion and thereby created order. But by narrowing the circle of activity to God, men, and angels, they introduced a robust theo – and anthropocentrism quite foreign to Aristotle.

Since we are heirs to the dualism of the mechanists, the degrees of soul posited by their seventeenth-century opponents require a bit of explaining in order to make them conceivable, if not plausible. The ablest exponents of the anti-Cartesian position on this issue (all of them readily endorsed other aspects of Cartesian natural philosophy) were the Cambridge Platonists Henry More and Ralph Cudworth, and the neo-Aristotelian Leibniz. Although these authors certainly did not agree on all points, it is possible to extract a common, minimal position on ensouled nature from their writings. In all cases, their departure point was a dilemma posed by the Cartesian ontology of passive matter and active soul: either the world order arose from the interplay of chance and blind necessity in the manner sketched by Epicurus, *or* God would be obliged to recreate or at least conserve the world order by direct agency at every instant. The first alternative was rejected out of hand as absurd; the second smacked of a perpetual miracle, sentencing God to perpetual and indecorous labor, and condemning any natural philosophy based on secondary causes to failure from the outset. Anti-mechanist critics diagnosed the problem to be the brutalization and pacification of matter, and proposed ensouled nature as the way out of this dilemma.

More (*The Immortality of the Soul*, 1662), Cudworth (*The True Intellectual System of the Universe*, 1678), and Leibniz ("De ipsa natura," *Acta Eruditorum*, 1698; *Specimen Dynamicum*, Part 1, *Acta Eruditorum*, 1695) all appealed to a staple set of counterexamples in order to justify their assumption of an ensouled nature. How could the mechanical philosophy explain sympathetic cures, musical instruments that vibrated in unison, and the power of the maternal imagination? How could it deal with the marvelous instincts of animals, which constructed the spider's web and the bee's honeycomb? How could it explain why any organism was structured and functioned as an integrated whole, as opposed to a heap of parts? Only something like "plastic powers" or "spirit of nature" or "indwelling active principles" could account for these phenomena, they argued. However, they were all at pains to insist on the inferiority of the soul of nature to that of humans, much less to God. Leibniz warned that a fully anthropomorphized nature would revive "heathen polytheism";[23] Cudworth admitted that human actions may not be "carried on with such Constancy, Eaveness and Uniformity, as the Actions of *Nature* are;" but pointed out that we surpass nature in being able to act "*Electively* and *intendingly*, with *Consciousness* and *Self-perception*." Yet ensouled nature was still elevated above the stupid matter of the Cartesians, for it possessed *"Self-Activity"* and the power to *"Artificially direct* its own Motion."[24]

Perhaps nothing is more revealing of how the anti-mechanists understood ensouled nature than their recurring analogies to the kind of labor it performs. Far from rivalling God, ensouled nature was his "servant," his "Drudging Executioner," the "Manuary Optificer" to God's "Architect." Just what it meant to be a servant in the seventeenth century is made painfully clear by Cudworth's eleboration of the latter analogy: "We account the Architects in every thing more honourable than the Manuary Optificers, because they understand the reason of the things done, whereas the other, as some Inanimate things, only do, not knowing what they do."[25] The labor of nature is drudging labor, and the knowledge of nature is tacit knowledge, akin to the unconscious "habits" that sustain the art of the musician and the dancer. It is at the level of manual labor and bodily habits that Cudworth hazards a genuine, albeit selective, anthropomorphism: here was familiar proof that it was possible to act *"For the Sake of Ends,* and *Regularly* or *Artificially,* and yet be it self devoid of *Knowledge* and *Understanding.*"[26]

Nature was not the only entity to be endowed with half a soul. Anti-mechanists such as David-Renaud Boullier (*Essai philosophique sur l'âme des Bêtes,* 1728) made a similar plea on behalf of animals, adducing the evidence of comparative anatomy (why would God have bothered to endow animals with brains and perceptual apparatus so similar to ours if they were not designed for a similar function?), and of instincts too intricate to be the mere routine of an automaton. Animal souls surpassed that of nature in being capable of sensation and emotion, but did not achieve the understanding and will of the human soul, much less the still more distinct ideas and wider-ranging volition of angels.[27] Once again, intelligence and activity (or rather, stupidity and passivity) are tightly intertwined with a hierarchy of labor and an ethics of exploitation. To those who might worry how a benevolent God could allow humans to enslave animals capable of suffering, Boullier cooly tested their consistency with this all-too-precise parallel: "I do not doubt that the Negroes who serve me are not pure automata, given the marts of intelligence they give me daily; The Creator if too benevolent to have given them a soul which in such bodies can only be miserable; he replaces it with a system of springs [*jeu de ressorts*] whose aim is my comfort and convenience."[28] Robert Boyle seemed to have had similar moral misgivings in mind when he argued that excessive admiration of nature's art might be a "discouraging impediment to the empire of man over the inferior creatures of God,"[29] or when he cheerfully praised automata, which for him encompassed all of nature except humans, because "inanimate engines may be so contrived, as to act as we please; whereas angels and human souls are endowed with a freedom of acting, in most cases, as themselves please."[30] Stupidity – be it that of matter, beasts, or servants – was not simply an ontological category in the mechanical philosophy; if often carried social and ethical associations of legitimate subordination and exploitation as well.

This is one reason why ensouled nature might be deemed dangerous as well as wrongheaded: it would be ethically inconvenient for humans to respect the feelings of these "drudges"; the anthropocentrism of exploitation entailed the antianthropomorphism of the mechanical philosophy. However, the chief reason advanced in the most sustained attack on ensouled nature, Robert Boyle's *A Free Inquiry Into the Vulgarly Received Notion of Nature* (comp. 1666, publ. 1686; abridged Latin version *Tractatus de ipsa natura*, 1688) was impiety rather than loss of labor. To grant nature even a smattering of autonomy, much less to accord nature the status of God's plenipotentiary, was an act of *lèse majesté* committed against God. At its worst, too great a veneration of nature could lead to the heresies to polytheism, and Boyle devoted many pages to pagan religion and the worship of heavenly bodies in order to make this risk vivid. Yet even the orthodox Christian, wholly untempted by the rites of the sun and moon, might err in admiring nature's works too fervently: Boyle cautioned that not only had scripture told us that God was "jealous"; there was moreover no Hebrew word for "nature" in the Old Testament. If God indeed required subordinates "for the management of the world," better to case angels rather than nature in that role.[31] Here at least the motives for the "disenchantment of the world" are patently Judeo-Christian rather than secular, and Aristotle is the enemy as much because he was a pagan as because of his false teachings in natural philosophy.

Like Pascal, Boyle presents these teachings in crudest caricature, mocking the "axioms" of Aristotle and the scholastics as childish anthropomorphisms. How, for example, can nature seek the shortest path (*Natura semper agit per vias brevissimas*), without knowledge of "the exigency of particular circumstances,"[32] or seek to regain its natural state without memory? Boyle conceded the anti-mechanists' point that there existed as yet no mechanical explanation for such phenomena, but he was

Whereas Boyle considered it the duty of nature (is such there be) to be ceaselessly nurturing and vigilant, he intimates that God's labor began and ended with the free (indeed, arbitrary in every sense) act of creation. Here he found himself skewered on the second horn of the anti-mechanists' dilemma, for he was no more friendly to an Epicurean explanation of the world order by chance and necessity than they were. How to account for the regularity or "artificiality" of nature if it consisted only of brute, passive matter? Descartes had faced this difficulty squarely, and concluded that God conserves His creation at every moment, albeit always according to the fixed rules of motion He ordained at the outset.[36] Earlier English mechanists, such as Walter Charlton, had followed their master in this point, denying that it was beneath divine dignity to attend to "the most minute, and seemingly most trivial and contemptible transactions on this great exchange of the world,[37] and Malebranche grasped the nettle of occasionalism.[38]

Although Boyle took too lofty a view of God's exalted station to be able to countenance so much divine labor, he was also loathe to allow God servants, for this would lead willy-nilly to an ensouled and potentially usurping nature. Boyle's solution was to claim that not only was nature artifice rather than artist; it was an artifice of a peculiar kind, an engine rather than a lavishly ornamented luxury object. Appealing over and over again to the Strasbourg clock and to the automata that swam like real ducks or tooted like real flutists, Boyle argued that the world was nothing but a "great automaton," composed of still other smaller automata, in the manner of Chinese nested boxes, and that God was the most ingenious of all engineers. Referring obliquely to Aristotle's analogy in *On the Heavens* to the deity as puppeteer, Boyle suggested that it would be still more decorous if the divine artificer were not "fain (by drawing sometimes one wire or string, sometimes another) to guide and oftentimes over-rule the actions of the engine," but rather to arrange for "all things to proceed, according to the artificer's first design, and the motions of the little statues [of the Strasbourg clock], that at such hours performs these or those things, [land] do not require, like those of puppets, the peculiar interposing of the artificer, or any intelligent agent employed by him. . . ."[39] There is some irony in Boyle's contrast of automaton to puppet, for in Aristotle's view "automata" *were* puppets. However, the aim of the analogy is clear, if not cogent: automata hovered between the passivity of matter and the activity of soul, somehow conserving within their springs and weights the initial impulse imparted by the engineer but not genuinely self-moving. Boyle thus hoped at one stroke to eliminate the need for servants, in particular "an intelligent and powerful being, called nature,"[40] *and* to keep God's hands clean of demeaning labor. Deep within the mechanical philosophy and its critique of anthropomorphism was buried the dream of the perfect servant, industrious and orderly but without uppity ideas or, indeed, any ideas whatsoever.

Natural Laws

Alas, the dream was indeed an illusion, for neither Boyle nor any of his fellow-mechanists was able to explain how stupid matter could "obey" – and the anthropomorphism is telling here – even the initial impulse imparted by divine activity, much less to sustain it in orderly fashion. All, including Boyle, admitted that "natural laws" could only be understood in a metaphorical sense, for "to speak properly, a law being but a *notional rule acting according to the declared will of a superior*, it is plain, that nothing but an intellectual Being can be properly capable of receiving and acting by law."[41] Cudworth triumphantly concluded from such admissions that "their *Laws* of *Nation* concerning *Motion*, are really nothing else, but a *Plastick Nature*;"[42] Leibniz constituted his monads into a moral community so that they would be capable of understanding and obeying these laws.[43] Boyle himself speculated that the question of "how senseless matter, to whose nature motion does not at all belong, comes to be put into motion, and qualified to transfer it according to determinate rules, which itself cannot understand"[44] might belong to the inscrutable divine mysteries. He thus confirmed the anti-mechanists' gloomiest prediction that the mechanical philosophy, for all its talk about intelligible causes, would drive natural philosophy to abandon explanation altogether in favor of miracles.

Why did Boyle and the other mechanists hold so firmly to the confessed anthropomorphism of natural laws, despite its glaring incompatibility with their ontology? Once again, the principal motives were theological. Boyle, ever worried about saving divine labor, thought they might free God's time up for minding ensouled creatures, especially the hyperactive and wilful angels and demons, for "the government of one daemon, may be as difficult a work, and consequently may as much declare the wisdom and power of God, as the government of a whole species of inanimate bodies, such as stones and metals, whose nature determines them to a strict conformity to those primordial laws of motion, which were once settled by the Great Creator, and from which they have no will of their own to make them swerve."[45] Boyle's angelology was perhaps idiosyncratic, but not his emphasis on the act of creation: as John R. Milton has pointed out, only a theology with a creation myth could conceive of God as legislator at all. The notion of natural law was almost unknown in the natural philosophy of antiquity; indeed, the conventional Sophist opposition between *physis* and *nomos* made it almost an oxymoron. As applied to nature, the term came into its own in the theology of the high and late Middle Ages, particularly in the radical voluntarism of the fourteenth-century nominalists. Many seventeenth-century mechanists, including Mersenne, Gassendi, Descartes, and Boyle, were deeply indebted to this voluntarist tradition.[46] It was their near obsessive concern with the divine fiat, with God's completely unfettered legislative power (Boyle even went so far as to suggest that it lay within God's prerogatives to create other worlds in which matter *was* animate), that recommended the notion of natural law to these natural philosophers, despite its obvious drawbacks given their ontology. If the "disenchantment of the world" meant ridding nature of agency and understanding, this did not imply a retreat from Judeo-Christian theology: by making nature the other, the seventeenth-century mechanists Christianized (or perhaps judaicized – they were inordinately partial to the Old Testament) natural philosophy to a degree comparable or even surpassing the Thomist synthesis of the thirteenth century.

In the final analysis, it is indeed dubious whether the mechanists could have ever succeeded in purging natural philosophy of anthropomorphism in its entirety, even if they had been willing to jettison the crashing exception of natural laws. By eliminating Aristotle's final and formal causes in nature as anthropomorphic, but retaining his categories of activity and passivity, along with all the values that saturated these terms, they trapped themselves. For the principal explanandum of natural philosophy was what seventeenth-century writers revealingly if paradoxically called the "artificiality" of nature, i.e., its ubiquitous regularities. The very word "artificiality" (and its cognates in Latin and several other European languages) simultaneously meant that which is contrived, which shows evidence of intelligent labor, and that which displays regularities, intricate structure, or complex organization. Boyle and others tried to divert the artistry from nature to God, thus inaugurating a long-lived natural theology of divine final causes that endured until Darwin: nature was no longer artisan, but artifice; and the art of nature was no longer the opulent luxury item, but the finely tuned engine.

But the mechanists never succeeded in explaining how so stupid a nature could manage to function so smoothly as an engine, without constant divine intervention. Instead, they covertly transferred the intelligence of the artificer back into the artifice. Despite the claim that the mechanical philosophy naturalized the artificial, it would be closer to the mark to say that it artificialized the natural. Boyle himself admitted that all his examples of natural phenomena were in fact "artificial bodies,"[47] and Leibniz thought that natural machines were even more artificial than human ones, for "the machines of nature, living bodies, are still machines in their smallest parts, into infinity."[48] For all their protests that divine art was incomparably greater than human art, the seventeenth-century mechanists nonetheless understood nature – "the art of God" – almost entirely on the analogy of human handiwork, and only one narrow class of human handiwork at that. In order to rid natural philosophy of one kind of anthropomorphism, they arguably introduced a still greater anthropomorphism into their conceptions of God as engineer, as David Hume was later to argue in his *Dialogues Concerning Natural Religion* (comp. ca. 1755, publ. 1779). Nature may have become the other, but she was still all too recognizable in her artificiality.

Conclusion

Nature was also – and remains to this day – all too monolithic. What, after all, does it mean to say that nature had become "the other'? It can only mean other than us – in the case of the seventeenth-century version of the anthropomorphism taboo, other than human intelligence and purposefulness. Only from a staunchly anthropo*centric* perspective does "nature" appear as a single, integrated entity, as opposed to a plethora of kinds and individuals that make up the nonhuman universe. There may arguably be natural kinds without humans, but not nature, that great miscellany of the non-us. In the late twentieth century as well as in the late seventeenth century, our antianthropomorphic insistence that nature is the other – the very linguistic usage that allows us to take the census of the universe in one short work, "nature" – rests at least in part upon a profound anthropocentrism.

Notes

1. Michel Foucault, *Les Mots et les choses* (Paris: Editions Gallimard, 1966).
2. Alexandre Koyré, "The Significance of the Newtonian Synthesis," in his *Newtonian Studies* (Cambridge, MA: Harvard University Press, 1965), pp. 3–24.
3. Max Weber, "Wissenschaft als Beruf" [1919], translated as "Science as a Vocation," in H. H. Gerth and C. Wright Mills (eds.), *From Max Weber* (New York: Oxford University Press, [1946] 1958), pp. 129–156.
4. August Comte, *Cours de philosophie positive* [1830], 5th edn., 6 vols. (Paris: Société Positiviste, 1892–94), Vol. 1, Ire Leçon, pp. 1–11; Vol. 5, LIIIe-LIVe Leçons, pp. 92–93.
5. Upholders of this historiography tend to treat the most obvious exception, Germany *Naturphilosophie* of the early nineteenth century, as a short-lived "repalse" into earlier thought patterns. For a more nuanced and broader view of the phenomenon, see Andrew Cunningham and Nicholas Jardine (eds.), *Romanticism and the Sciences* (Cambridge: Cambridge University Press, 1990).
6. Friedrich Schiller, "Die Götter Griechenlands," version of 1798/1800.
7. See for example, Charles T. Fritsch, *The Anti-Anthropomorphism of the Greek Pentateuch* (Princeton: Princeton University Press, 1993).
8. On the themes of analogy and mimicry in Renaissance natural history see Jean Céard, *La Nature et les prodiges. L'Insolite au XVIe siècle en France* (Geneva: Droz, 1977), pp. 230–235 *et passim*.
9. Thomas Nagel, *The View from Nowhere* (Oxford: Oxford University Press, 1986), especially pp. 60–66.
10. Thomas Browne, *Religio Medici* [1642]. Facsimile reprint of 1643 edn. (Menston: Scolar Press, 1970), Vol. 1, pp. 16, 35.
11. All Aristotle translations are from Jonathan Barnes (ed.), *The Complete Works of Aristotle. The Revised Oxford Translation*, 2 vols. (Princeton: Princeton University Press, 1984).
12. Robert Boyle, *A Disquisition About the Final Causes of Natural Things* [1688] in Thomas Birch (ed.) [1772], 6 vols. Facsimile Reprint (Hildesheim: Georg Olms, 1966), Vol. 5, p. 397.
13. Paolo Rossi, *Francesco Bacone: Della magia alla scienza* [1957] (Torino: Einaudi, 1974), pp. 39–40; Francis Bacon, *De augmentis scientiarum* [1623] in Basil Montagu (ed.), *The Works of Francis Bacon*, 16 vols. (London: WIlliam Pickering, 1825–34), Vol. 8, pp. 90–91; cf. René Descartes, *Principia philosophiae* [1644] in Charles Adam and Paul Tannery (eds.), *Oeuvres*, 10 vols. (Paris: Cerf, 1897–1908), Vol. 8, Part IV, Art. 203, pp. 325–326.
14. Paula Findlen, "Jokes of Nature of Jokes of Knowledge: The Playfulness of Scientific Discourse in Early Modern Europe, "*Renaissance Quarterly* **43** (1990), 292–331.
15. Ambroise Paré, *Des Monstres et prodiges* [1573] Jean Céard (ed.), (Geneva: Droz, 1971), p. 139.
16. Guillaume Rondelet, *Histoire entière des poisons* [1558], quoted in Céard 1977, *op. cit.*, Note 8, p. 298.
17. On the *Wunderkammern* see Julius von Schlosser, *Die Kunst – und Wunderkammern der Spätrenaissance: Ein Beitrag zur Geschichte des Sammelwesens* [1908]

(Braunschweig: Klinkhardt & Biermann, 1978); David Murray, *Museums: Their History and Their Use*, 3 vols. (Glasgow: James MacLehose, 1904); Oliver Impey and Arthur MacGregor (eds.), *The Origins of Museums: The Cabinet of Curiosities in Sixteenth- and Seventeenth-Century Europe* (Oxford: Oxford University Press, 1985).
18. Lorraine Daston, "The Factual Sensibility," *Isis* **79** (1988), 452–470, esp. 464–465.
19. Francis Bacon, *New Organon* [1620] Fulton H. Anderson (ed.) (Indianapolis: Bobbs-Merrill, 1960), II. 29, p. 178.
20. Isaac Newton, *Mathematical Principles of Natural Philosophy* [1687], trans. Andrew Motte, rev. Florian Cajori, 2 vols. (Berkeley/Los Angeles/London: University of California Press [1934], 1962, Bk. III; "General Scholium"; Vol. 2, pp. 544–545.
21. Ralph Cudworth, *The True Intellectual System of the Universe* [1678] in Bernhard Fabian (ed.), *Collected Works of Ralph Cudworth*, 2 vols. (Hildesheim/New York: Georg Olms, 1977), Vol. 1, pp. 157–159.
22. Blaise Pascal, *Récit de la grande expérience de l'équilibre des liqueurs [1648]*, quoted in Tetsuya Shiokawa, Pascal et les miracles (Paris: A.-G. Nizet, 1977), pp. 57–58.
23. Gottfried Wilhelm Leibniz, Letter to Thomasius (20/30 April 1609), quoted in Catherine Wilson, "*De Ipsa Natura*: Sources of Leibniz's Doctrine of Force, Activity and Natural Law," *Studia Leibnitiana* **19** (1987), 148–172.
24. Cudworth *op. cit.*, 1678, Note 21, p. 162.
25. *Ibid.*, p. 156.
26. *Ibid.*, p. 157.
27. David-Renaud Boullier, *Essai philosophique sur l'âme des Bêtes* [1728], 2nd edn. [1737] (Paris: Fayard, 1985), p. 343.
28. *Ibid.*, p. 122.
29. Robert Boyle, *A Free Inquiry Into the Vulgarly Received Notion of Nature* [1686] in *Works op. cit.* 1772, Note 12, Vol. 5, p. 165.
30. Robert Boyle, *Of the High Veneration Man's Intellect Owes to God* [1685] in *Works, op. cit.* 1772, Note 12, p. 141.
31. Boyle, *op. cit.*, 1868, Note 29, pp. 183, 172, 173.
32. *Ibid.*, p. 224.
33. *Ibid.*, p. 162.
34. *Ibid.*, p. 278.
35. *Ibid.*, p. 197.
36. Gary Hatfield, "Force (God) in Descartes' Physics," *Studies in History and Philosophy of Science* **10** (1979), 113–140.
37. Walter Charlton, *The Darknes* [sic] *of Atheism Dispelled by the Light of Nature* (London: William Lee, 1652), p. 121.
38. Nicholas de Malebrance *Recherche de la vérité* (Paris, 1674).
39. Boyle, *op. cit.*, 1686, Note 29, p. 163.
40. *Ibid.*, p. 164.
41. *Ibid.*, p. 170.
42. Cudworth, *op. cit.*, 1678, Note 21, p. 151.
43. Wilson, *op. cit.*, 1987, Note 23, p. 162.
44. Boyle *op. cit.*, [1685], Note 30, p. 150.
45. *Ibid.*, p. 142.

adamant on a priori, and largely theological grounds that such anthropomorphisms paid nature too great a compliment, bordering on idolatry. Although Boyle saddled the Aristotelians with the most extreme anthropomorphic interpretations of these maxims, to which Aristotle himself had certainly never subscribed nor more than a scant handful of his disciples, there is internal evidence that it was the Renaissance naturalists whom Boyle actually had between his sights. It is otherwise difficult to explain his preoccupation with extraordinary phenomena and anomalies of all sorts, for these were always marginal to Aristotelian natural philosophy, though of central concern to the Renaissance naturalists, as we have seen. Boyle complained: ". . . it does not a little darken the excellency of the divine management of things, that, when a strange thing is to be accounted for, men so often have recourse to nature, and think she must extraordinarily interpose to bring such things about; whereas it much more tends to the illustration of God's wisdom, to have so framed things at first, that there can seldom or never need any extraordinary interposition of his power."[33]

Boyle's treatment of anomalies – new stars, earthquakes, plagues, inundations, monstrous births – is a tortured and tortuous one, and I cannot explicate it in detail here. His position is a precarious balancing act, for he wished simultaneously to claim that *natura naturans* or ensouled nature does not exist, *and* that such anomalies are proof of its incompetence; that although such anomalies are a slur on the benevolence of *nature*, they are nonetheless proof of *God's* wisdom; that although divine providence plans ahead with infinite precision and care, miracles are not only possible but actual. I want only to remark in passing on the strikingly gendered double standard that keeps this position from toppling under the strain of its own contradictions. Because nature is, for Boyle, self-evidently maternal, "a nursing mother to the creatures,"[34] any injury to human, animal, or plant can be held against her as dereliction of duty. God, in contrast, is not only paternal but patriarchal: His calamities punish the wicked, for His justice (indeed, even His concern for uniformity in the world) always trumps His benevolence in cases of conflict. Nor did His creatures have any right to better treatment. As creator, "he was not bound to make, or administer, corporeal things after the best manner, that he could, for the good of things themselves; among which those, that are capable of gratitude, aught to praise him for having vouchsafed them so much as they have, and have no right to except against his having granted them no more."[35]

46. John R. Milton, "The Origin and Development of the Concept of the 'Laws of Nature', *"Archives of European Sociology* **22** (1981), 173–195; see also Francis Oakley, *Omnipotence, Covenant, and Order: An Excursion in the History of Ideas from Abelard to Leibniz* (Ithaca/London: Cornell University Press, 1984), pp. 82–113; Matthias Schramm, "Roger Bacons Begriff von Naturgesetz," in Peter Weimar (ed.), *Die Renaissance der Wissenschaften im 12. Jahrhundert* (Zürich/München: Artemis, 1981), pp. 197–209; Jane E. Ruby, "The Origins of Scientific, 'Law'," *Journal of History of Ideas* **47** (1986), 343–359.
47. Boyle, *op. cit.*, 1686, Note 29, p. 231.
48. Gottfried Wilhelm Leibniz, *Monadology* [comp. 1714, publ. 1840] in Robert Latta (trans. and ed.), *The Monadology and Other Philosophical Writings* [1898] (New York/London: Garland, 1985), Section 64, pp. 254–255.

THE MANIFEST AND THE SCIENTIFIC

RAPHAEL FALK[1]
Hebrew University of Jerusalem

Biology has evidently borrowed the terms "heredity" and "inheritance" from every-day language, in which the meaning of these words is the "*transmission*" of money or things, rights or duties – or even ideas and knowledge – from one person to another or to some others; the "heirs" or "inheritors."[2]

In his well-known, though less well-read book *Gödel, Escher, and Bach*, Douglas Hofstadter tells the story of the history of non-Euclidean geometry. Euclid attempted to give definitions of ordinary, common words such as "point," "straight line," "circle," and so forth.

How can you define something of which everyone already has a clear concept? The only way is if you can make it clear that your word is supposed to be a technical term, and is not to be confused with the everyday word with the same spelling.... Well, Euclid did not do this, because he felt that the points and lines of his *Elements* were indeed *the* points and lines of the real world.... But this is where he slipped up, for an inevitable consequence of his using ordinary words was that some of the images conjured up by those words crept into the proofs which he created.[3]

Hofstadter suggests that the clue to the realization of the discoverers of non-Euclidean geometry was the uncoupling of their thinking from preconceived notions. The proposition of the non-Euclidians are only

"repugnant to the nature of the straight line" if you cannot free yourself of preconceived notions of what "straight line" must mean.... If, however, you can divest yourself of those preconceived images, and merely let a "straight line" be something which satisfies the new propositions, then you have achieved a radically new viewpoint.[4]

Without going into the dispute of the accuracy of Hofstadter's historic presentation of the details, I wish to claim that this description highlights an important aspect of scientific creativity. Scientists creatively use analogies from concepts of daily life for their hypotheses and theorems, but in the process become captive of the manifest image of the man-in-the-world, inherent in such analogies. As a result, a creative procedure may eventually turn into a handicap for scientific intuitions and creativity.

The creative role of analogous inductive thinking has been well established, and hardly needs further argumentation to convince. In the present paper I wish to discuss the constraining effect of the relation between the manifest and the scientific image, i.e., between the images of the world constructed on the personal experiences of individuals encountering themselves in society, and the striving for images based on impersonal reasoning, which may involve imperceptible conceptual entities. I shall follow primarily two examples from the early history of genetics. The one is the story of the objects of Darwinian natural selection, which is best known through the clash between biometricians and Mendelians in the first decade of the century, and its conceptual (though not actual) resolution through the introduction of the terms phenotype and genotype by Wilhelm Johannsen in 1909.

The other is, in a sense, a more confined conflict, about the meaning of the genetic linkage maps of the chromosomes, as developed in the 1910s by T. H. Morgan and his students. In both cases the manifest image provided the framework within which the scientific image unfolded, but eventually constrained the independent development of the scientific image. Finally, I will also mention a reverse inductive process, namely the relaxation of the constraints of the manifest image of the man-in-the-world in the face of the show of authority of the scientific image, through the case of the apparent authority that genetics conferred on eugenics.

The Manifest and the Scientific Images of the Man-in-the-World

I borrow the juxtaposition of the manifest versus the scientific image of man from the elaboration of Wilfrid Sellars in his *Science, Perception and Realtiy*.[5] The manifest image of man-in-the-world is the framework in terms of which man encounters himself.[6] It is a world constructed of personal experiences of the individual with himself and of his encounter with the immediate objects of his environment, primarily other persons. Although the manifest image encompasses elements such as discipline and critic, it does not include the type of scientific reasoning that "involves the postulation of imperceptible entities, and principles pertaining to them, to explain the behaviour of perceptible things."[7]

According to Sellars, one thing that is common to all philosophical traditions is that they accept the manifest image as the real. "They attempt to understand the achievements of the theoretical science in terms of this framework, subordinating the categories of theoretical science to its categories."[8] Such a subordination of the categories of theoretical science to those of the manifest image had been, according to Hofstadter, the fate of geometry from the days of Euclid until the introduction of Non-Euclidean geometries, two thousand years later. This is also that which is captured in the quotation from Johannsen in the beginning of this paper. Although we note that the scientific image is methodologically dependent on the world of common sense, it "purports to be a complete image, i.e., to define a framework which could be the whole truth about that which belongs to the image. . . . From its point of view the manifest image on which it rests is an 'inadequate' but pragmatically useful likeness of a reality which first finds its adequate (in principle) likeness in the scientific image."[9] In the last part of this paper I will touch on the conflict between the inevitable and the unattainable, namely between the need to draw consequences form the inevitable interrelationship of the manifest and the scientific image and the unattainable ideal of the independent, objective scientific image.

The Controversy on What is Inherited

Transmission of physical and behavioral traits from parents to their offspring has been envisioned since ancient times as analogous to the man-made traditions and rules of property inheritance. Thus, it is not surprising to find that when students of the life sciences endeavored to develop a theory of biological inheritance they took their clues from the manifest image of material inheritance, i.e., they developed theories of the transmission of the perceptible traits or properties of living organisms.[10]

In 1866 Gregor Mendel published his work that could have laid the foundation for the disengagement of the transmission of properties from that of the inheritance of the *factors* that provide for a potential for traits. However, this opportunity for the uncoupling of the scientific image of inheritance from a manifest image was appreciated neither by Mendel's contemporaries, nor by the "rediscoverers" of Mendel at the turn of the century. At the beginning of the twentieth century, when the modern science of genetics was founded, not only were events channeled by the manifest image of heredity, but to some extent, its long shadow has continued to exert its impact on the scientific image up to the present day. Scientists, geneticists included, never gave up the concept of the inheritance of traits. Although the Gordian knot of traits and inheritance was cut by Johannsen and by Morgan in 1910, it is a common practice even nowadays to relate to "genetic traits,"[11] and to assign specific genes to traits which are defined according to various criteria.

After the publication of the *Origin of Species* in 1859, nothing in biology was the same again. In the decades that followed, even biologists who accepted Darwin's theory of evolution by natural selection of preexisting variants, were profoundly divided. There were those who accepted Darwin's conception of gradual evolutionary changes within species that eventually accumulated to such an extent that new species were established, and those who accepted Darwin's mechanism of evolution in principle, but denied that the evolution of new species was a direct extension of intraspecific variation. Thomas Huxley, although an ardent supporter of Darwin, adhered to the concept of saltational evolution of new species, as did Francis Galton. Being captives of the manifest image of heredity, they justly questioned how selection among organisms that varied in some traits would lead to the hereditary establishment of these traits in the selected populations. The difficulties were both at the phenomenological and at the empirical levels. In spite of some examples of variation that spanned the interspecific borders, and of the fuzzy nature of the borders of delineation of some species, in the majority of cases assigning an organism to one species rather than to another was not difficult. Furthermore, according to existing hereditary notions (primarily that Darwin's pangenes), interbreeding of selected organisms with others would "swamp out" the achievements of selection within a few generations to such an extent that no known mechanism could replenish them. Thus, although Darwin provided a metaphor for the material basis of the process of evolution, his central thrust of reducing the origin of species to the origin of intraspecific variation, was challenged by these scientists.

This sentiment of many biologists of the last decade of the nineteenth century can be illustrated by William Bateson's switch from

The Manifest and the Scientific 61

research in embryology to the study of natural history, or, as Coleman called it, "his rebellion against the morphological tradition."[12] Bateson was interested in the problems of evolution of species. Following Darwin, he, as many of his contemporaries, first approached the study of evolution as embryologists would. His disappointment in this path is captured most clearly in his (unsuccessful) application letter for the post of deputy to the Linacre professor at Oxford in June 1890, where he pointed out that "since the promulgation of the doctrine of Descent" a new class of facts is sought and a new kind of question is asked.

Now, the readiest method of answering these questions seemed to be the embryological method. By using the so-called Law of von Baer, that the history of the individual is the history of the species, it should be possible to determine the pedigree of the form by examining the manner of its development. . . . but at the outset, embryological evidence as to the history of Evolution is open to the objection that it rests on an error in formal logic. For the stages through which a particular organism passes in the course of its development can be taken as indications of its pedigree only when it shall have been proved as a general truth that the development of individuals does follow the lines of which the species was developed. But the proof of this proposition rests entirely on the observation that in the development of animals stages are passed through in which they resemble other forms: from this resemblance it is inferred that they are descended from those forms. Hence the truth of the general proposition is established assuming it true in special cases; while its applicability to special cases rests on its being accepted as a general truth.[13]

In the introduction to his treatise of 1894, Bateson noted that according to Darwin's theory of evolution by continuous modification

specific diversity of form is consequent upon diversity of environment, and diversity of environment is thus the ultimate measure of diversity of specific form. Here then we meet the difficulty that diverse environments often shade into each other insensibly and form a continuous series, whereas the Specific Forms of life which are subject to them on the whole form a discontinuous series.[14]

Bateson saw the problem of evolution as one of selection of traits, and could not visualize how a continuously acting variable (the environment) would be operational in shaping a discontinuous fauna of distinct species. He was, of course, well aware of the long and complex process of the development of these traits. Having been an embryologist himself, his conceptual view of development was primarily that of an epigeneticist, believing that all the new structures of the organism developed from an originally undifferentiated mass of living material in the course of embryonic development. Yet, the details of the developmental process seemed not too relevant to the process of evolution, and accordingly he felt no need to make a conceptual distinction between the traits and the imperceptible entities that provided the potential for the development of the traits.[15] Thus, he could not envision that there might be a difference between selection *of* the trait and selection *for* the trait, namely between *selection of objects* and *selection for properties*.[16] Had he obtained such insight it would have been quite feasible for him to realize that while the one variable ("diverse environments") varied continuously, the other ("specific forms") varied discontinuously. This lack of distinction was grounded in the mechanistic reductionist philosophy that prevailed in the life sciences of most of the nineteenth century. It was a direct expression of the manifest image and its transformation into the scientific image. Man-made machines dominated not only the economy and social life of the century, but also the scientific concept of the world.[17] Living organisms were assemblies of "unit characters," which were integrated at the morphological as well as at the physiological and behavioral levels. Thus there was a one-to-one relation between the unit characters and the factors (or *Anlagen*, the embryological primordia) of these units.

Biologists, who for ages disputed whether traits were already preformed in the fertilized egg or whether they unfolded only during embryogenesis, found themselves in agreement on the existence of some determinative structures or functions in the shaping of properties. And with the encroachment of the physical, mechanistic-reductionist thinking into biology, discrete properties or traits of organisms became realities, not just heuristic devices of efficient research.

Although speculations on the existence of particular determinants of traits were put forward by Darwin as well as by his successors, all variants of such pangene theories made no concpetual distinction between the particles and the traits that the particles were replacing. The gemmules of Darwin, or the pangenes of Weismann or de Vries, did not demand a change of the image of the organism as an assembly of unit-characters, on which natural selection exerted its effect. When de Vries in the first months of 1900 finally grasped the potential hidden in Gregor Mendel's paper for his notion of heredity, he did not waste time. He opened his "rediscovery" papers of 1900 not with Mendel, but with de Vries:

According to pangenesis the total character of a plant is built up of distinct units. These so-called elements of the species, or its elementary characters, are conceived of as tied bearers of matter. A special form of material bearer corresponding to each individual character.[18]

Mendel's work provided Bateson and de Vries with the necessary rationale for a reduction of the theory of the origin of species to that of the origin of new traits that are not "swamped out" long before a new species could be established: traits and their primordia were really discontinuous as shown by Mendel's laws of segregation. For Bateson and for de Vries, Mendel provided the theoretical framework within which they could establish their manifest image of Darwinian evolution in a context of a scientific image.

Their opponents, primarily Weldon and Pearson, did not contest the manifest notion of selection or that of the inheritance of traits: traits were the entities of inheritance, and selection directly affected their frequencies. Pearson, the logical positivist, would have no man-in-the-world images pollute the autarchic scientific image that was based on directly observed data. Notwithstanding, when he adopted Quetelet's and Galton's notion that the statistical parameters of populations represented the true "types" of populations, Pearson did impute the manifest image on the biological meaning of the data.

It was Johannsen who exposed this inconsistency in Pearson's logic. Johannsen's observations of breeding experiments led him to conclude that "supposedly races of cultivated plants can, even when they *appear* completely homogeneous, *consist* of several independent forms with different proportions."[19] Thus Johannsen was eventually led to conclude that an explicit distinction should be made between the level of selection of *objects*, namely the selection of *Phenotypes*, and selection for *properties*, namely the selection for *genotypes*. Johannsen's analysis suggested that both saltationists and biometricians were barking up the wrong tree. Johannsen's explication of the two levels established a conceptual framework in which students of heredity could have formulated their assertions without committing themselves to the inheritance of traits. Johannsen thus raised the problem of "Heredity and Variation" – as de Vries called it – from a manifest to a scientific level, or put differently, he allowed the young science of genetics to free itself from the grasp of the manifest image of heredity that has been preserved in the conception of post-Darwinian evolutionism.[20]

As to the "heredity" of "Heredity and Variation," geneticists studied the transmission of imperceptible conceptual entities, the genes, by following traits that served as their markers. By and large they maintained a heuristic separation between the question of the transmission of these genetic entities and the question of their manifestation during the development of the individual. On the other hand, as to "variation," R. A. Fisher[21] and his successors formalized the link between the theory of evolution by natural selection of continuous traits and the theory of *selection of* discontinuous genetic determinants of traits. Yet, although the uncoupling of a scientific image of inheritance and variation from a manifest image was a decisive step in the establishment of genetics as an independent discipline with its own type of questions and methodology, the grasp of the manifest image on the scientific image was never relaxed. The concept of a "genetic trait" has not disappeared from the genetic discourse and it haunts geneticists of today no less than it misleads the man in the street.[22]

Conceptual Versus Material Maps

As I endeavored to show elsewhere,[23] one of the immediate consequences of Johannsen's differentiation between phenotype and genotype was to pave the way for Thomas Hunt Morgan to separate problems of inheritance as transmission from those of inheritance as embryological development.[24] Starting in 1910, Morgan made peace with the "evidence" for the theory of inheritance provided by cytologists[25] and readily took his clues from their images. Already on September 10, 1911, Morgan submitted to *Science*[26] a short note in which he "venture[d] to suggest a comparatively simple explanation" for the phenomenon of coupling, that is, for the increasing number of observed deviations from the independent segregation of genetic markers predicted by the Mendelian laws of inheritance. Morgan did not accept Bateson's formal interpretation for the phenomenon discovered by him. He adopted an overt manifest physicalistic image of the chromosomes as strings or cords of beads of genes, and chromosome pairing in meiotic prophase was perceived as intimate twisting of pairs of these strands around each other. Such a manifest image was receptive to the notion that mechanical torsion forces eventually disrupted the strand. Once the torsion force was relieved as a consequence of the disruption, some strands could rejoin with the "wrong" partners. "If the materials that represent these factors are contained in the chromosomes, and if those factors that 'couple' be near together in a linear series," and if cytological observations "that during strepsinema stage homologous chromosomes twist around each other," then we may come up with a consistent image. We may interpret the observed deviations form the Mendelian expectations as consequences of the mechanical forces acting on twisted strands. The differences in coupling frequencies observed between various traits (i.e., the variation in the deviations from the expected Mendelian random segregation of the factors) was a consequence of "the linear distance apart of the chromosomal materials that represent the factor. . . . The results are a simple mechanical result of the location of the materials in the chromosomes."[27] The physical-mechanical image of chromosomes in meiosis as semirigid tangible cords of everyday life (based on microscopical observations of fixed cytological material) were the source of the induction that the genetic factors are physical entities arranged along the chromosomes. Phenomena that were identified by the interpreting terms "coupling" and "repulsion" (anchored in Bateson's nonparticulate vibrational concepts of matter)[28] were given a simple mechanistic manifest image of linkage of genetic matter, and the data were redescribed by the interpretative term "crossing over." Sturtevant succinctly summarized the integration of the observations into the theory by noting that these "could happen only if 'crossing over' is possible."[29]

Although Morgan was aware of the limited deductive power of the metaphor of the semirigid entities of the man-in-the-world to the results of breeding experiments, he and his students saw in the linkage wording of the phenomenon a most powerful image that, as a matter of fact, fueled all their research programs over the next thirty years. The nonfree association of factors

> means, on the chromosome view, that the chromosomes, or at least certain segments of them, are more likely to remain intact during reduction than they are to interchange materials. On the basis of these facts Morgan has made a suggestion as to the physical basis of coupling. He uses Janssens' chiasmatype hypothesis as a mechanism.[30]

The fertility of this image was revealed in Sturtevant's comment: "It would seem, if this hypothesis be correct, that the proportion of 'crossovers' could be used as an index of the distance between any two factors."[31] Upon this insight Sturtevant constructed the first linear map of a chromosome, in which '[t]he unit 'distance' is taken as a portion of the chromosome of such length that, on the average, one cross-over will occur in it out of every 100 gametes formed."[32] Note, however, that Sturtevant wrote 'distance' in quotation marks. The cytological image provided only a manifest framework for the scientific genetic image. As he hurried to explain: "Of course there is no knowing whether or not these distances as drawn represent the actual relative spatial distances apart of the factors." Yet, the material, mechanical transfer concept was too lucrative, and he continued the same sentence in explicit physicalistic language" . . . but what we do know is that a break is far more likely to come between C and P than between B and C."[33] The interfield transfer[34] from cytology to genetics goes here beyond that of theories; it is extended to encompass the nonscientific manifest images of mechanical twisting of two cords. although the linkage maps were formally nothing but abstractions described in terms of linear arrangements of the genes along the chromosomes, i.e., they were accepted in the framework of the scientific image, the manifest image of material chromosomes in which exchange takes place in proportion to the physical distances along the chromosomes was from early on overwhelming.

How Far Does the Image Go?

The intertwining of the two images is uniquely exposed by the assault of William Castle on the claim for the linear arrangement of the genes in the chromosome, and the responses of Morgan and his students, notably that of H. J. Muller to that assault. Castle reiterated Morgan's insight: "Morgan has suggested that what binds or links two characters together is the fact that their genes lie in the same body within the cell-nucleus."[35] He agreed that the evidence for Morgan's conclusion that such bodies are the chromosomes "is very strong." Castle also accepted that Morgan had established that "the genes within a linkage system have a very definite and constant relation to each other." However, he was doubtful of the image of the twisted cords and its culmination that "the arrangement of the genes within a linkage system is strictly linear."[36] It went beyond both established scientific knowledge and available sense data, and depended on accessory hypotheses that were nothing but *ad hoc* assumptions nourished by the manifest mechanistic image of the chromosomes as cords twisted around each other. First, the material counterpart of the image, the chromosome, does not comply with what we know about the characteristics of matter. "It is doubtful . . . whether an elaborate organic molecule ever has a simple string-like form."[37] Second, the model holds only within a restricted range of values even as a predictive working hypothesis. "In reality it has been found that the distances experimentally determined between genes remote from each other are in general less than the distances calculated by summation of supposedly intermediate distances."[38]

Morgan's model was no doubt aesthetic and simple, but to account for its discrepancy Morgan adopted "certain subsidiary hypotheses, of 'coincidence,' 'double crossing over,' etc."[39] Castle, on the other hand, took the data "exactly as they stand," and found it possible "to make a very complete and on the whole self-consistent reconstruction of the architecture of the sex-chromosome from the linkage relations of its genes."[40] After considering the data, presumably without any preconceived notion of a manifest image, he came up with "a roughly crescentic plate longer than it is wide, and wider than it is thick" (Figure 1). Castle asserted rhetorically, "What, it might be asked, does this reconstruction signify? Does it show the actual shape of the chromosome, or at any rate of that part of it in which the observed genetic variations lie? Or is it only a symbolical representation of molecular forces?"[41] Castle tries to resist the intuitively attractive, yet "uncritical" traffic of conceptions between the manifest and the scientific image of the mapping procedure. Even though the scientific model was conceived on a manifest image, there must be a limit to further adoption of ad hoc images, merely to maintain a scientific model pleasing to the man-in-the-world image of "reality." To Sturtevant and his colleagues' claim that "this curvature is due to the phenomenon of double crossing over,"[42] Castle answered that the image of "double crossing-over need not be assumed as an explanation of the observed regrouping."[43] Lucrative as it may be to visualize genes as beads on a one-dimensional chromosome string, we should be careful not to impute more meaning to the linkage data than they actually allow. He offered his three dimensional map as a possible alternative for the explanation of the data. However, upon reading his explanations it becomes obvious that he could no more avoid imputing "scientific reality" to his three-dimensional image than Morgan's group could with their one-dimensional linear image.

It is no coincidence that the most explicit antagonist to Castle was Muller, who early on claimed that genes are real material entities – the atoms of inheritance – rather than constructs of the experimenter that can be treated as if they were real, as professed by Morgan and Sturtevant.[44] Muller examined one by one the assumptions and the concepts of Castle's "declaration" which was "so widely at variance with the conclusions jointly agreed upon by all Drosophila workers."[45] Muller agreed that Castle's argument was an argument without a flaw, *"once the premises are admitted."*[46] However, these premises were what Muller challenged. Muller endeavored to explicate the relations between sense data, the manifest image of the observer-scientist as a man-in-the-world, and the scientific image as it should be endorsed, in a style that is rare in scientific literature.

Muller first presented a manifest image, "a chain," that the sense data impose on us. However, only the essential bipolar linkage characteristic of the image has relevance for the scientist.

Whether or not we regard the factors as lying in an actual material thread, it must on the basis of these findings be admitted that the forces holding them linked together – be they physical, "dynamic" or transcendental – are of such nature that each factor is directly bound, in segregation, with only two others – in bipolar fashion – so that the whole group, dynamically considered, is a chain.[47]

As a matter of fact, the scientist should keep aloof of any preconceived manifest images or scientific biases.

This does not necessarily mean that the spatial relations of the factors accord with these dynamic relations, for it is conceivable a priori that factor A might be far off from B, in another part of the cell, or that both might be diffused throughout the cell, and that they might nevertheless attract each other, during the segregation division, by some sort of chemical or physical influence.[48]

Manifest images were only crutches that helped to overcome a handicap in our ability to abstract, with no material analogy intended. The scientific image is strictly a mental abstraction.

In the discussion that follows, no implication is intended when the term "linear series," "distance," etc., are used; these will refer only to the relations existing between the points in the linear map, which may be regarded merely as a mathematical mode of representation of the data themselves.[49]

The *abstract* scientific image relates, however, to a real world, and as such it may be meaningfully juxtaposed with it, notwithstanding its intuitive origins in a manifest image.

It will be shown, however, at the conclusion of this article, that when the various conditions which have to be fulfilled at segregation are taken into consideration, any other explanation for these peculiar linear linkage findings than an arrangement of genes in a spatial, physical line proves to be hazardously fanciful.[50]

Observations lead us to imagine relations like those that we are acquainted with in daily life (such as that of a chain and its links). When we form a scientific theory we should disengage ourselves and abstract from such manifest images.

Muller agreed that there was no compelling evidence to make the linkage map model concrete. Furthermore, he insisted that "[i]t has never been claimed, in the theory of linear linkage, that the per cents of crossing over are actually proportional to the map [physical] distances: what has been stated is that the per cents of crossing over are calculable from the map distance – or, to put the matter in more mathematical terms, that the per cents of crossing over are functions of the distances of points from each other along a straight line."[51] Muller also admitted that Castle's abstract model accommodated his chemical perceptions. The premise that made him reject it was his physical model, founded on the intuitive manifest image. "[T]here is an a priori objection . . . to accepting a curved line as an explanation to the linkage relations, in that it is very difficult to imagine a plausible set of conditions in the chromosome which would hold the factors rigidly in this curved line."[52] It was to uphold this physicalistic model, that double crossovers were suggested to justify the nonlinearity of "distantly located genes" as well as the absence of crossover values exceeding 50%. And since these accessory hypotheses were not enough, an additional one, that of "interference" between crossing-over events, was superimposed to explain the absence of double cross-overs predicted by the previous accessory hypothesis.

Hard as he tried, Muller could not be satisfied with such a biased argument for the physicalistic model, hazardously dependent on manifest images, as a refutation of Castle's premises. The only rational way to bolster his a priori manifest image argument was by providing crucial empirical evidence for it. "[Q]uite aside from a priori reasons, there is an experimental result absolutely fatal to the curved line 'explanation'; this consists in the finding of those classes which are termed by the *Drosophila* workers 'triple crossovers.'"[53] Indeed, it was the extensive empirical evidence, intertwined with the penetrating rational analysis of its consequences and of the alternatives, that made Muller's work[54] a masterpiece of scientific inquiry, and convinced that genetic community in the heuristic value of Morgan's manifest image of the chromosome. Fifty years later, L. C. Dunn noted in his *Short History of Genetics* that "Muller's study was important in showing how far theory could be carried by purely genetical methods."[55] The model, which had its foundations in the manifest image of the twisted cords, got further support in the 1930s, from Curt Stern and by H. B. Creighton and Barbara McClintock, when cytological techniques were developed that proved the correlation of genetic and cytological exchange. "However, so thorough had been the genetical experiments, like those of Muller, that Stern's demonstration seemed anticlimactic."[56]

Muller's discrimination between the intuitive heuristic value of manifest images and the rational-empirical analysis of the scientific images, did not reduce the impact of the manifest image on the creative thinking of geneticists. Just the opposite happened. Belling's model[57] of copy-choice replication notwithstanding, the image of twisted cords that rupture by mechanical tensions prevailed well into the molecular era. Phenomena that did not tally with it, such as deviations from the one to one segregation of the products of meiotic events in which exchange took place[58] (later known as "gene conversion") were summarily declined. However, with the increasing dissatisfaction of mechanistic models as explanations at the cellular and molecular level, and with the increasing insights of the power of molecular-chemical mechanisms, the constraints of the mechanical attraction and dissociations and of the torsion forces were eventually replaced by models of specific chemical bonds and enzymatic reactions.

Eugenics, the Authority of the Scientific Image

No less impressive than the effect of the manifest image on the scientific image, and the mutual enrichment of the manifest and the scientific image, is the impact of the scientific image on the manifest image. To this contributed both the ideal of the scientific image as a sublime, more objective image, and the "proof" of the truth of such an image, through the accomplishments of the industrial (and the ensuing scientific) revolution.

Darwinism had been rapidly connected to established social concepts after the publication of the *Origin* in 1859. Similarly, the establishment of Mendelian genetics quickly became entangled in social and racial prejudices. As early as 1908, it was claimed by Raymond Pearl, that the science of genetics is approaching the state "when it will be possible for a person to give an order to a breeder specifying that he would like to have a plant or an animal embodying such and such characters, and after a due lapse of time, to receive from the breeder precisely what he ordered, just as though he were dealing with a manufacturing mechanist."[59] At the face of such "truly wonderful advance in man's ability to control nature" would anybody hesitate to apply science to "the amelioration of man himself"?[60] Thus Pearl set out to incorporate his conceptions of society and culture into the new scientific image relating to biological abilities. Of the social movement

which we commonly call "the advance of civilization,' . . . society, acting principally through both Church and State, has continuously endeavored to make man better.
. . . these lines of endeavor, successful as they have been in their respective fields, have not touched one possibility of amelioration of the human species. They . . . have nothing directly or effectively to do with his character or quality *at* birth.[61]

Now that we as scientists know better, wouldn't it be beneficial to apply science to the social and moral properties of man? "[I]f this is a good way in which to improve animals and plants in general, is it not worth considering in connection with the problem of the betterment of the human race?"[62] Asking "Are mental and moral qualities likewise inherited, and if so with what degree of intensity?" Pearl, the expert in animal breeding, refers to Karl Pearson, who "spent several years in the collection and reduction of data on the degree of resemblance between brothers and sisters . . . [which] included such things as 'ability,' 'temper,' 'vivacity,' 'assertiveness,' 'conscientiousness,' and other similar attributes."[63] Thus, Pearl sublimed to the precarious way by which imperceptible genes were assigned to well-defined qualitative biological traits, and, under the hallowedness of the scientific image, extended the generalizations of genetics, not only to quantitative traits definable in biological terms, but also to human traits defined in behavioral and social terms. From the elevated pedestal of the authority of science, he asserts: "the evidence indicates that mental and moral characters are inherited, and with about the same degree of intensity as physical characters."[64]

In 1916 Castle published his textbook for students of biology, *Genetics and Eugenics*, which became the most widely used college text in the field, going through four editions in fifteen years.[65] In the introduction he wrote:

since man is zoologically merely one of the higher animals, it is evident that his reproduction is a very special case falling under the general laws of genetics, and before we can properly understand this special case we must know something of the general laws of genetics. . . . To determine what are the general principles of genetics and to what extent man is subject to them are primarily biological problems, but to determine how far these are socially controllable is a problem for the sociologist, and one which I shall not attempt to answer without help from sociologists."[66]

The position that Castle takes, which is typical of students of the sciences, is that the study of the biology of man is a pure scientific problem, and that the social aspects, though not the business of the scientist, must take primary notice of the scientific-biological determinations. In other words, the scientific image is above, and independent of the manifest socio-cultural image of man-in-the-world. It is significant, however, to note how Castle on the first page of the chapters dealing with eugenics, in spite of his declaration of the supremacy of the scientific image on the manifest image, and its independence of it, accepts the power of social considerations to check the scientific claims. "The statement is often made that mixed races are feeble, but if this is ever true it is not because they are mixed, but because the specimens that mix are feeble. Mating out of the race, when mates within the race are available, is *prima facie* evidence that the individual so mating is a social cast."[67] And some lines further on: "But human racial crossing in general is a risky experiment, because it interferes with social inheritance, which after all is the chief asset of civilization."[68]

Less careful and less scrupulous about the extension of scientific generalizations to the social aspects and the license to exert it, was Castle's former teacher, the first American who published on Mendelism in 1901,[69] Charles B. Davenport. Davenport opened the preface to his 1915 book, *Heredity in Relation to Eugenics*, noting the great advances in our knowledge of heredity that have revolutionized the methods of agriculturists. He continued, "It was early recognized that this new knowledge would have a far-reaching influence upon certain problems of human society – the problems of unsocial classes, of immigration, of population, of effectiveness, of health and vigor."[70]

Davenport called attention to the biases that the manifest-image-directed scientific conceptions were led to in the past. This tendency should be stemmed and reverted: "Modern medicine is responsible for the loss of appreciation of the power of heredity. It has had its attention too exclusively focused on germs and conditions of life. It has neglected the personal element that helps determine the course of every disease."[71] A scientific conception that liberated itself from the impact of the manifest image of the uniqueness of human being, must view man as "an organism – an animal; and the laws of improvement of corn and race horses hold true for him also. Unless people accept this simple truth and let it influence marriage selection human progress will cease"[72] The extension of the biological generalizations to the social aspects of human life established the supremacy of the biological scientific images. It is from such a perspective of the authortiy of the scientific image that he urged not to be content with the "euthenical" standpoint (i.e., that of improving the human environment as a means of managing human biological resources), and proposed to impose the eugenic program: "The general program of the eugenist is clear – it is to improve the race by inducing young people to make a more reasonable selection of marriage mates; to fall in love intelligently."[73] Once it has been accepted that the scientific image is superior to the manifest one, the eugenic program "also included the control by that state of the propagation of the mentally incompetent" although, for the time being, it fell short of implying "the destruction of the unfit either before or after birth."[74]

This arrogance of the scientists about the primacy of the scientific image was by no means limited to "racists" and their like. In August 1932, H. J. Muller, delivered what was considered to be a "leftist political" speech to the Third International Congress of Eugenics in New York. In it, according to Carlson, [75] Muller gave air to his becoming "increasingly incensed over the shallow thinking of eugenicists." Many who attended the talk felt that Muller had leveled a mortal blow at the eugenics movement,[76] and in the printed text of it Muller added a footnote that referred to "the lack of justification for the assumption made by most eugenicists, that our present social stratification is positively correlated with genetic worth."[77] Yet, the second paragraph of Muller's talk states "That genetic imbeciles should be sterilized is of course unquestionable."[78] For Muller it was immoral to fail to apply scientific knowledge to human affairs. What was true was also just. However, once scientists took license to be the arbitrators of "knowledge," of how things are, and the brokers of the primacy of such an image of science over that of any manifest image of the man-in-the-world, the stage was set for things to come. Categorical statements, such as form the outset there could be no doubt whatever that mental racial differences exist," in the authoritative text on *Human Heredity* of Baur, Fischer and Lenz in its English translation[79] must have been accepted with the same seriousness that they have been uttered. We may raise an eyebrow to assertions such as these:

Least of all human beings is the Nordic, the slave of the moment, for he excels the members of all other races in constancy of will and foresight. Owing to his Promethean character, he inclines to subordinate his sensual impulses to a remote aim. . . . These mental peculiarities of the Nordic race are obviously connected with the northern environment. . . . Persons who wish to keep themselves alive in the northern environment must be able to build durable houses and seaworthy ships. That is why most technicians and inventors to-day belong to the Nordic race.[80]

Yet, this only reflects the prevailing optimistic vein of confidence in the achievements of scientists, believed to have been obtained by the sublimation of the scientific image over the manifest image and its uncoupling from it. As is so explicitly expressed in the above mentioned quotation from Davenport,[81] the liberation of man from the determinism of "euthenics" – of the fate imposed by social and political determinants – must be expected to be accompanied by the acceptance of subjugation to the inevitable determinants of "eugenics.

The Inevitable and the Unattainable

The scientific image has for centuries been grounded in the manifest image of man-in-the-world. From there it derived its conceptual frameworks, its intuitions and metaphors. The manifest image has always been the fountain of its models. In reciprocation the scientific image constantly modified the manifest image. The relationship between the two has always been one of inevitable interdependence. Yet, this inevitable interdependence has never been accepted with acquiescence by thinking persons. The constant need for asking questions and demanding explanations precedes the elements of Cartesian skepticism and Baconian empiricism as the hallmarks of the scientific image. These are the elements that generate the scientific image, that can establish it beyond the trivial, the obvious, and the daily. These, however, are also the elements that nurse the concept that although the scientific image gets its clues from the manifest image, it must sublimate the manifest image and become independent of it, in order to develop meaningful concepts and interrelations of its own. Euclidean geometry was constructed directly on a manifest image, non-Euclidean geometry was established only after the scientific image freed itself from that manifest image. The science of heredity started with the manifest image of inheritance, but it could develop independently only after it freed itself of the manifest image of the inheritance of traits. Also, the conceptual framework of the theory of evolution was given new impetus when genetic theory relieved it from the manifest image of the heredity of traits. Castle cast doubts at the linear model of the chromosomal maps because it was directed too much, to his taste, by the manifest intuitive image of chromosomes. Muller claimed that the scientific image was inevitably construed according to the manifest image, and, since any other interpretation was hazardously fanciful, he tried the unattainable, a rational-empirical foundation of a scientific argument, independent of the manifest image.

Philosophers of science were divided between rationalists who constructed the scientific image as one that confronts and scrutinizes the manifest image of the man-in-the-world, and empiricists, who constructed a scientific image that should have superseded the manifest image. Logical positivism, that hybrid of rationality and empiricism, was probably the latest attempt to establish the unattainable, the myth of a science both objective and rational, which examines truth, the whole truth and nothing but the truth. In short, it craved for a normative scientific framework founded on solid rational analyses of experience. With such a conception, the scientific image could deal only with the context of justification, and shunned any relations with intuitions, inductions, or metaphysical speculations of the context of discovery. Its scientific image was that of reduction, because the only objective way to scrutinize and supersede the manifest image was by accepting science as a methodology that strips problems and phenomena to their elementary entities, and then reconstructs these elements by applying nothing but elementary objective principles.

It is an alarming fact that the (so far) final disillusion from this naively optimistic image of the scientific needed the high price it had to pay. Social-Darwinism and eugenics were enshrined in the name of the scientific hallowedness. As such they upheld the impact of social and racial injustices that raged in many of the most developed industrial-scientific societies. This led also to the adoption of a totalitarian regime by a scientific community, in the hope to be able to realize that image of the scientific, such as happened in Nazi Germany. The unassailable belief in the independence of the scientific image, and accordingly in its possession of "The Truth," presumptively provided scientists with the possibility, even the need, to deal with the issues of scientific matters objectively, without the interference of subjective or moral issues of the manifest image. It is ironic that while scientists in Germany hoped that the Nazi regime would provide them with the opportunity to enforce their scientific truths, many of the scientists who actively contributed to the Allies' war effort during World War II testified that they did so out of the sincere conviction that the truths of the scientific image were *not* detached from the manifest image. It is important to rehearse these issues today, when the belief in the power of science has again attained such colossal dimensions.

Notes

1. I wish to thank Dvora Kamrat-Lang and Sara Schwartz for their penetrating and thoughtful comments.
2. Wilhelm Johannsen, "The Genotype Conception of Heredity," *The American Naturalist* **45** (1911), 129–159, here, p. 129.
3. D. R. Hofstadter, *Gödel, Escher, Bach, an Eternal Golden Braid* (New York: Basic Books, 1979), p. 90.
4. *Ibid.*, p. 92.
5. Wilfrid Sellars, *Science, Perception and Reality* (New York: Routledge and Kegan Paul, 1963).
6. *Ibid.*
7. *Ibid.* p. 7.
8. Sellars, *op. cit.*, 1963, Note 5, p. 19.
9. *Ibid.*, p. 20.
10. In such a context it is, however, difficult to understand the common belief that once a person has inherited "genes" for some trait – especially "genes for some defect" – he or she is destined to harbor the trait, come what may, as if it were an inevitable fate determined by some laws of nature. We do not treat the laws and rules of society (at least, not any more) as inevitable deterministic dooms, and this pertains not only to crooks and habitual law breakers! I suspect that this belief relates essentially to the meta-image of the manifest and the scientific. Whereas the deterministic will of God has to a large extent been progressively eliminated from the rules, traditions, and laws of manifest images of the individual and of the social man-in-

the-world, it has not been eliminated at all from the image of the man-in-the-world of science – or, more appropriately, Science, with a capital "S." In this respect, Science merely took over much of the former rôle of the manifest image of God.
11. See R. Falk, "The Dominance of Traits in Genetic Analysis", *Journal of the History of Biology* **24** (1991), 457–484; and especially Fred Gifford, "Genetic traits," *Biology and Philosophy* **5** (1990), 327–347; and Kelly C. Smith, "The New Problem of Genetics: A Response to Gifford," *Biology and Philosophy* **7** (1992), 331–348.
12. W. Coleman, "Bateson and Chromosomes: Conservative Thought in Science," *Centaurus* **1** (1970), 228–314, here, p. 242.
13. Beatrice Bateson, *William Bateson, F.R.S.* (Cambridge: The University Press, 1928), pp. 31–32.
14. W. Bateson, *Materials for the Study of Variation* (London: Macmillan, 1984), pp. 2–5.
15. This identity between the unit-character and its genetic determinant was also expressed in Bateson's "presence and absence" theory, that he devised after the "rediscovery" of Mendel's work, of gene function: The variability of a Mendelian trait was directly the function of whether the proper gene was present or absent. There were, of course, interactions that resulted in deviations from the Mendelain ratios. But these were due to interactions between the unit-traits. Such was the case when chicken with a "pea" comb were mated to chicken with a "rose" comb, and gave progeny that had a "walnut" comb.
16. See E. Sober, *The Nature of Selection* (Cambridge, MA: MIT Press, 1984), p. 97ff.
17. For a stimulating discussion of the prevailing "context of justification" as provider of fuel to the "context of discovery" see G. Gigerenzer, "From Tools to Theories: A Heuristic of Discovery in Cognitive Psychology," *Psychological Review* **98** (1991), 254–267.
18. Hugo de Vries, "Das Spaltungsgesetz der Bastarde," *Berichte der deutschen botanischen Gesellschaft* **18** (1900), 83–90. Translation quoted from Curt Stern and Eva R. Sherwood, *The Origin of Genetics* (San Francisco: Freeman, 1966), p. 107.
19. W. Johannsen, *Über Erblichkeit in Populationen und reinen Linien* (Jena: Gustav Fischer, 1903), p. 176 of the translation of excerpts in *Selected Readings in Biology for Natural Sciences* **3** (Chicago: University of Chicago Press, 1955), pp. 172–215.
20. See R. Falk, "Mendel no de Vriesian," submitted for publication, 1993; R. Falk, and S. Schwartz, "Morgan's Hypothesis of the Genetic Control of Development," *Genetics* **134** (1993), 671–674.
21. R. A. Fisher, "The Correlation Between Relatives on the Supposition of Mendelian Inheritance," *Trnasactions of the Royal Society of Edinburgh* **52** (1918), 399–433.
22. See Falk, *op. cit.*, 1991; Gifford, *op. cit.*, 1990; Smith, *op. cit.*, 1992, Note 11.
23. Falk, *op. cit.*, 1993, Note 20.
24. Coleman, *op. cit.*, 1970, Note 12. pp. 239–241.
25. Thomas Hunt Morgan, "Chromosomes and Heredity," *The American Naturalist* **44** (1910), 449–496.
26. T. H. Morgan, "Random Segregation versus Coupling in Mendelian Inheritance," *Science* **34** (1911), 384.
27. *Ibid.*
28. Coleman, *op. cit.*, 1970, Note 12, pp. 263ff.

29. Alfred H. Sturtevant, "The Linear Arrangement of Six Sex-linked Factors in Drosophilia, as Shown by their Mode of Association," *The Journal of Experimental Zoology* **14** (1913), 43–59.
30. *Ibid.*
31. *Ibid.*
32. *Ibid.*
33. *Ibid.*
34. Lindly Darden and N. Maul, "Interfield Theories," *Philosophy of Science* **44** (1977), 43–64.
35. W. E. Castle, "Is the Arrangement of the Genes in the Chromosome Linear?", *Proceedings of the National Academy of Sciences, Washington* **5** (1919), 25–32, here, p. 25.
36. *Ibid.*, p. 26.
37. *Ibid.*, p. 26. Castle asked for scientific material-chemical evidence for the physicalistic-mechanistic manifest intuitions of Morgan and his students. Unfortunately not much was known about chemical polymers at the time. In hindsight it is pathetic to realize how wrong he was about molecules that have "simple string-like form." L. C. Dunn, in *A Short History of Genetics* (New York: Mcgraw-Hill, 1965), pp. 114–115, comments that although it was Janssens's interpretation of the events during meiosis in the amphibian oocyte in terms of breakage and reunion that Morgan seized upon in constructing his theory of linkage with crossing over, it was the opinion of E. B. Wilson and other cytologists, such as C. D. Darlington, that "the cytological evidence provided by Janssens was not sufficient to prove that exchange (crossing over) did in fact occur." Yet, "the importance of Janssens' work lay not in what it proved cytologically but in what it suggested for testing by genetical methods."
38. Castle, *op. cit.*, 1919, Note 34, p. 26.
39. *Ibid.*, p. 26.
40. *Ibid.*, p. 30.
41. *Ibid.*, p. 30.
42. A. H. Sturtevant, C. B. Bridges, and T. H. Morgan, "The Spatial Relations of Genes," *Proceedings of the National Academy of Sciences, Washington* **5** (1919), 168–173.
43. Castle, *op. cit.*, 1919, Note 35, p. 31.
44. Coleman, *op. cit.*, 1970, Note 12, p. 235. See also R. Falk, "What is a Gene?" *Studies in the History and Philosophy of Science* **17** (1986), 133–173, here pp. 148–151.
45. Herman J. Muller, "Are the Factors of Heredity Arranged in a Line?" *The American Naturalist* **54** (1920), 97–121, here, p. 97.
46. *Ibid.*, p. 98 (italics added).
47. *Ibid.*, pp. 100–101.
48. *Ibid.*, p. 101.
49. *Ibid.*
50. *Ibid.*
51. *Ibid.*, p. 98.
52. *Ibid.*, p. 105.
53. *Ibid.*, p. 105.

54. H. J. Muller, "The Mechanism of Crossing Over," *The American Naturalist* **50** (1916), 193–221; 284–305; 350–366; 421–434.
55. Dunn, *op. cit.*, 1965, Note 37, p. 147.
56. *Ibid.*
57. J. Belling, "A Working Hypothesis for Segmental Interchange between Homologous Chromosomes in Flowering Plants," *University of California Publications in Botany* **14** (1928), 283–291.
58. See Carl C. Lindegren, "Chromosome Maps of *Saccharomyces*," *Proceedings of the 8th International Congress of Genetics. Hereditas* (Supplement), p. 338.
59. R. Pearl, "Breeding Better Men," *The World's Worker* (January 1908), 9818–9824.
60. *Ibid.*
61. *Ibid.*, pp. 9818–9819.
62. *Ibid.*, pp. 9819.
63. *Ibid.*, pp. 9820–9821.
64. Pearl, *op. cit.*, 1908, Note 59, p. 9821.
65. D. J. Kevles, *In the Name of Eugenics* (Harmondsworth: Penguin, 1985).
66. W. E. Castle, *Genetics and Eugenics* (Cambridge: Harvard University Press, 1916). pp. 3–4.
67. *Ibid.*, pp. 233–234.
68. *Ibid.*
69. A. H. Sturtevant, "The Early Mendelians," *Proceedings of the American Philosophical Society* **109** (1965), 199–204, here, p. 202.
70. Charles B. Davenport, *Heredity in Relation to Eugenics* (New York: Henry Holt, 1915), p. iii.
71. *Ibid.*, p. iv.
72. *Ibid.*, p. 1.
73. *Ibid.*, p. 4.
74. *Ibid.*
75. E. A. Carlson, *Genes, Radiation, and Society. The Life and Work of H. J. Muller* (Ithaca: Cornell University Press, 1981), p. 178.
76. *Ibid.*, p. 180.
77. H. J. Muller, "The Dominance of Economics over Eugenics," *The Scientific Monthly* **37** (1933), 40–47.
78. *Ibid.*
79. E. Baur, E. Fischer, and F. Lenz, *Human Heredity* (New York: Macmillan, 1931), p. 625.
80. *Ibid.*, p. 656.
81. Davenport, *op. cit.*, 1915, Note 70. p. 4.

THE NEXUS OF ANIMAL AND RATIONAL: SOCIOBIOLOGY, LANGUAGE, AND THE ENLIGHTENMENT STUDY OF APES

ROBERT WOKLER
University of Manchester

The Place of Primatological Linguistics in Sociobiology

Scientific exactitude is no mere literary device, and experimental and laboratory researchers in pursuit of it have accordingly been largely unreceptive to studies of their figurative use of language. On the other hand, it is just with respect to those sciences which aspire to explanations of human nature and behavior that the symbolism, imagery, and metaphor in part constitutive of that behavior should appear most pertinent to its description. It is thus a commonplace of the social sciences — whatever might be claimed about their objectivity — that the terminology employed by investigators to account for the conduct of human affairs unavoidably overlaps with and reflects the terminology of the subjects under scrutiny. The social sciences are in a sense translations from one vocabulary to another, and they seek to render human patterns of life intelligible precisely because they do make sense, which it is thought can be construed by persons who may be motivated by quite diverse and even discordant judgments and values. But for those who approach human behavior within the framework of the natural sciences, the ascription of deliberate sense and articulated meaning to what people do or say will be insufficient to account for their behavior in a satisfactorily causal way, since it appears to grant privileged status to the intentions of individuals who alone can report what they truly had in mind, as expressed and embellished in a dense medium of communication that is moreover unique to one species alone. Social scientists' undue emphasis upon language tends to render obscure the biological place in nature that men and women occupy, it has been argued, and the peculiar attribute that makes it possible for us to invoke the imagery of myth and metaphor must not shroud in eternal mystery the genetic determinants of our strategies of reproduction and interpersonal relations, as well as our collective prospects of survival.

Nowhere are these tensions between two perspectives on the use of language more conspicuous than at the interface of the biological and social sciences, which the newly synthesized discipline of sociobiology has attempted to cement over the past couple of decades. The history of this discipline, and of its practitioners' extraordinary reliance upon just those social sciences whose metaphorical language they hold to be unscientifically opaque, has often been recounted. Sociobiologists' ideas of kin selection are derived from anthropology; their conceptions of inclusive fitness are based upon game theoretical models along lines earlier mapped out by economists; while selfishness, altruism, and aggression draw their meaning from the social behavior of individuals, of which genes can only be thought replicators by a mighty and wholly artistic leap of the imagination. As many commentators have remarked, it has always been so. The terms "struggle for existence," "survival of the fittest," and even "evolution" were gratefully borrowed by Darwin from the field of sociology which his latter-day disciples would have us expel from our minds, if they could just come to believe that we had any.[1] Alas, having attributed to genes properties that are unintelligible except as expressions of mind, Edward Wilson and his disciples regard all reference to mental activity and criteria as "a retreat into mysticism." But it is not Wilson's mindless ants who are slavish followers of their queen, so much as Wilson himself of his dogmas, which, instead of explaining the workings of the human machine without the ghost, leave us with a ghostly coded endoplasm that appears to have human purposes but never takes human form.

Sociobiology's critics have frequently remarked that the subject is not of such contemporary vintage as its upholders maintain, despite the recent invention of some of its terminology. If it was at the height of the Cold War when a number of its approaches, especially to the allegedly genetic sources of aggression, came to achieve their widest popular acclaim, these approaches were established in the human sciences long before they became fashionable with the general public. Indeed, their recent celebration during the Cold War, and now perhaps their decline in public notice with its apparent ending, in many ways recapitulate the rise of social Darwinism – itself the "sociobiology" of the late nineteenth century in an age of particularly expansive European imperialism.

Of course a scientific doctrine is not discredited by an explanation of its popular appeal, and its standing is not undermined just because it can be demonstrated to have been put to political uses, as the contributions to this collection of Peter Bowler, Gregg Mitman, Antonello La Vergata and Peter Weingart amply demonstrate. Some of the members of the sociobiology study group of Science for the People in particular have gone too far in objecting to the methodology of this newly synthesized science, through their accusations of sexism in Wilson's approach and their suggestion that the whole discipline stemmed from the same eugenic doctrines that, at worst, "led to the establishment of gas chambers in Nazi Germany" or, at best, merely buttressed "the institutions of their society by exonerating them from responsibility for social problems."[2] But while objections to their political bias may have been unnecessarily abrasive and misplaced, Wilson and his followers have thus far failed to meet the challenge of those who, like Steven Rose, Richard Lewontin, and Leon Kamin in *Not in our Genes* and elsewhere, have claimed that dominance, aggression, and other social qualities imputed to our species as a whole are not natural objects but historically and ideologically conditioned terms which describe human conduct in certain contexts only. In constructing a world of intrinsically prescribed and naturally selected social traits that they deem universal, sociobiologists have all too commonly just reified types of social phenomena so as to explain them, mistakenly supposing the possessive, self-aggrandizing, and individualist dispositions characteristic of entrepreneurial economic activity to be innate, genetically coded, behavioral properties, sometimes attributed to so-called "culturgens," which are never shown, except by an imputed necessity, to have any bearing at all upon mankind's cognitive development.[3]

Well-merited censure from such quarters might have been less severe in tone and substance if sociobiologists themselves had only been more modest in their claims to innovation, less determined to relegate earlier attempts to account for the social nature of man to the realms of fantasy, pseudo-science, or prehistory. Greater familiarity with previous distinctions drawn between human biology and culture, and with other conceptions of the relation between mankind and animals, might have usefully tempered the extravagance of some of their ideas. And they might have learned a thing or two about the presuppositions of their own enquiry even from such ancient works as Plato's *Gorgias*, Aristotle's *Generation of Animals*, or Lucretius's *De rerum natura*. Too seldom are discussions of the future of sociobiology punctuated by reflections on its past; and it is then only its quite recent past, when under-laborers are shown to have run some fraction of the course, that ever appears to be thought worthy of explanation. Wilson's case on behalf of a "biologized" science of ethics could have been presented less farcically than it seems to most reviewers if he had taken stock of the dismal fate that befell its earlier formulations, such as that of Spencer at the hand of Hobhouse or Lalande. Since so much of sociobiology pertains to analogies of behavior understood to be homologies across animal species, it is unfortunate, too, that its contemporary practitioners have learned so little of the complexities regarding man's place in nature that confronted their biological precursors of the seventeenth and eighteenth century – such as Ray, Leibniz, Buffon, and Bonnet. In this regard it is curious to note, for instance, how Trivers, in his seminally important and widely reprinted essay on "Reciprocal Altruism," challenges the (genetic?) tendency of social scientists who have failed to differentiate nonaltruistic from reciprocal benefits, including among the culprits "Baier 1958" and "Rousseau 1954,"[4] thus, by citing a modern edition of an eighteenth-century author, conflating two hundred years of that putative tendency in a single line.

My principal aim here will be to address a particular problem in seventeenth- and eighteenth-century speculation about human nature, which sociobiologists have largely ignored at their peril. This problem is essentially that of the linguistic capacity of the great apes, in so far as these animals appear most akin to mankind biologically but do not display evidence of what was and is taken, today no less than in the Enlightenment, to be our most distinctive capacity for culture. Because of their greater proximity to our species in the *scala naturae* or any other taxonomy of natural organisms, apes are better candidates for the testing of sociobiological hypotheses about the genetic determination of human behavior than are ants. Because human social systems depend so remarkably upon a kind of natural language that no other species have apparently mastered, linguistic performance forms the most notably striking divide between mankind and the rest of the animal world. In the Enlightenment, questions to do with the cultural uniqueness of our species and its biological links with all other creatures were, I mean to show, addressed above all to the linguistic aptitude of apes, and the relative neglect of such questions by contemporary sociobiologists has proved one of the principal factors undermining the credibility of their discipline. I shall also argue that primatologists, who have indeed frequently addressed them, have nevertheless in too many instances failed to do so with the perspicuity of their more speculative seventeenth- and eighteenth-century forebears. In both cases, I shall try to establish that the conceptual problems raised by Enlightenment scientists and philosophers were much the same as the problems underlying sociobiological and ape language studies today. But the diverse ways in which they sought to settle such problems have not been recapitulated by many current experimental scientists, who have adopted a less sceptical and less critical outlook in their understanding of what may be man's place in nature or in their grasp of biological and physiological factors bearing on the evolution of languages.

My focus will thus be on the transfer of images and metaphor between biology and the social sciences in two different senses. On the one hand, through the ape language debates themselves, I shall be concerned with the articulation of imagery and metaphor as the putative divide between biological and cultural explanations of human nature, in so far as the scientific status of sociobiology depends upon plausible explanations of the absence of such interspecific transfers between man and ape. On the other hand, in examining a particular theme in speculative anthropology, I mean to show that certain notable seventeenth- and eighteenth-century scientific metaphors have been conspicuously overlooked by modern commentators, whose own research has been conceptually weaker for that absence. The Enlightenment study of apes and language, I hope to show, provides an important illustration in the history of science of a problem whose empirical methodology would be well-served by attention to fundamental principles about the description and classification of the patterns of mankind's behavior. Such principles were once raised by conjectural historians of human nature and of the evolution of language, who had precious little first-hand information about the activities of apes but who were nevertheless well-versed and knew much about the purpose of scientific inquiry.

The Contributions of Tyson, Rousseau, and Other Enlightenment Thinkers

As long ago as 1699, Edward Tyson, in his *Orang-Outang* (actually a chimpanzee), published what is still today the best available study of the anatomy of a nonhuman primate compared in detail with the anatomy of man.[5] Tyson was drawn to his subject by many precursors who had commented upon chimpanzees and other apes and monkeys (often employing the generic term "orang-utan," a Malay expression equivalent to the Latin *homo sylvestris*, or "man of the woods"). From such iconographical descriptions as could be found in Bernhard von Breydenbach's *Peregrinatio in terram sanctum* of 1486, Conrad Gesner's *Historiae animalium* of 1551, Nicolaas Tulp's *Observationum medicarum* of 1641, or Jacob Bontius's *Historiae naturalis* of 1658, he inferred that there had been a confusion of several different creatures, some of which bore only a passing resemblance to the animal he dissected, although – because his specimen had already died by the time be first set eyes on it – he imagined, like so many before him, that it was an upright biped, perhaps assisted in its gait by the use of a walking stick. From Claude Perrault's *Suite des mémoires pour servir à l'histoire des animaux* of 1676, moreover, and ultimately from Descartes's *Traité de l'homme* of 1664, Tyson adopted a dualistic perspective on the nature of man, which inspired his belief that the brain, larynx, and other organs of the orangutan were just vessels and pipes, lacking the animation of a spiritual faculty that enabled humans, possessing remarkably similar organs, to both think and speak.[6]

His contribution to the study of human nature by way of the comparative anatomy of man and chimpanzee was to have a profound impact upon Enlightenment and subsequent reflection on this subject, for it established two central principles around which debate would henceforth revolve. The first is that the closest connection between mankind and animals in the natural world must be that formed by anthropoid apes, on account of the striking physical resemblance of their organs and features to ours; the apparently sublime intelligence of elephants, or the noble countenance of lions, which had so impressed earlier commentators, would henceforth warrant no special place for those animals next to man in the order of nature. If it was not yet supposed that we had descended, or ascended, from apes, after 1699 it could at least no longer be doubted that, among all earthly creatures, apes or orangutans were our next-of-kin. Second, Tyson's work made it plain that the most fundamental questions about the status of the human race among animals must be focused ultimately on our mastery of language and the apparent linguistic incompetence of apes. He was convinced that the closer the similarity he could establish in physical terms between chimpanzee and human organs of speech, the more it must be evident that our exclusive use of speech (for he supposed apes to be inarticulate, if not mute) must be due to a spiritual faculty unique to man. Thus, while apes could be shown to occupy the next link to ourselves in the chain of being, they did not challenge our predominance over all nature, manifested in our command of language, which reflected our possession of mind and reason. As was remarked by Buffon, on this subject Tyson's most illustrious eighteenth-century disciple, the gulf between the organutan and man was one that nature herself "could not bridge."[7]

Many Enlightenment commentators were to take issue with both Tyson and Buffon, either through closer examination of the reported behavior and supposed capacities of apes and other animals (sometimes suggesting that they too must have minds), or through further investigation of the nature of language, particularly as it passed from primitive cries and utterances to refined and metaphorical figures of speech. The study of the evolution and transformation of language and other cultural institutions forms one of the most striking features of eighteenth-century sciences of the evolution of human physical attributes and of the transformations of species. The discipline of sociobiology, to be sure, presupposes such developments in the study of both culture and nature – especially the materialist critique of dualism already set forth in the Enlightenment – to the effect that there must be a physical (or genetic) foundation to human conduct if it is to be rendered intelligible within the natural world.[8] Yet, instead of building upon the doctrines of their precursors, sociobiologists have all too characteristically ignored them, leaving relatively unexplored, by contrast with their interest in invertebrate parallels to mankind's behavior, what ought properly to form the core of this discipline – a genetic account of the human capacity for language joined to studies of its social formation and variations in disparate environments. It is upon the study of language above all other cultural institutions that sociobiology's fate as a true science of human nature must ultimately depend.[9] Tyson perceived the physical and spiritual nexus of man with respect to apes, on the one hand, and language, on the other; sociobiologists ignoring these subjects do not so much bridge society's artifacts with our animal roots as open a new chasm between mankind and the rest of nature.

From Jean-Jacques Rousseau's reflections on *l'homme sauvage*, which he also sometimes termed an "orang-utan," practitioners of sociobiology might have learned more still. Drawing at length upon the sixteenth- and seventeenth-century African voyages of Andrew Battel, Olfert Dapper, and Girolamo Merolla, as recounted in the abbé Prévost's *Histoire générale des voyages* (1746–89), Rousseau, in his *Discours sor l'inégalité*, put forward the hypothesis that the large Congolese animal described by these travelers, and taken to be of the same family as the orangutan of the East Indies, might really be a human being like the rest of us. No less than Tyson and Buffon, Rousseau was impressed by this animal's alleged "human countenance" and by its strikingly "exact resemblance to man." For if it was true that the beast was in so many respects just like a man, then why, he wondered, should we not accept at least a *prima facie* case for its humanity? Adopting Buffon's biological definition of a species in terms of the fecundity of sexual unions, he insisted that we could only establish by experiment whether matings between humans and organutans might prove fruitful. While most eighteenth-century naturalists subscribed to Tyson's view that orangutans were beneath the level of humanity because they lacked the capacity to think and speak like men, Rousseau maintained that our savage ancestors were unlikely to have been wiser or more loquacious than orangutans, since both reason and speech, he claimed, were just "potential faculties,"[10] which must have undergone a long history of development in complex social settings before they could have become manifest in our behavior. We could not point to the languages of civilized peoples as proof of the subhumanity of orangutans, because linguistic competence must be mastered and in itself is not a natural characteristic of man. Hence, as La Mettrie had already demonstrated in his *Histoire naturelle de l'âme* of 1746, the apparently mute condition of the orangutan could be explained merely in terms of the creature's scant opportunity to employ and develop those vocal organs that it shared with the rest of us.

Of course, since Rousseau never actually saw a true orangutan, and since his account of the creature's behavior was drawn from statements of witnesses who disagreed among themselves, his reflections about its capacities must be treated with some scepticism. It was not until the 1770s that a sufficient number of live specimens came to be available in Europe for reliable studies to be undertaken, and it was not until after Rousseau's death that scientists came to agree that the animal was definitely a species of ape different from the chimpanzee, with which he and so many naturalists had confused it. But the originality of his comments about the humanity of orangutans was widely recognized in the late eighteenth century – inspiring the Scottish historian of languages, Lord Monboddo, for instance, and, on the other hand, alarming Bonnet, Blumenbach, and other scientists who could not accept such an unorthodox notion of primitive man.

Especially in the light of the Darwinian revolution of the nineteenth century, we ought to bear in mind that Rousseau's commentary on *l'homme sauvage* in his *Discours sur l'inégalité* forms perhaps the boldest hypothesis about the physical transformation of mankind in an age when most arguments about the chain of being remained fundamentally wedded to a belief in the fixity of species. Of course Rousseau must not be credited with a fully fledged theory of human evolution, for at least three reasons: first, because he hesitated to speculate on comparative anatomy and allowed that *l'homme sauvage* must have been formed physically rather like *l'homme civil*, upright and adroit; second, because he supposed that modern man could have developed from a species of animal still present in the contemporary world; and third, because he imagined that apes and men together were zoologically distinct from all other species, and that monkeys could not be counted among our progenitors because they lacked the attribute of *perfectibility*, a term he coined to account for man's capacity to become civilized, which only apes and humans shared. Nevertheless, Rousseau appears to have been the first Enlightenment figure to suppose that there might be a temporal and sequential relation between particular species in the natural chain, and the first, moreover, to conceive that the last link in that chain – the relation between apes and men – might be one of genetic continuity.

From the perspective of the contemporary science of sociobiology and the current state of primate studies, furthermore, it should be observed that Rousseau's portrait of the orangutan as a kind of savage in the state of nature happens to have been drawn with greater accuracy than any description of the animal's behavior for at least the next two hundred years. The same Rousseau whom Trivers attacks in 1971 for his misconception of animal altruism, produced in 1755 the best available description of the true orangutan of Southeast Asia until Trivers' own day. Rousseau followed other commentators in recognizing the orangutan's natural habitat to be the tropical forest, but while most of his contemporaries agreed about the creature's promiscuity, he alone in the eighteenth century inferred that the male of the species mated only occasionally and formed no lasting sexual bond with the female. While other naturalists imagined that orangutans were carnivorous in their diet and aggressive in their conduct, Rousseau had no doubt that they were frugivorous and generally peaceable animals. Not until the publication in 1869 of *The Malay Archipelago* by Alfred Russel Wallace (co-founder, with Darwin, of modern evolutionism) was it confirmed that orangutans are nomadic beasts without clearly defined territorial ranges, again as Rousseau had already perceived in the *Discours sur l'inégalité*. And not until the late 1960s was it established that he had been right all along in his guess that orangutans even lack any distinctive social system and that, apart from infrequent copulation, their lives are essentially inactive and hermitic. None of the other higher primates have shown themselves so introverted, private, and reclusive.[12]

John MacKinnon's *The Ape Within Us*, as well as his earlier *In Search of the Bed Ape*, provides a remarkable confirmation of Rousseau's account of the life of orangutans, which should bring pleasure to historians of ideas and comfort to armchair primatologists. MacKinnon's studies verify much the same patterns of orangutan behavior as were established independently by those of Peter Rodman, also dating from the late 1960s, and they have been largely confirmed as well by Biruté Galdikas since 1971, both in her observations of wild orangutans and in her rehabilitation centre for orangutan orphans, allowing, however, that Galdikas finds these animals somewhat less nomadic and more territorial, even if only peripatetically, over a sizeable area. Rousseau, in Paris, conjectured about these animals by stripping all the artificial influences of society away from natural man. MacKinnon, in the tropical forests of Borneo and Sumatra, spent months upon months tracking orangutans down, his blood drained by leeches, his strength by mud, and his patience by flies. Their conclusions, however, are much the same, and, from the field work of MacKinnon, Rodman, Galdikas, and others, we now know that Rousseau had been right all along in remarking upon these creatures' nomadic existence, vegetarian diet, infrequent sexual relations, and essentially solitary and indolent lives. It would seem that Rousseau's masterpiece of speculative anthropology is thus also a most remarkable contribution to empirical primatology.

Both his and contemporary findings on this species, it must be noted, nevertheless differ markedly from those of perhaps the world's best-known primatologist today, Jane Goodall, whose study of *The Chimpanzees of Gombe* is the fruit of twenty-five years' investigation in the forests around Lake Tanganyika. That notably long period of research is significant, since Goodall acknowledges that after the first ten years she had formed the impression that chimpanzees are far more peaceable than humans. But the fragmentation of their packs and communities which she and her colleagues came to witness gave rise to a level of violence (including even cannibalism) that surprised her and has so altered our understanding of chimpanzee behavior that we cannot but perceive this creature's evolution as strikingly different in its path – more social, more territorial, more aggressive (occasionally predatory) – from that of the Asian orangutan.

The main interest to sociobiologists of these animals – orangutans and chimpanzees alike, as well as gibbons and gorillas – ought to be their significance for an understanding of the genetic origins of human behavior, since all the evidence we have – from paleontology, immunology, comparative anatomy, and especially biochemistry – corroborates what most visitors to zoos will have guessed anyway: that the great apes share more of our physical traits and genetic material than do any other species. Yet their life patterns are all too rarely discussed by sociobiologists, even though such evidence of instinctive territoriality and aggression as they display, and which is supposed to be characteristic of mankind, too, seems far less clear and universal among them than is the case for ants. Orangutans in particular appear so little spurred by the promptings of territoriality and aggression that they lazily undermine all expectations of those traits by geneticists who remark on human evolutions, and for many sociobiologists they have therefore become a new missing link, losing their place in the synthesis.

As Enlightenment commentators perceived, moreover, the apparent lack of language of creatures bearing such a striking physical resemblance to man needs explanation. From the similitude of their bodily parts ought to arise a homology of their functions, it was supposed, and yet apes showed no outward sign of our most striking capacity for culture. Was it just a defect in their education or, alternatively, a consequence of the stunted growth of their vocal organs, for which some instruction in gestural languages might offer compensation? The fate of these creatures, not least their souls, was often deemed to be much like man's, so that if only they were able to display some command of language, even the merest semblance of figurative speech, no human accomplishments need prove beyond their grasp. "Parle, et je te baptise," confided the Cardinal de Polignac to the orangutan in the collection of the Prince of Orange, according to Diderot in his *Suite du Rêve de d'Alembert*.[12] Such a prospect of an introduction into our religious world might, of course, make any sensible creature pause. To show skill at the use of metaphorical speech might lead to the deliverance of an animal's soul, but only by imperiling the safety of its body. As Rousseau once remarked, following a seventeenth-century voyager who claimed to have overheard their conversation in the forest, it's no wonder that these animals, when confronted with the prospect of salvation, enslavement, or culture, wisely *pretend* to be mute.[13]

Communicating with Apes Today

Questions to do with the linguistic competence of apes were put by many eighteenth-century philosophers and natural scientists – Rousseau, La Mettrie, Monboddo, Camper, and Blumenbach among them – but the one indisputable material ingredient necessary to formulate an answer to them has only become available recently – i.e., the research grant. Without it, and the laboratory facilities and armies of assistants it pays for, the Gardners could not have undertaken to teach their celebrated chimpanzee, Washoe, the use of gestures and mimicry based on American Sign Language, nor the Premacks to teach Sarah to discourse with plastic chips. Experiments such as these, as well as others reported for a popular audience in *The Ape's Reflexion* by Adrian Desmond, and for more academic readers in *Speaking of Apes*, edited by Thomas Sebeok and Jean Umiker-Sebeok, point to the experimental and conceptual difficulties of communicating with lexical markers whose linguistic status is doubtful, as well as the hazards of expecting chimpanzees' solutions of forced-choice problems to resemble children's mastery of generative grammatical rules. No less than the sociobiology debate, whose metaphors came to be embroiled in the political diplomacy of Cold War international relations, the ape-language studies from the mid-1960s have recapitulated and reinforced the tensions evident particularly in the practice of American psychology, between empiricists or behaviorists on the one hand, and rationalists or mentalists on the other.[14] By and large, those who have remained resolutely optimistic about communicating with apes have presupposed the venerable principles of an unbroken continuity in nature's ladder joining man and animals, and, in the matter of ape languages, they have enlisted with the camp of psychological empiricists. The Gardners and Premacks, to be sure, have been cautious in their claims about the intelligence of apes, but a parasitic host of commentators, impressed by their research, has come to regard caution as a human sentiment best thrown to the winds, and language as a primatological form of chitchat that chimpanzees and gorillas have finally, albeit mutely, begun to copy from us. Such imitation of human behavior cannot be deemed of human design. Indeed, at this level, parroting parrots are more adept at articulating phonemes than are aping apes.

Much the most engaging book yet published on the subject is Herbert Terrance's *Nim* (that is, Nim Chimpsky, in backhanded honor of Noam Chomsky, who contends that linguistic aptitude is a uniquely human trait). Terrace embarked on his work excited at the prospect of communicating with another species, but he never let his enthusiasm take hold of his judgment, and after nearly four years of teaching his pupil a sign language, he concludes that Nim's successes show none of the spontaneity and complexity featured in the first sentences of children. Not only did Nim imitate, for the most part, rather than generate his signs, he seldom strung them together to form new meanings. Nor did he communicate for purposes other than to meet a compelling need and instead of conversing with his interlocutors, he generally interrupted them. Like La Mettrie and Monboddo in the eighteenth century (and perhaps also some disgruntled mothers of offspring at our local, understaffed schools), Terrace is persuaded that even the best-trained chimpanzees perhaps may not yet have realized their full linguistic potential. All the available evidence shows, he nevertheless reflects ruefully, that the gestural utterances of these creatures are shaped almost entirely by the dictates of their teachers.

And how many teachers are needed for the task! In the course of his education Nim had no less than sixty, most having him as their only common interest. Here, we find, if not the reserve army of capitalism, then at least the real band of nature's little helpers – mistakenly thought by Wilson to be homosexuals facilitating the transmission of some of mankind's more altruistic impulses – all gathered together in a country manor, whose lord Nim soon became, so that he could discover the ways of nature and Her helpers at his leisure. As Terrace reports, already in the 1970s this proved a costly exercise, with $70,000 spent on chimp sitters alone, and nearly $250,000 (a fair part of which came out of the author's own pocket) in all. Despite such expenditure and attention, Nim often found his teachers wanting, and no activity so dull as that of learning a language. He was better with paints.

Terrace's conclusions, to be sure, have been challenged by other scientific commentators, most notably Duane Rumbaugh and his collaborators in the Lana Project conducted with different chimpanzees in the 1970s and early 1980s, and by Francine Patterson in her education of Koko the gorilla from 1972. Rumbaugh, working with American Sign Language, and Patterson, adopting a lexigraphical language called Yerkish specifically designed for communication with nonhuman primates, both found their specimens to be rather more creative, innovative, and spontaneous than was Nim. When subjected to sufficiently sympathetic and patient tuition, it appears that apes may indeed display somewhat greater inventiveness and fractionally larger and better structured vocabularies than were drawn by Terrace from his less tractable pupil. More recently still, in the mid-1980s, Sue Savage-Rumbaugh turned an ever-optimistic faith in the outcome of such endeavors in yet another fresh direction, towards the bonobo (*Pan paniscus*), deemed to have a larger vocal repertoire and assortment of natural gestures and more sophisticated patterns of communicative eye contact than any other ape or monkey.[16]

But on the evidence thus far supplied, these persisting efforts to form an interspecific linguistic bridge between animals and man are misconceived. That apes are manifestly intelligent creatures is as indisputable as mankind's firmly rooted biological and physiological place in nature. Descartes' and Tyson's critics in the Enlightenment already recognized that animals think, plan, and convey their thoughts as well as express their feelings to one another, no less than humans do.[16] Apes can employ a range of symbols to represent what they mean, and perhaps even deliberately to obscure it, by misleading human tutors much as they might deceive other animals. Dolphins and dogs, as well as chimpanzees, exhibit remarkable intelligence and conceptual skills, and it may well be the case, as Irene Pepperburg has argued in her recent study of psittacine cognition, published in Sue Taylor Parker's and Kathleen Gibson's *"Language" and Intelligence in Monkeys and Apes* in 1990, that the African grey parrot can count. It is, after all, with reference to the behavior of such animals that humans, too, have come to ape and parrot. But what neither apes nor parrots have ever learned, even after the most skillful tuition or credulous appraisal, is how to use a language and thereby enter the still plainly exclusive domain of human culture. The more any real command of metaphor appears beyond their grasp, the more anthropomorphic seems our description of what they do, as we metaphorically impute our own intentions to them, in transference of images and purposes that other species do not naturally share.

Premack, in *Gavagai! or the Future History of the Animal Language Controversy*, presents a robust case for scientific scepticism about the linguistic capacity of apes, without having to suppose that mankind was created separately from all other animals, or is uniquely endowed with a spiritual and rational soul. Apes, he argues, for all their abundant communicative skills, have shown no capacity to master a natural language whose semantics may be mapped from its grammatical rules. Not only have they failed to learn a natural language in the laboratory; they appear, in the wild, even to lack a natural language of their own. They possess no skills remotely comparable to that of the human child in mastering syntactic formula which make the remarkable generation and development of its language possible. This distinction between apes and children in the refinement of their aptitudes ought to be the most conspicuous lesson of all the ape-language trials thus far conducted, unfortunately too often by experimenters who, in seeking to draw man back to nature, contrive to turn apes into children. While apes may employ abstract signs and gestures, they do not evoke or evince the kind of meanings that human culture has conveyed by virtue of articulation in diverse forms of language. No other animal apart from man shows any sign of having developed a natural language, still less a phonetic language formed from a voiced code of acoustic signals. The neural mechanisms making such communication possible in *Homo sapiens* must no doubt have been subject to a long evolutionary history of selective adaptation. Tyson at the end of the seventeenth century and Camper at the end of the eighteenth were doubtless each in error in focusing, respectively, upon the unanimated larynx and inflatable laryngial pouch of the orangutan, in their explanations of the animal's incapacity to enunciate intelligible sounds. If they had been students of modern morphology, they might have looked instead at the acoustic and filtering properties that have made the human supralaryngial vocal tract, uncharacteristic of other higher primates, including the great apes, so peculiarly well-adapted for speech.[17] But as distinct from a regrettably too large contingent of modern commentators, they had no doubt that out of the formation of speech had come the articulation of the images of human culture. As the child suddenly accelerates from infancy, so with the birth of speech must the human race have passed through a revolution that generated the artifacts from which civilization springs.

The Human Specifity of Metaphor

Sociobiologists and primatologists alike would do well to turn to the lessons learned from the excavations and discoveries of the human remains of the upper palaeolithic era. These point to the evolution of the supralaryngeal vocal tract only within the past 250,000 years, during a period that marked the demise of Neanderthal hominids and the ascendancy of modern *Home sapiens* perhaps just thirty to forty thousand years ago, when the remarkable transformations of toolmaking and stoneworking, together with the sudden appearance of art, may be taken as material signs of the first flourishing of culture, concomitant with the development of vocally encoded language.[18] No models of any consequence have been proposed as yet by sociobiologists to explain the genetic origins of language, and it is unclear what form such models could plausibly take, given that in any human population natural languages are perhaps the preeminent phenotypic characteristics acquired by individuals, in particular and distinct ways in the course of their upbringing. L. L. Cavalli-Sforza and M. W. Feldman, for instance, in their quantitative approach to *Cultural Transmission and Evolution*, suggest how rates of phonological change within populations and linguistic isolation as a factor of geographical distance might be measured in ways perfectly acceptable to evolutionary biologists. But they are the first to admit that mastery of particular languages is not attributable to genetic differences of linguistic propensity, and that cultural selection, such as in the learning of languages, may remain permanently inexplicable in terms of the natural selection of the physical properties and bases of human behavior. The sociobiology study group and other critics of the discipline maintain that there is no reason to believe that *any* of our behavioral traits (except, perhaps, a few pathological ones) are coded in our genes, and this contention, if seemingly somewhat implausible, has still to be overcome, however damaging it is in principle to all attempts to explain the biological roots of the way we live.

At least with regard to language, conjectures about the genetic coding of any deep structural grammar, even allowing for its uniquely human specificity, have so far proved of remarkably doubtful value, except perhaps in so far as transformational grammars of our languages may be symbolic analogues in culture for the configurations of genes that determine the patterns of our physical development.[19] Wilson himself appears to find the matter unproblematic, devoting only a few lines to the subject of language in *On Human Nature*, where he instead views religion, on which he includes a whole chapter, as constituting "the greatest challenge to human sociobiology."[20] But rather than speculate on unrelated parallels between cultural and organic evolution, sociobiologists need to identify those features of language, if any exist, that reflect molecular sources predisposing us towards the complex manipulation of symbols. Thus might they pose a centrally significant question of Enlightenment speculative anthropology in a new and contemporary idiom. On this subject, paradoxically, Wilson and Chomsky seem to agree, since despite their differing views of man's links with animals, they both believe that our linguistic competence must have a determinate genetic foundation, which behaviorists and environmentalists alike, misled by suppositions of our infinitely malleable nature, have failed to perceive.

One recent and notable work, Peter Wilson's *Man, the Promising Primate*, is specially exciting largely because it attempts to turn the subject inside out. Following arguments propounded by Gould in *Ontogeny and Phylogeny*, Wilson claims that our capacity for language depends on our plasticity and the fact that, compared with other species, the human infant comes into the world prematurely, relatively generalized in its morphology and dispositions, "unformed into anything specific."[21] This uniqueness does not derive from any genetic implantation of superior attributes but rather from the condition, in which each of us finds himself in infancy, of having to confront his problems of adaptation for himself. In the course of our development, Wilson claims, we acquire self-consciousness bound together with and expressed through language, such as other creatures, genetically adapted to particular environments, have neither need nor occasion to do. This thesis is built around relatively recent writings in physical anthropology as well as philosophy, with frequent commendatory references to Enlightenment ideas of human nature, including those of Rousseau. It might have drawn benefit from greater familiarity with contemporary biology and especially genetics – above all, the idea of allometric growth as an alternative to the notion of selective adaptation, for human culture, of specific organs, or other bodily attributes. Such an idea, indeed, resembles the contention of Rousseau and other Enlightenment thinkers that human history is made possible not by the programming of our conduct by instincts, but by our being characteristically underdetermined and relatively unprogrammed in our dietary, sexual, and, indeed, general behavior. Here

Wilson curiously overlooks the extraordinary resemblance of his view of mankind's plasticity to Rousseau's conception of perfectibility.

Wilson's book title and main theme trade on a striking ambiguity of meaning. We are promising primates, he suggests, because we have potentialities for self-development unique to our species, and equally because we fulfill those potentialities in the performative linguistic acts by which we give ourselves freely to our sexual partners and form the pair bonds upon which the history of our species depends. Human bondage is thus regarded as stemming from the undertaken word rather than the underlying gene, and language is understood as prescribing fixities beyond the power of nature to control. Metaphor and the manipulation of symbols curtail the prospects of our species' development no less than they liberate us from permanent subjection to biological constraint. In captivating us with words, culture imposes obligations upon us that are as deeply felt as any promptings we share with other creatures – indeed, perhaps more deeply, because we recognize such chains to be of our own making. If Wilson's argument is correct, and if by chance other apes come to realize their promise too, they might sensibly pause to wonder whether they ought to keep the secrets of sociobiology to themselves.

The ape language debates in contemporary primatology provide a salutary reminder that the plausibility of empirical research depends in large measure on the nature of the questions to which it is addressed. Speculative anthropologists of the Enlightenment formulated such questions along lines that remain implicit in the practice of sociobiology today. These commentators recognized that generalizations about human behavior must embrace traits that are apparently unique to man as well as others that are distributed more widely across the *scale naturae*. They concentrated in particular upon language as a mirror of mind or as a skill cultivated by the exercise of certain faculties. They perceived metaphor as a mark of specifically human ingenuity, and a measure of the divide that fundamentally separated man from ape. In hesitating to ascribe human purposes to other animals, they showed caution in the use of interspecific transfers of metaphor, often informed by a respect for the great diversity of nature, and the distinct identity of other animals, which is still worthy of scientific emulation today.[22]

Notes

1. Among the most comprehensive treatments of these transfers of terminology and metaphor are those of Antonello La Vergata's *Equilibrio e la guerra della natura. Dalla teologia naturale al darwinismo* (Naples, 1984); Robert J. Richards' *Darwin and the Emergence of Evolutionary Theories of Mind and Behavior* (Chicago, 1987); and Robert M. Young's *Darwin's Metaphor: Nature's Place in Victorian Culture* (Cambridge, 1985). They are elaborated further in the contributions of Bowler and La Vergata to this volume.
2. Elizabeth Allen, et al., "Against Sociobiology," in Caplan (ed.), *The Sociobiology Debate*, pp. 260, 264.
3. See Lumsden and Wilson, *Genes, Mind and Culture*, especially pp. 7, 26–30, 99–236.
4. See Robert L. Trivers, "The Evolution of Reciprocal Altruism," first published in *The Quarterly Review of Biology* **46** (1971), reprinted in Caplan (ed.), *The Sociobiology Debate*, p. 225.
5. The finest and most richly detailed discussion of Tyson's commentary on the chimpanzee remains Ashley Montagu's *Edward Tyson and the Rise of Comparative Anatomy in England*.
6. See Tyson, *Orang-Outang*, p. 55.
7. See the *Oeuvres philosophiques de Buffon*, p. 359. With regard to this link between Buffon and Tyson, see my "Tyson and Buffon on the orangutan."
8. See Yolton's *Thinking Matter* and my "Anthropology and Conjectural History in the Enlightenment."
9. This point has been stressed by evolutionary biologists and historians of biology, if all too seldom by sociobiologists themselves. See, for instance, Mayr, *The Growth of Biological Thought*, p. 622.
10. See Rousseau's *Discours sur l'inégalité*, Note 10, pp. 209–211.
11. For a fuller treatment of Rousseau's conception of the orangutan, from which these remarks have been compressed, see my "Perfectible Apes in Decadent Cultures."
12. See Diderot's *Rêve du d'Alembert, suite de l'entretien*, in Diderot's *Oeuvres complètes*, ed. H. Dieckmann, J. Varloot et al. (Paris, 1975–), Vol. XVII, p. 206.
13. Rousseau to David Hume March 29, 1766. See the *Correspondance complete de Rousseau*, ed. R. A. Leigh (Geneva and Oxford, 1965–), Vol. 29, p. 66. This suggestion stems from Richard Jobson's *The Golden Trade* of 1623.
14. A point neatly made by Premack in *Gavagai!*, p. 2.
15. See Savage-Rumbaugh, *Ape Language*, pp. 384–397.
16. These subjects are amply discussed by George Boas in *The Happy Beast in French Thought of the Seventeenth-Century*; Hester Hastings in *Man and Beast in French Thought of the Eighteenth Century*; Leonora Rosenfield in *From Beast Machine to Man Machine*; and Keith Thomas in *Man and the Natural World*.
17. With regard to the utility for an articulated language of the supralaryngial vocal tract, see especially Lieberman, *On the Origins of Language*, pp. 81, 84, 88, 159, and *The Biology and Evolution of Language*, pp. 256–286.
18. See especially Lieberman, *On the Origins of Language*, pp. 121–182, and *The Biology and Evolution of Language*, pp. 287–334.
19. I have in mind especially the work of Roger D. Masters. See, for instance, his *The*

Nature of Politics, Chap. 3, pp. 69–113, and with Glendon Schubert (eds.), *Primate Politics*.
20. See Wilson, *On Human Nature*, p. 175. There is also precious little attention paid to language in Eibl-Eibesfeldt's *Human Ethology*, Lockard's *Evolution of Social Behavior* or Lopreato's *Human Nature and Biocultural Evolution*.
21. Wilson, *Man, the Promising Primate*, p. 153.
22. These remarks pursue a line of enquiry to which I have turned occasionally over a number of years, most recently in "Paroles de singes," whose themes are herein elaborated and expanded with regard to a wider range of contemporary studies in primatology, sociobiology, and evolutionary linguistics. I am grateful to the appointed Bielefeld conference reader, whose scepticism not only of historical research but even of the utility of recent experiments, when they are deemed to have been superseded by the most up-to-date discoveries, gave me this opportunity to reaffirm my conviction that current scientists in these disciplines may sometimes have as much to learn from their precursors, including conjectural historians of human nature, as from each other.

Selected Additional Bibliography

Barnett, S. A. *Biology and Freedom: An Essay on the Implications of Human Ethology*. Cambridge, 1988.
Blumenbach, Johann Friedrich. *De generis humani varietate nativa*. Gottingen, 1775.
Boas, George. *The Happy Beast in French Thought of the Seventeenth Century*. Baltimore, 1933.
Bontius, Jacob. *Historiae naturalis*. Amsterdam, 1658.
Breydenbach, Bernhard von. *Peregrinatio in terram sanctum*. Monguntina, 1486.
Brown, Roger. *Psycholinguistics: Selected Papers*. New York, 1970.
Brown, Roger. *A First Language*. London, 1973.
Le Clerc, G. L., comte de Buffon et al. *Histoire naturelle cénérale et particulièra*. 44 vols. 1749–1804, selections in Piveteau, J., ed. *Oeuvres philosophiques de Buffon*. Paris, 1954.
Camper, Petrus. "Account of the organs of Speech of the Orang Outang, in a Letter to Sir John Pringle." *Philosophical Transactions of the Royal Society* **69** (1779), i. 139–159.
Camper, Petrus. "De l'orang-outang, et de quelques autres espèces de singes," in Camper's *Oeuvres*. Paris, 1803. Vol. I, pp. 5–196.
Caplan, Arthur L. ed. *The Sociobiology Debate: Readings on Ethical and Scientific Issues*. New York, 1978.
Cavalli-Sforza, L. L. and M. W. Feldman. *Cultural Transmission and Evolution: A Quantitative Approach*. Princeton, 1981.
Cheney, Dorothy L. and Robert M. Seyfarth. *How Monkeys See the World: Inside the Mind of Another Species*. Chicago, 1990.
Clutton-Brock, Timothy H. and Paul H. Harvey. eds. *Readings in Sociobiology*. Reading, 1978.

Dawkins, Richard. *The Selfish Gene*. Oxford, 1976, repr. 1981.
Descartes, René. *Traité de l'homme*. Paris, 1664.
Desmond, Adrian. *The Ape's Reflexion*. London, 1979.
Diderot, Denis. *Le Rêve de d'Alembart* in Diderot's *Oeuvres comolètes*. ed. H. Dieckmann, J. Varloot et al. Paris, 1975–. Vol. XVII, p. 206.
Ehara, A., T. Kimura, O. Takenata and M. Iwamoto. eds. *Primatology Today: Proceedings of the XIIIth Congress of the Primatological Society*. Amsterdam, 1991.
Eibl-Eibesfeldt, Irenäus. *Human Ethology*. New York, 1989.
Galdikas, Biruté M. F. "Orangutan adaptation at Tanjung Putting Reserve: Mating and ecology," in Hamburg, D. A. and E. R. McCown. eds. *The Great Apes*. Menlo Park, 1979.
Gardner, R. Allen, Beatrix T. Gardner, Thomas E. Van Cantfort. eds. *Teaching Sign Language to Chimpanzees*. Albany, 1989.
Gesner, Conrad. *Historia animalium*. 5 vols. Tiguri, 1551–1558.
Glass, Bentley, et al. eds. *Forerunners of Darwin: 1745–1859* Baltimore, 1959, 2nd ed. 1968.
Goodall, Jane., *The Chimpanzees of Gombe: Patterns of Behavior*. Cambridge, Mass., 1986.
Gould, Stephen Jay. *Ontogeny and Phylogeny*. Cambridge, Mass., 1977.
Gregory, Michael S., Anita Silvers and Diane Sutch. eds. *Sociobiology and Human Nature*. San Francisco, 1978.
Hasting, Hester. *Man and Beast in French Thought of the Eighteenth Century*. Baltimore, 1936.
Kevles, Bettyann. *Watching the Wild Apes: The Primate Studies of Goodall, Fossey and Galdikas*. New York, 1976.
Labov, William. *Sociolinguistic Patterns*. Philadelphia, 1972.
Lieberman, Phillip. *On the Origins of Language: An Introduction to the Evolution of Human Speech*. New York, 1975.
Lieberman, Phillip. *The Biology and Evolution of Language*. Cambridge, Mass., 1984.
Lockard, Joan S. ed. *The Evolution of Human Social Behavior*. New York, 1990.
Lopreato, Joseph. *Human Nature & Biocultural Evolution*. Boston, 1984.
Lovejoy, Arthur O. *The Great Chain of Being: A Study of the History of an Idea*. Cambridge, Mass., 1936.
Lumsden, Charles J. and Edward Wilson. *Genes, Mind and Culture: The Coevolutionary Process*. Cambridge, Mass. and London, 1981.
Lumsden, Charles J. and Edward Wilson. *Promethean Fire: Reflections on the Origin of Mind*. Cambridge, Mass. and London, 1983.
MacKinnon, John. *In Search of the Red Ape*. London, 1974.
MacKinnon, John. *The Ape Within Us*. London, 1978.
Marler, Peter R. and J. G. Vandenbergh. eds. *Handbook of Behavioral Neurobiology*. Vol. 3: *Social Behavior and Communication*. New York and London, 1979.
Masters, Roger D. *The Nature of Politics*. New Haven, 1989.
Masters, Roger D. and Glendon Schubert. eds. *Primate Politics*. Carbondale, Ill., 1991.
Mayr, Ernst. *The Growth of Biological Thought: Diversity, Freedom, and Inheritance*. Cambridge, Mass., 1982.
Midgley, Mary. *Beast and Man: The Roots of Human Nature*. Hassocks, 1979.

Monboddo, Lord. (James Burnett). *Of the Origin and Progress of Language*. 6 vols. Edinburgh, 1773–1792.
Montagu, Montague Ashley. *Edward Tyson and the Rise of Comparative Anatomy in England*. Philadelphia, 1943.
Montagu, Montague Ashley. ed. *Sociobiology Examined*. New York and Oxford, 1980.
Parker, Sue T. and Kathleen R. Gibson. eds. *"Language" and Intelligence in Monkeys and Apes: Comparative Development Perspectives*. Cambridge, 1990.
Patterson, Francine and Eugene Linden. *The Education of Koko*. London, 1982.
Perrault, Claude. *Mémoires pour servir à l'histoire naturelle des animaux*. Paris, 1676.
Pinker, Steven. *Language Learnability and Language Development*. Cambridge, Mass., 1984.
Premack, David. *Gavagail or the Future History of the Animal Language Controversy*. Cambridge, Mass., 1986.
Prévost, Abbé Antoine-François. ed. *Histoire generale des vovages*. 20 vols. Paris, 1746–89.
Reynolds, Peter C. *On the Evolution of Human Behavior: The Argument from Animals to Man*. Berkely and Los Angeles, 1981.
Rodman, Peter. "Population composition and adaptive organization among orangutans" in Crook, J. and R. Michael, eds. *Comparative Ecology and Behavior of Primates*. London, 1973.
Roger, Jacques. *Les sciences de la vie dans la pensée française du XVIII siecle*. Paris, 1963.
Rose, Steven, Richard Lewontin and Leon Kamin. *Not in our Genes: Biology, Ideology and Human Nature*. New York and Harmondsworth, 1984.
Rosenfield, Leonora. *From Beast Machine to Man Machine: The Theme of the Animal Soul in French Letters from Descartes to La Mettrie*. New York, 1940.
Rousseau, Jean-Jacques. *Discours sur l'inégalité*, in Gagnebin, B., M. Raymond, et al. eds. *Oeuvres complètes de Rousseau*. Vol. 3. Paris, 1964.
Rumbaugh, Duane. ed. *Language Learning by a Chimpanzee: The Lana Project*. New York, 1977.
Ruse, Michael. *Sociobiology: Sense or Nonsense?* Dordrecht, 1979.
Sahlins, Marshall. *The Use and Abuse of Biology: An Anthropological Critique of Sociobiology*. Ann Arbor, 1976.
Savage-Rumbaugh, E. Sue. *Ape Language: From Conditioned Response to Symbol*. Oxford, 1986.
Sebeok, Thomas A. and Jean Umiker-Sebeok. eds. *Speaking of Apes: A Critical Anthology of Two-Way Communication with Man*. New York, 1980.
Smellie, William. *Philosophy of Natural History*. 2 vols. Edinburgh, 1790–1799.
Snowdon, C. T., C. H. Brown, and M. R. Petersen. eds. *Primate Communication*. Cambridge, 1982.
Terrace, Herbert. *Nim*. London, 1980.
Thomas, Keith. *Man and the Natural World: Changing Attitudes in England, 1500–1800*. London, 1983.
Tinland, Franck. *L'Homme sauvage*. Paris, 1968.
Todt D., P. Goedeking and D. Symmes. eds. *Primate Vocal Communication*. Berlin, 1988.

Tort, Patrick. *La Pensée hierarchique et l'evolution*. Paris, 1983.
Tort, Patrick. ed. *Misère de la sociobiologie*. Paris, 1985.
Tyson, Edward. *Orang-Outang or, the Anatomy of a Pygmie*. (1699) facsimile reprint with an introduction by Ashley Montagu. London, 1966.
Vartanian, Aram. *La Mettrie's L'Homme machine: A Study in the Origin of an Idea*. Princeton, 1960.
Wallace, Alfred Russel. *The Malay Archipelago*. 2 vols. London, 1869.
Wilson, Edward O. *Sociobiology: The New Synthesis*. Cambridge, Mass., 1975.
Wilson, Edward O. *On Human Nature*. Cambridge, Mass. and London, 1978.
Wilson, Peter J. *Man, the Promising Primate: The Conditions of Human Evolution*. New Haven, 1980, 2nd edn., 1983.
Wokler, Robert. "Tyson and Buffon on the orang-utan," *Studies on Voltaire and the Eighteenth Century* **155** (1976), 2301–2319.
Wokler, Robert. "Perfectible Apes in Decadent Cultures: Rousseau's Anthropology Revisited," *Daedalus* (Summer, 1978), 107–134.
Wokler, Robert. "From Apes to Races in the Scottish Enlightenment: Kames and Monboddo on the History of Man" in Jones, Peter. ed. *Science and Philosophy in the Scottish Enlightenment*. Edinburgh, 1989.
Wokler, Robert. "Anthoropology and Conjectural History in the Enlighenment" in Fox, C., R. Porter and R. Wokler, eds. *Inventing Human Science*. Berkeley, 1993.
Wokler, Robert. "Paroles de singes," in Blanckaert, Claude, ed. "Les sciences humaines." *Autrement*. Hiver, 1993.
Yerkes, R. M. and A. W. Yerkes. *The Great Apes: A Study of Anthropoid Life*. New Haven, 1929.
Yolton, John W. *Thinking Matter: Materialism in Eighteenth-Century Britain*. Minneapolis, 1983.

II: "Struggle"

SOCIAL METAPHORS IN EVOLUTIONARY BIOLOGY, 1870–1930: THE WIDER DIMENSION OF SOCIAL DARWINISM

PETER J. BOWLER
Queens University, Belfast

The popularity of the term "Social Darwinism" highlights the problems lying in wait for anyone wishing to analyze the question of the transfer of metaphors between biology and social theory. Here is a term which, in its original meaning, denotes the application of metaphors that were highly successful in biology to social theory and to social policy. Darwin used the concepts of the "stuggle for existence" and the "survival of the fittest" to revolutionize biology through the introduction of a materialistic theory of evolution. The Social Darwinists then – illegitimately, in the eyes of most users of the term – argued that the same process of struggle and elimination of the unfit should be encouraged in society as a means of guaranteeing progress. But, as both Marx and Engels pointed out at a very early stage in the debate, Darwin's theory could itself be seen as the projection onto nature of the social values of laissez-faire capitalism. The transfer of metaphors works both ways, and historians of science are still debating the extent of the role played by social metaphors in the construction of evolutionary theory.[1] No one, however, doubts that models and metaphors, which have gained the kudos of being labelled "scientific," have then been employed by social thinkers.

The analysis that follows will certainly raise the question of the extent to which social metaphors are transferred into, as well as from, biology, but it is not my intention to offer a resolution of this controversial issue. My main purpose is to highlight the complexity of the situation by stressing the many different metaphors that have participated in the interaction. By using the single term "Social Darwinism" to denote a wide range of social applications of evolution theory, we tend to oversimplify the situation by focusing on the best-known evolutionary mechanism. Recent work in the history of science has stressed the extent to which non-Darwinian mechanisms of evolution were accepted in the late nineteenth and early twentieth centuries. To understand what might be best called "social evolutionism" we must take these alternatives into account. More confusing still, there are some biological theories that *seem* Darwinian and were actually used by Darwin and his followers, but which were unconnected with the basic mechanism of natural selection and thus were also used by anti-Darwinian biologists. The rival theories often had parallels, and perhaps conceptual sources, in rival models of human history. All theories – pro-, anti- and partly-Darwinian – reflect social values and raise complex questions about the interaction between the two areas of discourse.

The classic expression of the view that Darwin's theory had an impact on nineteenth-century social thought is Richard Hofstadter's *Social Darwinism in American Thought*.[2] According to this interpretation, Darwin's mechanism of the "survival of the fittest" in the "struggle for existence" was used by Herbert Spencer and his followers to legitimize the ideology of *laissez-faire* individualism that sustained the competitive ethos of capitalism. But Hofstadter himself complicates the issue by insisting that there are several other applications of the theory that also count as Social Darwinism. Later on in the nineteenth century the struggle metaphor was used to justify the extermination of "inferior" races in the age of imperialism. By a slight twist of the argument, the supporters of eugenics were then able to justify their calls for restricting the breeding of the "unfit" members of society by comparing such a process with artificial selection. The sheer range of social applications makes it difficult to take seriously the claim that they were all inspired by a single biological theory, especially when we take into account the many changes that took place in evolution theory during the post-Darwinian decades.

The assumption that all these policies can be attributed to the influence of Darwinism has been seized upon by those commentators who see that theory as the foundation-stone for a materialistic worldview in which all moral values are abrogated in the name of expediency and profit. Darwinism was assumed to be a bad theory because it destroyed the meaning of human life and encouraged us all to behave like animals. It also implied that every individual's character was rigidly fixed by biological inheritance, a point reinforced by the subsequent emergence of Mendelian genetics. By contrast, rival theories such as Lamarckism are supposed to convey a more humane message – Hofstadter used the sociologist Lester Frank Ward to show how Lamarckism could be used to sustain a social philosophy based on state invervention directing society (and the human race) toward a chosen goal. The late Arthur Koestler defended a similar view, following in the footsteps of Samuel Butler, George Bernard Shaw, and other literary opponents of materialism.[3] Socialists and antimaterialists thus concur in finding Darwinism a useful symbol for all that they dislike in modern biology.

The possibility that this negative image of Darwinism was an artifact of hindsight was raised by Robert Bannister.[4] For Bannister, Hostadter's intepretation was a myth created by stressing the apparently amoral character of Spencer's philosophy. The purpose of this myth was to legitimize the reformist policies of the early twentieth century by contrasting them with the harshness of what had gone before. Bannister suggested that Spencer's evolutionism was not, in fact, very Darwinian, and some critics argued that he was trying to defend the objectivity of science by driving a wedge between biological and social Darwinism. Bannister's real point, though, was that Social Darwinism was never as popular as Hofstadter had claimed, nor was it perceived as a totally amoral philosophy of life. This view has been sustained by later studies of Spencer and his influence. It has also been shown that the term "Social Darwinism" was used only in a perjorative sense and did not come into use until the early twentieth century.[5]

If Bannister did imply that Social Darwinism was not very Darwinian, we are now in a position to make sense of such a claim without implying that there was a rigid dividing line between social and biological evolutionism. Reinterpretations of the Darwinian Revolution have shown that nineteenth-century evolution theory was very different from its modern counterpart.[6] To imply that Spencer and his followers adopted a social policy based on the survival of the fittest among individuals subject to random genetic variation would be to make them Social Darwinists in a modern sense. But it is now clear that very few biological evolutionists – let alone social evolutionists – were in a position to take such a possibility seriously before the emergence of the modern genetical theory of natural selection.

To understand the social evolutionism of the late nineteenth century, we need to forget modern Darwinism and try to relate the various social policies to the scientific theories of the time. Once we recognize that even the biological Darwinism of the late nineteenth century had components that were non-Darwinian by modern standards, we can better appreciate the true character of Spencer's efforts to produce a universal philosophy of evolution. At the same time, we shall be able to see how some anti-Darwinian theories could be used to justify social policies that might easily be mistaken as Darwinian by someone who was not aware of the tensions existing within the biological sciences.

The most obvious non-Darwinian mechanism of evolution was the Lamarckian theory of the inheritance of acquired characteristics. It is seldom realized that Spencer himself was more a Lamarckian than a Darwinian in his biology. By recognizing this fact we can appreciate that much free-enterprise Social Darwinism depended on the claim that struggle was the best stimulus to individual self-development, not a means of eliminating the congenitally unfit. Both Darwinians and Lamarckians believed that progress was the result of the species' struggle against its environment, and both believed that if this struggle was evaded the species would degenerate. At this level the metaphor of the "struggle for existence" took on a significance transcending the scientific debate over the evolutionary mechanism.

If the concept of a struggle among the individuals making up a population could be used by both Darwinians and Lamarckians, there was another level of competition that became more important as the century drew to a close. Darwin and Wallace both believed that a newly-evolved species would tend to spread out from its point of origin, displacing or exterminating the more primitive forms that populated the surrounding territory. In the late nineteenth century this model of conflict between species was widely used to explain the geographical distribution of species, and after 1900 it was taken up by palaeoanthropologists trying to explain the succession of primitive human races in the Palaeolithic or Old Stone Age. The implications of this model for modern imperialism were not lost on its exponents, who thus founded yet another form of Social Darwinism. Yet conflict *between* species was not a consequence of the basic Darwinian mechanism, and the theory was taken up by many biologists and anthropologists who were not Darwinians in the sense accepted today. Indeed, many Lamarckians lent their support to racism by claiming that the "higher" races were evolved in the more stimulating environment of the North.

Those who see Darwin's theory as a product of the social values of his time usually point to the parallel between natural selection and the ethics of free-enterprise individualism. But Darwin's theory was also a product of his voyage around the world aboard H.M.S. *Beagle*, a classic expression of British naval imperialism. By the latter decades of the century, the imperialist ethic had become far more self-conscious, and it is not surprising that we should find the metaphors of conquest and colonization creeping into evolution theory and into palaeoanthropology. A study of other areas of nineteenth-century thought reveals, however, that there had been a long tradition of explaining European history and prehistory in terms of successive waves of invaders coming in from a source of cultural progress to the East. Darwinism (and evolutionism in general) thus drew its inspiration from both the liberal philosophy of continuous progress and the rival, more conservative view of history that stressed successive waves of conquest.

In these circumstances, tracing the direction of the interactions between biology and social theory becomes extremely difficult. Did the imperialist model of race-conflict advocated by the palaeoanthropologists come from biogeography or from the conservative model of history? Or must we follow Stephen Gould's suggestion that there are "eternal metaphors" that constantly reappear in our efforts to interpret the past, rival interpretations that spring into the mind whenever the evidence or the cultural environment creates an opening?[7] To have any hope of dealing with such issues we have to develop a much more sophisticated understanding of the changes taking place within science, within social theory, and also within society itself. My contention is that the oversimplified notion of "Social Darwinism" stands in the way of our efforts to achieve this goal.

Laissez-faire Social Evolutionism

The classic image of Social Darwinism as an amoral social policy based on the rejection of all traditional values stems from the identification of Darwinism with the kind of struggle that prevails in a free-enterprise society. The assumption seems to be that if one advocates leaving everything in society to the self-interest of individuals, then there can be no room for the hope or expectation that social development has a moral goal. We are all struggling against one another in the mere hope of survival, caring nothing for what happens to anyone else or to society as a whole. This emphasis on the struggle motif is favored by critics who maintain that the free-enterprise system is based solely on self interest. Those with wealth or ability seek only to enjoy the fruits of their good fortune without sharing any with others less fortunate than themselves.

Such a model of the free-enterprise society certainly seems Darwinian in the sense that good or ill-fortune is a matter of luck, and those who win in the lottery have no interest in modifying the natural outcome of the process by which the unfit are cast aside. Even if this is a valid assessment of captitalist ethics, it is not an accurate picture of the arguments actually used to defend the system in the nineteenth century. As Robert Richards has argued, it is a mistake to see Spencer and his followers as advocates of pure materialism.[8] The whole point of Spencerianism was that nature should be allowed to progress without interference because the ultimate goal of social development was a moral one. We may dismiss Spencer's arguments as pure verbiage designed to conceal the true selfishness of the commercial elite, but to assess the public impact of those arguments at the time we have to take them at their face value.

To the extent that Darwin's theory became associated, in the public mind at least, with Spencer's social philosophy, we must accept that it was being identified with a model of progress through struggle that does not correspond to the modern Darwinian theory of selection. In a strictly Darwinian mechanism, individuals are born with fit or unfit characters (as measured by the local environment) and can do nothing to influence their fate. Natural selection is a kind of genetic Russian roulette in which we all must participate, and no effort on our part can save us if we are born with harmful characters. But it was certainly not Spencer's purpose to invoke selection acting upon a random distribution of such rigidly determined hereditary characters. The whole purpose of the struggle that must take place within a free-enterprise society was not to eliminate the congenitally unfit, but to make everyone fitter. Liberal thinkers endorsed Samuel Smiles' vision of a society based on *Self Help* (his book of that title was published in the same year as the *Origin of Species*). Spencer merely generalized this emphasis on self-improvement to make it the basis of all social progress.

For Spencer, the great problem with socialism was that it protected everyone from the chief stimulus to self-development. If you could be sure of a living provided by the state, why work hard to develop your own potential? Only harsh necessity would force all members of society to exercise their abilities to the full. Anyone who made a mistake would be taught a lesson by suffering the consequent penalties – and this would teach them to do better next time. Having thus learned how to function effectively in society, such a person would do this or her best to pass the message on to their children, thus perpetuating the effect of the lesson in future generations.

> If to be ignorant were as safe as to be wise, no one would become wise. . . . Unpitying as it looks, it is better to let the foolish man suffer the appointed penalty of his foolishness. For the pain – he must bear it as well as he can: for the experience – he must treasure it up, and act more rationally in the future. To others as well as to himself will his case be a warning. And by multiplication of such warnings, there cannot fail to be generated in all men a caution corresponding to the danger to be shunned.[9]

Spencer admits that the policy looks pitiless, yet its ultimate purpose is to force everyone to become more intelligent and more adaptible. Nature is the best schoolteacher when it comes to endowing everyone with the Protestant virtues of thrift and industry.

Spencer never ruled out the possibility that some unfortunates might actually die as a consequences of their inability to cope. This is what he called an indirect way of bringing the species into equilibrium with its environment. It was to describe this mechanism – which is, of course, natural selection – that Spencer coined the term "survival of the fittest" in his *Principles of Biology*.[10] The fact that Spencer coined this notorious phrase has no doubt fueled claims that he was the original Social Darwinist, yet for him this was always a secondary mechanism. Far more important was the direct mechanism of adjustment by which the individual survived and learned from its experience, thus improving itself and transmitting the improvement to future generations. This was an explicitly Lamarckian process: struggle was the spur to self-development, and the evolution of society (or the species) was the sum total of generations of such individually acquired characters. Small wonder that Spencer went out of his way to defend Lamarckism against Weismann's attacks in the 1880s.[11] Far from being an expression of a rigid hereditarian policy, his was a theory that depended upon the individual having the ability to improve itself when suitably stimulated. For all that which he has been called a Social Darwinist, *Spencer was really a Social Lamarckian*. He justified a society based on struggle not by reference to natural selection, but by portraying struggle as the chief force encouraging self-development.

This Lamarckian component in Spencer's thought has been ignored because evolution theories have all too often been portrayed as rigidly polarized in their social implications. Lamarckism is identified with the social reformism of Lester Frank Ward and others, who sought to improve the human race by using education to foster the characteristics desirable in future generations.[12] This *was* a form of social Lamarckism, but it was not the only form. Spencer dismissed claims that we can control the future development of the human race because he thought that nature was too complex for us to be able to predict the results of our actions. Progress would only result from letting nature take its course; individual competition, not a state-controlled school system, was the best form of education leading to the improvement of the human race. This alternative version of social Lamarckism has been ignored because historians have allowed themselves to be influenced by an artificially rigid characterization in which Darwin's selection theory is identified as the only expression of the capitalist ethos in the biological realm.

If we accept that Darwin's theory became associated in the public mind with Spencer's philosophy of social progress, then it seems probable that many ordinary people simply failed to see the distinction between the Darwinian and Lamarckian modes of "progress through struggle." In the decades before the appearance of August Weismann's concept of the germ plasm – let alone the modern concept of the gene – it was easy to be so vague about the production of variations that no clear distinction was possible. Darwin himself accepted that variations were caused by an environmental disturbance of the reproductive process, and never ruled out the Lamarckian effect (i.e., directed rather than undirected variation). I believe that the language used by many late-nineteenth-century biologists and social theorists simply fails to distinguish between what we call Darwinism and Lamarckism. The idea that individual characters might by rigidly determined by heredity came in only with Weismann's theory of the germ plasm, and became uncontroversial only after the triumph of Mendelian genetics. When this kind of hereditarianism did emerge, its social expression was not the Spencerian philosophy of free-enterprise, but the eugenics movement's demand for state control of human reproduction.[13]

Struggle Against the Environment

The liberal philosophy of continuous progress though the accumulation of individual acts of self-improvement depended on the notion of competition between the members of a population, but it also invoked the metaphor of the population as a whole struggling against its environment. The claim that all human beings should work together to help in the struggle against nature has become known as "reform Darwinism."[14] The fact that Darwinism could be invoked to defend an ethic of cooperation gives us a clear warning of the dangers inherent in the notion of a unitary Social Darwinism. But my main interest here is the way in which the notion of the struggle against the environment was taken up by both biologists and social theorists. Implicit in Spencer's whole philosophy was the assumption that progress is stimulated by an *external* challenge, and what could be more of a challenge than a harsh environment? In an easy climate where food was plentiful, there would be no need to struggle and hence no stimulus toward self-improvement. Over many generations of exposure to such a soft environment, a population would simply fall behind in the race towards progress, when compared to another group whose environment provided a constant challenge.

Biologists took up the theme that the main cause of progress throughout the history of life had been exposure to a challenging environment. This is evident in the work of the Darwinist E. Ray Lankester, but equally evident in the Lamarckian embryologist E. W. MacBride. Lankester also noted that there was a reverse side to the coin: a stimulating environment gave progress, but the adoption of a less active lifestyle such as parasitism would cause not merely stagnation but actual degeneration towards a more primitive level of organization.[15] MacBride developed a philosophy of evolution in which the various invertebrate types were offshoots from the main stem of evolution, which led upwards towards the vertebrates. The main stem had kept to a stimulating environment in the open sea, but the invertebrates had stepped off the escalator and degenerated by taking up a mode of life based on crawling or burrowing on the bottom.[16]

The human implications of this were not lost on the biologists. Lankester compared biological degeneration to the fate of the Romans after they had conquered the known world and expressed concern over the future of modern mankind.[17] MacBride became a racist, convinced that the Irish – who had evolved in a softer, Mediterranean environment – were an inferior branch of the white population that should be subject to compulsory sterilization. The Anglo-Saxons, who had evolved in the harsh climate of Northern Europe, were the dominant race upon which the empire depended for its success. He thus supported eugenics from a Lamarckian rather than a Darwinian or a Mendelian perspective.[18]

The implications of this approach for the race question were already apparent even in Spencer's writings. He had realized that biological evolution applied to the human races and extended the concept of a cultural hierarchy into a hierarchy of racial abilities. Anthropologists such as E. B. Tylor extended the liberal model of progress to other cultures by supposing that all branches of mankind had advanced at different rates through the same hierarchy of cultural stages. Australian aborigines were still stuck at the level reached by white Europeans back in the Old Stone Age. This was reinforced by the work of archaeologists such as Sir John Lubbock in Britain and Gabriel De Mortillet in France, who also compared the successive levels of Palaeolithic culture with various modern "savages."[19]

As J. W. Burrow and George Stocking have shown, the initial form of this cultural evolutionism rejected the racism that was beginning to be promoted in the middle decades of the century.[20] All races were supposed to have the same *capacity* to advance, even if they had been subjected to different levels of stimulus. But Spencer's theory threatened this older liberal interpretation by suggesting that cultural and intellectual evolution went hand in hand. Those races that had lagged behind culturally must also have lagged in intellectual development; they were now permanently fixed at a lower level of the intellectual and moral hierarchy. By the later decades of the century it was virtually impossible for a white European to escape the sense that he stood at the head of an evolutionary ladder of racial development. Liberals and conservatives alike agreed that the "lower" races were fixed in an inferior position by virtue of their evolution in a less stimulating environment.

Racial Conflict

The evolutionists were convinced that, despite their common origin, some races were more advanced than others. This immediately raised the question of what would happen when two races, originally separated by a geographical barrier, came into contact. Since the lower could not raise itself sufficiently by self-development, the mental differential would persist, and it was only a short step to visualizing the subsequent events in terms of the struggle for existence. Extending the level of conflict from the individual to the racial level provided yet another form of Social Darwinism, – yet, like the Spencerian philosophy, this too must be analyzed with care to determine how much was genuinely Darwinian.

It is certainly true that Darwin used competition between closely related varieties and species as an explanatory tool, especially in biogeography. He believed that a newly, and hence more highly evolved species would inevitably spread out from its home territory, displacing the more lowly forms that had preceded it. A successful new form would thus occupy an expanding geographical area; previously existing types would be wiped out unless they were able to survive in remote refuges protected by some barrier from competition with the later products of evolution. This phenomenon explained many otherwise puzzling features of geographical distribution, such as the marsupials of Australia.

This approach was developed in A. R. Wallace's classic *Geographical Distribution of Animals* of 1876. Wallace, like Darwin, assumed that the continents had always occupied their present positions, and from this proceeded to explain the faunas of the different regions in terms of successive migrations of higher types from a centre of progressive evolution in the northern regions, especially Eurasia. In the last decades of the century, geographical distribution blossomed into a major field of study.[21] Wallace's image of successive waves of more highly evolved invaders spreading out from Eurasia (and to a lesser extent North America) was extended by the Canadian palaeontologist W. D. Matthew in his "Climate and Evolution" of 1914.[22] The imperialist language of this school of thought is quite explicit: species have their "headquarters," but can "invade" or "colonize" another territory and displace or exterminate it inhabitants. The whole theory was sustained by a conviction that the harsh environment of northern Eusrasia provided a constant supply of more highly evolved types, generating successive waves of invasion sweeping across the rest of the world.

To be fair, there were many biogeographers who rejected this model. In the era before the popularization of continental drift, many pointed to now sunken landbridges between continents in earlier geological periods to explain how species migrated from one area to another. Advocates of the land-bridge theory were not so likely to present northern Eurasia as the main center of progressive evolution, but they were equally convinced that any newly evolved form would spread out to occupy as much territory as it could gain access to, at the expense of the previous occupants.

It is important to note, however, that the metaphors of expansion and conquest could be used independently of the theory of natural selection. Talk of the "survival of the fittest" among species and races did not require belief in the natural selection of individual variations as the mechanism that actually produced the more highly evolved forms. Wallace and Matthew were Darwinists in the modern sense of the term, committed to natural selection as the only mechanism of evolution. But such a position was rare in the decades around 1900, and many who made use of the "waves of conquest" theory in biogeography were not Darwinians. Matthew's chief at the American Museum of Natural History, Henry Fairfield Osborn, accepted much of his theory but rejected Darwinism in favor of orthogenesis, or evolution by definitely directed variation.[23]

We normally think of Darwinism as a theory of continuous change, and so it is when confined to the evolution of a single population. But once it came to explaining the historical relationship between successively evolved forms, Darwin himself was forced to adopt a very different model of change. At any one point on the earth's surface, change was discontinuous, consisting of waves of invading forms sweeping across as they spread out from some distant homeland where continuous progress actually took place.

The social implications of such a model of geographical distribution in an age of imperial conquest were gradually made apparent. European technology was now breaking down the barriers that had protected the "lower" races in their geographically isolated refuges. Darwin himself did not advocate the deliberate extermination of "lower" races, but he assumed that for complex reasons such races gradually declined in numbers whenever confronted by invading white men.[24] By the end of the century the Darwinist Karl Pearson was openly advocating the replacement of inferior races by white populations wherever this was possible.

It is a false view of human solidarity, a weak humanitarianism, not a true humanism, which regrets that a capable and stalwart race of white men should replace a dark-skinned tribe which can neither utilise its land for the full benefit of mankind, nor contribute its quota to the common stock of human knowledge. The struggle of civilized man against uncivilized man and against nature produces a certain partial "solidarity of humanity" which involves a prohibition against an individual community wasting the resources of mankind.[25]

In a footnote to this passage Pearson warned against the brutality of simple extermination, but nevertheless rejoiced that whites were replacing the natives of America and Australia.

Darwinism supplied a rhetoric of struggle which could easily be translated from the individual to the racial level. Pearson was a staunch defender of natural selection acting within populations, but in the modern world he believed that human communities needed solidarity in order to resist external enemies. The struggle for existence had moved from the individual to the national and racial level. The rhetoric of competing nations became popular in the decades before World War I and was widely believed to have contributed to the ruthlessness of the German war machine. This was certainly the view of the American biologist Vernon Kellogg, who visited the invading Germans in Belgium and wrote his *Headquarters Nights* to warn against this version of Social Darwinism.[26] In the years following the war, many biologists argued strongly against the application of Darwinian metaphors to international affairs.

That Darwinism played a role in stimulating this atmosphere of national and racial rivalry cannot be denied. But we must remember that the "invasion and extermination" model of biogeography could be applied even by evolutionists who were *not* Darwinians. A good illustration of this can be seen in palaeoanthropology, which underwent a theoretical shift in the early twentieth century that brought the field into line with the model of geographical distribution advocated by Wallace and Matthew. In the late nineteenth century, extinct human types such as the Neanderthals had been added to the bottom of the supposed hierarchical ladder of modern races leading from the apes to white Europeans. Neanderthals were seen as an early stage in human evolution. In the years before World War I, this model of continuous progress was challenged by a new image of human evolution in which the Neanderthals were seen as a separate branch of the human family tree, which had been wiped out when modern humans invaded Europe from the East.[27]

The pioneer in the movement was the British geologist W. J. Sollas, whose *Ancient Hunters* of 1991 identified various modern "savages" in different stages in the European archaeological record. But instead of arguing for indigenous development, Sollas insisted that the modern savages had been displaced from Europe by higher types invading from elsewhere. The Eskimos were the remnants of the people who had occupied Europe during the Magdelenian period of the Old Stone Age, now marginalized in the North. The Australian aborigines were Neanderthals who had found a refuge in that remote corner when the rest of the world was taken over by Cro-Magnon man. Sollas soon had to abandon this last claim, however, as Arthur Keith and Marcellin Boule spearheaded a move to brand the Neanderthals as so primitive that they could not be indentified with even the lowest forms of *Homo sapiens*. They were a distinct and more primitive species of mankind, wiped out by our own ancestors.

The claim that the Neanderthals and other ancient human types had been displaced by more highly developed invaders expanded the theory of race conflict onto a wider stage and provided yet more rhetoric to back up the ideology of imperialism. Sollas clearly recognized that his new model of prehistory had broader implications. The principle of "might is right" was nature's way of ensuring the success of more highly evolved types.

What part is to be assigned to justice in the government of human affairs? So far as the facts are clear they teach in no equivocal terms that there is no right which is not founded on might. Justice belongs to the strong, and has been meted out to each race according to its strength; each has received as much justice as it deserves. What perhaps is most impressive in each of the cases we have discussed is this, that the dispossession by a new-comer of a race already in occupation of the soil has marked an upward step in the intellectual progress of mankind. It is not priority of occupation, but the power to utilise, which establishes a claim to the land. Hence it is a duty which every race owes to itself, and to the human family as well, to cultivate by every possible means its own strength: directly it falls behind in the regard it pays to this duty, whether in art or science, in breeding or in organization for self-defence, it incurs a penalty which Natural Selection, the stern but beneficient tyrant of the organic world, will assuredly exact, and that speedily, to the full.[28]

The imperialist message of this passage is obvious enough, and it would be tempting to dismiss Sollas as the worst kind of Social Darwinist, obsessed with natural selection as the vehicle of progress. His words seem to parallel those of Karl Pearson quoted above. Yet elsewhere in his book he ridiculed the Darwinian mechanism as impotent to explain the generation of higher characters.[29] Like many other critics of Darwinism, Sollas was prepared to accept natural selection as a mechanism for weeding out the types left behind in nature's progress – but he confessed himself unable to explain how the higher types were, in fact, evolved.

The expulsion of the Neanderthals from human ancestry was widely accepted during the 1920s and 1930s. It was also widely assumed that central Asia was the source of the successive waves of more advanced humans who spread around the world. This view had been adopted by A. R. Wallace and was now championed by the palaeontologist H. F. Osborn and by Davidson Black, the discoverer of Peking man.[30] Only in the 1930s did opinion begin to shift back to Darwin's view that Africa was the cradle of mankind.

The ideological overtones so evident in the passage quoted from Sollas above were apparent to other supporters of the theory. Michael Hammond has shown that Boule was motivated by a desire to challenge the political message that the socialist de Mortillet had associated with his linear progressionism.[31] Keith, like Sollas, clearly saw the theory of invasion and extermination as a justification for imperialism. He had become acutely conscious of the racial differences among modern humans, and extending the antiquity of the human species enabled him to explain the appearance of such fundamental variations in the non-Neanderthal stock. He favored the view that racial interaction would lead to struggle and extinction: "What happened at the end of the Mousterian period we can only guess, but those who observe the fate of the aboriginal races of America and Australia will have no difficulty in accounting for the disappearance of *Homo neanderthalensis*. A more virile form extinguished him."[32] In later years Keith went on to develop a whole theory of human evolution based on racial competition.

Keith, again like Sollas, was by no means a Darwinian in the modern sense of the term, confirming that an emphasis on *racial* struggle need not have an origin in the Darwinian theory. Some Darwinians certainly did translate the struggle for existence from the individual to the racial level, and no doubt everyone made use of the rhetoric of "struggle" and "survival" encouraged by Darwinism. But the idea that races must come into conflict to determine their destiny was part of a Darwinian biogeography that flourished independently of the selection theory and that was based on a model of history quite different than Spencer's progressionist evolutionism. If the theory of Neanderthal extinction drew some of its scientific plausibility from its congruence with the Wallace-Matthew school of biogeography, it also drew upon a long tradition in

European historiography that had provided an alternative to the liberal vision of cumulative progressive evolution owing to individual human activity.

I have called this alternative model the "cyclic" theory of progress because it stressed the noncumulative and episodic nature of human cultural development.[32] Philologists such as Max Müller had explained European prehistory in terms of waves of Aryan invaders marching in form the East, each contributing a new language and adding new cultural traditions to those already developed by their predecessors. The archaeologists who created the "three-age" system assumed that the new bronze and iron technologies were introduced by more advanced peoples invading from the East. Historians such as Charles Kingsley saw the Greeks, Romans, and Teutons as the basis for successive waves of cultural development, each conquering people picking up from where its predecessors left off. Progress was discontinuous and depended on something more than individual effort – each race and each culture made a unique contribution to human development, but each in its turn became exhausted and had to pass the torch on to a new race so that the process could be renewed.

The amazing thing is not that palaeoanthropologists eventually interpreted earlier types such as the Neanderthals in the light of this theory, but that it took them so long to make the connection. There had, in fact, been earlier efforts in this direction. Sollas had drawn upon the writings of the geologist W. Boyd Dawkins, whose *Early Man in Britain* of 1880 had suggested that the Palaeolithic Eskimos were pushed out of Europe by invaders. Dawkins drew directly upon the mythology of an Eastern center of progress from which the higher types had spread out, arguing that "the Neolithic peoples migrated into Europe from the South-east, from the myterious birthplace of successive races, the Eden of mankind, Central Asia."[34] This view of the Palaeolithic had been largely ignored during the late nineteenth century because interpretations of human evolution had been dominated by the supporters of the continuous model of progress in which it was more convenient to see the Neanderthals as missing links in a linear pattern of development.

Far from proposing something new, Sollas, Keith, and Boule were merely extending into the Palaeolithic a rival, more conservative model of history that had long been available in other fields. What is more interesting in this context is the link between this cyclic model of human progress and Matthews,' in some respects, very Darwinian theory of biogeography. The liberal Spencerian model of continuous development, of which Darwinism is often seen as an extension, stressed competition between individuals but saw the overall march of history as a continuous process. Yet when Darwinian biologists tried to explain the succession of events that generated the modern distribution of species, they too were forced to adopt a model in which waves of more successful species spread out across the globe, exterminating or marginalizing the earlier products of progressive evolution. It would seem that by the end of the century, the barriers between the rival models of historical development were breaking down. But was this because the model of continuous evolution no longer suited the ideology of an age of imperialism – or because the evidence from biogeography was intrinsically more difficult to explain without invoking discrete events such as episodes of migration?

Our survey of the various forms of Social Darwinism has forced us to conclude that the transfer of metaphors from biology to social thought is a far from straightforward process. Darwin's particular theory of evolution by natural selection was certainly not the only source for the metaphor of the "struggle for existence." Without wishing to minimize the extent to which Darwin's theory did reflect (and could hence be used to endorse) the ideology of *laissez-faire* individualism, we have seen that other versions of evolutionism, especially Spencerian Lamarckism, could equally well be interpreted as the biological analogs of that social philosophy. The concept of species struggling against their environment was exploited by many biological thinkers, both Darwinian and non-Darwinian, who thus seemed to endorse the claim that social progress was stimulated by the creative response of the human mind to environment challenge. When applied to nations or races via the image of migration and conquest, the "struggle" metaphor may have been the product of a conservative view of human history, which was belatedly incorporated into palaeoanthropology and biology only at the end of the nineteenth century. The question of why the various models were invoked in certain areas only at certain times awaits more careful analysis of the complex interactions involved.

In these circumstances, any effort to define what is Darwinian about Social Darwinism is fraught with danger. Without wishing to deny the interaction between biology and social thought, we must be aware of the complexity of the situation in both biology and historical studies, and of the changes that took place in both these fields in the course of the late nineteenth and early twentieth centuries. Darwinism alone cannot be seen as the source of the struggle metaphor – indeed some versions of what might look like Social Darwinism turn out to be based on philosophies of development whose origins lie in a worldview that had always stood opposed to that of *laissez-faire* individualism. Darwin's conversion of the world to evolutionism may have served as the means by which a whole range of ideas about change were transferred into biology – and then reused to endorse theories about human relationships. But the ideas that participated in these interactions included many that were not directly associated with Darwin's theory. To ascribe any use of the "struggle" metaphor to the influence of Darwinism is clearly a misuse of history. Recognizing the complexity of the situation surrounding the popularity of this metaphor poses a whole series of historical questions about its transfer from one domain to another, questions which have been concealed by our readiness to take the term "Social Darwinism" at face value.

Notes

1. See for instance Robert M. Young, *Darwin's Metaphor: Nature's Place in Victorian Culture* (Cambridge: Cambridge University Press, 1985) and John C. Greene, *Science, Ideology and World Wiew* (Berkeley: University of California Press, 1981). Recent books on Darwin confirm that this debate is ongoing. Ernst Mayr, *One Long Argument: Charles Darwin and the Genesis of Evolutionary Thought* (Cambridge, Mass.: Harvard University Press, 1991), denies any input from social values. Robert J. Richards, *The Meaning of Evolution: The Morphological Construction and Ideological Reconstruction of Darwin's Theory* (Chicago: University of Chicago Press, 1991), advocates a progressionist interpretation of the theory, but Adrian Desmond and James Moore, *Darwin* (London: Michael Joseph, 1991), see the origins of the theory in the more pessimistic social values of the 1830s.
2. Richard Hofstadter, *Social Darwinism in American Thought* (revised ed., New York: George Braziller, 1959).
3. Robert C. Bannister, *Social Darwinism: Science and Myth in Anglo-American Social Thought* (Philadelphia: Temple University Press, 1979).
4. Arthur Koestler, *The Ghost in the Machine* (New York: Macmillan, 1967); for earlier literary attacks on Darwinism, see Samuel Butler, *Evolution Old and New* (London, 1879) and the preface to George Bernard Shaw's *Back to Methuselah*.
5. Donald C. Bellomy, "'Social Darwinism' Revisited," *Perspectives in American History*, n.s. **1** (1984), 1–129.
6. See Peter J. Bowler, *The Eclipse of Darwinism: Anti-Darwinian Evolution Theories in the Decades around 1900* (Baltimore: Johns Hopkins University Press, 1983)

and *The Non-Darwinian Revolution: Reinterpreting a Historical Myth* (Baltimore: Johns Hopkins University Press, 1988).
7. Stephen Jay Gould, "The Eternal Metaphors of Paleontology," in A. Hallam (ed.), *Patterns of Evolution* (Amsterdam: Elsevier, 1977), pp. 1–26.
8. Robert J. Richards, *Darwin and the Emergence of Evolutionary Theories of Mind and Behavior* (Chicago: University of Chicago Press, 1987).
9. Herbert Spencer, *Social Statics: or the Conditions Essential to Human Happiness Specified* . . . (London, 1851), pp. 378–379.
10. See Spencer, *Principles of Biology*, 2 vol. (London, 1864), I, p. 444.
11. Spencer, *The Factors of Organic Evolution* (London, 1887) and "The Inadequacy of Natural Selection," *Contemporary Review* **43** (1893), 153–166 and 439–456.
12. See Hofstader, *Social Darwinism*, Chap. 4; also George W. Stocking, Jr., "Lamarckianism in American Social Science," *Journal of the History of Ideas* **23** (1962), 239–256.
13. A recent survey of the history of eugenics is Daniel Kevles, *In the Name of Eugenics: Genetics and the Uses of Human Heredity* (New York: Knopf, 1985).
14. See Eric Goldman, *Rendezvous with Destiny* (New York: Knopf, 1952). Greta Jones, *Social Darwinism in English Thought* (London: Harvester, 1980), also stresses the many different uses of the theory.
15. E. Ray Lankester, *Degeneration: A Chapter in Darwinism* (London, 1880); see Peter J. Bowler, "Development and Adaptation: Evolutionary Concepts in British Morphology, 1870–1914," *British Journal for the History of Science* **22** (1989), 283–297.
16. E. W. MacBride, *Textbook of Embryology: Invertebrates* (London, 1914), e.g., p. 662.
17. Lankester, *Degeneration*, p. 33. Lankester's views almost certainly influenced H. G. Wells' vision of a degenerate future for mankind in *The Time Machine* of 1895; see Bowler, "Holding Your Head Up High: Degeneration and Orthogenesis in Theories of Human Evolution," in James R. Moore (ed.), *History, Humanity and Evolution: Essays for John C. Greene* (Cambridge: Cambridge University Press, 1989), pp. 329–353.
18. See Peter J. Bowler, "E. W. MacBride's Lamarckian Eugenics and the Social Construction of Scientific Knowledge," *Annals of Science* **41** (1984), 245–250.
19. On the history of archaeology in this context see Peter J. Bowler, *The Invention of Progress: The Victorians and the Past* (Oxford: Basil Blackwell, 1989).
20. See J. W. Burrow, *Evolution and Society: A Study in Victorian Social Theory* (Cambridge: Cambridge University Press, 1966) and George W. Stocking, Jr., *Race, Culture and Evolution: Essays in the History of Anthropology* (New York: Free Press, 1968) and *Victorian Anthropology* (New York: Free Press, 1987); see also Bowler, *The Invention of Progress*.
21. See Janet Browns, *The Secular Ark: Studies in the History of Biogeography* (New Haven: Yale University Press, 1983) and Martin Fichman, "Wallace, Zoogeography, and the Problem of Land Bridges," *Journal of the History of Biology* **10** (1977), 45–63.
22. See William Diller Matthew, *Climate and Evolution* (New York, 1939), which reprints the original article of 1914; see Ronald Rainger, "Just before Simpson: William Diller

Matthew's Understanding of Evolution," *Proceedings of the American Philosophical Society* **130** (1986), 453–474.
23. On Osborn see Bowler, *The Eclipse of Darwinism* and Ronald Rainger, "Vertebrate paleontology as biology: Henry Fairfield Osborn and the American Museum of Natural History," in Rainger et al. (eds.), *The American Development of Biology* (Philadelphia: University of Pennsylvania Press, 1988), pp. 219–256.
24. Charles Darwin, *The Descent of Man*, 2nd ed. (London, 1885), pp. 181–192.
25. Karl Pearson, *The Grammar of Science*, 2nd ed. (London, 1900), p. 369.
26. Vernon Kellogg, *Headquarters Nights: a Record of Conversations and Experiences at the Headquarters of the German Army in France and Belgium* (Boston, 1917); on the reaction against Darwinian metaphors see Gregg Mitman, "Evolution as Gospel: William Patten, the Language of Democracy, and the Great War," *Isis* **81** (1990), 446–463.
27. See Peter J. Bowler, *Theories of Human Evolution: A Century of Debate, 1844 1944* (Baltimore: Johns Hopkins University Press and Oxford: Basil Blackwell, 1986).
28. W. J. Sollas, *Ancient Hunters and their Modern Representatives* (London, 1911), p. 383.
29. *Ibid.*, p. 405.
30. See Bowler, *Theories of Human Evolution*, esp. pp. 174–181.
31. Michael Hammond, "The Expulsion of the Neanderthals from Human Ancestry," *Social Studies of Science* **12** (1982), 1–36.
32. Arthur Keith, *The Antiquity of Man* (London, 1915), p. 136.
33. See Bowler, *The Invention of Progress*.
34. W. Boyd Dawkins, *Early Man in Britain* (London, 1880), p. 306.

"STRUGGLE FOR EXISTENCE":
SELECTION AND RETENTION OF A METAPHOR

PETER WEINGART
University of Bielefeld

Functions and Dysfunctions of Metaphors in Science

"The entire Darwinian theory of the struggle for existence is simply the transfer of the Hobbesian theory of *bellum omnium contra omnes* and the bourgeois-economic one of competition as well as Malthus' demographic theory from society into organic nature. After having accomplished this trick . . . it is easy to transfer these theories back from natural history into the history of society . . . and to claim one had proved this thesis as an eternal natural law of society." This famous interpretation of Darwin's theory by Friedrich Engels has been repeated by Nietzsche, Spengler, and countless lesser scholars to this day. It also underlies the general understanding of Social Darwinism and contains a description of the origin and function of metaphors in science. It provides an example of three different ways that metaphors are used as media of exchange of meaning: 1) from everyday language to scientific language; 2) from scientific to scientific language, and 3) from scientific to everyday language. The history of science is full of examples of each of these cases, much has been written on them, and it can no longer come as a surprise that metaphors reflect the links between scientific, social, and political discourses, and therefore corroborate that science is very much a social activity rather than anything else.

The issue of metaphors in science arises partly because science attempts, among other things, to control and standardize the use of language, and thus to construct an artificial language removed from that of everyday use. At least that is normally the case when the objective is to facilitate communication free from ambiguities and subjective interpretations. One could speak of a second order language contrasted to a first order language, and wherever concepts are used that originate in a different context (either first order to second order or one discipline to another) this seems metaphorical or analogical. This also pertains to the problem of taking a concept from one discipline to another, usually as the transfer of models.

Debates over the permissability of the use of metaphors in science are futile, since the flow of concepts from everyday language to scientific language, or generally between different contexts is inevitable. The problem is primarily which functions and dysfunctions certain metaphors have in a concrete case. Thus, there is, indeed, no reason to be "afraid of metaphors."[1]

Among the functions of metaphors one may mention that they illustrate, exemplify, focus attention, and by way of cross-fertilization lead to new hypotheses and insights. Among the dysfunctions are the danger of reification, the implicit and unnoticed redefinition of the subject matter of the research field in which the metaphor is used by way of a transfer of meaning. Clearly, the use of metaphors is neither generally positive nor generally negative. As Bono correctly points out, the use of the computer metaphor in brain science is productive (at least to a point) while in the case of race and gender analogies the suggested links are "misleading mistaken, embarassing, or worse."[2] It depends on the specific context and on the preferences of the observer. However, I am not concerned with that part of the much discussed metaphor issue.

One of the potential dysfunctions of metaphors occurs with the transfer back from science into nonscientific contexts, or in Engels' words, from "natural history" into the "history of society." This transfer is problematic if and insofar as the authority of science is carried by or attributed to the metaphor. It is taken as representing a proven theory, an "eternal natural law," and not just as an illustrative analogy that could easily by replaced by another one. This recourse to the authority of science, be it implicit or explicit, by no means requires a precise or "justified" application of the metaphor. In fact, the metaphor may be misused and/or abused, i.e., when a different meaning in the new context or a normative meaning is given to it. In these cases ideologies, entire worldviews, *Weltanschazuungen*, are elevated to the level of scientific truths. This is obviously an aspect of the general process of "scientification," but while the large majority of transfers involves single concepts with very limited connotations, some cases stand out where a few concepts represent a self-contained theory and develop into a coherent worldview.[3]

The focus here is on one of the most famous cases: the diffusion of a metaphor from science, i.e., Darwin's theory of evolution, into other areas of social life as well as its repercussions on the reception of Darwinism in the historical and social sciences. The metaphor in question is Darwin's use of the metaphor "struggle for existence," and I will limit myself almost exclusively to its German translation as *Kampf ums Dasein* and its fate in the German context.

Why choose an atypical or at least extreme example? Because the *Kampf ums Dasein* metaphor has been identified as the core of Social Darwinism, which, in turn, is held responsible for the emergence of Nazism.[4] The latter association has probably had profound repercussions on the reception of biological thinking in post-World War II, on the relation between the biological and social sciences as well as on historical accounts of the Nazis' rise to power. The *Kampf* metaphor thus encompasses all of the possible types of impact mentioned before and illustrates the process by which metaphors become like "viruses of the mind," multiply all of a sudden, diffuse into many different contexts of meaning, and eventually, after having caused major changes of perception and interpretation fade away.[5]

The following study proceeds from a number of assumptions. Scientific metaphors, like *Kampf ums Dasein*, although phrased in everyday language, are highly condensed representations of complex theories or parts of theories. As will be shown, when such metaphors are absorbed into other "foreign" contexts of meaning, e.g., public debates on the relation between nations before the background of increasing nationalism, complex processes are set in motion. If these contexts prove "favorable," there is mutual reinforcement through particular meaning relations. Say, the metaphor seems to strike the cord of the current "Zeitgeist" and is used to describe political phenomena. But in the process and by way of repeated use meaning relations may change in time, new concepts and theories suggest new interpretations and may trigger the effect of dormant metaphors or lead to their loss of effects. Ultimately, the original meaning of the metaphor may be entirely lost or reversed.

The account to follow will show (thereby giving empirical support to some studies) that Social Darwinism was less prevalent in those contexts where it was claimed by historians to be strongest. More importantly it will also show that the transfer of the metaphor back from science into public and political discourse entails a loss of its original meaning, or more precise yet, the loss of any meaning relation. It will show further, that the meaning of the metaphor changes as its conceptual environment changes.

Kampf um's Dasein – Popularization and Contexts of Usage

Darwin's Usage and German Translation

In one form the metaphor already appears in the subtitle of the *Origin of Species* as "the preservation of favored races in the struggle for life." The third chapter of the book is entitled "Struggle for Existence," and in it Darwin spells out the bearing of the struggle for life on the process of natural selection and species differentiation. Here he also remarks on the status of the term "Struggle for Existence": namely, that he uses it "in a large and metaphorical sense, including dependence of one being on another, and including (which is more important) not only the life of the individual, but success in leaving progeny."[6] It should suffice here to refer to some of those studies that have looked very carefully at the origins of the metaphor. Young traces it primarily to Malthus' *Essay on the Principle of Population* and to Charles Lyell's *Principles of Geology* stressing the "common context of biological and social theory." Bowler has shown the discontinuities in detail in the use of the concept of struggle between Darwin and Malthus which centers around the difference between intraspecies competition and interspecies struggle.[7] Leaving the interpretation of the finer grain to the historians of Darwin one obvious conclusion is that a common term like "struggle" was used widely by scientists, was taken from everyday language, and employed consciously as a metaphor. An important aspect is pointed out by Bowler: regardless of the differences between them, Malthus and Darwin both took part in a period of intellectual change during which the view of nature and society as a harmonious system gave way to one characterized by the "law of struggle."[8]

Since the concern here is with the German career of Darwin's metaphor, the first issue is its translation. Albeit a story all by itself, a few remarks are in place. In the first translation of the *Origins* by H. G. Bronn, "favored races" was translated to "vervollkommnete Rassen" which actually means "perfected races." That did part of the damage (notwithstanding the correction in the second translation by Victor Carus in 1867). Darwin's interchangeable use of "struggle for life" and "struggle for existence" was translated to *Kampf ums Dasein*.[9] That implied more damage since "struggle for life" could be translated more adequately into *Kampf ums Leben* which would easily encompass both meanings Darwin had in mind, namely the struggle for survival of species in a certain environment of other species under particular ecological conditions as well as the individualistic struggle between members of the same species. *Kampf ums Leben* or perhaps even more adequately *Kampf ums Überleben* would suggest the unconscious, general struggle for survival in the natural environment while *Kampf ums Dasein* assumed the connotation of an individual, conscious, and ultimately lethal conflict. In this connotation the metaphor is suggestive of the Hobbesian vision of the *bellum omnium contra omnes.*

The damage was complete with the combination of "perfection" and the struggle metaphor to *Vervollkommnung durch den Kampf ums Dasein* (i.e., "perfection through the struggle for existence"), because this elevated the metaphor to the level of the normative and instrumental as was later exemplified by the eugenicists' strategy to "perfect the human race by eliminating the unfit."[10]

When speaking about "damage," a dividing line between neutral science and value laden political use of the Darwinian metaphor is implied. The very fact that the seemingly simple translation from one language into another could produce the change in meaning of which it is also known to have occurred in the transfer from one context (scientific) to another (political) within the same language demonstrates the elusive nature of metaphor: floating freely between contexts of use.

The Kampf-Metaphor and Reception of Darwinism in the Popular Press

The popularization of the metaphor is illustrated by its career in Germany's foremost encyclopedia. By 1872 it was already listed as a commonly used term in connection with Darwinism, but by 1898 it had achieved the elevated status as an independent entry in the famous "Brockhaus Lexikon." Beyond this the picture becomes more complicated. Studies of the public impact of scientific theories suffer from the well-known difficulty that usually little is known apart from the circulation figures of books and journals and more or less well informed guesses about their readers. From there it is still a long way to empirical evidence of their impact, i.e., by shaping opinions and attitudes.

With respect to Darwinism, numerous studies with diverging results have been undertaken.[11] It cannot be claimed here that the methodological difficulties have been overcome, but only that the evidence has been substantiated. We will first look at a selection of popular science journals, the assumption being that they had their readership primarily among the "Bidungsbürgertum" and, thus, among the opinion-leading elite of the German empire during the last quarter of the nineteenth century. Then, that analysis will be supplemented by a look at a number of political arenas with which Darwinism in its analogical use has most often been associated.

One of the first popular science journals to take up Darwin's ideas in Germany was the weekly *Ausland* (Abroad). Although sceptical at first, by 1865 it had become a fervent supporter of the "new biological theory." Its self-posed question – what makes Darwin so popular – the journal answered by stating that although the theory was not proved, and with respect to its claims about the past probably never could be proved, it represented a "unified principle of *Weltanschauung*." Because of its analogical character for many neighboring fields, it had brought about a new age. At the beginning of the 1870s, the journal's editor Friedrich von Hellwald, like many other authors at the time of the Franco-Prussian War, resorted to Darwinism as proof that war was a natural law. The theory of the "struggle for existence" that applied to societies must also be applicable to the relationship between peoples. The war between the civilized people (*Kulturvölker*) of Europe against the "primitive" people (*Naturvölker*) had to be understood as *Kampf ums Dasein*.[12]

During the decade of 1870–80, *Ausland* became the chief popularizing journal for Darwin's theory but, as Zmarzlik points out, its glorification of the *Kampf ums Dasein* ran counter to the ethical values held by the bourgeois public. Therefore, in the 1880s, it radically changed its editorial policy.[13] Von Hellwald, who had been the chief propagator of Darwinism in the journal, gave up his editorship in 1884, complaining to his friend Ernst Haeckel that his efforts had not been very successful now that the times had turned away from the theory of evolution. Seven years prior to that he had welcomed the appearance of another new journal that had taken over the lead in representing the Monist *Weltanschauung: Kosmos*.

Kosmos was a Monist journal "based on the theory of evolution in connection with Charles Darwin and Ernst Haeckel" which first appeared in 1877. As an explicitly Darwinian journal, it self-confidently anticipated resistance from the public and, thus, a *Kampf ums Dasein*.[14] Although at first embracing the humanities as well as the sciences, the journal shifted its focus to evolution after five years. Most prevalent were discussions on theories of heredity (notably Weismann's) and selection.

In the latter context a wide range of applications of the *Kampf* metaphor can be found. From the interpretation of human history to the theory of all living organisms, from the history of science and competition between ideas to the struggle between cells the metaphor was used as a descriptive device.

One theme which attracted special attention in the journal was also the prominent example of the Social Darwinist debate. This concerned the question whether Darwin's theory was socialist or aristocratic, an issue that arose out of the famous dispute between Virchow and Haeckel in which Virchow had attacked Darwinism as a theory that could be adopted by Social Democrats and was, thus, dangerous. Haeckel had responded by pointing out that Darwinism was "aristocratic" rather than democratic and even less socialist because of its principle of the "survival of the fittest" which had been translated into "victory of the best." During the first five years of its existence, *Kosmos* was a Social Darwinist journal propagating the aristocratic orientation of Darwin's theory of selection. But in transporting political messages with Darwin's theory, it was an exception at the time.

In the early 1880s the increasing number of critical voices indicated that the enthusiasm for the Darwinist fad had subsided. The *Kampf* metaphor came into disrepute as a "vague and ambiguous term with which so many Darwinists in lack of a definite category were befogging themselves."[15] After ten years of publication, the editor of *Kosmos* announced its demise, claiming that the theory of evolution had not only gained unquestioned authority in biology but had also become a self-evident and irrevocable prerequisite in other disciplines.[16] Contrary to this evaluation, the journal had obviously lost the *Kampf ums Dasein* which it had anticipated at its birth.

In other periodicals Darwinism played a much smaller role, but its diffusion and popularization followed very similar patterns. One of the most widely read journals at the time, the *Deutsche Rundschau*, until the turn of the century carried relatively few articles on Darwin's theory and almost none with a Social Darwinist flavor, with the notable exception of Oscar Schmidt's *Darwinismus und Socialdemokratie*.[17] More prevalent were philosophical essays on the implications and limitations of the theory of evolution for ethics, the history of ideas, etc.

Another widely circulated review for politics, literature, and art, *Die Grenzboten*, which until the formation of the empire had been the most influential organ of the liberal bourgeoisie, had by 1871 allied itself with the National Liberal and anti-Catholic camp of Bismarck allies. At the same time it was decidedly anti-Darwinian, and by 1900 the journal declared that the twentieth century did not want to have anything to do with Darwinism. The axiom of the *Kampf ums Dasein* was attacked as the "same sophistry which had led to imperialism and anarchy."[18]

By the early 1890s, authors even commented on the loss of attention to Darwinism. It had become common knowledge rather than being limited to science, and in this process it had lost its attraction. Even the theologians had recovered from their shock. At the same time science, i.e., biology, had calmed down, the revolution had gone into its second generation, specialization set in and for the layman it became ever more boring.[19] Throughout this time the debate over the aristocratic vs. the democratic or socialist character of Darwin's theory continued, demonstrating not only the suitability of the theory as a *Weltanschauung*, but also its ambiguity with respect to conflicting political positions. The ultimate issue was whether society was based on a selective struggle for existence with the survival of the "best" or whether that struggle led to their degeneration and eventual demise.

The relationship of Marxist and Social Democratic authors to Darwinism must be seen in the context of social, political, and economic events occurring in the last quarter of the century. In 1873 the "great depression" set in and initiated a change in the political climate, notably a discrediting of political liberalism and enthusiastic Manchester capitalism. The dramatic structural changes, industrialization, and urbanization, brought the working class into politics and created widespread fear of the socialist movement among the middle class and aristocracy. While the defense and legitimation of power interests on the part of the ruling classes is usually taken as the basis of Social Darwinism, the reception of Darwinism by the political Left is overlooked.

During the election campaign for the *Reichstag* in 1877, August Bebel attacked Prussian militarism in a brochure, pointing out that war and the military system had degenerating effects on the population. In support of his thesis, he cited a passage from Haeckel's *Natürliche Schöpfungsgeschichte* which Haeckel himself omitted from later editions of the book. This episode is typical for the ambiguity of the *Kampf* metaphor with respect to its evolutionary or degenerating effects, which recurs in the shift of arguments from eugenicists before and after World War I. Typically, Marxists and Socialists denied the interpretation – referring to Darwin – that the "struggle for existence" had to be an inevitable conflict between individuals; they focused on the negative results of individual competition and pointed to the superiority of mutual help. Since the difference between the individualistic stance of Darwinism and the collectivistic philosophy of socialism seemed to represent a contradiction between them, a flood of publications was devoted to the defense of the commensurability of socialism and Darwinism.[20] Socialist authors used Darwin's theory as proof for the materialist alternative to religious explanations of man's creation and the role of such explanations in legitimating the existing social order, notably the monarchy. The scientific authority of evolutionary theory was seen to coincide with the claims to the scientific nature of Marx's theory of social evolution. Evolution in nature counted as scientific proof that social progress had to take place with the same law-like necessity. This association of socialism with Darwin's scientific theory was carried on through the 1890s and explains the impact on and fascination of Socialist authors like Kautsky and Bebel with eugenic schemes, in spite of the fact that they were much more explicitly selectionist and Social Darwinist than the various applications of the *Kampf* metaphor.

The same association had motivated Virchow already in 1877 to criticize Haeckel and the attempts of his followers to include Darwin's theory in school curricula. In his famous talk before the fiftieth convention of the *Naturforscher und Ärzte* he insinuated in calculated vague terms a relation between Darwinism, socialism and the Paris Commune, most likely because he feared that if the same were done by reactionaries the relative freedom gained by science after the unification of the empire could be jeopardized.[21] It did not help Haeckel to insist that if any political tendency were to be attributed to Darwin's theory it could only be an "aristocratic" and hardly a socialist one.

In summary, scanning the range of major periodicals attached to various political positions, two points stand out: the reception of Darwinism follows the pattern of fashion, i.e., it declines as it becomes more widely diffused; and its use as an ideology is not limited to one group whose interests it matches, but rather is utilized by many groups who interpret it as they choose. This ubiquitous usage, if any at all, renders it problematic to assume a specific impact of the theory, be it instrumental or legitimating. The common denominator seems to be another aspect: taking recourse to the authority of science and natural laws in order to give credibility to other ideas, opinions, and arguments.

Darwinism in Political Arenas

War and the Military: It comes as no surprise that, given its origin in common language, the *Kampf* metaphor assumed different meanings in different contexts and at different times. The most conspicuous example for the ambiguous use of the metaphor came to dominate public and scientific discussions about the impact of Darwinism: the association of the metaphor with war between peoples and nations. The connection may have contributed to the idea that war means progress although even that idea was not new. A look at the popular scientific literature in which the metaphor appears shows that from the late 1860s onward, the metaphor is used extensively and with increasing radicalism to justify war as a natural principle.[22] After the turn of the century a radical author like Klaus Wagner writes in Darwinian terms of the "selective struggle for new space" (*Auslesekampf um Neuland*), the "great cultural significance of the selective struggle for existence" (*hohe Kulturbedeutung des auslesenden Daseinskampfes*), the "selection of peoples as natural selection" (*Völkerauslese als natürliche Auslese*), and propagates explicitly the enslavement of the "lower peoples" in what was or was to become the colonies.[23]

The identification of the *Kampf* metaphor with the German war ideology became a major topic of English and French popular writings during World War I and served to associate Social Darwinism and German militarism. This helped to put the blame for both on the specific German use of the term and to overlook two important aspects: "war Darwinism," as La Vergata calls it, had adherents elsewhere, and it added little or nothing to the militarist defense of war other than a new metaphorical repertoire and its scientific prestige as well as giving evidence of the almost unlimited variety of interpretations of Darwin's theory.[24]

The usurpation of the *Kampf* metaphor by saber-rattling militarists makes it seem as if there were no other voices. However, the Social Democrats gave a different reading of Darwin. For Alfred Dodel-Port, Darwin had not taught that the evolution of the human race was to be sought in bloody fight between peoples but in the care and strengthening of intellectual powers and social instincts. And Karl Kautsky attacked "vulgar jurists, economists and historians who did not have any clue about the sciences as well as scientists and medical doctors who did not know anything about the social conditions and developments for giving any struggle in history the same name, no matter how it originated: to them they were all *Kämpfe um's Dasein*."[25]

Analysts within the peace movement deplored the popularity that the metaphor and the theory about the struggle for existence had assumed because of the Franco-German war and attacked the mindless analogy between nature and human society. The same critical stance can be identified after the First World War when Müller-Lyer branded Social Darwinism as a "cultural zoology" that had become an "intellectual plague" after the war of 1870–71. The meaning of "struggle" had become so broad that it no longer differed from other concepts such as "competition." Thus, one reads about peaceful Eskimos fighting the struggle for existence in the Arctic, and the millions of Chinese doing the same while living from their gardening, "even though it could only be a struggle with their vegetables."[26]

Since the association of war with the *Kampf* metaphor has been dealt with extensively elsewhere, it need not be pursued here further. Instead, a related arena may be examined in which the language of "struggle" and Social Darwinism may be expected to have played a role: the build-up of the German Navy.

The So-called *Flottenfrage*: The build-up of the German Navy became a prominent political issue in the last decade and a half before the turn of the century. At the time of rising nationalism and intensifying competition with leading imperialist power Great Britain, support of the navy assumed an integrative function for the bourgeoisie, which was directed against the rising influence of Social Democracy. The Imperial Naval Office (*Reichsmarineamt [RMA]*) set up a special news bureau commissioned to mobilize mass support with popular and scientific propaganda. At the same time a number of organizations took part in these efforts, among them the All German Association (*Alldeutscher Verband*), the German Naval Association (*Deutscher Flottenverein*), and the Free Association for Naval Treaties (*Freie Vereinigung für Flottenverträge*). In all of these, academics, professors, teachers, artists, and writers played a leading role, and the obvious focus of the RMA's propaganda was the *Bildungsbürgertum*. The speeches, articles, and communications give a fairly accurate picture of the positions of German scholars on the topic, and if Social Darwinist thinking was present it might be expected to be seen there.

However, a review of the RMA's annual publication *Nauticus*, of the collection of speeches and opinion statements from German university teachers on the significance of the *Flottenfrage*, and of a survey conducted by the *Münchner Allgemeine Zeitung*, comes to a disappointing result. The *Kampf* metaphor had been generalized and segregated from its Darwinism context. The "competitive struggle" (*Konkurrenzkampf*) between classes, nations, peoples, shipping lines, for food and world markets, was the dominating terminology. An exception was Max Sering arguing explicitly against the theory of the *Kampf ums Dasein*, which he branded as nonscientific and only a pragmatic political argument used by the British Prime Minister. Repeatedly, and in spite of an abundance of *Kampf* and *Dasein*, in this discourse one can find explicit rejection of Darwin's theory as British and not in line with German interests. Among the fifty odd scholars answering to the survey of the *Münchner Allgemeine Zeitung* in 1897 were such renowned names as Karl Binding, Felix Dahn, Hans Delbrück, Max Weber, and the Darwinians Ernst Haeckel and August Weismann. Haeckel was the only one to speak of a "German *Kampf ums Dasein* next to other major powers in Europe."[27]

Thus, Steinberg may be right in a very general sense in saying: "The permeation of German academic thought by Social Darwinian conceptions led to widely-held views about the nature of politics as a struggle for survival. The state was a living organism engaged in a life and death battle."[28] But here, when analyzing the rhetoric of struggle, it is advisable to heed Kelly's advice to distinguish between "those who occasionally appropriated a Darwinian phrase or two and those who undertook a sustained and detailed application of Darwinism to human society. The first group – those vast ranks of saber-rattlers, socialist-baiters, and self-righteous rich who happened to live in a Darwin-conscious age – can be called Social Darwinian only in the loosest sense of the word."[29]

Colonialism, Commerce and Trade: Very much the same applies to related policy arenas: colonialism and commerce and trade where Social Darwinism is expected to be strongest. Especially in the area of trade, the almost complete absence of any allusions to Darwinism alongside the self-confident assertion of the right of the entrepreneur and the denial of equality to the workers is conspicuous. The only notable exception is Alexander Tille, who called himself in Haeckelian fashion a "social aristocrat" and sought to turn Darwinism into a social ethic along the lines of early eugenic arguments. In his treatise "Struggle for the Planet," he also followed the common racist topos that the "lower races" were less efficient and would thus be pushed off the face of the earth.[30] Tille, when asked by Haeckel to collaborate in the Krupp-Prize (see below) turned down the offer saying that to apply evolutionary theory to social theory the most elementary building blocks had not yet been assembled.[31] The discourse on colonialism was primarily concerned with the imagined dangers of interracial marriages, and well into the first decade of the new century, a biological concept of race in the strict sense can hardly be detected in colonialist writings. Purity of race was a distant derivative of the topos of the "struggle between nations" insofar as it was considered an important prerequisite for success in that struggle, but "race" was mostly used in a humanist connotation prevalent in anthropology at that time.

Friedrich Ratzel, the founder of Bio- or Anthropogeography, who was later claimed by the Nazis as one of their own, in his more than 30 monographs and some 1,240 articles was far more sophisticated on issues such as the origin of the Arian race and the demarcation between and the unity of races than was palatable for them. Thus, he had to be heavily censored in order to serve ideological purposes.[32] Ratzel started out as a Darwinian influenced by Haeckel, but by 1875 had turned against Darwin's selectionism, characterizing it as the "crude hypothesis of the survival of the fittest in the struggle for existence."[33] Ratzel is often identified as a Social Darwinist because of his coining of the term *Lebensraum* which became the catch metaphor of colonialist and later "Greater German" expansionism. Actually, he reinterpreted the Darwinian *Kampf* metaphor in Malthusian fashion: to a large extent Darwin's often misunderstood and abused "struggle for existence," in his opinion, had to be a struggle between living organisms for space.[34]

Thus, wherever one looks before 1990 – and the list could be easily extended[35] – various public discourses, which are usually associated with Social Darwinism, actually show very little, if any, impact of it. They do reveal a very loose usage of Darwinian thought and terminology, i.e., of Darwinian metaphors, the most widespread of which is the *Kampf* metaphor. Kelly's conclusion is corroborated: "Social Darwinism, in whatever form, never achieved a mass popularity." And the few popular writers who talked a lot about struggle and race had such tenuous and ambiguous relationships to Darwinism that "it would be absurd to call them Social Darwinists."[36] This is not to be confused with the fact that the *Kampf ums Dasein* metaphor had become a highly inflated currency, together with elements of evolutionism and selectionism, that was being traded in all kinds of contexts with absolutely no relation to Darwin's theory. Popular Darwinism, in contrast to Social Darwinism, had reached millions of readers by World War I not least through the writings of Wilhelm Bölsche, whose reading of Darwin was that nature not only showed brutal struggles for survival but also cooperation and love, especially on the higher levels of evolution.[37]

A Change of Context: Weismann and the Krupp-Competition

Weismann's Significance and Impact

During the last decade or so of the nineteenth century, the discussion on race, especially the future of the Arian race and degeneration, began to emerge. In an increasingly nationalistic mood a number of political associations devoted to the promotion of nationalistic, Germanic, anti-Semitic or racial thinking were founded. One of them, the *Gobineau-Vereinigung* aimed to "solidify the idea of race against all doubts and assaults." The association, which was founded in 1894 and had a maximum of 360 members just before the outbreak of the war, managed the translation of Gobineau's *Essai sur l'inegalite des races humaines* into German. Only 1,000 copies of the four thick volumes, which had originally appeared in 1853–55, reached the German market between 1897 and 1901, and it is safe to assume that even these were not widely read. However, Gobineau's ideas on race received considerable attention and enthusiasm and to a great extent shaped the discourse on race among the public with *Völkisch* leanings. But, to be sure, that discourse had nothing in common with Darwinism nor with Social Darwinism, even though by virtue of the common fear of degeneration and defeat in the struggle for existence, cross-fertilization between them began to take place in the 1890s.

It is this context in which Weismann published in 1885 and in 1892 his speculative theory about the continuity of the germ plasm.[38] The theory had a complex history in Weismann's own work, which is not at issue here but has been well documented both at the time and later by historians of science.[39] Weismann's theory postulated that Lamarck and with him Darwin were wrong to assume that acquired traits could be inherited and that, instead, the hereditary material, i.e., the germ plasm, remained unaffected by the environment. Thus, Weismann had given the first radical formulation of a theory of heredity which diverged from the many explanations that had been current since Spencer's theoretical speculations about "physiological units" and Darwin's pangenesis hypothesis in one crucial aspect: the assumption that the substance of heredity was contained in the cell core and as such was unalterable.

Weismann's work, which had taken him from being an adherent of Darwin's view of personal selection and Lamarckism to being a strict selectionist, moved gradually to locating the hereditary material on the cellular and subcellular level. Essentially, Weismann's theory had extended Darwin's selection principle to the sphere within the germ plasm and called it "germinal selection." The selection processes taking place on this level of the smallest (hypothetical) units of life were given the greatest consequences since they were the basis of all variation and, thus, evolution. Weismann had taken the principle of the *Kampf ums Dasein* from the level of individual selection for which it was originally formulated to the level of the smallest units.[40]

When Weismann's *Keimplasma* appeared, attention to it was necessarily limited to the already highly specialized community of biologists working on heredity. There, Weismann's ideas were still highly controversial since the empirical evidence was contested. But their significance for the debate over the different interpretations of Darwin's evolutionism and also the *Kampf* metaphor must have been readily apparent to all those engaged in it. If Weismann was right, every individual was predetermined by his or her hereditary material and human evolution bore no hope of moral and cultural progress by way of changes in the social environment.

Although Weismann's work did not achieve any popularity outside the academic world and could not win over "environmentalist" popularizers of Darwinism, it gave support to a radicalization of Social Darwinism. Within that frame of reference all doubts about possible effects of the environment on heredity could now be cast aside, the primary objective had to be the preservation of good hereditary stock. Thus, the fears of degeneration, whether couched in terms of race or hereditary health, were intensified. Although race anthropologists or "anthroposociologists," as Schallmayer called them, and eugenicists stood in opposition to one another with respect to the biological foundations that could be claimed for race theory, they nevertheless pulled at the same end of the rope when it came to propagating those fears and drawing conclusions for policy making.[41]

References to Weismann's theory of heredity grew steadily in eugenic writings throughout the 1890s and after the turn of the century. The authority of science was called upon once again to give new lifeblood to the extension of Darwinism to society. Although the issues of heredity were far from being settled in the scientific context, social interpretations ran ahead of them.

The Krupp-Prize-Competition

This was accentuated most clearly by a prize competition issued in 1900 under the cumbersome title, "What can we learn from the principles of the theory of descent with respect to the internal development and legislation of states?"

The anonymous donor of the prize of 30,000 marks, later increased in view of the more than 60 essays received, was revealed to be the industrialist Alfred Krupp. Haeckel counts as the instigator. In retrospect the competition crystallized interests in and efforts to apply theories of heredity and evolution to social and political problems. The founding of biologistic and eugenic journals as well as the beginning of the eugenic movement in Germany can be traced back to Krupp's and Haeckel's initiative, even if it cannot be claimed to have been their prerequisite. The competition was to serve the "progress of science in the interest of the fatherland." Essays were to be accessible to laymen, and heredity, adaptation, and tradition, as well as political tendencies, had to be taken into account. Heredity was to be understood as inborn intellectual capacities and character traits which could be changed in the long term through selection or through the transfer of acquired traits. Adaptation was to be understood as adaptation of man to the changing economic, scientific, cultural, political, and legal conditions. While allegiance to either the Weismannian theory of heredity or to Lamarckism was not a criterion of exclusion, essays that questioned the importance of heredity or selection altogether could not be considered for the prize.

The question of the competition would seem to have invited Social Darwinist essays from hard-core proponents. The ten prize winning treatises, as well as some others which were published subsequently in a series entitled "Nature and State," not only reveal a remarkable range of different ideas attributable to the different disciplinary backgrounds of the authors, but they also show some striking similarities. Almost all of the authors accepted Weismann's theory and thus put the respective emphasis on the principle of selection as a force shaping society. This implied that most of them believed the *Kampf* metaphor to be applicable to modern societies, be it as the mechanism that retained the valuable and destroyed the inferior, be it as the necessary prerequisite for progress. The *Kampf* metaphor was not so much given new meaning but rather more focus: it was turned from a descriptive metaphor to a normative one as was exemplified most clearly in Schallmayer's prize-winning eugenic essay. The metaphor itself had almost disappeared but the whole book was dedicated to the question of how the state could control the selective process in such a way as to ensure that the hereditary quality of the population would be safeguarded from degeneration. His concern was that culture had rendered natural selection ineffective, that the negative effects of this on the germ plasm outweighed the positive ones and that this development posed a threat to hereditary quality and had to be reversed.[42]

Albert Hesse, another prizewinner, explicitly rejected the applicability of the *Kampf* metaphor as a natural law to society. Law and morality had always set limits on an outright Darwinian struggle of all against all. Thus, the principle of natural selection could only in rare cases be the basis of political measures. At best a selection of marital partners according to criteria of physical and intellectual heath could assure good offspring. Utopian plans of human breeding could then be discarded.[43]

The majority of entries, thus, leaned toward a very differentiated and careful adaptation of Darwin's theory to society. Some advocated versions of state socialism, mechanisms to protect the weak, and institutions of social welfare. Schallmayer's eugenics was explicitly antiracist and, compared to authors like Title and Woltmann, was moderately Social Darwinian. Kelly is correct in his judgment that most of the authors, despite their biologistic thinking remained "committed to humanitarian values. This is an important point to keep in mind when drawing parallels between Social Darwinism and Nazism."[44]

The Normative Turn of the Kampf Metaphor

True as that may be when reading the essays of the Krupp competition, Kelly underestimates the fact that, after 1900, Social Darwinism appeared in a different disguise: the relatively vague *Kampf* metaphor had been translated into a fairly precise normative scheme of selectionist demographic and eugenic policy. To Alfred Ploetz, eugenics, or "racehygiene" as he called it, was essentially a program to determine an optional organization of the internal *Kampf ums Dasein*, i.e., the struggle for reproduction and living conditions between individuals of one race, and of the "external" struggle, i.e., between races.[45] In the racehygienic discourse, coupled with Weismannism and shortly after the turn of the century with a crude Mendelism, the notion of the "fit" and the "unfit" in the *Kampf ums Dasein* was translated by 1913 into a normative economic calculation of the "costs of the inferior to the state" and by 1920 into the supposedly humane suggestion to "destroy" life unworth living.[46]

In the context of Nazi-ideology "inferior" assumed two meanings, both of which were apparent in the eugenic/racehygienic movement from its outset: to the anthropologically oriented wing of the movement (then to be called "racebiology"), "inferior" pertained to the non-Arian races including Jews; to the medically and social hygiene oriented wing, its meaning combined supposedly hereditary diseases like alcoholism and forms of behavior judged to be "antisocial."[47]

Although neither Social Darwinism in the narrow sense of the word as it was represented by the authors of the Krupp competition nor popular Darwinism like Haeckel's or Bölsche's were officially accepted by Nazi propaganda, this does not warrant the conclusion that Darwin's ideas had lost their effect or were ostracized. The National Socialist *Weltanschauung* was shaped from a garbled mixture of biological holism equating notions of organism, race, and *Volk*, of evolutionism and hereditary theory. This mixture was declared to be the truly "German Biology" and was peddled by the biological profession as the core of public education.[48] At the same time the Nazis themselves made repeatedly clear that National Socialism was a political and not a scientific movement, and that it drew a sharp distinction between what science had determined as "reality" and the various research areas and theories of individual scientists. Thus, neither Lamarck, Darwin, or Haeckel, regardless of their importance for the progress of science, nor any of their followers or opponents, could be equated with the movement.[49]

The leadership strategy of calling on the authority of science without becoming involved in ongoing debates and academic quarrels, and thereby retaining the charismatic authority of the leader, had consistently been employed by Hitler himself. In his notorious book *Mein Kampf*, a whole chapter was devoted to *Volk* and *Rasse*. Hitler never cited any of the sources he used, but it is obvious that he must have read much of the literature on racehygiene and race biology at the time (i.e., 1924), as Fritz Lenz later proudly claimed. The chapter opens with an account of Darwin's principle of the struggle of existence and selection, without a mention of Darwin, without even using the *Kampf* metaphor in full, and analogizing the struggle between species and the struggle between races. *Kampf*, he concluded, is always a means to promote the health and vigor of the species and thus the cause for its advancement. What follows is the murky melange of racehygiene and race biology in which the state is to take an active role in the selection of the "fit" and the preservation of racial purity.[50]

Conclusion: The Metadiscourse on the Metaphor

In the introduction the choice of the *Kampf* metaphor for this case study was justified with its central position in Social Darwinism and the connection it had to the explanation of National Socialism. That Social Darwinism was in reality much less prevalent in the public press and in pronouncements in various political arenas with which it had hitherto been associated runs counter to conventional wisdom in the historical and social sciences. In doing so, it supports Bannister's revision of Hofstadter's seminal work on Social Darwinism in the United States and Kelly's for Germany. Bannister's intention was to uncover the "myth of social Darwinism – the charge, usually unsubstantiated or quite out of proportion to the evidence, that Darwinism was widely and wantonly abused by forces of reaction."[51] Originally directed against that interpretation of the Anglo-Saxon variety of the myth, Kelly comes out with a similar result for Germany. Here, the "myth" of Social Darwinism as a precursor of National Socialism was spread by historians of science like G. Mann, and H. Conrad-Martius, by sociologists like G. Lukacs and

H. Plessner and by historians like Fischer, H. G. Zmarzlik, and H. U. Wehler. It would require a separate study to trace the important differences between them as D. C. Bellomy has done.[52]

It is obvious, though, that this revision is largely based on two things: a very loose definition of Social Darwinism on the part of those held responsible for creating the myth, and a much narrower definition of it on the part of their critics. The metadiscourse on Darwinism thus initiated replicates something that was found to be true for the use of Darwinism all along: "Everyone took the liberty to use Darwin as he pleased . . ."[53] Is it implausible to assume that scientific disciplines develop their own interests with respect to the perception of their subject matter and their own history? Bannister at one time came close to alleging that Social Darwinism "had been made up, initially by partisans, later by historians."[54] Bellomy sees an interest on the part of some historians by broadening the influence of Social Darwinism "to spread the guilt for imperialism around and establish a broader responsibility for Hitler and Auschwitz."[55] It is probably no accident that historians of science come out in defense of Darwin and try to clear him from the stigma of being implicated with conservative political thought or much worse with Nazism, wholly aside from the plausibility of their case. However, this leaves the question of the function of scientific theories or more narrowly of metaphors transferred from science into popular discourse entirely unanswered.

The strongest position on this issue is taken by a Marxist sociology of knowledge and by the social history which it inspires. It denies any independent role to ideas in shaping actions, and thus any meaningful function to the history of ideas in explaining actions. To this position, Darwin is the first Social Darwinist. But that theory remains too foggy for today's tastes; instead Social Darwinism "can best be understood out of a dialectical context of a specific social totality. . . ."[56] The crude assumption of "interests" emerging from the situation in the production process as the materialist basis of ideas and ideologies is no longer a sufficient base for explaining much of anything, and even Karl Mannheim was already ambivalent about applying the base/superstructure mechanism to scientific knowledge.

The puzzle of the transfer and diffusion of metaphors remains untouched by this scheme, but the evidence of it as presented here and in similar analyses clearly contradicts it. Rather than committing scientific metaphors to a secondary reflection of social processes, they, together with the theories from which they emerge, must be taken seriously in their own right.

I want to suggest that scientific theories are instrumental primarily in two ways: they directly inform specialized areas of social practice and they shape worldviews. Darwinism is an example of the latter type. In a more general sense than the directly effective kind, worldview-type theories affect notions of man's place in nature and society, the potentials and limitations of his activities. Old limitations are pushed aside and new ones are erected, orientations which were unquestioned before are replaced by others which are hypothetical and limited. Social relations and institutional arrangements can be profoundly affected by the emergence of such new theories.

Why a theory assumes the function of a worldview is sociologically unexplored. That it has achieved that status can be ascertained when its derivatives become effective on different levels of action. This was the case when and to the extent that the threat of degeneration, the need to care for the hereditary stock or the race became elements of the public belief system in the early decades of the century. Depending on its particular contents the theory will unfold its influence in different contexts of experiences and activity. It is unlikely, though, that in that coupling very rational calculations play a role. Even in the scientific context, the succession of theories and the reception of new knowledge does not follow very rational patterns. In everyday life discourses this can be expected even less.

The example of the reception of Darwin's theory demonstrates the extremely loose connection between the original theory and the meanings given to it in different contexts and at different times. The metaphors that are taken from the theory, such as the *Kampf ums Dasein*, even accentuate this. They are the small change of the currency. While the interpretation of the theory and the use of the metaphor are not narrowly determined by the theory, they are not completely independent from it either: at least the semantics remain a limiting parameter, and often the meaning relations, too. Therefore, it is more plausible to assume a model in which both social practice and scientific theories are on the same level, the latter being descriptions of the former. Metaphors mediate between them, crystallizing both. Applying this picture to the *Zeitgeist* of 1990s, one might say that the long and involved *Kampf ums Dasein* has ended in "Chaos."

Notes

1. See S. Maasen, in this volume, who presents the major positions and gives an innovative answer.
2. For a discussion of different position see J. J. Bono, "Science, Discourse, and Literature: The Role/Rule of Metaphor in Science," S. Peterfreund (ed.), *Literature and Science, Theory and Practice* (Boston: Northeastern University Press, 1990), pp. 59–89, 81. Bono refers to N. Stepan, "Race and Gender: The Role of Analogy in Science," *Isis* **77** (1986), 261–277.
3. On the "scientification" of everyday language see U. Pörksen, *Deutsche Naturwissenschaftssprachen* (Tübingen: Gunter Narr Verlag, 1986), p. 202.
4. Scholarship from the late 1970s has corrected this image and related convictions that are usually attributed to R. Hofstadter. *Social Darwinism in American Thought, 1860–1915* (Philadelphia: The University of Philadelphia Press, 1945); see R. C. Bannister, *Social Darwinism, Science and Myth in Anglo-American Social Thought* (Philadelphia: Temple University Press, 1979); and for Germany: A. Kelly, *The Descent of Darwin, The Popularization of Darwinism in Germany, 1860–1914* (Chapel Hill: The University of North Carolina Press, 1981).
5. I am only cautiously alluding to Hofstadter's use of a metaphor: D. R. Hofstadter, "Metamagical Themas, Virus-like sentences and self-replicating structures," *Scientific American* **248** (1983), 14–22. Also, Dawkins' use of his "memes" comes to mind but because of its problems will not be the favored metaphor. For a discussion of both see W. Durham, P. Weingart, *Culture as an Assembly of Units or a Seamless Whole* (unpubl. ms.).
6. C. Darwin, *The Origin of Species* (Harmondsworth: Penguin, 1859, 1968), p. 116.
7. R. M. Young, *Darwin's Metaphor, Nature's Place in Victorian Culture* (Cambridge: Cambridge University Press, 1985), p. 43 et passim. Also, R. M. Young, "Malthus and the Evolutionists: The Common Context of Biological and Social Theory," *Past and Present* **43** (1969), 109–145. P. J. Bowler, "Malthus, Darwin, and the Concept of Struggle," *Journal of the History of Ideas* **37** (1976), 631–650. For the somewhat surprising story of the related metaphor "survival of the fittest" see D. B. Paul, "The Selection of the "Survival of the Fittest," *Journal of the History of Biology* **21** (1988), 3, 411–424.
8. P. J. Bowler, *op. cit.*, 1976, p. 644.
9. On Darwin's difficulties with his own concept and the translation into German, see U. Pörksen, *op. cit.*, 1986, pp. 132–140.
10. See H. Schmidt, *Der Kampf ums Dasein* (Jena: Urania Verlagsgesellschaft, 1929).
11. A. Ellegard, Darwin and the General Reader, The Reception of Darwin's Theory of Evolution in the British Periodical Press, 1859–1872 (Göteborg: *Acta Universitatis Gothoburgensis*, Göteborg's Universitets Arsskrift, 1958), LXIV: D. Kohn (ed.), *The Darwinian Heritage* (Princeton: Princeton University Press, 1985), Part III. For Germany, see A. Kelly, *The Descent of Darwin* (Chapel Hill: The University of North Carolina Press, 1981).
12. "Was macht Darwin populär?" *Ausland* **44** (1871), 813–815; F. v. Hellwald, "Der Kampf ums Dasein im Menschen- und Völkerleben," *Ausland* **45** (1872), 103–106. As a specific comment on the Franco-Prussian War in terms of the "Kampf metaphor"

see g. Jäger, "Naturwissenschaftliche Betrachtungen über den Krieg," *Ausland* **43** (1870), 1161–1163.
13. H. G. Zmarzlik, "Österreichische Sozialdarwinisten. Ein Beitrag zur Brutalisierung des politischen Denkens im späten 19. Jahrhundert," *Der Donauraum* **19** (1974), 147–163.
14. "Prospekt," *Kosmos* **1** (1877), 1–3.
15. M. Wagner, "Darwinistische Streitfragen. Teil III.," *Kosmos* **14** (1884), 55–362, 356.
16. B. Vetter, "Abschiedsworte des Herausgebers," *Kosmos* **19** (1886), 82.
17. O. Schmidt, "Darwinismus und Socialdemokratie," *Deutsche Rundschau* **17** (1878), 278–292.
18. See "Ethik und Politik," *Die Grenzboten* **59** (1900), 249–257, 51.
19. See W. Bölsche, "Wankt unsere moderne naturwissenschaftliche Weltanschauung?" *Neue Rundschau* **3** (1892), 62–72.
20. The central periodical in this respect was *Die neue Zeit*, a monthly founded in 1883. See apart from O. Schmidt's article cited above, "Darwinismus contra Sozialismus" *Neue Zeit* **8** (1890), 326–333; K. Kautsky, "Darwinismus und Marxismus," *Neue Zeit* **13** (1895), 709–716; E. Aveling, "Charles Darwin und Karl Marx," *Neue Zeit* **15** (1897), 745–757; K. Kautsky, "Darwinismus und Sozialismus," *Der Sozialist* **34** (April 24, 1879).
21. See R. Virchow, "Die Freiheit der Wissenschaft im modernen Staatsleben" in K. Sudhoff (ed.), *Rudolf Virchow und die deutschen Naturforscherversammlungen*, (Leipzig: Akademische Verlagsanstalt, 1922).
22. Among the more moderate authors one finds M. Jähns, *Krieg und Frieden, Theorien und Praxis* (Berlin 1868): W. Preyer, *Der Kampf um das Dasein. Ein populärer Vortrag* (Bonn, 1869); Friedrich Hellwald, the editor of the journal *Ausland*, was a representative of the radical interpreters of the *Kampf*-metaphor, especially in connection with the war of 1870–71. F. v. Hellwald, "Zur Geschichte der germanischen Race, Teil I.," *Beilage zur Allgemeinen Zeitung* (October 15, 1870). Likewise, G. Jäger, "Naturwissenschaftliche Betrachtungen über den Krieg," *Ausland* **43** (1870), 1161–1163.
23. K. Wagner, *Krieg* (Jena, 1906).
24. A. La Vergata, *Biology and War*, 1870–1918.
25. A. Dodel-Port, "Charles Robert Darwin. Sein Leben, seine Werke, und sein Werke, und sein Erfolg," *Neue Zeit* **1** (1883), 105–119; K. Kautsky, "Kamerun," *Neue Zeit* **6** (1888), 13–27.
26. A. H. Fried, *Handbuch der Friedensbewegung, Teil I.: Grundlagen Inhalt und Ziele der Friedensbewegung* (Berlin/Leipzig, 1911), pp. 42–46; F. Müller-Lyer, *Der Sinn des Lebens und die Wissenschaft, Grundlinien einer Volksphilosophie* (München, 1919), p. 95.
27. See *Außerordentliche Beiträge zur Allgemeinen Zeitung*, Nr. 1/1-11–Nr. 23/3-6 (1898).
28. J. Steinberg, "The Kaiser's Navy and German Society," *Past and Present* **28** (1964), 102–110, esp., 108.
29. A. Kelly, *op. cit.*, 1981, p. 102.
30. A. Tille, "Der Kampf um den Erdball," *Nord und Süd* **80** (1897), 68–96.

31. See W. Schungel, *Alexander Tiller (1866–1912): Leben und Ideen eines Sozialdarwinisten* (Husum, 1980), p. 10.
32. See K. Haushofer (ed.), *Friedrich Ratzel, Erdenmacht und Völkerschicksal. Eine Auswahl aus seinen Werken* (Stuttgart, 1941).
33. J. Steinmetzler, "Die Anthropogeographie Friedrich Ratzels und ihre ideengeschichtlichen Wurzeln," *Bonner Geographische Abhandlungen* **19** (1956), 86–88.
34. F. Ratzel, *Der Lebensraum* (Tübingen, 1901), p. 51.
35. In the project on Social Darwinism from which this material is drawn, systematic scanning of the popular literature also covered the political associations like the *Alldeutscher Verband, Hammerbund, Deutschbund*, the Anti-Semitic and vegetarian movements, as well as the *Wandervögel*-movement.
36. A. Kelly, *op. cit.*, 1981, p. 109.
37. W. Bölsche, *Stirb und Werde. Naturwissenschaftliche Plaudereien* (Jena: Gustav Fischer, 1913).
38. A. Weismann, *Die Continuität des Keimplasmas* (Jena: Gustav Fischer, 1885); A. Weismann, *Das Keimplasma. Eine Theorie der Vererbung* (Jena: Gustav Fischer, 1892).
39. See F. Rohde, *Über den gegenwärtigen Stand der Frage nach der Entstehung und Vererbung individueller Eigenschaften und Krankheiten* (Jena: Gustav Fischer, 1895); E. Gaupp, *August Weismann: Sein Leben und sein Werk* (Jena: Gustav Fischer, 1917); F. B. Churchill, "August Weismann and A Break From Tradition," *Journal of the History of Biology* **1** (1968), 91–112.
40. E. Gaupp, *op. cit.*, 1917, pp. 249–253.
41. The most outspoken critic of Gobineau and anthropological race theories among the eugenicists was probably Schallmayer. See W. Schallmayer," Gobineaus Rassenwerk und die moderne Gobineauschule, "*Zeitschrift für Socialwissenschaft* N.F. **1** (1910), 553–572.
42. W. Schallmayer, *Vererbung und Auslese im Lebenslauf der Völker* (Jena: Gustav Fischer, 1903).
43. A. Hesse, *Natur und Gesellschaft. Eine kritische Untersuchung der Bedeutung der Deszendenztheorie für das soziale Leben* (Jena: Gustav Fischer, 1904), p. 172.
44. A. Kelly, *op. cit.*, 1981, p. 108.
45. A. Ploetz, *Die Tüchtigkeit unserer Rasse und der Schutz der Schwachen* (Berlin, 1895), p. 22.
46. See I. Kaup, "Was kosten die minderwertigen Elemente dem Staat und der Gesellschaft? *Archiv für Rassen- und Gesellschaftsbiologie* **10** (1913), 723–748; H. Binding and A. Hoche, *Die Freigabe der Vernichtung lebensunwerten Lebens* (Leipzig, 1920).
47. This duality found its way into Nazi-racehygiene legislation in the parallel notions of the protection of the "blood" (race) and heredity. See P. Weingart, et. al. *Rasse, blut und Gene. Geschichte der Eugenik und Rassenhygiene in Deutschland*. Frankfurt: Suhrkamp Verlag, 1988) p. 494.
48. See documentation from the journal *Der Biologe* in Ä. Bäumer (ed.), *NS-Biologie* (Stuttgart: Wissenschaftliche Verlagsgesellschaft, 1990); esp. H. J. Feuerborn, "Das Kernstück der deutschen Volksbildung: die Biologie, *Der Biologe* **4** (1935), 99–101.
49. See G. Hecht, "Biologie und Nationalsozialismus," cited in Ä. Bäumer, *op. cit.*, 1990,

p. 120. Hecht represented the position of the *Rassenpolitisches Amt* whose director Walter Groß held the same position. W. Groß, *Nationalsozialismus und Wissenschaft* (Berlin, 1937).
50. A. Hitler, *Mein Kampf* (München, 1933); Verlag Franz Eher Nachf., pp. 313, 475–480.
51. R. C. Bannister, *op. cit.*, 1979, p. 9.
52. D. C. Bellomy, "'Social Darwinism' Revisited," *Perspectives in American History*, New Series **1** (1984), 1–130.
53. A. Kelly, *op. cit.*, 1981, p. 8.
54. R. C. Bannister, *op. cit.*, 1979, p. 8.
55. D. C. Bellomy, *op. cit.*, 1984, p. 10.
56. H. U. Wehler, "Sozialdarwinismus im expandierenden Industriestaat" in I. Geiss and B. J. Wendt (eds.), *Deutschland in der Weltpolitik des 19. und 20. Jahrhunderts* (Düsseldorf, 1973), pp. 133–142, 134.

III: "Evolution" and "Organism"

THE IMPORTANCE OF THE CONCEPTS OF "ORGANISM" AND "EVOLUTION" IN EMILE DURKHEIM'S DIVISION OF SOCIAL LABOR AND THE INFLUENCE OF HERBERT SPENCER

PETER M. HEJL
University of Siegen

> How is it possible that the individual, in becoming more autonomous, depends more closely on society? How can it be at the same time more personal and more solidary?"[1]

According to Durkheim, the collective consciousness of a social group or society influences what its members take as evident, what they see as problematic, and what might constitute a solution in their eyes. This holds for science as well, both during its heuristic or (literally) creative phases as well as during those phases that Kuhn called "normal science."[2] The general goal of this contribution is to show that concepts like "organism" or "evolution" played and still play an important role in sociology. But the nature of this role does not seem to be a model that can be directly transferred from one domain to another to explain a class of preexisting phenomena. Even today, there are no generally agreed upon definitions of "social," "society," "individual," etc. At least for one root of modern sociology, the fuzzy concepts of "organism" and "evolution" allowed the pioneers to generate some order at the theoretical and hence empirical level. The metaphorical use of a concept or image depends, of course, on similarities that those who use the metaphor perceive between different phenomenological domains. Their perceptions are, in turn, shaped by historically, socially, and theoretically contingent experiences and contexts. This holds true both for early sociology and modern theorizing. Therefore, it seems adequate to understand the role of metaphors mostly as a constitutive one.

If we look at early sociology, this constitutive role of metaphors becomes even more salient. The metaphors "organism" and "evolution" played a central role in this respect. They helped the cognitive construction of the emerging social sciences by providing concepts sufficiently differentiated to permit a variety of interrelated questions and possible answers. At the same time, these concepts were flexible enough to allow for context-specific modifications. It is this interplay between metaphors and the actors-in-context who use them that leads to the meaning attributed *ex post* to a metaphor.

The constitutive role of the metaphors "organism" and "evolution" shall be shown through a close analysis of why and how they are important in Durkheim's sociology, and through a short, contrasting look at Spencer's position and, more specifically, Durkheim's reading of it.[3] I will argue that Durkheim's background (just as in Spencer's case) was of importance in his theorizing; show that Durkheim needed the theoretical input Spencer could offer (the organism as a model plus the theory of evolution) and that, therefore, Durkheim had to take the organismic model more seriously than Spencer in that he had to transform it.[4]

Durkheim's background allowed him to be acquainted with several cultures, and confronted him with a number of problems that were difficult to avoid. He was born in 1858 into a Jewish family which had provided rabbis to their community over eight generations. Originally, Durkheim was designated to the same office. By the age of thirteen he had learned Hebrew in a rabbinic school and was familiar with the Pentateuch and the Talmud. But then he refused to become a rabbi, and between the baccalaureate and his entrance into the Ecole Normale Supérieure, he became agnostic (Filloux, 1976: 259). In 1879, after three years of preparation in Paris, he passed the entry examination to the Ecole Normale. He left the Ecole Normale in 1882 with the *agrégation*.

It is important to briefly outline the social and political context of Durkheim's youth and studies. It is true that the political situation (Mayeur, 1973; 1984) became somewhat stabilized between the 1877 general election and the end of what became known as the "Dreyfus affair," but the fundamental social and political problems of these years were continuously present in the public debate throughout Durkheim's lifetime.[5]

Durkheim was born not only a Jew, but an Alsatian Jew. In this border region, where Jews were less well integrated than in the rest of France

(Marrus, 1972: 47f), a long tradition of anti-Semitism continued to exist.[6] Although Durkheim's hometown of Epinal remained French, Durkheim's family and Durkheim himself belonged culturally to those Alsatian Jews who opted for France, where Jewish emancipation had been most advanced since the French Revolution.[7] But this option for France, which meant that many Alsatian Jews chose to move into French territory, had important consequences. Through it, the Alsatian Jews joined those Jews within France and their position in sociopolitical life. Factually, they opted for assimilation into French culture and, if not directly for secularization, at least for a secular state (Marrus, 1972: 108). Most of the French Jews reacted overwhelmingly positively to the possibility of integration. Often they became fervent patriots – serving as officers or civil servants – and were, by the same token, ready to accept the surrounding culture.

If this characterizes the socioreligious subgroup to which Durkheim belonged, at least during his youth, the wider context is important as well. French political life had vacillated since the revolution between a democratic and a more autocratic, if not authoritarian, form of government (Nicolet, 1982). Even though France was still a country with a large agricultural sector in the second half of the last century, there was a growing bourgeoisie that clung to the liberal and democratic – not the proletarian – tradition of the French Revolution. Liberalism and parliamentarianism were, of course, conceptually linked to individualism. But there was another reason why the relation between the individual and society could attain such importance, which made it one of the main issues in Durkheim's first major work.

The defeat against Germany in 1871 brought a parliamentary government into power. It seemed not only more or less provisional, but was forced upon the country by the course the war took. Hence, there was a popular support for liberal democracy of unknown dimensions, many were still undecided as to what they wanted politically, and the political forces were still unorganized. In this situation, the old debates were revived between those who defended parliamentarianism and those who proclaimed the need for a unification of the country under a nonpartisan leadership. France's defeat was often understood in this debate as the defeat of "French individualism" by "German collectivism."[8] In order to prevent a similar debacle in the future, conservatives of all kinds[9] fought against the definite establishment of a parliamentary system. The pressure abated only in 1877, when general elections brought a strong republican majority to the parliament. This allowed the establishment of the republican regime between 1877 and 1881. The years after the election saw a bitter struggle over the separation of church and state. In order to give the republican system a firm and lasting foundation, many *Républicains* found it necessary to eliminate not only the still existing clerical control of the educational system, but also to laicize it and reform it completely. This reform was linked to another proposed explanation of France's defeat in the war against Germany, which was sometimes presented as a victory for the German teacher.[10] It was in the context of these projected educational reforms and hence of the teaching profession, that Durkheim, following his own proposition, was sent (during the academic year 1885–86) to Germany to become acquainted with the teaching of "moral sciences" in German universities.[11] In the same context, he was appointed to a professorship that was created especially for him in education and sociology in Bordeaux (Filloux, 1963: 69).

The fact that these debates had an impact on the young Durkheim can be linked to his time at the Ecole Normale Supérieure. The school is one of a number of elitist institutions where students are admitted exclusively upon the results of an entry examination. Situated above the level of universities, these institutions still today produce an elite group that governs the country by occupying the top positions in politics, the economy, the universities, and culture.[12] It was in the Cagne (specialised classes to prepare for the entry examination for the Ecole Normale Supérieure) and in the Ecole Normale that heated discussions on intellectual and political topics took place,[13] and it was there that Durkheim became friends with the later socialist leader Jean Jaurès.[14]

Given Durkheim's background, one may speculate about its influence on his fundamental, intellectual decisions taken after a period of hesitation. The main decision[15] was, of course, to contribute to the establishment of social science in France, and hence to become a sociologist (although he used this term only at a later stage). This goal perhaps seems too ambitious for a young man who had just finished his academic training. But one has to consider the situation of a highly centralized country like France since the beginning of French absolutism, and of an intellectual who belongs to the elite of his country through his educational background, or who, at least, has a greater chance to join it. Both factors together make this decision seem less pretentious.[16]

His second important decision concerned the relation between individuals and society. I will turn below to examine this relation and the role played by the metaphors "organism" and "evolution" in establishing it. But there are two significant features of Durkheimian sociology, present from the beginning and consistent in Durkheim's texts. One was the concept of holism.[17] Durkheim took the idea that "the whole is more than the sum of its parts" from Renouvier, then one of the most influential neo-Kantian philosophers (Digeon, 1959: 106ff) and Durkheim's teacher at the Ecole Normale. The second concept stems from another of his professors, Boutroux, and postulates that a true scientific discipline should have a subject matter of its own such that its explanations can not be reduced to that of another discipline.[18] Both ideas are, of course, complementary. They can be seen as Durkheim's conditions, which every candidate had to fulfill for a potential contribution to the social science in *statu nascendi*.

To sum up: one can reasonably assume that there was a combined effect of the experiences of his youth, his student years, and the political debates of these years, that led him to think of a theoretical concept that integrated rules or norms (both of human action and of their scientific explanation), social integration, individualism, holism, and disciplinary independence. I will now try to show that Durkheim needed what Spencer had to offer, but he had to transform it first.

What is the problem Durkheim attempts to solve? In 1885, when discussing the theoretical work of the influential jurist Ihering, Durkheim criticizes him for claiming that ancient Roman law was more respectful of personal independence than contemporary law. Durkheim says:

. . . this affirmation is . . . absolutely wrong. With progress going on, the human person distinguishes himself more and more from the physical and social environment . . . and becomes conscious of himself: the liberty he enjoys increases together with his social obligations. *There is a phenomenon difficult to understand and apparently contradictory, which, to our knowledge, has not yet been elucidated* (Durkheim, 1975c: 292, my emphasis).

Here in a nutshell is the problem Durkheim tries to solve. It is not until the *Division* that he is able to propose a solution. The answer to the question that arises from this observation (cf. the introductory citation) – and in fact the question itself – distinguish Durkheim not only from Spencer, but from other predecessors as well. With this question and the answer he proposes, Durkheim has no difficulties in drawing extensively on Spencer's work. At the same time, asking how personal freedom and social coherence can be achieved simultaneously is tantamount to rephrasing Durkheim's political problem of combining individualism with political unity.

The question as to why Durkheim draws so heavily[19] on Spencer when opting for societies as organism-like entities, can be answered by pointing at the broad use of the organismic analogy and by Spencer's already established reputation. With respect to the theory of evolution, Darwinism was, at Durkheim's time, and even during much of the twentieth century, rather unpopular in France.[20] Moreover, Spencer's account of the evolution of organisms and superorganisms presented Darwin's theory as a specific case and proved, not the least for its encompassing claims, inspiring to those who tried to understand society in a scientific way. But Spencer's appeal turns out to be limited,[21] Durkheim differs from Spencer with respect to important aspects of the mechanism that produces social evolution.[22]

That there should be important differences between Durkheim and Spencer becomes plausible if we look briefly at Spencer's social and intellectual setting. He came from a nonconformist background and was, as Burrow (1970: 1983ff) argues, both conscious and proud of it. The cultural tradition of Protestant dissent was linked to the philosophy of the Enlightenment and embedded in a long history of indivdual rights (Thompson, 1978: 28ff). Its social basis was the provincial educated middle class (Turner, 1992: 197). Also successful during the Industrial Revolution, this segment of the population grappled with an understanding of the economic and social process congruent to their own social and cultural situation between the working class, to whom they felt superior, and the established upper class, who had persecuted them for religious and political reasons. Some aspects of Spencer's general philosophy, as well as some of his key terms, suggest that his view of society and his individualism remained rooted in this context.[23]

Naturally, it is impossible to do justice to Spencer's pioneering work in a passage of an article about certain aspects of Durkheim's sociology. I must confine myself, therefore, to some remarks that allow us to situate where and why Durkheim agreed or disagreed with Spencer.

Spencer understood society analytically and as an organism.[24]

... regarding society as a thing, what kind of thing must we call it? It seems totally unlike every object with which our senses acquaint us. Any likeness it may possibly have to other objects, cannot be manifest to perception, but can be discerned only by reason ... Between a society and anything else, the only conceivable resemblance must be one due to *parallelism of principle in the arrangement of components.*

There are two great classes of aggregates with which the social aggregate may be compared ... Are the attributes of a society ... like those of a not-living body? Or are they ... like those of a living body? Or are they entirely unlike those both?

The first of these questions needs only to be asked to be answered in the negative. ... The second question, not to be thus promptly answered, is to be answered in the affirmative. ... (Spencer, 1893: 436, Spencer's emphasis).

Spencer saw societies, not individuals, as *analytical* entities. Although this difference is somewhat pushed aside owing to those long and almost monotonous passage where he develops the analogy between organisms and societies (for example in *The Principles of Sociology*), individuals are, according to Spencer, *natural* entities. Of course, there are passages where he also underlines that man is the product of society (Spencer, 1972: 56). However, his extreme liberalism and his position (with respect to the relation society-individual) show clearly that such a statement has to be placed within the functional role that society has to play in human evolution.

Despite his publications on biology, psychology, sociology, and on many political and economical issues, Spencer is a philosopher. His main interest is not so much scientific as it is political and moral. As La Vergata convincingly argues elsewhere in this volume, Spencer looks for general principles to guide action. Spencer seems to find these principles through the application of what he takes to be the general law of evolution to all domains in which he is interested. But these evolutionary principles are, at a general level, present in Spencer's work prior to his closer reading of biology in 1852–53. The concept of evolution is based upon his conviction that societies and humans undergo a change that can be labeled progress. Its goal is a state of harmony among all humans, and between humans and their physical environment: a state of perfect adaptation. This idea of progress and its goal can be understood as hardly secularized theological convictions. They even have a Calvinist flavor, if one considers Spencer's moral emphasis when he defends his liberalism, and when he discusses the suffering it might cause (Spencer, 1896: 232ff). This becomes quite evident in *Social Statics* where he argues that "all evil [sic] results from the non-adaptation of constitution to conditions." Then Spencer affirms as "true" that evil perpetually tends to disappear. In virtue of an essential principle of life, this non-adaptation of an organism to its conditions is ever being rectified; and modifications of one or both, continues until the adaptation is complete. . . . Man exhibits just the same adaptability (Spencer 1972: 8f).

Man's "goal" of adaptive change is the "social state":

It requires that each individual shall have such desires only, as may be fully satisfied without trenching upon the ability of other individuals to obtain like satisfaction (*op. cit.* 11).

The reason for the imperfect adaptation is that man "yet retains the characteristics that adapted him for an antecedent state . . . he is fitted for his original predatory life" (*op. cit.* 11, 21). This allows Spencer to define progress as "successive steps of the transition" to the social state. Progress is therefore brought about due to the universal law of adaptation that works via the diminishment of faculties not fully used, and to the complementary increase in faculties that are needed and used (*op. cit.* 12). Spencer concludes:

Progress, therefore, is not an accident, but a necessity, and ends quite dramatically: As surely as the tree becomes bulky when it stands alone, and slender if one of a group, [follows half a page of comparable examples] . . . so surely must the human faculties be moulded into complete fitness for the social state; so surely must the things we call evil and immorality disappear; so surely must man become perfect (*op. cit.* 13).

Spencer's "survival of the fittest" in this context is much more of a moral and social category rather than a biological concept (Bowler, this volume). The fittest is either the society or the individual most advanced on the road to progress. Spencer's views of society as an organism and of social evolution as progress towards the social state have to be placed within this concept of general evolution.

Given Spencer's moral and political convictions, the metaphor of the organism, as well as the biological concept of development/evolution, could not be as congruent with his "conclusions" as he intended. Critics during his lifetime already underlined the contradiction between society as an evolving organism and Spencer's extreme individualism (Rumney, 1934: 274ff, cf. below). In addition, von Baer's concept of evolution (understood by Spencer as "a change from an indefinite, incoherent homogeneity, to a definite, coherent heterogeneity") was primarily intended to describe the development of an organism from an embryonic to an adult state. For the analogical use of the biological concept of evolution, this required more than simply a direct parallelization. A further problem of consistency that arises from Spencer's theological background is that, despite his reductionist position,[25] he seemingly did not conceive of the telos of evolution by analogy to the concept of entropy.

A closer look at Spencer's view of societies and social evolution confirms what already appears at a more general level. Societies have their origin in voluntary associations of preexisting individuals who decide to cooperate in order to satisfy their interests.

The motive for acting together, originally the dominant one, may be defense against enemies, or it may be the easier obtainment of food . . . the units pass from the state of perfect independence to the state of mutual dependence and as . . . they do this they become united into a society rightly so called (Spencer, 1967: 63).

According to Spencer, societies, once established, exist as undifferentiated hordes. They are assumed to have lived in continuous war with other hordes, which caused them to unite and hence to form greater and more complex societies. To coordinate society's actions, war leaders emerged, whose influence was confirmed by their supposed power to control the supernatural and to ensure its help. As societal growth continued, the war leader became a permanent chief or king. His officers formed the nucleus of the developing state administration. With further growth, kingship became hereditary, and political assemblies and agencies underwent a process of institutionalization that established them on a permanent basis – the modern state appeared. From the beginning, and well into our times, the government has focused its attention on military activities, conquests, and territorial gains. At a more advanced stage, government activities shifted to the development of industry. In the modern state, Spencer thought the military state would progressively be replaced by the industrial state. In the latter, the government would limit its activities to protecting the citizens against assaults on their freedom by internal and external enemies. But this is not the end of social evolution.

The ultimate state to be hoped for is one in which the resources of a developed industrialism may be turned toward the perfection of human character in the higher and more truly socialized aspects of moral conduct, thus bringing into being the ethical state (Barnes, 1948: 123).

In Spencer's account of societal development, the concept of the organism and of evolution appear in two ways. Firstly, social evolution is but one aspect of cosmic evolution. Secondly, societies are seen as evolving organisms, which, in the course of their growth, must differentiate and solve problems linked to that differentiation, just like biological organisms. In this sense, societies must obey laws that cannot be altered through the intervention of governments or other social actors.

What Spencer had to offer was a detailed parallelization between societies and organisms, as well as a stage model for which he claimed it would give an account of how societies evolved from small and homogeneous to large and heterogeneous unities. In short, it was the model of a system and a description of its dynamics.

Societies can be understood as systems in several ways. They are all centered around the relation between individuals and society. This relation was already puzzling Durkheim when he reported on his German experience. The problem was implicitly defined as a dualism: individual(s) versus society. This certainly immediately suggested the traditional way of coping with dualisms: one postulates the dominance of one side of the dualism over the other. This way of conceptualizing the relation between individuals and society characterized the situation of sociology in Durkheim's eyes. In 1885, in a long and careful review of Schaeffle's *Bau und Leben des sozialen Körpers*,[26] Durkheim wrote, "Sociologists are divided into two schools, depending on whether they subordinate society to the individual or the individual to society" (Durkheim, 1975b: 375). Both positions assume a one-directional causal relation. Either society can be reduced to human nature. Then social phenomena are seen as mere consequences of the (by definition: nonsocial) properties of interaction humans. Or societies are conceptualized as systems of class relations, functions, or communications, and individuals become conceptually the strange beings assumed in what has been called the "Standard Social Science Model" (SSSM): cognitively and behaviorally undetermined but completely moldable because of their learning capacities.[27]

If one wished either to maintain the dualism without deciding on the relation between its sides, or if one had a solution to propose, the organismic model would be most convincing. When used as a metaphor, its function was to provide a plausible model and a source of inspiration for those who wished to conceptualize entities that consisted of interacting parts. The main advantage of thinking about society in terms of the organism is that one has to deal with an entity that exists in our empirical world and that can be analyzed accordingly. At the same time, it consists of components that, despite a certain independence, have to interact to survive individually (at least at the level of organs) as well as all together, and hence the whole organism survives. The organism is a system where components interact in such a way that they form a network of interactions.[28] This network functions as an internal environment of specific interactions, through which components contribute to each other's maintenance and hence to their own maintenance. This means that the organism can be seen as a system where different components cooperate both in their own interest and in that of their fellow components. Therefore, it constituted a natural model with which to understand society, both in Spencer's and Durkheim's eyes.[29]

Clearly enough, to opt for the organismic model is merely a first step. As Durkheim had already analyzed in his articles and book reviews, taking the organism as a model invariably implied taking a position with respect to the society-individual dualism. Either one gives priority to the societal level and ends with an authoritarian attitude like the socialists of the chair,[30] or one "denigrates excessively the individual" from an abstract idealistic position (*op. cit.* 329) like Wundt in his *Völkerpsychologie*. If one gives priority to the individual, the concept of society as an entity finally breaks down. For this reason, Durkheim criticizes the utilitarians and economists, and for this reason as well, he criticizes Spencer (*Division*: 180f). If it is true that conceptualizing the relation between individual and society in a dualistic form invariably leads to a priority of one part of the dualism over the other, then the fundamental problem is the dualism itself, because it creates the wrong alternatives.[31]

However, there is still another "dimension" of the problem. How can sociologists take inspirations from biology beyond the heuristic level provided by the study of composite unities? Durkheim's most important condition was that society had to be understood as a natural phenomenon. But what did he mean by "natural"? "Society as a natural phenomenon" to Durkheim did not mean that society was created by individuals who designed and then realized it while pursuing some goals. This, of course, again excluded utilitarian explanations which took society simply as a mean to individual happiness. It additionally excluded all variants of the radical German distinction between nature and culture. In this line of reasoning, Durkheim criticized Schaeffle for extensively using the organismic metaphor on one hand, and taking society as a man-made artifact on the other (*op. cit.* 285). In which sense then does society as an organism-like, composite, natural entity differ from biological organisms? Durkheim's answer constitutes part of the *Division's* core argument.

The title of Durkheim's famous first book, *On the Division of Social Labor*, is misleading to a certain extent. It underscores the procedural or dynamic aspects of his argument. The innovation he brings with respect to older discussions, and with respect to Spencer as well, is that he overcomes the obstacle of the society-individual dualism. By posing that human individuality resulted from social differentiation[32] *and* that society consisted of individuals, Durkheim replaced the traditional dualism with a relation of "circular causality."[33] This shift, which will be outlined below, forced Durkheim to look for an explanation for social change. The only convincing offer at hand was, in fact, Spencer's account of evolution.[34]

If Durkheim were not interested in a systematic account of social change, his contention concerning the relation between society and the individual would have been inconsistent. In fact, saying that society consists of individuals suggests that individuals are independent and even prior to society, as Spencer claimed. But maintaining at the same time that individuals are, to a great extent, products of society certainly implies a priority for society. The contradiction is solved by Durkheim's evolutionary model of the transformation from traditional to modern societies. The main features of the model developed in the *Division* are as follows:
a) the claim that societies are natural entities,
b) the claim that, for humans, the social environment is more important than the physical environment,
c) the replacement of the dualistic relation between individual and society by a circular and causal relation of interactions in time, which therefore leads both to the modification of societies and to that of individuals, and
d) the modification of the notions both of the individual and of society from a more "monist" to a more "pluralist," or network-like, concept.

What are the details of Durkheim's model? How is it influenced by the organismic metaphor and by Durkheim's reading of Spencer?

Taking societies as natural entities allowed Durkheim to replace the individual in the traditional sense as the basic natural entity in social theory, and to then transform sociology into a discipline of "truly" or, as he insisted, *sui generis* social objects. This takes place at two levels: the epistemological and the conceptual.

At the epistemological level, Durkheim demands that sociologists should consider social facts as "things." But despite his claims to build a positivistic social science, he never understood it in the naive sense in which positivism is sometimes understood. For our purpose it is sufficient just to point out his critique of traditional empiricism as "irrationalism" (Durkheim, 1985: 19f), and to see his insistence that "to treat the facts of a certain class as things is not to attribute them to this or that category of the real; it is to take a certain mental attitude with respect to them" (Durkheim, 1983: XIII).[35] This not only allows, but requires, a position that takes social phenomena, like those of other domains, not as given, but as analytically defined. One outcome was Durkheim's developing interest in the social origin of categories and knowledge.

At the conceptual level, Durkheim had to prove that social phenomena follow regularities that are socially produced. In modern terms, he had to show that social unities are self-organizing and self-regulating, hence they have properties that define "natural" phenomena since antiquity (Hejl, 1992b). Spencer had already defined societies – not individuals – analytically, and he certainly offered a conceptually rich and already widely accepted model with his structural and functional parallelization between organisms and societies, as well as a treasure of illustrative details. The organismic model makes clear that, according to the question one asks, one might look either at the level of the whole organism, or that of organs, of cells, or of cell components, i.e., at the cell membrane, etc. But in every case, and no one would doubt it, one is confronted with "natural" phenomena. According to the explanatory goal, one might then look at the relation between the level in focus and the natural phenomena "above" or "below" that level.

As already mentioned, Spencer took his evolutionary principle from the embryologist and evolutionist, von Baer.[36] Despite the importance that Spencer attributed to social processes as causes for change in individuals (adaptation leading to the "social state"), he took social evolution as being ultimately an adaption to the physical environment (social evolution as part of the process of cosmic adaptation). Durkheim saw this very clearly (*Division*: 247, 333f). In his eyes, such an explanation was incompatible with the phenomenal independence of societies, which means with their character as natural entities as well as their importance compared to that of man's physical environment. After having discussed "internal" reasons of individuals, which might lead to the division of labor, he concluded:

> It is therefore outside him [the individual], which then means in his environment, where the determining causes for social evolution are to be found. If societies change and if he [the individual] changes, it is because the environment changes. On the other hand, as the physical environment is relatively stable, it cannot explain this succession of changes. As a consequence, it is in the social environment that one has to look for the true conditions. Variations occurring here cause those variations which societies and individuals go through (*Division*: 231f).

As this passage shows, Durkheim diverges from Spencer's view that man adapts ultimately to his physical environment. He instead suggests that social life constitutes for humans an environment that is not only different from, but also more important than the physical environment.[37]

Durkheim solves the ambiguity of the relation between the individual and society in two ways. The first is that the dualism is replaced by a relation of interaction. Secondly, and consequently, individuals are made more social and societies are more related to individuals.

Durkheim's explanation of social evolution is definitely inspired by Spencer's organicism and evolutionism, although there is a substantial modification linked to the shift from the physical to the social environment. Durkheim takes from Spencer what Spencer took from von Baer's embryology: social evolution is the transition of an organism-like entity from an undifferentiated homogeneous to a differentiated heterogeneous state. The overall process is modeled after the development of a zygote to an adult organism. Furthermore, Durkheim and Spencer both believe that the importance of individuals grows with social evolution, but Durkheim claims to see the whole process so differently from Spencer, that finally he is more opposing than supporting him (*Division*: 169f).

On the basis of the available ethnographic and historical material, Durkheim assumed early human communities were characterized by segmentary social differentiation. Consequently, the members of these communities are assumed to be more or less equal. Living under the same social conditions, there are no extrabiological differences that lead to individualization for social reasons. The access to practical, social, religious, etc. knowledge is suggested to be more or less open to all members of these societies.[38] Already at this starting point of social life, Durkheim differs from Spencer. Where Spencer sees the origin of societies as a cooperation between individuals who decide to cooperate while pursuing their own interests, Durkheim conceives of this "starting point" in a significantly different way.

> A society, in the scientific sense of the word" says M. Spencer . . . exists only if cooperation is added to the [already existing] juxtaposition of individuals." We saw that this alleged axiom is the opposite of truth. On the contrary, it is evident, as said by Auguste Comte, "that cooperation, far from having produced society, supposed necessarily its spontaneous formation." What brings humans together are mechanical causes and impulsive forces like the affinity of blood, bonds to the same soil, ancestor-worship, shared habitudes, etc. It is only once the group has been formed on these foundations that cooperation is organized (*Division*: 262).

The central concept to describe the solidarity of these early societies is, of course, that of the "collective consciousness." In the present context, this concept is important because it illustrates both Durkheim's difference with Spencer and his factually greater propinquity to the organismic model. As is well known, Durkheim calls "collective consciousness" the consciousness the members of a society have developed as a result of those more or less egalitarian interactions he thought to be typical for undifferentiated societies. When shared throughout such a society, it allows its members to communicate and to interact successfully. It is a consciousness that characterizes not isolated individuals, but a given society. Hence it plays at the social level an analogous role to the consciousness of individuals, as already pointed out by Filloux (1979: 143). But this concept is not only nearer to the organismic model in taking societies as organisms with their proper consciousnesses; placing the concept of the collective consciousness at the very beginning of societies in fact establishes an analogy between the coalescence of two cells into a zygote and the formation of a social unity or protosociety. In both cases, a new unity emerges.[39]

Therefore, we have a manifest difference between Durkheim and Spencer in the way they conceive of societies and their formation. With respect to Spencer, Durkheim not only maintains his antireductionism, but raises the discussion from a behavioral to a cognitive level.[40] Whereas Spencer argues on the basis of analogies between animal colonies and organisms on one side and societies on the other side, Durkheim adds the level of representations. Whereas Spencer discusses structural and functional consequences of societal growth, Durkheim, without giving up Spencer's line of reasoning and the illustrations it provides, tries to shift the discussion of its consequences to the levels of individual and social knowledge repertories and/or systems.[41] This is illustrated by the much more explanatory than analogical model he develops in order to describe the shift from undifferentiated to differentiated societies.

Modern readers who are familiar with evolution theory normally do not associate von Baer or Spencer, but Darwin with the term "evolution." This would have been the contrary in Durkheim's and Spencer's time. Moreover, approaches inspired by modern evolution theory are normally situated at the level of utility-maximizing individuals and/or on a biology-based version of "human nature" that often turns out to be no less utilitarian.[42] Now, as Durkheim refuses the combination of utilitarianism and individualism that characterized Spencer's evolutionary thinking, he turns to Darwin. This allows him to introduce Darwin's "principle of divergence,"[43] which means to hold to the concept of differentiating development, without having to adopt Spencer's assumption about the role of individuals at the same time. A look at the details will show the importance of the shift.

Durkheim assumes that at some point in time a "progressive condensation" (*Division*: 238ff) of social life occurred in three ways: through a growing population, the formation of towns, and better transportation opportunities. Then Durkheim criticizes Spencer's view that division of labor occurs spontaneously as an adaptation to different external conditions. According to Durkheim, this does not explain why the members of a group accept to become different. Spencer explains how evolution continues, once it has appeared, but he does not "tell us what is the spring that produces it" (*Division*: 248). Next, Durkheim cites the passage from the *Origin of Species* already mentioned, where Darwin reports of observations that demonstrate how increased competition and a higher rate of differentiation coincide. Furthermore, Durkheim refers to a similar argument from Haeckel and affirms that the competition between different tissues of an organism is eased because they feed on different substances. Finally, Durkheim concludes, "Man undergoes the same law," and ends up with a number of illustrations from the social domain (*Division*: 249f).

Durkheim's urge not to assume, but to explain the occurrence of the division of labor is systematically linked to the conceptualization of the societal *statu nascendi*. Whereas Spencer takes individualization and division of labor to exist from the very beginning, Durkheim tries to explain them and show why they do not lead to the kind of unlimited competition Spencer praises as necessary and Durkheim thinks to be detrimental. Referring to his concept of the collective consciousness (to the representational level), Durkheim argues that, owing to the already existing social bonds (from emotional to institutional ones), the outcome of the "struggle for life" is not a generalized conflict, but a growing division of labor (*Division*: 253).[44] The social "mechanism" that Durkheim describes in two chapters starts from the assumption that the social control typical of traditional societies or communities is reduced during societal growth. This is due to the declining influence of families, traditions, and religion occurring within the developments mentioned. At the same time, the collective consciousness (in more modern terms, socially-shared realities as well as action programs and related norms) becomes more abstract. As a result, individual members of such societies must increasingly decide for themselves in which situation they are and how to act. Tasks such as the specification of situations and the choice of action, which used to be fulfilled socially, are taken over more and more by individual members, albeit within the "given" social framework. This necessarily leads to differentiations, i.e., a development of diverse perceptions and acts. Finally, the members of the "societies" in *statu nascendi* (in contradistinction to the former "communities") increasingly perceive themselves as those who act and who are primarily concerned by their acts. The community members who, according to the concept of community, differ only slightly as far as their cognition and their actions are concerned, gradually become individuals who are autonomized vis-à-vis the emergent societies.

With growing social differentiation, members of the evolving societies become less involved in interactions that resemble each other, both social interactions and interactions with their material environment. The changes produced by the initial causes finally themselves become causes for further changes. The process finally perpetuates itself. Its basic pattern can be described as follows: individualization $\Rightarrow$ increase of social differentiation $\Rightarrow$ individualization $\Rightarrow$ increase of social differentiation, and so on. This reciprocity between the level of components and the level of organization results in changes in both levels, which can be reconstructed in modern terms as self-organizing (Hejl, 1990; 1992c). The prerequisite to conceive of this self-organizing mechanism is an analytical approach that allows one to differentiate between the social and the individual level without giving one side of the traditional dualism a causal priority over the other. In Durkheim's context, this obviously meant to argue primarily for the (relative) independence of the social level and against a position that takes individuals fundamentally as evident basic entities.

But what exactly does Durkheim mean by "individual" and by "society"? Durkheim gave the word "individual" several distinct meanings. Two of them are needed here to understand his theorizing: the individual of individual psychology and the individual in the sociological or socio-psychological sense.

Durkheim defines one kind of individuals as living systems that are characterized by biological differences. He considers biological differentiation the "first (and most original) basis of any individuality" (*Division*: 175).

... there is a domain of psychic life ... which varies from each individual, regardless of the state of development of the collective type involved. This area consists of representations, emotions and tendencies which relate to the organism and the states of the organism; this is the world of internal and external feelings and perceptions, and the movements directly linked to it (*ibid.*).

This concept covers what Durkheim takes to be the individual in the psychological sense of the term. When he criticizes the use of "individuals" as building blocks of a social science, he refers to the generalization of this isolated individual (the *homo oeconomicus* is just another variant).

The individual in the modern (sociological) sense enters the stage with development of the division of labor. Its formation results from processes already mentioned. Durkheim conceives of modern man as an individual being because he is less dependent on traditions and religion, and is under a perhaps not smaller but more general social control than his precursors (*Division*: 395f). The "individual" as a unit of analysis, therefore, is transformed by Durkheim from a culturally-produced evidence into an analytical concept.

But Durkheim's rejection of the individual of individual psychology did not prevent him, as outlined earlier, from taking individuals, conceived as social individuals, as the basic units of societies.

Although sociology defines itself as the science of societies, it cannot, in reality, deal with human groups which are the immediate object of its research, without finally touching upon the individual, the ultimate element of which these groups are composed. For society can only constitute itself on condition of penetrating individual consciousness and fashioning it "according to its image and resemblance": without dogmatizing too much, one can therefore say for certain that many of our mental states, and some of the most important ones, are of social origin. Here it is the whole, which to a large extent, makes up the part; accordingly, it is impossible to try to explain the whole without explaining the part, at least as a repercussion (Durkheim, 1987d: 314).

If one tries to investigate in greater detail what especially characterizes the modern individual, one finds a striking similarity between the concept of the individual and that of society: both are considered networks that act as unities. If we look first at the individual, we find Durkheim's famous concept of the "dualism of human nature" as a basic description. For Durkheim, human beings differ from other animals because of their larger mental capacities. It is thought to consist basically of two (analytically separated, *Division*: 99) parts.

There are in us two consciousness: one contains the states which are personal to each of us and through which we are characterized. The states included in the other are common to the whole society. The first represents only our individual personality and constitutes it; the second stands for the collective type and, as a consequence, stands for the society without which it would not exist (*Division*: 74).

Yet this simple dualism exists only at the general level where the psychic and the social level are distinguished. In fact, in a footnote attached to the passage cited above, Durkheim adds:

To simplify the exposition, we suppose that the individual belongs only to one society. As a matter of fact, we belong to several groups at the same time and there are in us several collective consciousnesses; but this complication does not modify the relation we are establishing (*Division*: 74).

As a consequence, Durkheim sees a fundamental parallel between individuals and societies: both are systems modeled on the organism.[45] But whereas individuals consist of networks of "several collective consciousnesses" including (dualism!) "representations, emotions, and tendencies which relate to the organism and the states of the organism," societies consist either of the network formed by the set of all representations of social origin or of the network of interacting individuals (who are defined via these representations). In one case, the nodes of the network are the representations, or consciousnesses (the plural is important!) that characterize individuals, whereas in the other, the nodes are either the interacting individuals, or, at a higher level of aggregation, the interacting social subsystems characterized by the specific sets of representations typical for science, or the economy, or the political system, etc.

Consequently, there is no contradiction when Durkheim underlines sometimes that society consists of individuals and sometimes that social life consists of representations (Durkheim, 1983: XI). It is the importance of the representational level, its differentiation and distribution, which constitutes the difference between biological organisms and societies.

Representations in differentiated societies differ from those in traditional ones, of course. The collective consciousness becomes generalized and, therefore, forces the singular members of differentiating societies to decide for themselves and to accept the ensuing consequences. The social subsystems which are formed in this transformation are each characterized not by a collective consciousness and hence by collective representations, but by "particular representations": the knowledge and norms of *specific* social subgroups. Individuals are individuals in the social domain because they become autonomized centers of decision and action *and* because they belong to different combinations of social subsystems. The consciousness of the individual in differentiated societies thus presents the same type of differentiation that exists in the society as a whole, although obviously the consciousness of no individual represents the factual differentiation of its society.

To conceptualize societies as organisms (as networks with a boundary) requires one to ask how the interactions of the components are regulated such that a behavior is generated at the level of the organism. To complete the discussion of Durkheim's use of the organismic metaphor and of his difference with Spencer, it is important to look at Durkheim's view on regulation. This will bring us, at the same time, back to his sociopolitical context.

One important reason why the organism interests those engaged in making sense of the functioning of composite unities is that different parts cooperate in a seemingly harmonious way. But how is this harmony achieved? Spencer assumed that individuals would so adapt to each other that at some point, institutions and the state would be of little importance. This corresponded to his concept of the evolution from military states to laissez-faire market societies. Durkheim criticizes Spencer in two respects. First, he disagrees with Spencer's view that the course social evolution takes is circular, because it starts with individuals who form societies with a state that will then disappear again. He critically asks, "The movement of history would therefore be circular and the progress would consist of a turn backwards?" (*Division*: 171)

Second, using ethnographic and historical evidence, Durkheim argues that Spencer's account of social evolution is not only historically wrong (*Division*: 170ff), but an absurdity. Durkheim's main argument is that historically, the role of the state as the regulating organ increased continuously (*Division*: 199ff). Hence there is broad evidence for the parallel development of social control agencies and of individualism and individual freedom.[46]

It is in discussing this problem of the state's role for societal regulation that Durkheim fundamentally criticizes Spencer's conclusions on the formation of societies. Interestingly, Durkheim's point had been made in 1877 by another Frenchman, H. Marion.[47] According to Spencer (who answers Marion's criticism in a postscript to part II of his *Principles of Sociology*), Marion criticizes him because Spencer argues that the more organisms are developed, the more their central nervous system becomes important, whereas he maintains the opposite position for societies when defending liberalism. Spencer's answer to this criticism not only resumes his position marvelously, but also clearly shows the difference to Durkheim.

I regret that when writing the foregoing chapters I omitted to contrast the lives of the individual organisms and of social organisms in such a way as to show the origin of this seeming incongruity. It is this: Individual organisms . . . have to maintain their lives by offensive or defensive activities, or both: to get food and escape enemies ever remain the essential requirements. Hence the need for a regulating system by which the actions of senses and limbs may be coordinated. Hence the superiority that results from a centralized nervous apparatus . . . It is otherwise with societies. Doubtless during the militant stages of social evolution, the lives of societies, like the lives of animals, are largely . . . dependent on their powers of offense and defense; and during these stages, societies having the most centralized regulating systems can use their powers most effectually, and are thus, *relatively to the temporary requirements,* the highest. . . . Increases of industrialism and decrease in militancy, gradually brings about a state in which the lives of societies do not depend mainly on their powers of dealing . . . with other societies, but depend mainly on those powers which enable them to hold their own in the struggles of industrial competition. . . . In animals, then, the measure of superiority remains the same throughout, because the ends to be achieved remain the same throughout; but in societies the measure of superiority is entirely changed, because the ends to be achieved are entirely changed (Spencer, 1893: 586ff).

Durkheim, again taking a position closer to biology than Spencer's, does not accept the difference between societies and organisms that exists at this point in Spencer's view. Durkheim says, "Even the biological comparisons on which Mr. Spencer likes to base his theory of the free contract are much more its refutation" (*Division*: 195). According to Durkheim, Spencer concludes that, owing to the existence of lower regulatory centers in the organism, the central regulatory system loses its importance in more differentiated organisms. In contradiction, Durkheim underlines that, although decentralized, there is regulation, and that it might well be one of the economical problems of his time, that intermediary centers of regulation are lacking.

Interwoven with and backed up by his historical analysis of the development of law from repressive and criminal law to private law, Durkheim argues that there is a continuity and probably even an increase in regulatory activities. But whereas they were formerly based on the repression of any attempt on the societies and their collective consciousness, with differentiation and individualization, the collective consciousness partly loses in importance, and hence law as the main regulatory device changes as well. It is now directed towards the growing individualization of its members through laws that protect the individual, as well as describe and prescribe what are considered "individual rights." Besides human and civil rights, it is the development of contractual law that marks this change. Durkheim thinks of regulatory activity, which the whole exerts on its parts, in no way as the intervention of an autonomous center. True, he outlines the growth of central government (*Division*: 200ff), but he argues in the same context that there are activities which he calls "special functions," which stay and should stay outside the influence of the state. His main example is the economy, which he parallels with the digestive system. At the same time, even the capital of a country can be replaced if conquered by an enemy, and the government itself is seen as being dependent on the governed.[49]

If one looks at Durkheim's sociopolitical background and at his work on the division of social labor, one finds a striking parallelism between them. We have the "cultural heritage" of holism from Renouvier and the idea that a true science should have its own domain, which he took from Boutroux. The sociopolitical conflict focused on the problem of liberal democracy versus some more authoritarian regime. This discussion was linked to the dualism of individual versus collective/society. Given Durkheim's situation as member of a minority in assimilation, one can reasonably assume that he valued both collectivism (integration, assimilation) and individualism (without individualism and secular liberal politics, integration would be much more difficult). Hence the dualism of individual versus society coincided in the personal and the public domain. In the context of the Cartesian and state tradition of France, the necessity of institutions and regulations were moreover evident.

Why was Durkheim nearer to the "organismic model" he mainly took from Spencer? The simple answer is that Durkheim needed it more than Spencer. Given the premises of the character of a "true science" and of holism, and adding the interest to get over the dualism of individual versus society, Durkheim needed a model that could help him understand the functioning of a composite unity. At this point, the systematic problem appears: how one should deal with the obvious inconsistency when society is defined as that which consists of, as well as "produces," individuals. The problem can be solved if the relation is seen as an interaction in time. This strongly suggests that one should look for a model to explain the dynamic behavior of composite entities. For this reason, one might conclude that Durkheim "had to" choose Spencer's evolutionary theory. Only the developmental version of organismic evolution allowed one to view social change as a process of differentiation that occurred inside a system. Therefore, it brought about a transformation at the systems level and at that of the components, and finally, systematically connected the aspects in which Durkheim was interested. As a consequence of his reformulation of the relation between individual and society, he had to solve problems that appeared as a result of this decision. This contributed to his concept of a secularized dualism of human nature, which foreshadowed his sociology of knowledge and religion, a major interest in the following years.

Why was Durkheim more successful in comparison to Spencer? As Durkheim probably correctly points out, although Spencer was a productive, leading, and influential author in the field of sociology, he was less interested in sociology than in his evolutionary philosophy. In a time of rapid social change from industrialization and professionalization, Durkheim was more professional and opened up new insights with an approach that offered concepts useful to many domains, but which does not pretend to produce a complete answer. Moreover, and this does not have to be underscored, Durkheim was part of the academic system, whereas Spencer remained outside.[50]

Some final remarks on the "organismic model" and its functioning as a metaphor are helpful. A close look on the justification of the term "mechanical solidarity" shows how the "organismic model" works as a metaphor. In contradistinction to Spencer, Durkheim believes that early societies have developed a high degree of social cohesion not despite, but because of, their homogeneity. There is a strong social cohesion which goes together with low social individualization. Actors who have the same knowledge and use the same cognitive mechanisms will behave in the same way as well (*Division*: 105): the social cohesion that results from this "collective consciousness" is called "mechanical solidarity." It is "mechanical" because the behavior is produced by actors thought to be very similar, but who are, of course, organisms (*Division*: 74). The biological examples Durkheim (*Division*: 167 ff) uses are undifferentiated colonies of invertebrates and the more or less identical annular segments of worms. Drawing on biological literature, he argues that although the parts of colonies or worms can be considered living entities themselves, it is the higher level of the whole colony or animal that has to be taken as an individual as well, although of a different kind.[51] But why exactly is the "solidarity" between the parts required for that step? According to Durkheim, the basic units in his examples are so linked physically that any movement or action of some of their parts affects the others mechanically ("mechanical" is used here in its literal sense). This is, of course, not the case in what he calls "primitive societies." The physical links between the members of animal colonies that guarantee their solidarity (at least up to a certain point) become, in the case of societies, the shared knowledge of the world and of the society itself: the collective consciousness. In fact, Durkheim sees the formation of this consciousness as a mechanical process, owing to the absence of social differentiation and consequently of differential experience. But even then, there can be no doubt that the concept of mechanical solidarity is based on a complete modification of what served as a starting point. Because he has to explain how physically-independent individuals form a kind of superorganism, Durkheim has to replace the physical bonds by cognitive processes that have a similar behavioral effect: the collective consciousness. Because he believes that cognitive processes can be explained causally and that they are generated by physical systems on which they depend (*Division*: 217), he has no problems in understanding them as "mechanisms" (*Division*: 317), just as modern cognitive psychologists do. But it is not the concept of animal colonies that suggests societies as networks of interacting individuals, who, as a result of these interactions, produce very similar consciousnesses and hence act cooperatively as a group. This is suggested much more by the a priori acceptance of the individual and the social levels and of the necessity to establish a causal link between these levels. The biological metaphor serves in the case of mechanical solidarity more as an example to demonstrate indubitably that there are entities of the type claimed for societies, than as a model to conceptualize them.

The situation is different in the case of differentiated societies with organismic solidarity. This concept conceives of actions at the social level as results from different, but complementary contributions, and the organism can then be a much more profitable example of how composite entities might function. Nevertheless, there are problems that prevent too close of a parallelization. In contradiction to the cells or organs of an organism, individuals (even less than social subsystems) are neither biologically nor socially determined to fulfil certain functions. Moreover, individuals can change from one occupation to another, as Durkheim underlines (*Division*: 319ff). But individuals have access as well to the environment of their group or society, which is evidently not, or in a different sense, the case for the components of an organism.

To sum up, it is highly doubtful to speak of a "model" in the sense of a "selective description of a biological 'original'" in the case of an organism, as model theory requires (Stachowiak, 1973). Canguilhem (1970: 319ff) argues that the modern theory of the cell and the organism, which understands the organism as consisting of interacting cells, was itself largely inspired by the use of economic and political models. And, in fact, as mentioned earlier, there is a tradition of understanding societies in terms of the human body prior to any precise knowledge of the organism.[52] Therefore, the "organismic metaphor" might be better understood as the biological version of a more general picture used to describe composite entities.[53] This picture, in fact, belonged to the standard concepts in social philosophy, economics, biology, and sociology. As the "organismic metaphor" circulated between the disciplines, it served their explanatory ends. During this process, it eventually became modified in one discipline in a way understood as a gain in precision in another. The "organismic metaphor" then seems to be like a chameleon. Sometimes it becomes the "economic metaphor of the division of labor," sometimes the "metaphor of society," and sometimes a "biologistic concept."

Finally, some critical remarks are in order concerning the effects the metaphorical use of the concepts "organism" and "evolution" had on the development of the social sciences. The concept of societies as superorganisms led sociologists and anthropologists to construct societies to a large extent as developing organisms – as entities that differentiate from internal causes (Tenbruck, 1992). Despite its importance, this view tends to overstress the character of societies as unities. Although analytically defined, sociologists in this tradition tend to take this character as a "true" descriptor. As a result, they neglect characteristics no biologist would neglect if speaking of the development and/or evolution of organisms: development and evolution of whatever entity take place in an environment, and the ontogenetic development (the heritage of von Baer's embryology) comes to an end either with the adult stage or with the death of the organism considered. As societies are mostly treated as eternally existing and differentiating unities, changes that result from interactions with other societies are neglected. Nevertheless, such interactions do lead to effects both at the level of organisms and of societies. These effects might vary from death in the physical sense, to cultural extermination, to additive growth, to integration, and to the resulting emergence of a social and cultural variety. Therefore, we have a kind of differentiation and internal variety that, although not independent of, is not exclusively determined from internal causes. Of course, Spencer's and Durkheim's theories allow one to conceptually integrate external and internal causes of social change. But Spencer's combination of individualism with the origin of social growth and Durkheim's critique of Spencer together with Durkheim's emphasis on collective consciousness, caused later discussions to be kept in this very dichotomy instead of utilizing the integrational approach.

Acknowledgments

I have the pleasure of thanking R. Boyd and T. Shinn for numerous helpful comments and suggestions on an earlier draft of this contribution. My gratitude goes as well to S. Schultz who helped me with her linguistic capacities.

Notes

1. See Durkheim, 1986, *De la division du travail social*, XLII, hereafter referred to as *Division*.
2. For a striking example of how socially-shared evidences are constitutive for apparently robust, theoretical concepts, see Gigerenzer, 1991, who shows the impact statistical concepts have on theorizing in cognitive psychology.
3. My intention, therefore, is not to describe and systematically assess Spencer's influence on Durkheim, nor to evaluate the "accuracy" of Durkheim's reading of the English philosopher.
4. Of course, here is not the place to discuss theoretical problems of historical analysis, but is should be clear from the introduction that I do not subscribe to a view that pretends that "facts," "problems," and "solutions" impose themselves or can be copied from data in a positivistic sense. Any analysis is based on and propelled by prior experiences and concerns. In the present case, this background is an interest in systems theory and self-organizational "properties" of biological and social systems, and the hypothesis that "grand theories," and hence the founding fathers of sociology, had to grapple with these problems. That a systems theoretical posture with respect to Durkheim's work is quite helpful has been recognized by different authors, starting with T. Parsons in 1968. As will be argued below, the reason is simply that the organismic metaphor is a precursor of systems theory, see in this sense Birnbaum, 1971.
5. On cultural change parallel to the formation of Durkheimian sociology, see Tiryakian, 1979.
6. Cobban, 1978: 50. On the general history of anti-Semitism in France, see Winock, 1982: 83ff.
7. The Declaration of Human and Civil Rights in 1789 included Jews. There was a drawback after the Restoration (1814/15). Finally, in 1848, French Jews were granted full civil rights. On the general development of Jewish emancipation, see Greive, 1982: 149ff.
8. On the intellectual debates in France after 1871 and their relation to Germany, see Digeon, 1959 and Guitton, 1968.
9. Monarchists, backed up by the Church, high-ranking officers, but conservative intellectuals as well, and later on, Bonapartists.
10. See Cobban, 1978: 24, and as a case study of the establishment of sociology within the context of the other disciplines, Weisz, 1979.
11. On the importance of educational reform in France after 1871, and especially on a reform of the universities and of scientific research, see Digeon, 1959: 364ff. On Durkheim, see especially p. 379ff, and on Fustel de Coulanges p. 237ff.; on Fustel's influence on Durkheim, also with respect to Durkheim's critique of Spencer, see. Prendergast, 1983/84. Fustel started his career as a professor in Strasbourg. In the context of the struggle between French and German science, he aimed to fight against German historical science, which he accused of having a nationalist bias. This international context of the development of science is often neglected, although it plays a role in the transformation not only of educational and scientific institutions, but also of what is taught and studied, including the foundation of disciplines, or

shifts of importance between them. This seems particularly true for the case of France and Germany where such mutual and quite direct influences can be observed from the German defeats during the Napoleonic wars on (foundation of the University of Berlin as part of a larger reform movement inspired by von Humboldt) until today's general discussions of the relation between the two countries (see Meyer, 1979), or for example during the 1992 French referendum on the further development of the European political community (treaty of Maastricht), when France's relation to Germany became a major point of reference both for the partisans of the "Oui" and those of the "Non."

12. For the importance of the Ecole Normale Supérieure from the time of Durkheim on, see Bourgin, 1938, especially on Durkheim, p. 215ff. On the role that the Ecole Normale played as a mean of social mobility, see Karady, 1972.
13. The lasting importance these discussions had on Durkheim is demonstrated in a short passage in the article through which he took position to the Dreyfus affair and that, not by chance, was titled "Individualism and the intellectuals." With respect to the long struggle for full civil and political rights, and hence for democracy, he says: "The men of my generation remember how enthusiastic we were when, some twenty years ago [around 1878], we finally saw the last barriers fall that contained our impatience" Durkheim, 1987c: 276.
14. Although Durkheim did not become a party member, he was not only near to socialist positions, but in fact influenced the formation of the central political positions of the reformist French socialists led by Jaurès. On the importance of the "social question" and of socialism for Durkheim, see Durkheim, 1971; Filloux, 1963; Filloux, 1977; Birnbaum, 1971.
15. Filloux speaks in this respect of "the fundamental decision of 1881–1882" (1963: 67) as a "mission" Durkheim takes upon himself. That Durkheim in fact conceived of sociology as a contribution to a better praxis is made clear in *Division*: XXXIX.
16. That Durkheim and the Durkheimians in fact had the social competence to be successful is analyzed in detail by Karady, 1979. On the formation of the Durkheimian school, see Besnard, 1983, and Nandan, 1977. For a general analysis of the French university system and of the emergence of the social sciences, cf. Clark, 1973. For those who are not French, it is difficult to understand that the majority of leading French intellectuals are civil servants either in the Grandes Ecoles or in the universities of Paris. This means that, having passed their entry examinations, they enjoy a rather comfortable situation that encourages a production that often is an expression of this context. This holds true, as Bourdieu, 1992, points out, for movements like deconstructivism as well as for Durkheim: "one can see . . . in Durkehim's texts all that his thinking on the state owed to the fact that he was a functionary of the state" (*op. cit.*, 40).
17. Although the term was only used after Smuts, 1926.
18. On these influences, see Durkheim, 1975e; Durkheim, 1975f. In 1893, Boutroux presided over the jury when Durkheim defended his doctoral thesis in Bordeux. For the broader influence of Renouvier, both on Durkheim and in France, see also Lukes, 1975: 54ff.
19. The factual importance Durkehim accorded to Spencer's work can be demonstrated by quantitative analysis of the citations in the *Division*. There is a total of 156 authors

cited in the text (without the Preface added to the second edition). If we look at the citations (counting names only once per page of appearance), we have an overall number of 384 citations. Only 54 out of the total of 156 authors are mentioned more than once. A subset of 16 authors is quoted five times and more. But this group provides 45.3% of all citations. By far the most frequently mentioned author is Spencer with 52 citations (13.5%). Comte, who comes second, is far less important with only 18 references (4.7%). The other authors of this group are Tarde, Thonissen 12 (3.1%), Waitz 10 (2.6%); Perrier 9 (2.3%); Levasseur 8 (2.1%); de Candolle, Fustel de Coulange 7 (1.8%); Aristoteles, Morgan, Rein, Voigt 6 (1.6%); and Marquardt, Schmoller, A. Smith 5 (1.3%). On the sociological importance of Spencer in Durkheim's view, see Durkheim, 1987b: 91ff.
20. See Boesinger, 1980, esp. 319, and Limoges, 1980.
21. If we differentiate between those quotations where Durkheim either cites and/or approves of Spencer and those where he criticizes him, we find a ratio of 20 approving/neutral to 32 critical references to the English philosopher.
22. The widely-shared reading of Durkheim as an author who focused on the normative aspects of social life tends to neglect the philosopher of science interested in "mechanistic," i.e., "causal" or "rational" explanations. It is a misunderstanding to take Tönnis" (1979) dualism "mechanistic" – "organic," which Durkheim adopted and transformed, as referring necessarily to different types of explanations, or even to incompatible philosophical attitudes. Although this was the case when Descartes' mechanism declined (Tocanne, 1978: 44ff), or during the debates between defenders of industrialization and defenders of the established "organic order" (On the German debate that just illustrates these discussions, see Sieferle, 1984). Durkheim leaves this debate behind, certainly at the level of explanations. For him the organism, and hence societies, are evidently mechanisms! In this respect, and there are others, he definitely belongs to modern science, see for example *Division*: 53, 202, 258, 327ff, esp. p. 331.
23. Spencer's individualism is part of the more general phenomenon of Anglo-American individualism. Just as its counterpart, continental collectivism, it should be considered a culturally produced evidence, an expression of mentality (see the presentation and discussion of the concept in Le Goff/Chartier/Revel, 1978, and Vovelle, 1985), and hence should be distinguished from deliberate reductionism. An additional factor for English individualism (see particularly Macfarlane, 1978) can be seen in the British colonies, which could absorb those who were unsatisfied with religious, political or economic conditions in England itself (Turner, 1992: 184). In the United States then, this role was taken by the Western frontier.
24. The concept of the organism allowed Spencer to adopt and/or develop mainly from "evolutionary" (at his time: developmental) biology concepts still widely used in today's sociology, e.g. "function," "structure," and hence "structural differentiation and functional differentiation." He furthermore proposed much less-used biology-based typologies of societies, and undertook an immense work of comparative sociology not unlike work in comparative anatomy for example. See Spencer's *Descriptive Sociology*, published as a series of monographs under his direction, beginning in 1874.
25. Itself similarly based on metaphysical reasons: the eternal laws of matter and motion had to replace God's law, see La Vergata, this volume.

26. On the relation between Durkheim, Schaeffle, Tönnies, and Simmel, see Gephart, 1982.
27. See Tooby/Cossmides, 1992: 24ff. I share the authors' view of what they call the SSSM but claim that their criticism of Durkheim is based on a highly limited reading of his work and ignores his attempt to escape the traditional alternative of "society versus individual." Because the authors place their work on singular humans, naively taken as natural entities called "individuals," they fail to realize that Durkheimian sociology can be read more profitably as bridging the gap between the "social" and the "individual" in the older tradition.
28. Although a modern term, it is not only for analytical reasons that it seems adequate to use the concept of "network" to describe Durkheim's idea of societies (and organisms), see below. The abstract network concept makes it easier to grasp the problem of regulation in systems like organisms and societies.
29. That this insight had to grow during the years prior to the *Division* can be seen from the difference between the distant way in which he speaks of the organism as a model for sociology in 1885 (Durkheim, 1975b: 373), and the extensive use he makes of the organismic analogy in 1893, when the *Division* appeared.
30. See Durkheim, 1975b: 280f. On his relation to the socialists of the chair, see Assoun, 1976 and Lacroix/Landerer, 1972.
31. On the logical problems of conceptualizing operationally more or less closed systems with our binary logic, see Gunther, 1976.
32. I use the now more common terms "social differentiation" and "differentiated societies," although Durkheim made a sharp distinction between "differentiation" and "division of labor." The first is linked to processes of disintegration, whereas the second creates social cohesion, see *Division*: 344. Durkheim uses the term himself quite often in the modern sense, despite his criticism, see for instance pp. 24, 124, 157, 200.
33. To employ this definitely more recent term is justified by Durkheim's factual and conscious use of the corresponding concept, see Durkheim, 1983: 95f, and the discussion below.
34. One could reasonably say that Durkheim was interested in social change because of his historical background and because there was competition between history and the new science of sociology. On Durkheim's relation to Fustel de Coulange, who taught History at the Ecole Normale, and on Durkheim's critique of historical science, see Gottlieb, 1979.
35. It would be worthwhile to fully analyze the consistency of Durkheim's writings on epistemological and knowledge issues and his critique of Pragmatism with a look at modern constructivism, see Durkheim, 1955, and on his critique of pragmatism Allcock, 1982; Cuvillier, 1958; Deledalle, 1959; Gaudemar, 1969; Joas, 1985.
36. "Evolution" at that time meant more or less what today would be called "development," see Bowler, 1975; see Gould, 1977: 112, who speaks of von Baer as Spencer's Malthus. On von Baer and his position in the debate on evolution, see Lenoir, 1982 and Gould, *op. cit.*
37. Durkheim does not deny that environmental factors play an important role, see *Division*: 245f, but he denies that they are the most important ones, and that their importance remains unchanged during human evolution. The important argument is

that without the social solidarity achieved prior to any differentiation, a great struggle for life would lead to social breakdown. As to the changing importance of the physical environment, he argues that with growing differentiation, society itself becomes such a complex environment that its importance leaves that of the physical environment far behind. It can be shown (Hejl, 1993) that Durkheim's sociology of knowledge factually foreshadows the actual discussion among primatologists about the "social intelligence hypothesis," see Byrnewhiten, 1988.

38. With his ideal type of a "primitive society," Durkheim assumed a degree of cultural uniformity and stability that we consider oversimplified today. The high degree of conflict in hunter-gatherer societies demonstrates this oversimplification. Durkheim certainly was aware that social differentiation, and hence individualization, existed from the very beginning of societies onward, see *Division*: 146. But for systematic reasons, and because he had to emphasize the social factor in individualization, he held to this dualism instead of conceptualizing societal evolution as a change from a lower to a higher level of differentiation.

39. It is in this context that the "effervescence créatrice" Durkheim describes later in *Les formes élémentaires der la vie religieuse* (The Elementary Forms of Religious Life, see Durkheim, 1985: 307ff, 611) gets its full meaning as a kind of (mutual) fecundation, in which a new society is born or an existing one recreated.

40. In Spencer's description of the analogy between the organism and society, language and thought play in societies the role physico-chemical interactions play in organisms. Their role is dominantly a functional one, see Spencer, 1893: 449f.

41. There is a tendency to consider culture as a system of representations, for a short overview, see Durham, 1991: 7. Although Durkheim seems to hold a similar position, if we look at his concept of collective consciousness, the distinction he makes between collective consciousness and social consciousness points in a direction which is more adequately described by a concept of culture that defines cultures as repertories of "ideational" entities or representations with embedded knowledge systems, see Hejl, 1992a.

42. For work close to biological reproduction, see Thornhill, 1991, and as an example for the use of biological and evolutionary considerations on a topic apparently far from biology, see Masters, 1989. For an approach that avoids the biological fallacy to a great extent, although inspired by evolutionary theory, see Boyd/Richerson, 1985. For a critical assessment of the use of Darwinian evolutionary theory to explain social evolution, see Hallpike, 1986.

43. See Darwin, 1963: 160. Durkheim does not use Darwin's term where he cites Darwin's observations, but speaks later on (*Division*: 259) "Darwin's . . . law of the divergence of characters." On the "principle of divergence" in the context of Darwin's theory, see Rieppel, 1989: 148.

44. Durkheim seems to ignore the accelerating effect that the transition from a hunter-gatherer life to an agricultural economy has on social development. One can even argue that the fundamental tripartite differentiation that Dumezil, 1958, as well as Oexle, 1988, took as a specific trait of Indo-European societies is a universal phenomenon of agricultural societies once they have reached the required productivity. See Falkenstein, 1976: 22ff for early history, and Johnson/Earle, 1987, for an anthropological account of the interplay between environment, individuals and culture.

45. La Vergata (in this volume) hypothesizes with respect to Spencer that he was influenced by phrenology. Although difficult to substantiate directly, the same influence can be assumed for Durkheim. It might have worked as an additional factor that strengthened the evidence of the organismic metaphor. Durkheim neither cites Gall nor phrenology, but one has to remember that Gall worked in France, and despite errors of early phrenology, eventually stimulated neurophysiology by providing a first draft of the localization theory of brain functions (see Changeux, 1984: 25ff). Durkheim was acquainted with the theory through the work of Wundt and rejected it accordingly to a certain extent (*Division*: 322). Although Durkheim never cited Broca with respect to the brain, he knew Broca (*Division*: 104; Durkheim, 1975g: 76) who demonstrated the existence of the speech center in the brain, which is now named after him.
46. In this context, Durkheim develops his famous argument of the non contractual conditons of contracts (*Division*: 184ff).
47. In the *Division* (p. XLIV) Durkheim refers to Marion only in a way that suggests that he was unacquainted with Marion's criticism of Spencer on that point.
48. Although he otherwise accepts the differences between organisms and societies brought out by Spencer, see Spencer, 1893: 448ff and *Division*: 319f.
49. This had happened during the 1870/71 war, when part of France, including Paris, was occupied and the French Government moved to Bordeaux.
50. For a discussion of Spencer's personality and the tensions that arose from his being one of the last amateurs in a world of growing academic specialization, see Burrow, 1970: 180ff.
51. Even today biologists discuss the concept of the organism and how its parts constitute the "higher individual" of the organism. The relevance of this discussion becomes clear, if one remembers that since Weisman, biologists assume that there is continuity of the germ line and hence separation between heritability of traits on one side and non heritable variations occurring during ontogeny on the other. In contradistinction to Darwin, who believed in a Lamarckian type of heritage, modern evolution theory mainly followed Weismann and hence became disinterested both in the organism and in societies. See Buss, 1987, and for evidence of Lamarckian heritage and the necessity to rethink evolutionary theory, see Falk/Jablonka, 1994. Obviously, these developments in genetics constitute a serious challenge to sociobiology as well.
52. See Dohrn-Van Rossum, 1978, and Bockenforde, 1978.
53. On the interdisciplinary context in which biology and its basic concepts developed in France, see Canguilhem, 1970, who underlines (p. 71) that between 1848 and 1880, every biologist or physicist had to refer directly or indirectly to Comtian concepts if he wanted to relate his work to the intellectual debates.

Selected Additional Bibliography

The years of publication, written in parantheses, are those given by Lukes, 1975: 561 ff; they indicate either the first publication or the first publication of the cited version. Titles in square brackets are from the editors.

Allcock, John B. 1982. "Emile Durkheim's encounter with pragmatism." *Journal of the History of Sociology* **4**(1): 27–51.
Assoun, Paul-Laurent. 1976. "Durkheim et le socialisme de la chaire." *Revue Française de Science Politique* **26**(5): 957–982.
Barnes, Harry E. 1948. "Herbert Spencer and the Evolutionary Defense of Individualism." In Barnes, Harry E. (ed.) 1948. *An Introduction to the History of Sociology*. Chicago, London: Univ. of Chicago Press, pp. 110–137.
Besnard, Philippe (ed.) 1983. *The Sociological Domain*. The Durkheimians and the founding of French Sociology. Cambridge: Cambridge University Press and Paris: Maison des Sciences de l'Homme.
Birnbbaum, Pierre. 1971. "Préface." In Durkheim, Émile. 1971. *Le socialisme* Sa définition, ses débuts, la doctrine Saint-Simonienne, Introd. by M. Mauss. Paris: PUF.
Bockenförde, Ernst-Wolfgang. 1978. "Organ, Organismus, Organisation, politischer Köper (VII-IX)." *Geschichtliche Grundbegriffe. Historisches Lexikon zur politisch -sozialen Sprache in Deutschland.* Bd. **4**: 561–622.
Boesinger, Ernest. 1980. "Evolutionary Biology in France at the Time of the Evoluntinary Synthesis." In Mayr, Ernst and William B. Provine (eds.) 1980. *The Evolutionary Synthesis. Perspectives on the Unification of Biology*. Cambridge, Mass., London: Harvard University Press, pp. 309–321.
Bourdieu, Peirre. 1992. "Thinking about Limits." In Featherstone, Mike (ed.) 1992. *Cultural Theory and Cultural Change*. London, Newbury Park, New Delhi: Sage, pp. 37–49.
Bourgin, Hubert. 1938. *De Jaurès à Léon Blum*. L'École normale et la politique, Paris: Arthème Fayard.
Bowler, Peter J. 1975. "The Changing meaning of 'evolution'." *Journal of the History of Ideas* **36**: 95–114.
Boyd, Robert and Peter J. Richerson. 1985. *Culture and the Evolutionary Process*. Chicago, London: University of Chicago Press.
Burrow, J. W. 1970. *Evolution and Society: A Study in Victorian Social Theory*. London: Cambridge Univ. Press.
Buss, Leo W. 1987. *The Evolution of Individuality*. Princeton: Princeton University Press.
Byrne, Richard W. and Andrew Whiten. 1988. *Machiavellian Intelligence: Social Expertise and the Evolution of Intellect in Monkeys, Apes, and Humans*. Oxford: Clarendon Press.
Canguilhem, Georges. 1970. *Etudes d'histoire et de philosophie des sciences*. Paris: Librairie Philosophique J. Vrin.
Changeux, Jean-Pierre. 1984. *Der neuronale Mensch*. Wie die Seele funktioniert – die Entdeckungen der neuen Gehirnforschung. Reinbek b. Hamburg: Rowohlt.
Clark, Terry N. 1973. *Prophets and Patrons: The French University and the Emergence of the Social Sciences*. Cambridge, Mass.: Harvard Univ. Press.
Cobban, Alfred. 1978. *A History of Modern France*. Bd. 3: France of the Republics 1871–1962.
Cuvillier, Armand. 1958. "E. Durkheim et le pragmatisme." In Cuvillier, Armand. 1958. *Sociologie et Problémes actuels*. Paris: Vrin. 69–87.
Darwin, Charles. 1963. *Die Entstehung der Arten*. Stuttgart: Reclam.
Deledalle, Gérard. 1959. :Durkheim et Dewey: un double centenaire." *Les Études philosophiques* n.s. **4**: 493–498.

Digeon, Claude. 1959. *Le crise allemande de la pensée française (1870–1914)*. Paris: PUF.
Dohrn-Van Rossum, Gerhard. 1978. "Organ, Organismus, Organisation, politischer Körper (I–VI)." *Geschichtliche Grundbegriffe. Historisches Lexikon zur politisch-sozialen Sprache in Deutschland*. Bd. 4, Stuttgart: Klett, pp. 519–560.
Dumezil, George. 1958. *L'idéologie Tripartite des Indo-Européens*. Brüssel: Latomus.
Dumont, Louis. 1985. "A modified view of our origins: the Christian beginnings of modern individualism." In M. Carrithers, St. Collins and St. Lukes (eds.) 1985. *The Cetegory of the Person. Anthropology, philosophy, history*. Cambridge, London, New York: Cambridge Univ. Press, pp. 93–122.
Durham, William H. 1991. *Coevolution: Genes, Culture and Human Diversity*. Stanford: Stanford Univ. Press.
Durkheim, Émile. 1955: (1955a). (55) *Pragmatisme et sociologie*. Cours inédit prononcé à la Sorbonne en 1913/1914 et restitué par Armand Cuvillier d'après des notes d'étudiants. Preface by A. Cuvillier. Paris: Vrin.
Durkheim, Émile. 1969 (1938a). *L'évolution pédagogique en France*. With an introduction by M. Halbwachs. Paris: PUF, 2nd edn.
Durkheim, Émile. 1971 (1928a). *Le socialisme*. Sa définition, ses débuts, la doctrine Saint-Simonienne. Introd. by M. Mauss. Paris: PUF.
Durkheim, Émile. 1975a. *Textes*. Bd. a: Eléments d'une théorie sociale. Ed. by V. Karady. (Coll. "Le sens commun") Paris: Editions de minuit.
Durkheim, Émile. 1975b (1885). "[Organisation et vie du corps social selon Schaeffle]" (56). In Durkheim, Émile 1975a: 355–377.
Durheim, Émile, 1975c (1887c). "La science positive de la morale en Allemagne." In Durkheim, Émile. 1975a: 267–343.
Durkheim, Émile. 1075d (1900c). "La sociologie et son domaine scientifique." In Durkheim, Émile 1975a: 13–36.
Durkheim, Émile. 1975e (1913a). "[Controverse sur l'influence Allemande et la théorie morale]." In Durkheim, Émile 1975a: 405–407.
Durkheim, Émile. 1975f (1907b). "[Lettre sur l'influence Allemande dans la sociologie Française. Réponse à Simon Déploige]." In Durkheim, Émile 1975a: 402–405.
Durkheim, Émile. 1975g (1895). "L'état actuel des études sociologiques en France." In Durkheim, Émile 1975a: 73–108.
Durkheim, Émile. 1983 (1901c). *Les régles de la méthode sociologique*. Paris: PUF. 21. edition.
Durkheim, Émile. 1985 (1912a). *Les formes élémentaires de la vie religieuse*. Le système totémique en Australie. Paris: PUF. 7. edition.
Durkheim, Émile. 1986 (1902b). *De la division du travail social*. Etude sur l'organisation des sociètès supèrieures. Paris: PUF. 11 edition (referred to as *Division*).
Durkheim, Émile. 1987a. *La Science sociale et l'action*. Ed. and introduced by J.-C. Filloux. (Coll. Le Sociologue) Paris: PUF. 2nd edition.
Durkheim, Émile. 1987b (1888a). "Cours de science sociale. Leçon d'ouverture." In Durkheim, Émile 1987a: 77–110.
Durkheim, Émile. 1987c (1898c). "L'individualisme et les intellectuels." In Durkheim, Émile 1987a: 261–278.
Durkheim, Émile. 1987d (1914a). "Le dualisme de la nature humaine et ses conditions sociales." In Durkheim, Émile. 1987a: 314–332.
Falk, Raphael/Jablonka, Eva. 1994. "Inheritance: Transmission and Development." *Final*

report of the working group 'Biological Foundations of Human Culture', Universität Bielefeld, Zentrum für interdisziplinäre Forschung. (In prep.).
Falkenstein, Adam. 1976. "Die Ur- und Fruhgeschichte des Alten Vorderasien." In *Die Altorientalischen Reiche I. Vom Paläolithikum bis zur Mitte des 2. Jahrtausends.* Ed. by E. Cassin/J. Bottéro/J. Vercoutter. 1976. (Fischer Weltgeschichte. Vol. 2), pp. 13–56.
Filloux, Jean-Claude. 1963. "Durkheimism and Socialism." *The Review* 5(2): 66–85.
Filloux, Jean-Claude. 1976. "Il ne faut pas oublier que je suis fils de rabbin." *Revue Française de Sociologie* 17(2): 259–266.
Filloux, Jean-Claude. 1977. *Durkheim et le socialisme.* Paris/Genf: Droz.
Filloux, Jean-Claude. 1979. "Durkheim et l'organicisme." *Revue Européenne des Sciences Sociales et Cahiers Vilfredo Pareto* 17(47): 135–148.
Gaudemar, Paul de. 1969. "Les Ambiguités de la critique durkheimienne du pragmatisme." *La Pensée* 145: 81–88.
Gephart, Werner. 1982. "Soziologie im Aufbruch. Zur Wechselwirkung von Durkheim, Schäffle, Tönnies und Simmel." *Kölner Zeitschrift für Soziologie und Sozialpsychologie* 34(1): 1–25.
Gigerenzer, Gerd. 1991. "From tools to theories. A heuristic of discovery in psychology." *Psychological Review* 98 (forthcoming).
Gottlieb, Alma. 1979. "The social theories of Fustel and Durkheim: toward an analysis of a neglected relationship." *Anthropology* 3(1–2): 139–153.
Gould, Stephen J. 1977. *Ontogeny and Phylogeny.* Cambridge, Mass.: Belknap Press of Harvard Univ. Press.
Greive, Hermann. 1982. *Die Juden.* Grundzüge ihrer Geschichte im mittelalterlichen und neuzeitlichen Europa. Darmstadt: Wissenschaftliche Bunchgesellschaft.
Guitton, Jean. 1968. *Regards sur la pensée française: 1870–1940.* Leçons de captivtié. Paris: Beauchesne.
Günther, Gotthard. 1976. *Beiträge zur Grundlegung einer operationsfähigen Logik.* 2 vols. Hamburg: Meiner.
Hallpke, Christopher R. 1986. *The Principles of Social Evolution.* Oxford: Clarendon Press.
Hejl, Peter M. 1990. "Self-regulation in social systems." In Krohn, Wolfgang, Günter Küppers and Helga Nowotny (eds.) 1990. *Selforganization – Protrait of a Scientific Revolution.* (Sociology of Sciences. A Yearbook, Vol. 14) Dordrecht, Boston, London: Kluwer Acad. Publ. pp. 114–127.
Hejl, Peter M. 1992a. "Culture as a Network of Socially Constructed Realities." In Fokkema, Douwe and Ann Rigney (eds.) 1992. *Trends in Cultural Participation in Europe since the Middle Ages.* Amsterdam, Philadelphia: J. Benjamins (In print).
Hejl, Peter M. 1992b. "Die zwei Seiten der Eigengesetzlichkeit. Zur Konstruktion natürlicher Sozialsysteme und dem Problem ihrer Regelung." In Schmidt, Siegfried J. (ed.) 1992. *Kognition und Gesellschaft.* Neue Wege des Konstruktivismus. Frankfurt a.M.: Suhrkamp, pp. 167–213.
Hejl, Peter M. 1992c. "Selbstorganisation und Emergenz in sozialen Systemen." In Krohn, Wolfgang and Günter Küppers (eds.) 1992. *Exergenz: die Entstehung von Ordnung, Organisation und Bedeutung.* Frankfurt a.M.: Suhrkamp, pp. 269–292.
Hejl, Peter M. 1993. "Soziale Konstruktion von Wirklichkeit." In Merten, Klaus, Siegfried J. Schmidt and Siegfried Weischenberg (eds.) 1993. *Medien und Kommunikation.* (in print).
Joas, Hans. 1985. "Durkheim und der Pragmatismus. Bewußtseinspsychologie und die

soziale Konstitution der Kategorien." *Kölner Zeitschrift für Soziologie und Sozialpsychologie* **37**: 411–430.
Johnson, Allen W. and Timothy Earle. 1987. *The Evolution of Human Societies. From Foraging Group to Agrarian State.* Stanford, Cal.: Standford University Press.
Karady, Victor. 1972. "Normaliens et autres enseignants de la Belle Époque: note sur l'origine sociale et la réussite dans une profession intellectuelle." *Revue Française de Sociologie* **13**: 35–38.
Karady, Victor. 1979. "Strategies de réuissite et modes de faire-valoir de la sociologie chez les durkheimiens." *Revue Française de Sociologie* **20**(1): 49–82.
Lacroix, Bernard and Béatrice Landerer. 1972. "Durkheim, Sismondi et les Socialistes de la chaire." *Année Sociologique* **23**: 159–182.
Le Goff, Jacques, Roger Chartier and Jacques Revel (eds.) 1978. *La Nouvelle Histoire.* Paris: CEPL.
Lenoir, Timothy. 1982. *The Strategy of Life. Teleology and Mechanics in Ninetheenth Century German Biology.* Dordrecht, Boston, London: Reidel.
Limoges, Camille. 1980. "A Second Glance at Evolutionary Biology in France." In Mayr, Ernst and William B. Provine (eds.) 1980. *The Evolutionary Synthesis.* Perspectives on the Unification of Biology. Cambridge, Mass.; London: Harvard University Press, pp. 322–328.
Lukes, Steven. 1975. *Émile Durkheim.* His Life and Work. A Historical and Critical Study. Harmondsworth: Penguin.
Macfarlane, Alan. 1978. *The Origins of English Individualism: The Family, Property and Social Transition.* Oxford: Basil Blackwell.
Marrus, Michael R. 1972. *Les Juifs de France à l'époque de l'affaire de Dreyfus, Assimilation à l'épreuve.* Preface by Pierre Vidal-Naquet. Paris: Calman Lavy.
Masters, Roger D. 1989. *The Nature of Politics.* New Haven, London: Yale University Press.
Mayeur, Jean-Marie. 1973. *Les débuts de la IIIe République. 1871–1898.* Paris: Seuil.
Mayeur, Jean-Marie. 1984. *La vie politique sous la Troisième République 1870–1940.* Paris: Éditions du Seuil.
Meyer, Michel. 1979. *Le Mal Franco-Allemand.* Paris: Denoel.
Nandan, Y. 1977. *The Durkheimian School.* Connecticut: Greenwood Press, Westport.
Nicolet, Claude. 1982. *L'idée republicaine en France (1789–1924).* Essai d'histoire critique. Paris: Gallimard.
Oexle, Otto G. 1988. "Die funktionale Dreiteilung als Deutungsschema der sozialen Wirklichkeit in der ständischen Gesellschaft des Mittelalters." *Ständische Gesellschaft und soziale Mobilität.* Ed. by W. Schulze. (Schriften des Historischen Kollegs. Kolloquien 12) München: Oldenburg, pp. 19–51.
Parsons, Talcott. 1968. *The Structure of Social Action: A Study in Social Theory with Special References to a Group of Recent European Writers.* Vol. 1. New York: Free Press. 1. Paperback Edition.
Prendergast, Christopher. 1983/84. "The Impact of Fustel de Coulanges' 'La Cité Antique' on Durkheim's Theories of social morphology and social solidarity." *Humboldt Journal of Social Relations* **11**(1): 53–73.
Rieppel, Olivier. 1989. *Unterwegs zum Anfang.* Geschichte und Konsequenzen der Evolutionstheorie. Zürich, München: Artemis.

Rumney, J. 1934. *Herbert Spencer's Scoiology: A Study in the History of Social Theory.* To which is appended a bibliography of Spencer and his work. With a preface by Morris Ginsberg. London: Williams and Norgate.
Sieferle, Rolf P. 1984. *Fortschrittsfeinde? Opposition gegen Technik und Industrie von der Romantik bis zur Gegenwart.* München: Beck.
Smuts, Jan C. 1926. *Holism and Evolution.* London: McMillan.
Spencer, Herbert. 1893. *The Principles of Sociology.* Part 1. London, Edinburgh: Williams and Norgate.
Spencer, Herbert. 1896. *Einleitung in das Studium der Sociologie.* 2. Teil. Leipzig: Brockhaus. 2. Edition with a postscript. First published 1873 as "The Study of Sciology." New York: Appleton.
Spencer, Herbert. 1967 [1876, 1882]. *The Evolution of Society.* Selections from Herbert Spencer's "Principles of Sociology." Edited and with an Introduction by Robert L. Carneiro. Chicago, London: University of Chicago Press. First published in "Principles of Sociology", part 1 (1876) and part 2 (1882).
Spencer, Herbert. 1972 [1851–1908]. *Herbert Spencer on Social Evolution.* Selected Writings. Edited and with an Introduction by J. D. Y. Peel. Chicago, London: University of Chicago Press.
Stachowiak, Herbert. 1973. *Allgemeine Modelltheorie.* Wien, New York: Springer.
Tenbruck, Friedrich. 1992. "Was war der Kulturvergleich, ehe es den Kulturvergleich gab?" In *Zwischen den Kulturen?*, Soziale Welt. Sonderband 8. Ed. by J. Matthes. Göttingen: O. Schwartz, pp, 13–35.
Thompson, E. P. 1978. *The Making of the English Working Class.* Harmondsworth: Pelican. 2nd Edn., 7th reprint.
Thornhill, Nancy W. 1991. "An evolutionary analysis of rules regulating human inbreeding and marriage." *Behavioral and Brain Sciences* **14**: 247–293.
Tiryakian, Edward. 1979. "L'École Durkheimienne à la recherche de la scoiété perdue: la sociologie naissante et son milieu culturel." *Cashiers internationaux de Sociologie* **66**: 97–114.
Tocanne, Bernard. 1978. *L'idée de nature en France dans la seconde moité du XVIIe siècle.* Contribution à l'histoire de la pensée classique. Paris: Klincksieck.
Tooby, John and Leda Cosmides. 1992. "The Psychological Foundations of Culture." In Barkow, Jerome H., Leda Cosmides and John Tooby (eds.) 1992. *The Adapted Mind: Evolutionary Psychology and the Generation of Culture.* New York: Oxford University Press, pp. 19–135.
Turner, Bryan S. 1992. "Ideologies and Utopia in the Formation of an Intelligentsia: Reflections on the English Cultural Conduit." In Featherstone, Mike (ed.) 1992. *Cultural Theory and Cultural Change.* London, Newbury Park, New Delhi: Sage, pp. 183–210.
Vovelle, Michel. 1985. *Idéologies et Mentalités.* Paris: Maspero.
Weisz, George. 1979. "L'ideologie républicaine et les sciences sociales. les durkheimiens et la chair d'histoire d'économie sociale à la Sorbonne." *Revue française de sociologie* **XX**: 83–112.
Winock, Michel. 1982. *Nationalisme, antisémitisme et fascisme en France.* Paris: Éditions du Seuil.

HERBERT SPENCER:
BIOLOGY, SOCIOLOGY, AND COSMIC EVOLUTION

ANTONELLO LA VERGATA
University of Calabrio

Transfer or Common Origin?

This paper is a case study in the constitutive role of two metaphors, the social organism and the struggle for life, in the thought of Herbert Spencer. It deals with the relationship between biology and the social sciences; but only in a limited sense is it a study of the transfer of one idea ("the social organism") from biology to the social sciences, and of the transfer of another idea ("struggle or competition") from the social sciences to biology and back again. It is, rather, a study of the *common* roots of these images in Spencer's thought. Like any story, this one, too, has a moral: namely, not everything that looks like a metaphor is really a metaphor. What are metaphors to us or to one author are not necessarily so to another; in some cases, metaphors can perform their functions (illustrative, heuristic, constitutive, affective, persuasive, didactic, argumentative, etc.) exactly because the author who uses them presents them as *literal descriptions*, and believes them to be. Finally, to the numerous functions of metaphors recognized so far we should perhaps add another one: the *moral* function, as the role played by a metaphor in a discourse is often better explained by its moral context than its conceptual context. All this can seem obscure, but it will become clear if the reader is patient enough to read this paper through.

It is a given that the genesis of Darwin's idea of natural selection was influenced by his reading of Malthus' essay on population. Ever since the last century, many people have gone so far as to say that Darwin "derived" his idea of natural selection, and bodily took that of struggle for existence, from that book. To many, Darwin's debt to Malthus became a debt to political economy itself. So, whether or not one approved of this transference, there seemed to be nothing extraordinary in the fact that the opposite took place too. Social Darwinism was therefore seen variously as a revenge of history, the payment of a debt, or just one further episode in the continuous exchange between different fields and disciplines. The struggle for existence offers therefore a unique opportunity for an object lesson in the transfer of an image from one field to another, and vice versa. Critical evaluations of this transfer and retransfer have ranged from utter disapproval of what has been seen as an undue extrapolation to praise for all those who, after Darwin's model, had an interdisciplinary mind that enabled them to cross boundaries between different fields and to achieve a mutual fertilization of them. From this point of view, Social Darwinism appeared an inevitable consequence of Darwinism, all the more so because Darwin himself has been recognized as something of a Social Darwinist[1] (and possibly the only one who was a genuine Darwinian).

Fortunately the idea has spread in recent times that the label "Social Darwinism" refers to a much more complicated thing than it was previously thought. It is now more frequently recognized that there were many varieties of Social Darwinism, and the very concepts of "influence," "transference," and the like have been submitted to closer scrutiny and criticism. Furthermore, thanks to recent developments in Darwin studies, some attention has been paid to Darwin's (and, to a lesser extent, to "Darwinian") metaphors. It has become clear that the metaphor of the "struggle for existence" is very complex. Indeed, I would describe it as a bundle of metaphors.[2] Much Social Darwinism was based on mistaking Darwin's phrase for a literal description. Many Social Darwinists were in fact emphasizing one aspect of this "polymetaphor" (if I am allowed this term) to the exclusion of others (for instance, intergroup to the exclusion of intragroup struggle); in other words, they were presenting their own variety of Social Darwinism, which was only one of the many that were historically possible, as the only legitimate application of Darwin's ideas. Whether they emphasized competition or cooperation, whether they supported laissez-faire or social imperialism, they were spoiling and ossifying the richness, variety, and flexibility of Darwin's concept. The present paper, however, is not devoted to giving a résumé of what I have written elsewhere.[3] Rather, it focuses on another locus classicus of the discussions on social Darwinian imagery. One might legitimately ask, in fact, whether this revision of Social Darwinism has extended to the man who has been seen as the prototype of the Social Darwinist.

It has, but only to a limited extent. Ever since the last century, Herbert Spencer (1820–1903) has been considered (with few, but important, exceptions) the man who enlarged the sphere of application of Darwin's theories by giving full philosophical expression to the new view of nature they contained. To give one example only, the French literary critic Hyppolite Taine wrote that Spencer took the theory of evolution out of the special field where it had been enunciated, i.e., botany and zoology, and extended it to "the whole of reality;" Spencer was "the Darwin of philosophy," and Spencer's ideas "bolder applications of Darwin's law." Evaluations of this kind were echoed in Richard Hofstadter's class *Social Darwinism in American Thought*. By applying the theory of evolution by natural selection to society, Spencer and the Social Darwinists vindicated the origins of that theory, as the "survival of the fittest" was a biological generalization of the competition that prevailed in capitalist society and was approved of by laissez-faire economists; Darwinism itself was "a derivation from political economy." The same judgment occurs even in those who have more recently stressed the originality of Spencer's doctrine of evolution and have rightly argued that as early as 1850 he put forth most of the ideas he is supposed to have derived from Darwin. The American anthropologist Marvin Harris goes so far as to turn the traditional image on its head and to maintain that it is Spencerianism (by which he means the doctrine of progress through struggle) that lay at the basis of Darwinism, as "Darwin's principles were an application of social-science concepts to biology."[4]

In judgments such as these (and the same holds true of those made by his contemporaries), the emphasis is on the word "application." Now, this concept, too, is a problematic one, and not quite the most appropriate to study the origins and role of metaphors, as these are not always the result of a mere application or transfer. In Spencer's case the metaphor of "struggle" and of the "social organism" played neither a simply illustrative nor a heuristic, but a *constitutive* role, and so much so that one could safely go so far as to doubt that they were metaphors at all. (Quite apart from the fact that Spencer would have been very sensitive on the point of being supposed, even by well-meaning commentators and supporters, merely to "apply" someone else's ideas. He was a hypersensitive man and not quite ready to acknowledge intellectual debts, but in this case he would have been absolutely right.)

In this paper I shall argue that Spencer did not anticipate the concept of natural selection; his biology owes very little to Darwin; his view of organic evolution is different from that of Darwin; and that these differences are due to scientific, ideological, and political components of Spencer's thought, which leave their indelible stamp on his attempt at an all-embracing philosophy. Although Spencer has been considered a sort of prototype of the Social Darwinist, his alleged Social Darwinism was hardly Darwinian at all, since the roots of his view of evolution lay in those very elements of his thought that prevented him from being a Darwinian. Finally, it was physics, not biology, that represented the ultimate model of scientific certainty in Spencer's eyes. He tried to reduce biological explanations to physical terms. Therefore, to focus only on the relationship between his sociology and his biology is to deal with the problem at a somewhat superficial level.

Spencer has also been considered an apostle of free-for-all competition, and extolled or despised accordingly. But – paradoxical though it may seem – to many others he was the one who provided some of the best arguments for solidarity and altruism, and among his fans there were not few socialists,[5] which of course infuriated him (This frequently happens to philosophers with cosmic ambitions). Spencer was a philosopher for all seasons and all tastes.

He has also been considered the philosopher of the "social organism," a metaphor that does not seem to fit well with that of the "struggle for existence," for the latter stresses contrast, while the former stresses cooperation through differentiation and "division of labor" (a metaphor taken from economy to biology and then brought back to social science). The contrast between the two was pointed out very concisely and neatly by Thomas Henry Huxley, "Darwin's bull-dog":

Suppose that, in accordance with this view, each muscle were to maintain that the nervous system had no right to interfere with its contraction, except to prevent it from hindering the contraction of another muscle; or each gland, that it had a right to secrete so long as its secretion interfered with no other; suppose every separate cell left free to follow his own "interest," and *laissez-faire* lord of all, what would become of the body physiological?[6]

Strange though it may seem, to Spencer's mind this criticism was baseless. To show why it was so, in what follows I shall concentrate on Spencer's early work; this will enable us to see what the relationship between the metaphor of "struggle" and that of the "social organism" was from the beginning. I shall try to demonstrate that these metaphors were not simply an affair between sociology and biology, but the result of a more complicated web of influences and demands, in which moral and political components were far from unimportant. I shall have to discuss the context rather than the content of these metaphors. Finally, I shall argue that even in their respective fields of origin, Spencer's images of "struggle" and of the "social organism" were not unproblematical, literal, simple, ultimate elements of discourse; they were themselves, to use his terms, "derivative, not fundamental." This applies in particular to the image of "struggle."

The organic analogy and the "struggle for life" (or the Spencerian idea of competition and elimination of the unfit that was later to be identified with Darwin's "struggle for life") made their first appearance in the same work, in 1851. More important, they were used for the same basic purpose: a plea for noninterference with the spontaneous development of society, which is gradual, unalterable, and self-adjusting. Although the organic analogy was not developed as much as the idea of competition was, it is evident that they supported each other and were complementary. Spencer's 1851 work, therefore, already contains the answer to Huxley's subsequent objection. It was only later on, as he developed the two metaphors separately, that their coexistence could become a problem and a source of tension, although he does not seem to have realized this. What was originally united developed in different directions. This happened not only in Spencer's mind, but in the sociological literature as well, where the phenomenon was obviously more evident. Some sociologists stuck to the organic analogy, others to the idea of competition. When syntheses of the two were achieved, they produced very different results from what Spencer had expected. To give one instance only, Benjamin Kidd used the organic analogy in his fortunate *Social Evolution* (1894) as an argument for state intervention in social and economic questions, which was the opposite of what Spencer wanted. Kidd was faced with a problem that, as we shall see, did not exist for Spencer, that of reconciling the right of every individual to compete with others without limits, but "on a footing of equality," and the necessity of subordinating individual claims to the higher interests of the community. Far from sharing Spencer's faith in the spontaneous coincidence of the two, Kidd pointed to religion as the factor that ensured the subordination of the individual to the state and to the demands of general progress. Here, again, something less Spencerian could hardly be imagined. Most later writers on the organic analogy diverged from Spencer's individualism in their theory of the state, and tended towards the contrary notion that in the social organism, units exist for the good of the whole.[7] And here again socialists could turn Spencer's imagery and arguments against Spencerianism. By implying the subordination of the individual to the needs of the whole, of the peripheral to the central, the metaphor of the "social organism" could be seen to undercut natural-law arguments against social reform carried on from above. As Beatrice Webb, one of Spencer's most notable socialist followers, suggested,[8] that metaphor implied that the role of the state could be reconciled on the model of the brain, that is, of the organ that coordinates the integration of individuals into the social organism.

Some of the conclusions concerning the role of biological metaphors in Spencer's thought which will be reached in this paper can be anticipated now in the shape of a methodological caveat. From the beginning, Spencer's social and political arguments were indissolubly mixed with biological themes. However, this does not exactly mean that explanatory models and a language were being borrowed from biology as an established discipline. The fact is that Spencer saw the problem he tackled as common to social and biological sciences and thought it natural to resort to common conceptual tools in order to solve them. From this point of view, biology and sociology were not so sharply separated from each other as we consider them. And then there was Spencer's obsessive tendency to reduce discussion of every order of reality to ultimate terms which are supposedly not typical of any special discipline. If we fail to recognize this, Spencer appears to be one who cavalierly transplanted ideas, expressions, problems, and solutions from one field to another. We do not understand much of his evolutionism if we project our ideas of the relationship between present-day sciences onto it, that is, if we evaluate his views in the light of distinctions that did not exist for him or for most of this contemporaries. Many of those that are metaphors to us were not so to Spencer, or were so only in a limited sense. Talking of metaphors presupposes a clear-cut distinction between established disciplines. There can be no metaphors in no-man's-land, in overlapping areas, or in borderline districts.

Human Nature

In his first book, *Social Statics, or the Conditions Essential to Human Happiness Specified, and the First of Them Developed* (1851), Spencer attempted to correct, from inside, the political attitude of the Utilitarians and in particular the Benthamites. It was a typical product of the thinking that Spencer had developed in the course of his association with the radical clubs of his native Derby, and that he had first aired in his essays, *On the Proper Sphere of Government*, published in the radical organ *The Nonconformist* in 1842.

As the subtitle indicates, the central theme of *Social Statics* is happiness; the directly political commitment is evident and the approach is moralistic, since Spencer tackles from the outset the problem of finding a rule of conduct. The general attitude he assumes is marked by an extreme laissez-faire. Every piece of state intervention is declared harmful, whether it take the form of poor laws, national education, sanitary and medical services, postal and banking arrangements, government-sponsored colonization, commercial law, or any other. The whole book is a defence of the rights of the individual, the rights of women and children, the right of complete liberty of expression and opinion, the rights of property, including intellectual property, even the right to ignore the state if one does not recognize it. The "first principle" on which the whole argument is based is that "every man has freedom to do all that he wills, provided he infringes not the equal freedom of any other man."[9]

The free expression of the capacities of each, the free fulfillment of individual activities will in time lead to a growing harmony of occupations and satisfactions. What is happiness if not "a certain state of consciousness?" As such, it is the outcome of certain sensations, that is, of modifications of consciousness by external influences. These "affections of the consciousness" are filtered, as it were, by those receptive and reactive powers known as "faculties." The road to happiness starts from the stimulus of desire and passes through its due satisfaction by means of the due exercise of the faculties: happiness is "a gratified state of all the faculties. The gratification of a faculty is produced by its exercise." Spencer's man is a complex of functional relations that vary in response to varying circumstances, "a congeries of faculties, qualifying him for surrounding conditions."[10]

The whole of Spencer's argument rests on this concept of human nature. It is not the abstract and static conception from which the Utilitarians "deduced" moral corollaries; human nature is seen as mutable, incapable of creating once and for all a "universal standard of greatest happiness." Happiness, as we have seen, consists in the exercise of the faculties, but "to be agreeable that exercise must be proportionate to the power of the faculty; if it is insufficient discontent arises, and its excess produces weariness." But the world does not contain two individuals who possess the same "combination of elements," the same "balance of desires," and so the conditions of happiness "must vary indefinitely," not only from epoch to epoch and from place to place but also from individual to individual. It follows that the Utilitarians' "cannon of social morality," the "greatest happiness for the greatest number," loses all meaning, since, "if humanity is indefinitely variable, it cannot be used as a gauge for testing moral truth."[11]

Nor can the rules of conduct be established by the state. The state cannot have this authority: it is a transitory institution, necessary only as long as the majority of men need to be restrained from harming others. It is useful only while civilization is in that phase of development in which men still retain some of those characteristics of barbarity that from the very dawn of human society have made a repressive institution necessary. Prisons, police, courts, the whole administrative and governmental apparatus exists because evil exists. The state is a "necessary evil." As civilization advances, the pressure exerted by government diminishes. Spencer here expounds ideas that were to play a leading part in his thought from this point right up to *The Man* versus *the State* (1884) and the *Principles of Sociology* (1876–97); in the latter he compared "militant" societies, ruled by a rigid, stifling apparatus of government, with industrial societies, in which relations between individuals have less and less need of regulation by norms imposed by authority. But above all the state represents a harmful interference with the free play of individual faculties that ensures true happiness. By intervening, with the purpose of providing for the individual's needs, it prevents the faculties from being exerted and offers paltry palliatives pregnant with harmful consequences, since it separates the citizen from "the stem discipline of necessity."

The complete man is the self-sufficing man – the man who is in every point fitted to his circumstances – the man in whom there are desires corresponding not only to all the acts which are immediately advantageous, but to those which are remotely so. Evidently, one who is thus rightly constituted cannot be helped. To do anything for him by some artificial agency, is to supersede certain of his powers – is to leave them unexercised, and therefore to diminish his happiness. To healthily-developed citizens, therefore, state aid is doubly detrimental. It injures them both by what it takes and by what it does. By the revenues required to support its agencies it absorbs the means on which certain of the faculties depend for their exercise; and by the agencies themselves it shuts out other faculties from their spheres of action.[12]

The state should confine itself to safeguarding the freedom of the individual, protecting it from external dangers; if it goes beyond this, it abuses its power and becomes an aggressor. Furthermore, no institution can modify the nature of individuals; the opposite is more likely: it is human character that expresses itself in some institutions rather than in others. There can be no correct moral doctrine without a correct doctrine of society. But the nature of a society is determined by the nature of the individuals of which it is composed.

The characteristics exhibited by beings in the associated state cannot arise from accident of combination, but must be the consequences of certain inherent properties of the beings themselves. True, the gathering together may call out these characteristics; it may make manifest what was before dormant; it may afford the opportunity for undeveloped peculiarities to appear; but it evidently does not create them. No phenomenon can be presented by a corporate body, but what there is a pre-existing capacity in its individual members for producing.[13]

It is curious how, in saying this, the future philosopher of the analogy between society and the organism excludes every holistic consideration and prefers physical to biological similarities. He talks of atoms, he mentions Dalton, and says that, just as the gravity of a body is the sum of the gravity of its parts, so every social phenomenon has its origin in some property of the individual and so the "moral forces on which social equilibrium depends reside in the social atom – man." Hence, "if we would know the nature of those forces, and the laws of that equilibrium, we must look for them in the human constitution."[14] The analysis of the human constitution takes on, therefore, a fundamental importance.

Spencer knows that his arguments will have much more weight if they are founded on a demonstration that man is made like this and not in the way his adversaries believe. "The concept of human nature," it has been justly said, "sets limits to political possibilities."[15] Spencer criticizes the Utilitarians because their conception of human nature is unsatisfactory, not because it is erroneous in itself to appeal to human nature. The fact is that recourse to "the human constitution" is only one aspect of a more general appeal to nature and to inviolable natural laws for the purpose of tying an opinion to something solid and safe from human error. What is Spencer doing, for example, in order to deny the utility of state intervention, if not showing that it damages the "natural" means by which the individual achieves happiness? And his defense of laissez-faire was founded on the conviction that "human nature is difficult to manipulate."

In the attempt to define the "human constitution" in order to use it as the "natural" basis of his moral and political doctrines, Spencer, even before he found such a basis in the doctrine of cosmic evolution, turned to the findings and methods of a science that was then still in vogue: phrenology. R. M. Young has shown that – although Spencer never used the word "phrenology" and although "there is no obvious direct textual link" between his formulations and those of Gall's, Spurzheim's and Combe's – it is to phrenology that he owes, for instance, the idea that man is "a congeries of faculties," or the idea that organs and faculties develop with use and atrophy in disuse.[16] The latter is a very important idea: it was destined to characterize his later conception of organic evolution, and it would be a mistake to think it derived directly from Lamarck. Phrenology accounted for individual differences by tracing them to the diverse development of the original faculties with which each individual is to a greater or lesser extent endowed. For example, the plasticity of this endowment of basic faculties helped to explain moral feelings no longer as the result of simple associations of sensations but as the result of modification, owing to the pressures of external conditions, of original mental dispositions. On the other hand, the complexity of mental and moral manifestations could be traced to a single source. There was, of course, a danger in this of innatism, and it is on this very point that the *Principles of Psychology* (1855) represent a turning point in Spencer's development. In this work Spencer was to transform Gall's "original dispositions" into the psycho-physical endowment, which the individual inherits from the accumulated experience stored in the nervous fibers of a long chain of generations. Moreover, by locating the "moral sense" in the brain, it was possible to give a material basis, as it were, to morality and to all talk of "the nature of man." Abstract arguments about the "essence" of man gave way to analyses of the variable relations between the faculties and surrounding conditions. Physical man and moral man came to be seen as homogeneous aspects of a single physico-biological complex, the object of a true "moral physiology." Precisely because they are closely related to man's physical make-up, ethical laws can aspire to the same "constancy and universality" as physical laws.

Man's social behavior comes from the development of the faculties of the individual and is therefore ruled by laws of causation similar to those that operate in every other sector of the cosmos.[17]

By pointing out the laws necessary to mental processes, phrenology earned the approval of people who required science to demonstrate the deterministic character of human actions. J. D. Y. Peel has shown that such people were intellectual forces that currently held sway and significantly, derived from a Calvinistic tradition that went back to Jonathan Edwards and was represented by Joseph Priestley, David Hartley, Harriet Martineau, Henry Thomas Buckle, and the industrialist Charles Bray, an active supporter of liberalism. Among the members of this movement, which is important for an understanding of the context in which Spencer worked out his doctrine, Calvinistic necessitarianism did not mean fatalism but optimism, certainty of the outcome of history. The possibility of giving morality a scientific basis, after the crisis in traditional values marked by the industrial revolution, transformed the deterministic immanentism of these "non-conformists" into a "secular *certitudo salutis.*"[18] It is significant that the means of achieving a new rationalization of history and of social events were sought for, by Spencer as by others, first of all in phrenology and then, when that was abandoned, in cosmic evolution. Social change was the result of the operation of uniform, verifiable laws. Henry Buckle's *History of Civilization in England* (1857–1861) insists as much as the *Letters on the Laws of Man's Nature and Development* (1851) by H. G. Atkinson and Harriet Martineau, on the absolute determinism of the actions of individuals and society.[19] The idea of social development as gradual, slow organic growth, in harmony with the laws of nature, found one of its most effective expositions in the "address to working men" by the radical Felix Holt, hero of George Eliot's political novel (1866). He appealed to moral tension, good sense, concreteness and moderation simultaneously. If the workers wanted to improve their condition, they must act in such a way as not to violate "the gradual action of steady causes" which characterizes natural processes, and must have faith in the "wonderful slow-working system of things." They must therefore struggle to preserve public order: in short, they must obey nature.

. . . Taking the world as it is – and this is one way we must take it when we want to find out how it can be improved [. . .] The nature of things in this world has been determined for us beforehand [. . .] But now, for our own part, we have seriously to consider this outside wisdom which lies in the supreme unalterable nature of things, and watch to give it a home within us and obey it.

The reference to "changing conditions in a struggling world" paled before this reassuring vision. For Felix Holt, as for Spencer, the crucial concept in politics was the nature of man: "We are dealing with human nature in its entirety, and with nothing else."[20]

The search for a scientific basis for the principles of conduct, Spencer was one day to affirm, had been the ultimate aim of his work as a philosopher; the *Principles of Ethics* (1879–93) were to be regarded as crowning the immense labor of the "Synthetic Philosophy."[21] In *Social Statics* this "scientific ethics" is outlined by reinterpreting the doctrine of the moral sense reexpressed in the terms of phrenology. Although he keeps the Utilitarian criterion of pleasure and pain, Spencer revalues the "moral sense" which Bentham had condemned as "an anarchic and capricious principle, founded solely upon internal and peculiar feelings." The task was now to work out a system of ethics derived from the fundamental axiom (the principle of the greatest freedom) originating in the moral sense. Since the moral sense by itself cannot solve every moral problem it confronts, the argument shifts to the environmental conditions in which the original dispositions unfold. The history of man's manifestations is the history of the interaction between his faculties and his surroundings. In *Social Statics*, Spencer wavers between the two poles, original faculties and the influence of the environment, but environmentalism was later to become a fundamental feature of his conception of evolution.

Progress and Adaptation

The word "evolution" never occurs in *Social Statics*. The doctrine of organic evolution, although it was later to be worked out as an extension of some of the ideas contained in *Social Statics*, was to amplify and integrate a preexisting vision: Spencer's evolutionism is by its nature and origin essentially philosophical, not biological.

In the organic world the cause of change continues to lie in an "essential principle of life," the tendency to adapt progressively to circumstances. On this universal law Spencer bases his moral and political optimism as much as his vision of the cultural and social development of humanity. The goal of humanity's march will be the complete adaptation of the individual to the "social state." Trusting in the perfectibility of man, Spencer affirms that happiness is just around the corner, the second if not the first. He shows that his doctrines conform to the direction in which things are developing and finds his political faith sanctioned by the inevitable course of nature. His evolutionism emerges from a moral and political context.

The basic concept of adaption is introduced in a chapter of *Social Statics* in which Spencer confronts the problem of evil. Both physical evil and the anomalous moral behavior which is the object of what he calls "moral pathology" are no more than special cases of a problem to which the solution seems to him very simple: evil is something relative and transitory. In enunciating the doctrine of the progressive "evanescence of evil," Spencer enlarges the meaning of the term "evil" to such generality that it loses its most specifically moral connotations. Evil is every form of "non-adaptation of constitution to conditions;" the general process of evolution is a progressive shrinking of the sphere of evil. The moral problem thus becomes diluted into a cosmic philosophy, and physical and biological change assume ethical valence. From this point of view, morality and biology melt into one another. Just as a plant dies if it is deprived of light or taken to a climate in which "the harmony between its organization and its circumstances" is destroyed, so every kind of human suffering, whether physical or mental, is due to man's finding himself in situations in which his faculties do not make him fit. "No matter what the special nature of the evil, it is invariably referable to the one generic cause – want of congruity between the faculties and their spheres of action."[22] This definition confirms the influence of phrenology and is so all-inclusive that it embraces the extinction of an animal species and the damage resulting from tobacco abuse. Nevertheless, if evil is the temporary residuum of the incessant process of adaptation to circumstance, the extraordinary adaptive capacity of all that lives authorizes the rosiest hopes.

Equally true is that evil perpetually tends to disappear. In virtue of an essential principle of life, this non-adaptation of an organism to its conditions is ever being rectified; and modification of one or both, continues until the adaptation is complete. Whatever possesses vitality, from the elementary cell up to man himself, inclusive, obeys this law.[23]

Progress is the gradual attainment of a balance between the faculties and the environment, the progressive disappearance of what is excessive and what is wanting in the development of the faculties. In the ideal man, every faculty will be exactly commensurate with the demands of circumstance. That such a stage will be reached appears to Spencer "logically certain." He has found a literally physiological reason for believing in the perfectibility of man. Not only is progress necessary, it leads in the direction desired.

Progress, therefore, is not an accident, but a necessity. Instead of civilization being artificial, it is a part of nature; all of a piece with the development of the embryo or the unfolding of a flower. The modifications mankind have undergone, and are still undergoing, result from a law underlying the whole organic creation; and provided the human race continues, and the constitution of things remains the same, those modifications must end in completeness. As surely as the tree becomes bulky when its stands alone, and slender if one of a group; as surely as the same creature assumes the different forms of cart-horse and race-horse, according as its habits demand strength or speed; as surely as a blacksmith's arm grows large, and the skin of a labourer's hand thick, as surely as the eye tends to become long-sighted in the sailor, and short-sighted in the student; as surely as a clerk acquires rapidity in writing and calculation; as surely as the musician learns to detect an error of a semitone amidst what seems to others a very babel of sounds; as surely as a passion grows by indulgence and diminishes when restrained; as surely as a disregarded conscience becomes inert, and one that is obeyed active; as surely as there is any efficacy in educational culture, or any meaning in such terms as habit, custom, practice; – so surely must the human faculties be moulded into complete fitness for the social state; so surely must the things we call evil and immorality disappear; so surely must man become perfect.[24]

Notice how Spencer, under the influence of phrenology, means change in a Lamarckian sense: the blacksmith's arm develops in the same way as the giraffe's neck. Man is modified *directly* by circumstances.

Spencer's conception of social development is still firmly rooted in his conception of natural processes; in a continuous process of adaptation, the individual and society are destined to reach ever higher and higher states of equilibrium until they reach perfection. We have nature to thank for this. Spencer passionately avows his faith in the universal tendency of nature towards perfection. The passage that follows is worth quoting in full. The speaker feels himself to be the prophet of emerging intellectual powers and believes that history and nature are on his side; his words express a naturalist lay version of the old theodicy, the more authoritative for being based on an appeal to science. It cannot be said too emphatically how deeply the physico-moral necessitarianism that animates *Social Statics* conditions the whole of Spencer's subsequent conception of evolution. Whoever can interpret the "great laws of existence," Spencer writes,

> no longer regarding the mere outsides of things, has learned to look for the secret forces by which they are upheld. After patient study, this chaos of phenomena into the midst of which he was born has begun to generalize itself to him; and where there seemed nothing but confusion, he can now discern the dim outlines of a gigantic plan. No accidents, no chance; but everywhere order and completeness. One by one exceptions vanish, and all becomes systematic. Suddenly what had appeared an anomaly answers to some intenser thought, exhibits polarity, and ranges itself along with kindred facts. Throughout he finds the same vital principles, ever in action, ever successful, and embracing the minutest details. Growth is unceasing; and though slow, all powerful: showing itself here in some rapidly-developing outlines; and there, where the necessity is less, exhibiting only the fibrils of incipient organization. Irresistible as it is subtle, he sees in the worker of these changes, a power that bears onwards peoples and governments regardless of their theories, and schemes, and prejudices – a power which sucks the life out of their lauded institutions, shrivels up their state-parchments with a breath, paralyzes long-venerated authorities, obliterates the most deeply-graven laws, makes statesmen recant and puts prophets to the blush, buries cherished customs, shelves precedents, and which, before men are yet conscious of the fact, has wrought a revolution in all things, and filled the world with a higher life. Always towards perfection is the mighty movement – towards a complete development and a more unmixed good; subordinating in its universality all petty irregularities and fallings back, as the curvature of the earth subordinates mountains and valleys. Even in evils, the student learns to recognise only a struggling beneficence. But, above all, he is struck with the inherent sufficingness of things, and with the complex simplicity of those principles by which every defect is being remedied – principles that show themselves alike in the self-adjustment of planetary perturbations, and in the healing of a scratched finger – in the balancing of social systems, and in the increased sensitiveness of a blind man's ear – in the adaptation of prices to produce, and in the acclimatization of a plant. Day by day he sees a further beauty. Each new fact illustrates more clearly some recognised law, or discloses some inconceived completeness: contemplation thus perpetually discovering to him a higher harmony, and cherishing in him a deeper faith.[25]

Yet there are people who try by every possible means to obstruct this beneficent concatenation of cause and effect; they formulate minutely particularized moral rules or political programs intended to direct the course of events by force. Their folly borders on impiety. They cling to unnatural and useless schemes for guiding mankind and assign to governments the task of molding individuals to conform to arbitrary models. But "no human laws are of any validity if contrary to the law of nature."[26] If the state takes it upon itself to guide the process of adaptation, the individual ends by adapting himself to artificial conditions, not to "the natural requirements of the social state."[27] Or he is compelled to be idle because the state sees to all his needs, but if they are not exercised his faculties atrophy and that means losing the means of attaining happiness. The engine that drives social development is not social institutions but the inner nature of man and things. The institutions express this development and mark its stages, but they cannot be expected to alter its course because they are not capable of modifying the character of the individual or that of society. The development of institutions, therefore, is no more than the epiphenomenon of the moral evolution of individuals. The public apparatus, Spencer was to write, reflects the emotional nature of the social aggregate of which it is the expression. For this reason the types of political organization are not a question of deliberate choice. "Constitutions are not made, but grow."[28] On the other hand, "circumstances" impose unalterable limits on what men do. The idea that society develops "spontaneously" and not under human guidance (an idea that was to be expressed in the concept of the "social organism" and was to be central to the work of several so-called Social Darwinists) was present in Spencer's thought before Darwin's theories became well known. Moreover, the use of biological analogy in the study of social evolution was not peculiar to Social Darwinism. What follows is the longest passage devoted to the organic analogy in *Social Statics*. Characteristically, it is preceded by an example taken from physics.

Let it be again remembered that men cannot *make* force. All they can do is to avail themselves of force already existing, and employ it for working out this or that purpose. They cannot increase it; they cannot get from it more than its specific effect; and as much as they expend on it for doing one thing, must they lack of it for doing other things. Thus it is now becoming a received doctrine, that what we call chemical affinity, heat, light, electricity, magnetism, and motion, are all manifestations of the same primordial force – that they are severally convertible into each other – and, as a corollary, that it is impossible to obtain in any one form of this force more than its equivalent in the previous form. Now this is equally true of the agencies acting in society. It is quite possible to divert the power at present working out one result, to the working of some other result. You may transform one kind of influence into another kind. But you cannot make more of it, and you cannot have it for nothing. You cannot, by legislative manoeuvring, get increased ability to achieve a desired object, except at the expense of something else. Just as much better as this particular thing is done, so much worse must another thing be done.

Or, changing the illustration, and regarding society as an organism, we may say that it is impossible artificially to use up social vitality for the more active performance of one function, without diminishing the activity with which other functions are performed. So long as society is let alone, its various organs will go on developing in due subordination to each other. If some of them are very imperfect, and make no appreciable progress towards efficiency, be sure it is because still more important organs are equally imperfect, and because the amount of vital force pervading society being limited, the rapid growth of these involves cessation of growth elsewhere. Be sure, also, that whenever there arises a special necessity for the better performance of any one function, or for the establishment of some new function, nature will respond. Instance in proof of this, the increase of particular manufacturing towns and sea-ports, or the formation of incorporated companies. Is there a rising demand for some commodity of general consumption? Immediately the organ secreting that commodity becomes more active, absorbs more people, begins to enlarge, and secretes in more abundance. Instrumentalities for the fulfilment of social requirements – for the supply of religious culture, education, and so forth, are similarly provided: the less needful being postponed to the more needful; just as the several parts of the embryo are developed in the order of their subservience to life. To interfere with this process by producing premature development in any particular direction is inevitably to disturb the due balance of organization, by causing somewhere else a corresponding atrophy. Let it never be forgotten that at any given time the amount of a society's given vital force is fixed. Dependent as is that vital force upon the degree of adaptation that has taken place – upon the extent to which men have acquired fitness for a co-operative life – upon the efficiency with which they can combine as elements of the social organism, we may be quite certain that, whilst their characters remain constant, nothing can increase its total quantity. We may also be certain that this total quantity can produce only its exact equivalent of results; and that no legislators can get more from it; although by wasting it they may, and always do, get less.[29]

Spencer was convinced that there was no conflict of interest between the individual and society. The development of the individual and that of society are not incompatible, according to Spencer; on the contrary, each implies the other. Just as man is a complex of faculty-functions, so society is "a commonwealth of monads." These, like the faculties of an individual, each act independently of the rest in its own sphere and together conspire to create the harmony of the whole.[30] The fullest unfolding of the capacities of individual men will correspond to the fullest development of society: the perfect man in the perfect society. The citizen will then feel that his interests are in unison with those of society. The ensuing concord, rather than resulting from political decisions, will be a spontaneous outcome.[31]

Thanks to the conception from phrenology, that man consists of a complex of faculties and functions, which develop under the influence of external circumstances, and thanks to the organic analogy that made it possible to apply the same conception to society, Spencer had secularized and temporalized natural theology's pre-ordained harmony and had rooted it in the nature of man and things. Man realized the potential of his own nature in the course of social and cultural development, but he was also molded by circumstances. Through civilization, the members of society developed a nature capable of rising to the demands of circumstances. Spencer's environmentalism was to grow even stronger as his evolutionism matured. With his usual wit, Huxley was to write:

You appear to me to suppose that external conditions modify machinery [the structure of the organism] as if by transferring a flour-mill into a forest you could make it into a saw-mill.[32]

But does not this environmentalism contradict the idea that man's development is determined by basic components of his constitution? Is not Spencer saying that moral manifestations, "originating as these do in the facts of man's constitution, . . . are unalterable by the accidents of external condition" – that they are, in other words, universal? Does he not affirm that "only by giving us some utterly different mental constitution could the process of civilization have been altered?"[33] *Social Statics* is based on this ambiguity, and the organic analogy performs an important function here too. The history of civilization is not the unfolding of a divine plan, but it is, all the same, in some sense foreordained in that it is the development of man's latent powers. The development of *man* is assimilated to the development of *men*, the development of species to that of individuals. The preformed development of what is biologically programmed coincides with the epigenetic development of what is molded by circumstances: the transformation of an acorn into an oak tree, the growth of a child, the social evolution which takes primitive man to "perfect" man are similar processes.[34] The result is a necessitarian conception of social evolution, streaked with teleology. Man's nature contains potentialities which by developing will produce harmony with the evolution of things. What Spencer was anxious to demonstrate was that everything tended in the direction he was hoping for: the best direction. The conclusion seems to be, basically, that we may congratulate ourselves on the fact that men are not allowed to influence the cosmic process and that the best thing they can do is abstain from interfering – a conception poised precariously between optimistic determinism and pessimistic fatalism. If for whatever reason the certainty that progress was just around the corner weakened, it would not take much to tip the balance towards pessimism. So man's nature ceased to be the physiological guarantee of progressive adaptation and appeared a sort of ball to be kicked about by social progress. As time passed, an analogous change took place in Spencer himself, as his subsequent works on sociology and various private writings testify.

Sanguine of human progress as I used to be in earlier days, I am now more and more persuaded that it cannot take place faster than human nature itself is modified; and the modification is a low process, to be reached only through many, many generations.

There is no hope for the future save in the slow modification of human nature under social discipline. Not teaching, but action is the requisite cause. To have to lead generation after generation a life that is honest and sympathetic is the one indispensable thing. No adequate change of character can be produced in a year, or in a generation, or in a century. All which teaching can do – all which may, perhaps, be done by a wider diffusion of principles of sociology, is the checking of retrograde action. The analogy supplied by an individual life yields the true conception. You cannot in any considerable degree change the course of individual growth and organization – in any considerable degree antedate the stages of development. But you can, in considerable degree, by knowledge put a check upon those courses of conduct which lead to pathological states and accompanying degradations.[35]

Except that he flew off the handle when he was reproached with teaching an "enervating fatalism."

The Stern Discipline of Nature

Spencer's faith in the "inherent adjustment of things" had another, important consequence. If the perfecting of man is brought about by the continuous, free interaction of the faculties with circumstances, every form of open, full conflict, every situation in which the faculties are subjected to great pressure cannot but be good for man, whose unlimited malleability is thereby stimulated to produce more and more perfect forms of adaptation to external conditions. The optimistic presupposition is that a faculty left to itself is sure to respond positively to the environment. Nature not only subjects the organism and the faculties to unceasing exertion, so that they can never rest, but ruthlessly eliminates any ineffective response. Man's progress is quantitative and qualitative: society is composed of a greater and greater number of better and better individuals.

Beneath her apparent cruelty, therefore, nature is beneficent. It is necessary, Spencer writes, using the language and arguments of justification typical of natural theology, to raise oneself to "the highest point of view" from which one sees how nature, by eliminating the unfit, progressively eliminates evil in a continuous "purifying process." Spencer's description of this is, of course, no anticipation of natural selection, but merely another version of the old idea of "nature's broom" as a means to harmony and beauty.

Pervading all nature we may see at work a stern discipline, which is a little cruel that it may be very kind. That state of universal warfare maintained throughout the lower creation, to the great perplexity of many worthy people, is at bottom the most merciful provision which the circumstances admit of. It is much better that the ruminant animal, when deprived by age of the vigour which made its existence a pleasure, should be killed by some beast of prey, than that it should linger out a life made painful by infirmities, and eventually die of starvation. By the destruction of all such, not only is existence ended before it becomes burdensome, but room is made for a younger generation capable of the fullest enjoyment; and, moreover, out of the very act of substitution happiness is derived for a tribe of predatory creatures. Note further, that their carnivorous enemies not only remove from herbivorous herds individuals past their prime, but also weed out the sickly, the malformed, and the least fleet or powerful. By the aid of which purifying process, as well as by the fighting, so universal in the pairing season, all vitiation of the race through the multiplication of its inferior samples is prevented; and the maintenance of a constitution completely adapted to surrounding conditions, and therefore most productive of happiness, is ensured.³⁶

The same pattern of betterment holds good for nature and for society.

The development of the higher creation is a progress towards a form of being capable of a happiness undiminished by these drawbacks. It is in the human race that the consummation is to be accomplished. Civilization is the last stage of its accomplishment. And the ideal man is the man in whom all the conditions of that accomplishment are fulfilled. Meanwhile the well-being of existing humanity, and the unfolding of it into this ultimate perfection, are both secured by that same beneficent, though severe discipline, to which the animate creation at large is subject: a discipline which is pitiless in the working out of good: a felicity-pursuing law which never swerves for the avoidance of partial and temporary suffering. The poverty of the incapable, the distresses that come upon the imprudent, the starvation of the idle, and those shoulderings aside of the weak by the strong, which leave so many "in shallows and in miseries," are the decrees of a large, far-seeing benevolence. It seems hard that an unskilfulness which with all his efforts he cannot overcome, should entail hunger upon the artizan. It seems hard that a labourer incapacitated by sickness from competing with his stronger fellows, should have to bear the resulting privations. It seems hard that widows and orphans should be left to *struggle for life or death*. Nevertheless, when regarded not separately, but in connection with the interests of universal humanity, these harsh fatalities are seen to be full of the highest beneficence – the same beneficence which brings to early graves the children of diseased parents, and singles out the low-spirited, the intemperate, and the debilitated as the victims of an epidemic.³⁷

Here Spencer uses the expression "to struggle for life;" earlier in the same work he had used the expression "struggles for existence."³⁸ But the context is not specifically biological; nor is the term used in the technical sense it is to assume in Darwin. It is obvious that the image of human life as "struggle" and "competition" was already current before Darwin, but it is a mistake to suppose that Spencer in *Social Statics* enunciated the principle of natural selection. Even his idea of the "struggle for life" (but it would be better to call it "struggle *in* life") is rather a combination of a Calvinist view of a life of labor and strife and of a cosmic expansion of phrenology's idea of development through exertion and adaption. All this is evident in his attack on poor laws. They arrest and deflect the harsh process of adaptation through which man must pass. The "spurious philanthropists" and the ingenuous "paupers' friends" who would like to make charity compulsory defy the laws of nature.

Blind to the fact, that under the natural order of things society is constantly excreting its unhealthy, imbecile, slow, vacillating, faithless members, these unthinking, though well-meaning, men advocate an interference which not only stops the purifying process, but even increases the vitiation – absolutely encourages the multiplication of the reckless and incompetent by offering them an unfailing provision, and *dis*courages the multiplication of the competent and provident by heightening the prospective difficulty of maintaining a family. And thus, in their eagerness to prevent the really salutary sufferings that surround us, these sigh-wise and groan-foolish people bequeath to posterity a continually increasing curse.[39]

Furthermore, to impose beneficence on people by law means to prevent the real development of the all-important faculty of sympathy. It is this faculty that distinguishes civilized man from the savage, that originates the idea of justice, that renders society possible, and "of whose growth civilization is a history." But it is also "the very faculty above all others needing to be exercised." Of this faculty poor-laws partially supply the place. By doing which they diminish the demands made upon it, limit its exercise, check its development, and therefore retard the process of adaptation."[40]

Administered by the state, sympathy favors the multiplication of "those worst fitted for existence" and in consequence hinders the multiplication of "those best fitted for existence," to whom it leaves less and less space. In this way the world fills up with people to whom life means pain and grief and is emptied of those to whom living is a delight. Organized charity "inflicts positive misery and prevents positive happiness."[41] Spencer's concept of "survival of the fittest" therefore appears and develops in a context that is more social and moral than specifically biological. It refers to the moral betterment of man, and to his physical nature too, but only in so far as the faculties are located in the brain. It certainly does not refer to the formation of new species; harsh competition develops the best qualities in every man rather than selecting preexistent variations.

Malthus had attached great importance to "moral restraint." Spencer used the term "self-restraint" to indicate "the ability to sacrifice a small present gratification for a prospective great one." More self-restraint would avoid imprudent marriages and the growth of a poverty-stricken population. "And were there no drunkenness, no extravagance, no reckless multiplication, social miseries would be trivial." But only stern experience can teach this. "Those in whom this faculty [of self-restraint] needs drawing out must be left to the discipline of nature, and allowed to bear the pains attendant on their defect of character. The only cure for imprudence is the suffering which imprudence entails."[42] Spencer asserts, in a strikingly Malthusian spirit, that to encourage improvident marriages by state grants (and the extra taxation to pay for them) means drying up the main source of self-restraint, the sense of paternal responsibility (quite apart from being unjust to the unmarried). He goes further than Malthus: not only the poor law but state education, hospitals, and every other form of state intervention is a harmful barrier to nature's action. "The ultimate result of shielding men from the effects of folly, is to fill the world with fools."[43]

At every stage of social evolution except, obviously, the last, "there is bound up with the change a *normal* amount of suffering, which cannot be lessened without altering the very laws of life." All the same, as we saw previously, these setbacks are not only inevitable but actually beneficial; in every case, therefore, "the process *must* be undergone, and the sufferings *must* be endured." Humanity must pass through all the phases necessary to its completion. To this end humanity must train and temper itself, by passing through *all* the institutions and social evils necessary to it as "moral ordeals." In Spencer's evolutionary gradualism there is a Calvinistic, moral component which must not be overlooked in order to focus on his political motives.[44]

Physics and Morals

Spencer himself said one day that *Social Statics* contained two ideas that showed that he had already taken the path that would eventually lead him to cosmic evolution. One was the concept of the organ-function relationship, i.e., the "physiological division of labour," of which he had from the beginning made a law for all forms of organization, natural as well as artificial. (He had quoted a manufacture, mercantile society and language as instances of development through differentiation.) The other was the analogy between the organism and the "body politic," both being "communities of monads" whose components were each endowed with special functions but all subservient to the whole. Spencer had developed these two conceptions together. As the most elementary organisms are mere aggregates of like units, vegetatively replicated, so the primitive forms of society consist of the repetition of a single element.

As every part of the body of a polyp is at the same time stomach, skin, and lung, so in tribal communities every individual is hunter warrior, priest, etc. As we proceed to higher organisms and more evolved social organizations, the differentiation and reciprocal subordination of parts increase. This process of specialization is the result of a fundamental characteristic of life: the "tendency to individuation." Its progressive realization produces ever more intense, more complex, more functionally integrated degrees of life. To be endowed with a greater variety of organs and faculties means to be more characterized, to possess a more pronounced individuality. The tendency to a greater complexity is also a tendency to individuation, to "becoming a thing," as Spencer puts it. Tissues, for instance, become "individualized" into distinct organs, and it is by individuation that the nervous system is formed. The law of growing complexity applies also to the "correspondence" (this is the term used in the *First Principles* and the *Principles of Biology*) between internal and external factors; the more complex the organism, the more complex is this correspondence; and the more complex this correspondence, the more vitality does the organism possess and the less is it at the mercy of external agents.

Spencer applies the concepts of "complexity," "differentiation," and individuation to all sorts of evolution, inorganic, organic, and "superorganic" (that is, social). In 1852 his vocabulary was enriched by the term "heterogeneity": greater heterogeneity is a mark of higher organization, in the productions of nature as well as of man, in organisms as well as in languages, architectural styles, engines. Spencer's idea that the phenomena of all orders could be distributed in a scale of increasing complexity was strengthened in 1851 by his reading of the works (which, he characteristically does not say) of Henri Milne-Edwards, the zoologist who coined the expression *division du travail physiologique*. Milne-Edwards did not provide Spencer with any really new concept (the social division of labor already being a commonplace in social and economic science); rather, he confirmed that this law had great explanatory power in biology too. (And to Spencer's eyes this amounted to evidence of "all-embracing truth.") As he was to write in the *The Study of Sociology*, the concept of the physiological division of labor "obviously originates from the generalizations previously reached in Political Economy," but then biology "gives it back to Sociology greatly increased in definiteness, enriched by countless illustrations, and fit for extension in new directions."[45] Here, therefore, is an instance of the close relationship between biology and social science. No wonder then that, when discussing the *biological* concept of the division of labor in the *Principles of Biology*, Spencer gave the progressive differentiation of *social* roles as an instance of it.[46] That it was neither a mere rhetorical device nor a consciously instrumental, heuristic, and explanatory model is evident from the famous and notorious essay on *The Social Organism*, where the use of organic analogies is pushed to extremes, and, for instance, coins are compared to "blood-discs."[46] According to Spencer, there is "a real analogy" between an individual organism and a social organism: "certain necessities determining structure [. . .] govern them in common." Which these necessities are does not concern us here. What is important is that by "real analogy" he means something more than the words may first suggest.

Figures of speech, which often mislead by conveying the notion of complete likeness where only slight similiarity exists, occasionally mislead by making an actual correspondence seem a fancy. A metaphor, when used to express a real resemblance, raises a suspicion of mere imaginary resemblance; and so obscures the perception of intrinsic kniship. It is thus with the phrases "body politic," "political organization," and others, which tacitly liken a society to a living creature: they are assumed to be phrases having a certain convenience but expressing no fact – tending rather to foster a fiction. And yet metaphors are here more than metaphors in the ordinary sense. They are devices of speech hit upon to suggest a truth at first dimly perceived, but which grows clearer the more carefully the evidence is examined.[48]

In short, "we are not here dealing with a figurative resemblance, but with a fundamental parallelism in principles of structure" and a "fundamental kinship" between the two sciences.[49]

In 1851 Spencer also read the third edition (1851) *Principles of General and Comparative Physiology* (1839) of the physiologist William Benjamin Carpenter. In this work, among other things, the views of the great embryologist Karl von Baer on embryogenesis were expounded, and from them Spencer derived the idea that "the development of every organism is a passage from the homogeneous to the heterogeneous." But all in all he seems to have found in both Milne-Edwards and von Baer (neither of whom was an evolutionist) nothing more than a confirmation of ideas he already entertained, which were simply to receive a final formulation by that reading. In the essay *Progress: Its Law and Cause* (1857) the appeal to von Baer, Goethe, and the other important embryologist Caspar Friedrich Wolff as the sources of the idea that organic development is a progression towards differentiation and heterogeneity amounted to no more than the quotation of three names. What mattered to Spencer was above all to generalize those formulas, to give them an evolutionary meaning they did not have, and to encompass reality in one law. In this essay he characteristically took instances of the law of progress from all departments of the universe: from physics to linguistics, from astronomy to economics, from biology to sociology, from the arts to everyday life. The foregone conclusion was "the law of organic progress is the law of all progress."[50]

Is, therefore, Spencer's evolutionism a result of his application of biological concepts to all aspects of reality? And, more particularly, did he biologize society? The answer is only a *partial* "yes." For, to begin with, there could be some good reasons for arguing that the opposite was often the case. The very concept of adaptation was presented in the *Principles of Biology* as an example of the fact that "our conceptions of vital processes [are] made clearer by studying analogous social phenomena. From the laws of adaptive modifications in societies, we may [. . .] hope to get a clue to the laws of adaptive modifications in organisms."[51]

Second, and more important, is the fact that the biological level is not the ultimate one: it can itself be reduced to the physical. In the chapter of the *Principles of Biology* just referred to, after delving into a long and elaborate comparison between social and biological adaptation, Spencer concludes "that the phenomena of adaptation fall into harmony with first principles," by which he means the metaphysics of force, matter, and motion expounded in the *First Principles*.[52] The whole of the *Principles of Biology* were based on it. But while a biological phenomenon such as adaptation can be explained in physical terms, the contrary is obviously impossible. In a word, biological phenomena, and biological laws, are derivative, not ultimate. The unity of knowledge was due not to a particular science, but to a metaphysics full of concepts derived from physics. Spencer *does* write that "there can be no rational apprehension of the truths of Sociology until there has been reached a rational apprehension of the truths of Biology."[53] But in the very chapter of *The Study of Sociology* from which this quotation is taken (and which bears the title "Preparation in biology") he says clearly that this is only the first part of the sociologist's task, the second being to arrive "at the concept of the Social Science which alone fully affiliates it upon the simpler sciences."

Only when it is seen that the transformations passed through during the growth, maturity, and decay of a society, conform to the same principles as do the transformations passed through by aggregates of all orders, inorganic and organic – only when it is seen that the process is in all cases similarly determined by forces, and is not scientifically interpreted until it is expressed in terms of these forces; – only then is there reached the conception of Sociology as a science, in the complete meaning of the word.[54]

Already in the essay on progress Spencer argued that the cause of the passage from the homogeneous to the heterogeneous was a law of "multiplication of effects," that is, of the effects produced by a force.[55] Biological phenomena were reduced to their physical components and thereby explained by philosophical concepts. In the *First Principles* the pivot of the whole Synthetic Philosophy was identified in the "persistence of force," a concept clearly derived from physics. The point was then to derive all evolutionary phenomena from that basic concept.

The task before us, then, is that of exhibiting the phenomena of Evolution in synthetic order. Setting out from an established ultimate principle, it has to be shown that the course of transformation among all kinds of existences cannot but be that which we have seen it to be. It has to be shown that the re-distribution of matter and motion, *must* everywhere take place in those ways, and produce those traits, which celestial bodies, organisms, societies, alike display. And it has to be shown that in this university of process, is traceable the same *necessity* which we find in each simplest movement around us, down to the accelerated fall of a stone or the recurrent beat of a harp-string. In other words, the phenomena of Evolution have to be deduced from the Persistence of Force.[56]

Evolution is but an "implication" of the persistence of force. Every change is only a particular aspect of a single, fundamental transformation of matter. It is in this physico-metaphysical jargon that Spencer couches his much-quoted definition of evolution: "Evolution is an integration of matter and concomitant dissipation of motion; during which the matter passes from an indefinite, incoherent homogeneity to a definite, coherent heterogeneity; and during which the retained motion undergoes a parallel transformation."[57]

Both the *Principles of Biology* and the *Principles of Sociology* open with a review of "internal" and "external" factors, that is, of the "forces" whose interaction produces evolution. The cause of organic evolution is the instability of living matter, a property that Spencer demonstrates with chemical and physical arguments to be just one special instance of the "instability of the homogeneous" expounded in the *First*

Principles.[58] All of Spencer's biology consists in the attempt to explain the correspondence between external (environmental) changes and the internal changes that are produced to counterbalance them. The basic object of biology is the study of the continuous interaction between the organism and its environment. Life is the maintenance of a moving equilibrium of forces acting in accordance with the ultimate laws of the redistribution of matter and motion.

Only when the process of evolution of organisms, is affiliated on the process of evolution in general, can it be truly said to be explained. The thing required is to show that its various results are corollaries from first principles. We have to reconcile the facts with the universal laws of the re-distribution of matter and motion.[59]

The *Principles of Biology* teem with expressions such as "dissipation of force," "quantity of force," "applied force," "momentum," "expenditure," and "labor" and with mechanical instances. The phenomena of evolution are said to be "evolutions of force." Even the concept of function does not escape being considered in physical terms. The very definition of life is adapted to this approach: "The continuous adjustment of internal relations to external relations."[60]

From as early as 1857 Spencer "set forth the law of evolution as holding throughout all orders of phenomena, and, joyned with it, the statement of certain universal physical principles which necessitate its universality."[61]

While we think of Evolution as divided into astronomic, geologic, biologic, psychologic, sociologic, &c., it may seem to some extent a coincidence that the same law of metamorphosis holds throughout all its divisions. But when we recognize these divisions as mere conventional groupings, made to facilitate the arrangement and acquisition of knowledge – when we remember that the different existences with which they severally deal are component parts of one Cosmos; we see at once that there are not several kinds of Evolution having certain traits in common, but one Evolution going on everywhere after the same manner.[62]

Set against this gigantic attempt, Darwin's theory seemed to Spencer to be only one special contribution to the great edifice of the Synthetic Philosophy. Darwin's object was the question of species, not the evolution of the cosmos. Darwin felt admiration but also misgivings towards Spencer's ability to produce bold generalizations and sweeping views. And Spencer – who was very sensitive on these questions – was right in pointing out not only the priority of his views to those of Darwin, but their peculiarity and originality.

The distinctness of origin [of the theory of cosmic evolution and that expounded in the *Origin of Species*] might, indeed, have been inferred the work [*First Principles*] itself, which deals with Evolution at large – Inorganic, Organic, and Super-organic – in terms of Matter and Motion; and touches but briefly on those peculiar processes so luminously exhibited by Mr. Darwin.[63]

In *First Principles*, Spencer explained the production of diverging varieties from the species stock by his law of the "multiplication of effects."[64] In the preface to the fourth edition of the work (1880) he wrote that, if it had been written after the *Origin of Species*, that paragraph would have been phrased differently, as reference would have had to be made to natural selection, but the rest of the book would have remained the same; for Spencer's general purposes, the "special cause" alleged by Darwin could safely be left unaccounted for. Organic evolution was by no means reduced to the theory of natural selection, which explained only one part of the "entire transformation."

How utterly different the popular conception of evolution is from evolution as rightly conceived will now be manifest. The prevailing belief is doubly erroneous – contains an error within an error. The theory of natural selection is wrongly supposed to be identical with the theory of organic evolution; and the theory of organic evolution is wrongly supposed to be identical with the theory of evolution at large. In current thought the entire transformation is included in one part of it, and that part of it is included in one of its factors.[65]

It was not the theory of natural selection that provided the basis for an evolutionary world view, or the clue to it, but the contrary. Even if the theory of natural selection were disproved, the theory of organic evolution would remain firmly standing; its basis lay elsewhere, in the principles of cosmic evolution.

The Doctrine of Evolution, rightly conceived, has for its subject-matter not the changes exhibited by the organic world only, but also the changes which went on during an enormous period before life began, and the changes which have gone on since life rose to its highest form, and Man, passing into the associated state, gave origin to the endlessly varied products of social life. It has for its subject-matter the entire cosmic process, from nebular condensation down to the development of picture-records into written language, or the formation of local dialects; and its general result is to show that all the minor transformations in their infinite varieties are parts of the one vast transformation, and display throughout the same law and cause – that the Infinite and Eternal Energy has manifested itself everywhere and always in modes ever unlike in results but ever like in principle.

It is well known that Darwin adopted, on Wallace's suggestion, Spencer's phrase "the survival of the fittest" as a better alternative to his own "natural selection." To both Darwin and Spencer the former was more exact than the latter, and a more literal description of phenomena. But Darwin would not have gone so far as to grant with Spencer that it was a better term because it was more "physical"; indeed, he would have thought it meaningless to put the question in these terms. Spencer, on the contrary, thought it was absolutely necessary to reinterpret natural selection in "purely physical terms" in order to absorb it into his cosmic evolution ruled by laws of matter and motion. It was in order to do this that he came to adopt the expression "survival of the fittest." He has left two accounts of how this happened.

Organic evolution being a part of Evolution at large, evidently had to be interpreted after the same general manner – had to be explained in physical terms: the changes produced by functional adaptation (which I held to be one of the factors) and the changes produced by "natural selection," had both to be exhibited as resulting from the redistribution [sic] of matter and motion everywhere and always going on. Natural selection as ordinarily described, is not comprehended in this universal redistribution. It seems to stand apart as an unrelated process. The search for congruity led first of all to perception of the fact that what Mr. Darwin called "natural selection," might more literally be called survival of the fittest. But what is survival of the fittest, considered as an outcome of physical actions? The answer presently reached was this: The changes constituting evolution tend ever towards a state of equilibrium. On the way to absolute equilibrium or rest, there is in many cases established for a time, a moving equilibrium – a balanced set of functions constituting its life; and the overthrow of this balanced set of functions or moving equilibrium is what we call death. Some individuals in a species are so constituted that their moving equilibria are less easily overthrown than those of other individuals; and these are the fittest which survive, or, in Mr. Darwin's language, they are the select which nature preserves. And now mark that in thus recognizing the continuance of life as the continuance of a moving equilibrium, early overthrown in some individuals by incident forces and not overthrown in others until after they have reproduced the species, we see that this survival and multiplication of the selection, becomes conceivable in purely physical terms, as an indirect outcome of a complex form of the universal redistribution of matter and motion.[66]

It is difficult to see in what way "survival of the fittest" was more "physical" than "natural selection," but Spencer was satisfied, and Darwin's original idea, as it were, denatured.

One last implication of Spencer's fascination with redefining natural selection in physical terms should be noted. Those who brand him a typical Social Darwinist and charge him with being an apologist of progress through unrestrained competition should remember what he wrote in reply to the philosopher J. Martineau's attacks on evolution: "Under its rigorously-scientific form, the doctrine [of evolution] is expressible in purely-physical terms, which neither imply competition nor imply better and worse."[67] How peculiar for the alleged prophet of laissez-faire Darwinism to play down the importance of competition! The fact was that to him competition was less important than cosmical evolution.

To all these reasons that prevented Spencer from being a real Darwinian – and prevent us from labeling him a typical Social Darwinist – on last, but not least, must be added. To him, natural selection always remained a "secondary agency" of evolution,[68] and he tried as far as he could to diminish its importance. An obstinate Lamarckian, he stuck to what he called "use-inheritance", that is, the inheritance of functionally acquired modifications, and even fought a battle against Weismann, fearing that the triumph of the latter's views would have fatal consequences for social as well as biological science. Characteristically he renamed both the Darwinian and the Lamarchian mechanism of adaptation in physical terms. The Lamarckian he called "direct equilibration," by which he meant that "the individual organisms become modified when placed in new conditions of life – so modified as to re-adjust the power to the requirements,"[68] that is, reestablishes the balance between internal and external forces. "Indirect equilibration" was the process by which the different adaptive responses were sifted. Direct equilibration explains the modifications of the individual organism, indirect equilibration explains those of the species. The latter would not be possible without the former, which, therefore, is by far the more important and greatly restricts the sphere of action of indirect equilibration. It goes without saying that, as he saw any organic change as an adaptation – that is, in itself adaptive – Spencer could not see why Darwin made such a fuss about random or chance variation. Almost no variations were useless or neutral; most of the "functional deviations," which Darwin attributed to the selection of chance variations, were the result of the continuous, direct, material interaction between the organism and its environment, a result that was handed down ready-made to the offspring for these to readapt, if necessary, to new changes in the circumstances. Spencer's view of the plasticity of the organism was essentially Lamarckian (although he criticized Lamarck for believing in an internal tendency to organic perfection). Needless to say, he was convinced that function precedes structure.[70]

Although Spencer's Lamarckism explains his difference from Darwin, it does not explain everything. Spencer's stubborn insistence on use-inheritance and direct equilibration is only the most evident aspect of something deeper, something that was not derived from Lamarck, but was common to Lamarck, Spencer, and most of the neo-Lamarckians, and was the trunk of which they were all branches: the attempt to explain biological phenomena at any level according to a physical and determinist model. One striking consequence of this was the common rejection of chance variation. Universal determinism allowed no exception. Spencer's ambition was to leave no holes in his reconstruction of the universe in terms of matter, motion, and force. And his attempt at an all-embracing rationalization of nature and society was part and parcel of it.

He was not alone in this. As I have argued elsewhere,[71] the various forms of Lamarckism were expressions of some common assumptions and demands: they were the products of the persistence, in different disciplines and contexts, of some basic, *interdisciplinary* attitudes and patterns of thought. On the biological plane, one of these was the tendency to seek the cause of all vital phenomena – at all levels – in the chemical and physical laws and properties of living and nonliving matter. On the plane of the relationship between the biological and the human sciences, the Lamarckians shared the opinions that human efforts and achievements (good as well as evil ones) could influence the course of both organic and social evolution, not only indirectly, by orienting selection, but, above all, directly, through the inheritance of acquired characters. The more sanguine among them agreed that there was a firm basis in biology for the hopes of mankind. On the contrary, individual and racial improvement through education and the accumulation of experience were virtually impossible if Weismann was right. This is a point on which Spencer concentrated in his attack on Weismann. Society was materially linked to nature by the blood and nerves of people. The idea of a nature-culture continuum and of a beneficent solidarity between body and mind, the idea that mental, moral, and social phenomena were a continuation and were governed by the same laws as of biological and physical phenomena were modern versions of the old, reassuring natural theological idea of the beneficent economy of nature. What was reassuring now was not the idea of a creator, but faith in universal physical causation. A variety of biologism in sociology accompanied by biological reductionism inspired by physical determinism was the basis for a scientific, moral, and social optimism that took the form of the ambitious project of making the universe the subject of only one science based on few ultimate principles and finding a sanction in nature for effort, merit, and progress.

By helping to construe reality in terms of effort and strife, as well as spontaneous development and beneficent self-adjustment, the images of "struggle" and of "social organism" played an important role in Spencer's moral perception of man, society and the cosmos. Metaphors may be poor arguments, but in some cases they exist to perform some deeper function in their authors' mind than that of compelling assent in the reader.

Acknowledgments

The final version of this paper benefited much from the discussion at the Bielefeld seminar, for which I thank the convenors and all the participants, from Peter Weingart's and Sabine Maasen's criticisms and suggestions, and from subsequent research carried during my first months as a Humboldt fellow at the Lehrstuhl für Wissenschaftsgeschichte, Institut für Philosophie, University of Regensburg. I thank Christoph Meinel for his comments and the Alexander von Humboldt-Stiftung for their generosity. I am very grateful to Virginia Browne, who translated part of the text from the Italian and revised the parts I had written in English.

Notes

1. J. C. Greene, "Darwin as a Social Evolutionist," *Journal of the History of Biology* **10** (1977), 1–27.
2. Antonello La Vergata, *L'equilibrio e la guerra della natura. Dalla teologia naturale al darwinismo* (Napoli: Morano, 1990). On Darwin's metaphors see Eduard *The*

Young Darwin and His Cultural Circle. A Study of the Influences Which Helped Shape the Language and Logic of the First Drafts of the Theory of Natural Selection (Dordrecht-Boston: Reidel. 1978); B. G. Beddall, "Wallace's Annotated Copy of Darwin's *Origin of species*," *Journal of the History of Biology* **21** (1988), 265–289; D. B. Paul, "The Selection of the 'Survival of the Fittest," *Journal of the History of Biology* **21** (1988), 411–424.
3. La Vergata, *op. cit.* 1990, Note 2; La Vergata, "Biologia, scienze umane e 'darwinismo sociale'. Considerazioni contro una categoria storiografica dannosa," *Intersezioni* **2** (1982), 77–97; La Vergata, *Nonostante Malthus, Fecondità, popolazioni e armonia della natura, 1700–1900* (Torino: Bollati Boringhieri, 1990).
4. Hyppolite Taine, *Derniers essays de critique et d'histoire* (Paris: Hachette. 4th edn., 1909), pp. 188–201; Richard Hofstadter, *Social Darwinism in American Thought, 1860–1919* (Boston: The Beacon Press, Revised edn. 1955 (orig. 1944)), p. 38; Marvin Harris, *The Rise of Anthropological Theory. A History of the Theories of Culture* (London: Routledge and Kegan Paul, 1968), pp 122–123. For a different view see J. W. Burrow, *Evolution and Soceity: a Study of Victorian Social Theory* (Cambridge: Cambridge University Press, 1970 (orig. 1966). Recent works on social Darwinism include G. Jones, *Social Darwinism and English Thought: the Interaction between Biological and Social Theory* (Brighton: Harvester Press, 1980); R. C. Bannister, *Social Darwinism. Science and Myth in Anglo-American Social Thought* (Philadelphia: Temple University Press, 1979); D. C. Bellomy, "'Social Darwinism' Revisited," *Perspectives in American History* **1** (1984), 1–130. Critical overviews are Antonello La Vergata, "Images of Darwin. A Historiographic Overview," in D. Kohn (ed.), *The Darwinian Heritage* (Princeton: Princeton University Press in association with Nova Pacifica, Wellington, New Zealand, 1985), pp. 901–972; and J. R. Moore "Socializing Darwinism: Historiography and the Fortunes of a Phrase," in L. Levidow (ed.), *Science as Politics* (London: Free Association Books. 1986), pp. 38–80.
5. For instance Enrico Ferri, *Socialismo e scienza positiva (Darwin, Spencer, Marx)* (Roma: Casa editrice italiana, 1894).
6. T. H. Huxley, "Administrative Nihilism" (1871), in *Collected Essays* (London: Macmillan, 1893), Vol. I, p. 271.
7. This applies for instance to such writers as Lilienfeld, Schäffle and Fouillée. On the organic analogy in sociology see H. E. Barnes, "Representative Biological Theories of Society," *Sociological Review* **17** (1925), 120–130, 182–194, 294–300; René Worms, *Organisme et société* (Paris: Giard et Brière. 1905); E. T. Towne, *Die Auffassung der Gesellschaft als Organismus, ihre Entwicklung und ihre Modifikationen* (Halle a. S.: Kaemmerer. 1903); F. W. Coker, *Organismic Theories of the State, Nineteenth Century Interpretations of the State as an Organism or a Person* (New York: Columbia University. 1910); Paul Weindling, "Theories of the Cell State in Imperial German," in C. Webster (ed.), *Biology, Medicine and Society, 1840–1940* (Cambridge: Cambridge University Press. 1981), pp. 99–145; Renato Mazzolini, "Stato e organismo, individui e cellule nell'opera di Rudolf Virchow negli anni 1845–1860," *Annali dell'Istituto storico italo-germanico in Trento* **9** (1983), 153–293 (German transl. by Klaus-Peter Tieck, *Politisch-biologische Analogien im Fruühwerk Rudolf Virchows* (Marburg: Basilisken-Presse, 1988). A pithy criticism of Spencer's organicism is D. G. Ritchie, *The Principles of State Interference, Four*

Essays on the Political Philosophy of Herbert Spencer, J. S. Mill and T. H. Green (London: Swan Sonnenschein. 1891).
8. Beatrice Webb, *My Apprenticeship* (London: Longmans, Green, 1926), p. 192.
9. Herbert Spencer, *Social Statics, or the Conditions Essential to Human Happiness Specified, and the First of Them Developed* (London: John Chapman, 1851), p. 103.
10. *Ibid.*, pp. 5, 75–76, 280.
11. *Ibid.*, pp. 3, 5, 37.
12 *Ibid.*, pp. 280–281.
13. *Ibid.*, p. 17.
14. *Ibid.*, p. 18.
15. A. Ryan, "The Nature of Human Nature in Hobbes and Rousseau," in J. Benthall (ed.), *The Limits of Human Nature* (New York: Dutton, 1974), p. 13.
16. R. M. Young, *Mind, Brain and Adaptation. Cerebral Localization and Its Biological Context from Gall to Ferrier* (Oxford: Clarendon Press, 1970), pp. 157–158. On phrenology see D. De Giustino, *The Conquest of Mind: Phrenology and Victorian Social Thought* (London: Croom Helm, 1975); S. Shapin, "Homo Phrenologicus: Anthropological Perspectives on an Historical Problem," in B. Barnes and S. Shapin (eds.), *Natural Order. Historical Studies of Scientific Culture* (London: Sage Publications, 1979), pp. 41–71; Roger Cooter, *The Cultural Meaning of Popular Science. Phrenology and the Organization of Consent in Nineteenth-Century Britain* (Cambridge: Cambridge University Press, 1985).
17. Spencer, *op. cit.*, 1851, Note 9., pp. 50, 58.
18. J. D. Y. Peel, *Herbert Spencer: the Evolution of a Sociologist* (London: Heineman, 1971), p. 109.
19. See for instance Henry Buckle, *History of Civilization in England* (London: Parker, 1882 (orig. 1857–1861)), Vol. I., pp. 8–9, 18, 23–27.
20. George Eliot, *Felix Holt, the Radical* (Harmondsworth: Penguin Books, 1972), pp. 614, 616, 617, 626. The *Address to Working Men by Felix Holt* was published in "Blackwood's Magazine" in 1868, two years afer *Felix Holt, the Radical* (1866).
21. Herbert Spencer, *The Data of Ethics* (first part of *The principles of ethics*) (London: Williams & Norgate, 1879), pp. iii–iv.
22. Spencer, *op. cit.*, 1851, Note 9, pp. 58, 59.
23. *Ibid.*, pp. 59–60.
24. *Ibid.*, p. 65.
25. *Ibid.*, pp. 293–294.
26. *Ibid.*, p. 207.
27. *Ibid.*, pp. 281–281.
28. Herbert Spencer, "The Social Organism," in *Essays, Scientific, Political and Speculative* (London: Williams & Norgate, 1868–1874), Vol. I, p. 384. See Spencer, *op. cit.*, 1851, Note 9, p. 263.
29. Spencer, *op. cit.*, 1851, pp. 390–391.
30. *Ibid.*, p. 451.
31. *Ibid.*, pp. 442, 455–456.
32. Letter of March 22, 1886, in L. Huxley (ed.), *The Life and Letters of Thomas Henry Huxley* (London: Macmillan, 1900), Vol. II, p. 126.
33. Spencer, *op. cit.*, 1851, Note 9, pp. 297, 413.

34. *Ibid.*, pp. 186–187.
35. Letters of February 5, 1890 and January 10, 1895, in David Duncan (ed.), *The Life and Letters of Herbert Spencer* (London: Methuen and Co., 1908), pp. 355, 366.
36. Spencer, *op. cit.*, 1851, Note 9, p. 322.
37. *Ibid.*, pp. 322–323, my italics.
38. *Ibid.*, p. 228.
39. *Ibid.*, p. 324. See also, pp. 379–380: "A sad population of imbeciles would our schemers fill the world with, could their plans last. A sorry kind of human constituion would they make for us a constitution lacking the power to uphold itself, and requiring to be kept alive by superintendence from without – a constitution continually going wrong, and needing to be set right again – a constitution even tending to self-destruction. Why the whole effort of nature is to get rid of such – to clear the world of them, and make room for better. Nature demands that every being shall be self-sufficing. All that are not so, nature is perpetually withdrawing by death. Intelligence sufficient to avoid danger, power enough to fulfil every condition, ability to cope with the necessities of existence – these are qualifications invariably insisted on. Mark how the diseased are dealt with. Consumptive patients, with lungs incompetent to perform the duties of lungs, people with assimilative organs that will not take up enough nutriment, people with defective hearts that break down under excitement of the circulation, people with any constitutional flaw preventing the due fulfilment of the conditions of life, are continually dying out, and leaving behind those fit for the climate, food, and habits to which they are born. Even the less-imperfectly organized, who, under ordinary circumstances, can manage to live with comfort, are still the first to be carried off by epidemics; and only such as are robust enough to resist these – that is, only such as are tolerably well adapted to both the usual and incidental necessities of existence, remain. And thus is the race kept free from vitiation. . . . Nature just as much insists on fitness between mental character and circumstances, as between physical character and circumstances; and radical defects are as much causes of death in the one case as in the other. He on whom his own stupidity, or vice, or idleness, entails loss of life, must, in the generalizations of philosophy, be classed with the victims of weak viscera or malformed limbs. In his case, as in the others, there exists a fatal nonadaptation; and it matters not in the abstract whether it be a moral, an intellectual, or a corporeal one. Beings thus imperfect are nature's failures, and are recalled by her laws when found to be such. Along with the rest they are put upon trial. If they are sufficiently complete to live, they *do* live, and it is well they should live. If they are not sufficiently complete to live, they die, and it is best they should die.
40. *Ibid.*, p. 321.
41. *Ibid.*, p. 381.
42. *Ibid.*, p. 353.
43. Herbert Spencer, "State-tamperings with money and banks" (1858), in Spencer, *Essays, Scientific, Political and Speculative* (London: Williams & Norgate, 1868–1874), Vol. II, p. 349.
44. "I look upon despotisms, aristocracies, priestcrafts, and all the other evils that affect humanity, as the necessray agents for the training of the human mind, and I believe that every people must pass through the various phases between absolutism and

democracy before they are fitted to become *permanently* free, and if a nation liberates itself by physical force, and attains the goal without passing through these *moral ordeals*, I do not think its freedom will be lasting." (Spencer to E. Lott, October 14, 1841, in Duncan (ed.), *op. cit.*, 1908, Note 35, p. 41.
45. Herbert Spencer, *The Study of Sociology* (London: Henry King, 1873), p. 335.
46. Herbert Spencer, *The Principles of Biology* (London: Williams & Norgate, 1864–1867), Vol. I, p. 160.
47. Spencer, *op. cit.*, 1860, Note 28, pp. 414–415, quoting from the German chemist Liebig with approval.
48. Spencer, *op. cit.*, 1873, Note 45, p. 330.
49. *Ibid.* pp. 332–333, 334.
50. Herbert Spencer, "Progress: its Law and Cause" (1857), in Spencer, *Essays, Scientific, Political and Speculative* (London: Williams & Norgate, 1868–1874), Vol. I, pp. 3, 29–30.
51. Spencer, *op. cit.*, 1864–1867, Note 46, Vol. I, pp. 193–194.
52. *Ibid.* p. 199.
53. Spencer, *op. cit.*, 1873, Note 45, p. 334.
54. *Ibid.*, p. 329.
55. Spencer, *op. cit.*, 1857, Note 50, pp. 31–33.
56. Herbert Spencer, *First Principles* (London: Williams & Norgate, 7th ed. 1922 (orig. 1862)), p. 323.
57. *Ibid.*, p. 321.
58. Spencer, *op. cit.*, 1864–1867, Note 46, Vol. I, pp. 10–16, 153–158; Spencer, *op. cit.*, 1922, Note 56, Part II, Chap. XIX.
59. Spencer, *op. cit.*, 1864–1867, Note 46, Vol. I, p. 410.
60. *Ibid.*, pp. 80, 168.
61. Herbert Spencer, *Autobiography* (London: Williams & Norgate, 1904), Vol. I, p. 552.
62. Spencer, *op. cit.*, 1922, Note 56, p. 438.
63. *Ibid.*, pp. 159, 359–362.
65. Herbert Spencer, "Lord Salisbury on Evolution," *Nineteenth Century* **38** (1895), 757.
66. Spencer, *op. cit.*, 1904, Note 61, Vol. II, pp. 100–101. See also Duncan (ed.), *op. cit.*, 1908, Note 35, pp. 558–559.
67. Herbert Spencer, "Mr. Martineau on Evolution" (1872), in Spencer, *Essays, Scientific, Political and Speculative* (London: Williams & Norgate, 1868–1874), Vol. III, p. 241.
68. Spencer, *op. cit.*, 1864–1867, Note 46, Vol. I, p. 455.
69. *Ibid.*, p. 435.
70. *Ibid.*, p. 153. Spencer's criticism of Lamarck is on pp. 403–404.
71. Antonello La Vergata, "Il lamarckismo fra riduzionismo biologico e migliorismo sociale," *Intersezioni* **10** (1990), 87–108.

THE SUPERORGANISM METAPHOR: THEN AND NOW

SANDRA D. MITCHELL
University of California, San Diego

As John Maynard Smith[1] has said:

Our choice of models, and to some extent our choice of words to describe them is important because it affects how we think about the world . . . our choice of model decides what phenomena we regard as readily explicable, and which need further investigation. (p. 120.)

My paper is concerned with one such choice, namely the use of superorganism metaphor in biological theory. In particular I am interested in the recent arguments for the "revival" of the superorganism metaphor for the study of social insects. Briefly, superorganism metaphors and theories take the individual organism as a model of functional integration or cooperation of parts and extend that model to describe and explore social groups of individuals. Just as individual cells in the body cooperate in the development, maintenance, and reproduction of the organism, so too (it is suggested by the metaphor) do individuals cooperate in the development, maintenance, and reproduction of a colony or a society.

Metaphorical transfers between organisms, social insects, and human societies has had a long history. Yet each historical and scientific context has stamped its own character on the use of such language. After decades when mention of superorganism was anathema in social insect studies,[2] in the last ten years, there have been multiple pleas to "revive" the superorganism. This revival "movement" includes the following scholars:
- E. O. Wilson,[3] who has claimed that, although no one used superorganism language through the reductionistic and empirical trends in entomology from the 1950s until now, even then it had a significant, albeit "semi-conscious" role. Furthermore, "the time may be at hand for a revival of the superorganism concept."[4]
- Charles Lumsden,[5]
- Thomas Seeley,[6]

– David Sloan Wilson and Elliott Sober.[7]

The about-face concerning the use of the superorganism metaphor raises an important question: How does this "choice of models and choice of words" affect the way one studies social insects? In effect, what does a superorganismic revival for theories of social insects amount to? The answer to this analytic question is further complicated by two factors: one is that the words chosen are explicitly metaphorical, and hence require the explication of the transfer of conceptual content between the primary and secondary contexts of application. This could be called the "horizontal dimension of transfer." The second factor is that the plea seeks to revive a framework that had originally been voiced in a very different historical and scientific context. To see what the theoretical and empirical content of the so-called revived superorganism model is now, one must also investigate the "vertical transfer" of content from the antecedent application of the metaphor to its contemporary instantiation. The story of the revival of the superorganism metaphor thus concerns the transfer of language and content both from one scientific context (theories of the organism) to a second scientific context (theories of social groups) and from one historical period (early twentieth century United States) to another (contemporary United States). It is a metaphor that is made to do scientific work by offering a conceptual framework in which to raise questions, generate explanations, and explore testable consequences in the study of social insects.

Critical Theory of Metaphors

Earlier in this century, philosophers of science had tended to view metaphors and models as merely heuristic, nonessential, and hence dispensable for science. While perhaps relevant to the context of discovery, like hallucinations and other bogeymen of the peculiarly human psyche, it was thought that metaphorical language did nothing for explanation, justification, or the rational development of scientific theories. According to Bono,[8] the "standard" view, developed in the scientific revolution in the seventeenth century, is that metaphors "introduce inappropriate, not-literary meanings into science, contaminating ... precise and stable meanings." That metaphorical language "compromises scientific inquiry and is to be avoided" (p. 62). But this picture of science is being replaced by one in which metaphors are seen as ubiquitous, important, and powerful in scientific practice and hence demand critical analysis. Stepan argues that "we need a critical theory of metaphor in science in order to expose the metaphors by which we learn to view the world scientifically, not because these metaphors are necessarily 'wrong'; but because they are so powerful."[9] Philosophers have been engaged in the development of such a critical theory. The literature on this subject is vast, but I will describe one trend within it that motivates the discussion of the organism metaphor that follows.

Metaphors were once understood to make explicit, definite claims of similarity or analogy between the primary reference or context and the secondary one. On this, the "comparison" view, communication with metaphorical language was possible only when all parties agreed with the specific similarity claims. With an admission of the less than precise meanings necessary for communication, even for scientific discourse, many philosophers now defend a view of metaphor as more "open-ended" and "interactive." Included here are Max Black, Richard Boyd, and Thomas Kuhn.[10]

Even with the loosening up of the criteria for metaphorical understanding, there are differences within this modern camp. Black, for example, claims that metaphors are useful only in pretheoretical stages of science, in pedagogical interactions, or in nontechnical popularization. Boyd and Kuhn, however, find a place for metaphors in the construction and development of theory in mature science. Boyd refers to these as "theory constitutive" metaphors, whose inductive open-endedness actually invites scientists to explore similarities (rather than require explicit ones). Metaphors are attempts to accommodate our language to not yet discovered causal features of the world. Kuhn (though differing with Boyd on the realistic or constructivist interpretation of such endeavors) agrees with this basic approach. Both claim (in Kuhn's words) that "metaphor plays an essential role in establishing links between scientific language and the world. . . ."[11] It follows that metaphors should not be left unjustifiable and "semi-conscious". Rather, they, like any other theoretical postulate should be accepted critically. Boyd says:

One should employ a metaphor in science only when there is good evidence that an important similarity or analogy exists between its primary and secondary subjects. One should seek to discover more about the relevant similarities or analogies, always considering the possibility that there are no important similarities or analogies, or alternatively, there are quite distinct similarities for which distinct terminology should be introduced. One should try to discover what the "essential" features of the similarities or analogies are, and one should try to assimilate one's account of them to other theoretical work in the same subject area – that is, one should *attempt* to explicate the metaphor.[12]

In addition to the horizontal comparisons between the two domains of application of a metaphor, a critical analysis of a specific metaphor requires the comprehension of the sources and implications of the use of the metaphor both within and outside the confines of scientific discourse. Metaphors are both enabling and constraining. While they structure our perceptions, allowing us to "see" causal structures in new domains, via the similarity to such structures in known domains, they at the same time proscribe certain observations, blind us to certain descriptions and awareness. While they serve as a program for research, they also run the danger of being mistaken for reality. And finally, while performing all these onerous tasks in scientific theorizing and practice, metaphors at the same time reflect the social structure, scientific organization and cultural milieu in which they are invoked.[13] A critical analysis, ideally, should expose all of these aspects of a scientific metaphor. In this study, I will undertake only a small part of this larger task.

Horizontal Transfer

In general, scientific theories make two sorts of ontological commitments in explaining phenomenological experience: one concerning what entities exist, and the other concerning what forces acct upon those entities. The revival of the superorganism metaphor attempts to expand the ontology of Darwinian theory to include superorganisms as real, explanatory entities. I believe there are two different sources for the appeal to the reality of this ontological level, each resting on a difference set of biological processes. (See Table I) E. O. Wilson[14] and Lumsden[15] describe superorganisms as entities subject to sociogenesis, the analogue of morphogenesis, while Wilson and Sober[16] and Seeley[17] invoke superorganisms as entities subject to the forces of natural selection operating at a group level. Individual organisms are paradigmatic examples of both development and selection. They display ontogenetic processes of meiosis, mitosis, and cell differentiation and specialization; hence a collection of individuals may be seen as a superorganism if it is subject to a similar ontogeny. As E. O. Wilson says,

The workers of advanced insect societies are not unlike cells that emigrate to new positions, transform into new types, and aggregate to form tissues and organs.[18]

Individual organisms are also the paradigmatic subject of natural selection and the locus of adaptation, hence a collection of individuals may be seen as a superorganism if it is similarly subject to natural selection and its traits are adaptations at that level. Seeley thus claims,

> It seems correct to classify a group of organisms as a superorganism when the organisms form a co-operative unit to propagate their genes, just as we classify a group of cells as an organism when the cells form a co-operative unit to propagate their genes.[19]

While the first approach emphasizes the developmental processes affecting the organism, the latter is concerned exclusively with how natural selection operates on organism-like entities. In this paper I will focus on the latter trend, investigating the relationship between one contemporary selectionist superorganism model and the progenitor metaphor developed by William Morton Wheeler in the early part of this century.

Table I. Horizontal Transfer

	Organism	*Superorganism*
Entities	Cells Organism	Individuals Superorganisms
Organization	Weismann's preformationist development	Complete convergence of individual interests
Processes	Ontogenesis Selection on indiviudal only	Sociogenesis Selection on colony only

Vertical Transfer

The historical dimension of the investigation of the superorganism metaphor is required by the explicit desire on the part of contemporary scientists to "revive" the superorganism of W. M. Wheeler[20] of the 1910s and 1920s rather than to introduce a newly coined, or newly framed, metaphor born of the contemporary scene. I will suggest that there are significant differences between these periods which, I believe, are sufficient to cast doubt on the desirability of reviving the metaphor.

One way to compare the two scientific contexts is to see what elements make up the respective "contrast classes" for the superorganism defenders in each. By so doing, I do not intend to draw sharp dichotomies between naturalist and experimentalist, physiological and evolutionary questions,[21] or vitalism and Darwinism,[22] but rather to make apparent the range of views which allows for a middle ground. American biologists in particular tended to explicitly endorse the dissolution of dichotomies and the expansion of possibilities. For example, Charles Otis Whitman, a teacher and colleague of Wheeler's asserted that the tendency to dichotomise into mechanism versus vitalism is destructive and confuses the important question.[23]

When W. M. Wheeler invoked the organism as an appropriate metaphor for the study of social insects in 1911,[24] what views was it designed to replace? What implicit choices enabled the 1911 reader to situate Wheeler's theory? Similarly, when Wilson and Sober or Seeley defend the superorganism revival (though not without some qualifications) to what current distinct alternatives is this approach supposed to be superior? The metaphor appears to espouse the same transfer of structures from the primary domain of application of cellular organization of the body to the secondary domain of division of labor, and specialization in the functional organization of societies. However, though the metaphorical word used may be the same, they well may be invoking completely different meanings.

Let us look briefly at Wheeler first.[25] Wheeler began work in biology as a taxonomist and developmental biologist, publishing his first papers cataloguing flora and fauna in his native Milwaukee in the late 1880s. He received his doctoral degree at Clark University in 1892, writing a thesis on insect embryology, and immediately took a position at the new University of Chicago – both under the influence of C. O. Whitman. He spent the next academic year, 1893–94 in Europe where he divided his time evenly between Theodor Boveri's lab in Wurtzburg and at the zoological station in Naples. Wheeler was also a "regular" at the Marine Biological Laboratory in Woods Hole, Massachusetts, in its earliest days. In 1899, Wheeler moved to the University of Texas at Austin where he was to fall in love with the study of social insects, and with ants in particular. In Texas, Wheeler found himself virtually surrounded by numerous unidentified species of ant. He was to devote most of the rest of his career – with academic positions at the New York Museum and his long-term association with Harvard University – to the investigation of social insects. Indeed his observations are still cited as primary resources; his theoretical and conceptual frameworks have been less long lived.

Wheeler's superorganism theory was proposed to explain social insects.[26] Wheeler saw this conceptual schema as a corrective to Darwinism because ". . . the struggle for existence is not more than half the truth . . . To us it is clear that an equally pervasive and fundamental innate peculiarity of organisms is their tendency to cooperation or "mutual aid."[27] The appeal to "mutual aid" is, of course, to Kropotkin. It is worth investigation why the "mutual aid" or "cooperation" metaphors employed by Kropotkin found resonating voice in the United States in the 1920s, in lieu of Darwin's "struggle for existence" metaphor. Daniel Todes suggests that Darwin's "struggle for existence" competitive individualist metaphor reflects the "unsurprising fact that he (Darwin) shared the ideological outlook of his class, his circle and family, and that such language might identify the author as bourgeois Malthusian, or perhaps, typically British."[28] Todes goes on to argue that Russia, the context giving rise to Kropotkin's "mutual aid" metaphor, was not a land suffering from population pressure, but rather from the harsh elements against which people had to work collectively to survive. Perhaps the United States of the late nineteenth and early twentieth centuries, and especially the Texas landscape where Wheeler developed his devotion to the study of social insects, was more geographically and demographically like Kropotkin's Siberia than Darwin's London. However, ecological determinism is a strong thesis, and there is little evidence to suggest it played a major role in Wheeler's superorganic metaphor.

Wheeler's[29] definition of the concept of an organism, which would form the backbone of his theory of superorganisms, included three features: organization for nutrition, reproduction, and protection.

An organism is a complete, definitely co-ordinated and therefore individualized system of activities, which are primarily directed to obtaining and assimilating substances from an environment, to producing other similar systems, known as offspring, and to protecting the system itself and usually also its offspring from disturbances emanating from the environment. The three fundamental activities enumerate in this definition, namely nutrition, reproduction, and protection, seem to have their inception in what we know, from exclusively subjective experience, as feelings of hunger, affection, and fear, respectively.[30]

Wheeler argued that an animal colony is properly identified as an organism, and not merely an analogue of one, because it displays the following three features: 1) individuality – it behaves as a unitary whole, has identity in space, resists both dissolution and fusion with other substances; 2) duality of composition – it displays the Weismannian division of germ plasm and soma; and 3) ontogenetic and phylogenetic development. Wheeler's superorganism model provided him with a framework to describe and explain the observed cooperation of individual insects.

One can begin to define the boundaries of Wheeler's superorganism metaphor by describing what it was meant to exclude. First, he seems to be rejecting a *narrow reading of Darwin*. It was narrow by emphasising competition and struggle and ignoring cooperation and mutualism. Darwinian competition was between individuals, and most severe between individuals of the same species. In this sense Wheeler can be viewed as antiindividualistic when competition excludes cooperation. Of course, Darwin also was prompted by the phenomena of sterile insects to entertain ideas of individual sacrifice to group benefit: "If such insects had been social, and it had been profitable to the community that a number should have been annually born capable of work, but incapable of procreation. I can see no very great difficulty in this being effected by natural selection."[31] Appeal to cooperation, at least in the case of social insects, was not an instance of anti-Darwinism, but rather only of a narrow reading of Darwin.

This characterisation of Wheeler is further supported by his defence of Darwinism in arguments with Father Erich Wasmann, a Jesuit living in Holland. Wasmann studied insect parasitism and guest species (one of Wheeler's favorite subjects) and was an outspoken anti-Darwinist who thought natural selection was inadequate for explaining the relationships he observed. His solution was to propose new instincts and new forces in its stead. Wheeler did not reject a role for natural selection, rather he wished to expand Darwinism to include nonindividualistic perspectives as well.

Wheeler has been accused of another brand of anti-Darwinism, namely of being a "crypto-vitalist."[32] He merits this label by his academic association with both the University of Chicago and Harvard University which, according to Ghiselin, housed biologists, philosophers, and social theorists "who often explicitly denied that they were vitalists, but whose positions amounted to much the same thing."[33] This accusation is the result, I believe, of Ghiselin's strict dichotomization of views of the period. It seems that for him any view that is not individualistic, mechanistic, and neo-Darwinian in the sense of the impending synthesis falls into the category of vitalism. But this is unfair to Wheeler. Although he was attracted initially to Bergson, he later departed from vitalistic views, remaking that "the resort to such metaphysical agencies (as *elan vital* and others) has been shown to be worse than useless in our dealings with the inorganic world and it is difficult to see how they can be of any greater service in understanding the organic."[34] In 1928 he referred to *elan vital* and entelechy as "little more than fetishes."[35] Perhaps his most colorful rejection of vitalism appeared in 1911 in reference to Driesch's entelechy.

His angel child ... comes, to be sure, of most distinguished antecedents, having been mothered by the Platonic idea, fathered by the kantian Ding-an-sich, suckled at the breast of the scholastic *forma substantialis* ... but nevertheless, I believe that we ought not to let it play about in our laboratories, not because it would occupy any space or interfere with our apparatus, but because it might distract us from the serious work in hand. I am quite willing to see it spanked and sent back to the metaphysical house-hold.[36]

Though eschewing association with the vitalists, Wheeler also rejected the genetic reductionism born of the synthesis of Weismann, Mendel, and population genetics which makes external selection the only relevant cause for explaining biological diversity, adaptation, and evolution. Internal organization and development were left out of the synthesis and Wheeler's early work on embryology may have predisposed him to reject a Darwinism that ignored development. Writing in reference to the "exquisitely adapted" specialisation in termites and ants he says:

And it strains our credulity to be told that such forms arise either from peculiar genes popping out of nowhere into the germ plasm or develop gradually under the guidance of natural selection from forms which, so far as we can see, must have an equal or even greater survival value. When we encounter such *impasses* as the foregoing, instead of embracing the Aristotelian Entelecheia ... or joining the apostles of the survival of the fittest and forever croaking "natural selection!," it is surely more commendable to sit down in the laboratory or in the field and say nothing but "*ignoramus*" till we have made a much more exhaustive behavioristic and physiological investigation of the phenomena.[37]

Nevertheless, Wheeler did not reject Darwinism, he supported a broader interpretation of Darwinian theory, one that could redress the lacunae of developmental, internal processes made invisible by the genetics of his time.

In summary, Wheeler invoked the superorganism metaphor for social insects against a strictly external selectionist interpretation of Darwin. This put him in the company of many Darwinian critics of the time, those concerned with internal processes, like embryogenesis and other developmental considerations, who found that individual competition in response to external environmental conditions was insufficient to ground the kind of phenomena they were describing. Wheeler was, after all, a taxonomist constantly reminded of the variations distinguishing the thousands of species of ants, bees, wasps, and termites and an embryologist who saw complexity developing over the lifetime of the organism. While competition between genetic individuals may not have been sufficient, neither did it require embracing a vital force, like Dreisch's entelechy, or Maeterlinck's spirit of the hive, or Bergson's *elan vital*, an agency existing separately from matter which directs, selects, or arranges the structure of matter that constitutes life. Wheeler seems to have occupied a position that corresponds to what Timothy Lenoir[38] has described as "vital materialism," or a midpoint in Jane Maienschein's suggested continuum between mechanistic materialism and vitalism. Wheeler vehemently opposed vitalism, while at the same time rejecting the complete reduction of biology to a form of physics.

The Current Superorganism

Wilson and Sober[39] use the superorganism concept to promote a hierarchical theory of selection. Their interpretation of Darwinian natural selection logically entails the rejection of privileging the individual organism as the only level at which selection can operate. Again, one can locate their position in contrast to a narrow interpretation of Darwinism, the latter being what S. J. Gould has called the "hardening of the synthesis."[40] For them, higher levels of organization – groups, species, or superorganisms – are reasonable candidates for selection. Furthermore, Wilson and Sober argue that not only is it theoretically possible to have such levels of selection, but in fact the conditions for their realization are not overly restrictive (contra the early objections to group selection),[41] and, in fact, they do exist in nature. However, it is ironic that the strategy they use to reject the hegemony of a single individual level of selection has features that are similarly restrictive. Just as the individual organism prejudice tends to privilege a single level as the only one relevant for selection, blinding biologists to the other levels, Wilson and Sober's superorganism theory privileges a particular form of functional organization at the group level, blinding one to diversity in this realm. The individual single-level theory that they reject by invoking the superorganism metaphor is merely replaced by a group single-level theory, thereby obscuring the multiple levels at which selection can and does operate in social insect societies.[42]

For Wilson and Sober, a superorganism is "a collection of single creatures that together possess the functional organization implicit in the formal definition of organisms" (p. 339). They adopt the Random House dictionary definition of an organism: "a form of life composed of mutually dependent parts that maintain various vital processes" (p. 339). While this formal definition is consistent with Wheeler's articulation of three essential activities of nutrition, reproduction, and protection, (the various vital processes), Wilson and Sober's use of functional organization is further restricted to organization around reproduction. They summarize their superorganism model as follows:

i) A popoulation is subdivided into a number of groups.
ii) Groups vary in properties that affect the number of dispersing progeny (group fitness).
iii) Variation in group fitness is caused by underlying genetic variation that is heritable.
iv) No differences exist in the fitness of individuals within groups.

Wilson and Sober (pp. 340, 343) claim that functional organization is the key to identifying at which level natural selection is operating. Functional organization is that which allows individual organisms or superorganisms to successfully survive and reproduce. When the four conditions are met, they claim that natural selection will endow groups with the same properties of functional organization that we normally associate with individual organisms (p. 342).

Implicit in their development of a superorganism is a particular form of functional organization of the parts of the paradigmatic individual organisms – namely a Weismannian preformationist developmental schema. In this view there is a sharp division of germ plasm from soma and the germ line is sequestered so early in development that there is no opportunity for competition between somatic cells to have evolutionary consequences. The Weismannian legacy makes variation during the development or experience of the parts (soma or workers) irrelevant to evolution because it entails the complete, early separation of germ cells from somatic cells. Weismann's theory, by requiring the rejection of Lamarckianism, was a cornerstone of the modern synthesis.

However, Leo Buss[43] has convincingly argued that this very ideal image of the functional organization of cells in an organism fails to be approximated by most taxa or throughout most of geological time. Buss lists all the taxa and thus displays how rarely the Weismannian ideal is met in the primary domain of the biology of individual organisms. So too is the idealized Weismannian superorganism rarely approximated by the range of organization found in social insects.

The Variety of Functional Organization in Social Insects

The Wilson-Sober model of superorganisms requires that there be no differential "fitness" between units at levels lower than the colony, or between levels (their condition iv). Differential fitness is differential survival and reproduction for them, the conditions required for there to be evolutionary consequences of selection at a given level. However, competition at levels lower than the colony can occur among individual colony-mates, among males, or even among genes, for example, with meiotic drive or segregation distortion genes. Competition can even occur between workers and queens over determining the colony sex ratio. Any of these would make the colony something less than a superorganism. The superorganism ideal may be achieved by either eliminating all genetic variation among the subcolony units (thereby removing a necessary condition for selection to operate at that level) or by suppressing all competition among genetically variant subcolony units at all levels (i.e., by creating mechanisms, like meiosis, for making such variation effect no change in distribution of traits in offspring).

Rather than produce a complete table of social insects, listing those who do and those who do not qualify under the stringent condition of no variation in fitness at below colony levels, I will instead point to some of the sources of within-colony competition that result in most, if not all, social insects failing the criterion.

Polyandry, for example, describes the situation for many species of ants, wasps, and bees in which the queen of a colony mates with multiple males. This reproductive strategy can give rise to many different sources of within-colony competition. First of all, males must surely compete for matings with the queen on her nuptial flights, as they do compete for resources within the colony during their early development and outside it while they prepare for reproduction. The queen mates during a few days of her life, then forms a colony, and dispenses the stored sperm over her lifetime, which for honey bees can be several years. Thus competition between males may reflect fitness differences for the ability to mate at all, for quantity of sperm deposited, for longevity of sperm for use during the queens lifetime, and for the ability of sperm to fertilize queen eggs rather than worker eggs. There is evidence that sperm use by the queen is not random. Given that males differ genetically, such fitness differences, which obtain below the level of the colony, are likely to have evolutionary consequences.

In addition, given that multiple males father the "body" of female workers, different subfamilies are formed in a colony. (All workers share the same mother, but some will share the same father and differ from others in that respect). In the social Hymenoptera, females develop into queens or workers on the basis of what they are fed. If workers that feed larvae can recognize those that are more closely related to them, i.e., members of their own subfamily, then competition may arise among individuals of different subfamilies over which larvae are raised as queens versus workers. Competition can be eliminated if workers are unable to distinguish among potential queen larvae. Empirical work with honey bees has demonstrated that workers have the ability to discriminate among larvae under some experimental conditions.[44]

In order to eliminate polyandrous competition at these levels, the following conditions must be met: 1) mates of queens contribute equal numbers of sperm; 2) sperm will have equal viability and longevity; 3) sperm are used randomly to fertilize eggs; and 4) workers are unable to discriminate among larvae with different subfamily relationships. These are very unlikely conditions.

In addition, polygyny, or multiple queens, is widespread among termites, wasps, and ants.[35] Here competition within the colony can arise through differential egg laying by the different queens, and also between the subfamilies generated by the multiple queens.

Social insect colonies thus fail to be superorganisms by failing to meet condition iv, the complete suppression of competition at levels internal to the organism, the extension of the Weismannian ideal model to social groups. There are good reasons to believe that there is also reproductive competition among colony units at lower organizational levels.[46] The most likely candidates to qualify as a Wilson and Sober superorganism would have a single queen who mates with a single male and workers that lack functional ovaries. Admittedly, some ant species fit this ideal; however, they represent only a small fraction of the different kinds of colony organization displayed by social insects. Insect societies are rich in diversity with respect to how they are functionally organized for reproduction, just as Buss has shown that individual organisms are similarly diverse in their internal functional organization. The social insect colonies may, even when different in their reproductive organization, be similar with respect to organization around defence or nutrition, two aspects of a superorganism that Wheeler emphasized and Wilson and Sober ignore. That these strong conditions are intended by Wilson and Sober is shown by their claims that (p. 346) "social insect colonies really do cease to be superorganisms, to the extent that natural selection operates within single colonies" (p. 346) or "the essential criterion is absence of within-group selection, which may be accomplished either by creating genetically uniform groups, or by suppressing the differential reproduction of genetically diverse groups" (p. 348).

Wilson and Sober promote the revival of the superorganism, but it is not Wheeler's metaphor that they invoke. The major differences are that Wilson and Sober limit functional organization to organization around reproduction, ignoring the nutritional and protection organization which Wheeler included,[47] and while Wheeler found selective scenarios insufficient to explain superorganismic phenomena, Wilson and Sober make selection at the group level the defining characteristic.

The scientific context for Wilson and Sober, the contrast class in which to situate their superorganism, is the individualistic theories that "have dominated evolutionary biology for the last twenty years" (p. 353). The superorganism is supposed to provide a "radical departure" from this narrow interpretation of Darwinism by acknowledging multiple levels of selection. They promote a hierarchical picture of natural selection, one not restricted to a single level, yet one logically consistent with the teachings of Darwin. In these regards, Wheeler and Wilson and Sober are similarly fighting their respective hegemonic, narrow interpretation of Darwinism, offering radical departures that nevertheless do not step outside the bounds of what is "truly" Darwinian. The Wilson and Sober superorganism, however, depends on the Weismannian ideal organism and not the contemporary picture of a plurality of organismic organization. The horizontal transfer of content with respect to the organization of the whole from its parts seems to be oblique in their case. The Weismannian organism, when elevated to the Wilson and Sober superorganism, contravenes their very goal of supporting a broad and hierarchical Darwinian theory. The Wilson and Sober superorganism makes invisible the diverse array of truly hierarchical organization found in social insects on which hierarchical selection can operate. Their agenda is to rid biology of the myopic picture of selection operating only at the individual organism level; their explicit reasons for reviving the superorganism model concern the inconsistency of adopting a theoretical framework that focuses only at the individual level. And while in fact they admit within-organism selection (in the primary context of the metaphor) as well as within-colony selection (in the secondary context of the metaphor), their overly restrictive model of the superorganism obscures these admitted facts.

Conclusion

By exploring the transfer of content and contexts of the original superorganism metaphor for social insects developed by William Morton Wheeler and comparing it with the "revived" superorganism of Wilson and Sober, the complex ways in which metaphors acquire their meanings begins to be clarified. The appeal to ancestral authorities to inspire allegiance to a theoretical construct may have rhetorical success, but in this case can do so only by virtue of ignoring the differences that made appeal to the metaphor salient in its own historical and scientific milieu. The Wilson and Sober superorganism's credentials as a "revived" form of Wheeler's original metaphor can only be judged by such a comparison.

Whether the superorganism metaphor and model proposed by Wilson and Sober should be accepted, regardless of its pedigree, depends on the content and promise of such a conceptual framework. It's content is understood by means of the horizontal transfer of structure from the primarily context of application – the organism – to its new domain – social insects. On that score I have argued that, ironically, the model obscures our vision of just the sort of variability Wilson and Sober wish to highlight and it does this by transferring the Weismannian ideal structure of the organism to define the structure of the superorganism.

Notes

1. John Maynard Smith, "How To Model Evolution" in J. Dupre (ed.), *The Latest on the Best* (Cambridge, Mass.: MIT Press, 1987), pp. 119–131.
2. Edward O. Wilson, "The Superorganism Concept and Beyond" in M. Chauvin, M. Noiret and P. Grasse (eds.), *L" effet de groupe chez les animaux. Colloque internationaux du centre national de la recherche scientifique*, No. 173 (Paris, 1968), pp. 27–39; and *The Insect Societies* (Cambridge, Mass.: Belknap Press. 1971).
3. Edward O. Wilson, "The Sociogenesis of Insect Colonies," *Science* **228** (1985a), 1489–1495; and "The Principles of Caste Evolution" in B. Holldobler and M. Lindauer (eds.), *Experimental Behavioral Ecology and Sociobiology* (Sinauer, 1985b).
4. *Ibid.*, 1985b, p. 317.
5. Charles Lumsden, "The Social Regulation of Physical Caste: The Superorganism Revived," *Journal of Theoretical Biology* (1982), 749–781.
6. Thomas Seeley, "The Honey Bee Colony as Superorganism," *American Scientist* **77** (1989), 546–553.
7. David Sloan Wilson and Elliott Sober, "Reviving the Superorganism," *Journal of Theoretical Biology* **136** (1989), 337–356.
8. James J. Bono, "Science, Discourse, and Literature: The Role/Rule of Metaphor in Science," in Stuart Peterfreund (ed.), *Literature and Science: Theory and Practice* (Boston: Northeastern University Press, 1990), pp. 59–90.
9. Nancy Stepan, "Race and Gender: The Role of Analogy in Science," *ISIS* **77** (1986), 261–277 (quote p. 277).
10. Max Black, "More about Metaphor" (19–43), Richard Boyd "Metaphor and theory change: What is "metaphor" a metaphor for?" (356–408) and Thomas Kuhn "Metaphor in Science" (409–419) in A. Ortony (ed.), *Metaphor and Thought* (Cambridge: Cambridge University Press. 1979)
11. Kuhn, *ibid.*, 1979, pp. 415–416.
12. Boyd *op. cit.*, 1979, p. 406.
13. Gregg Mitmann, this volume.
14. Wilson, *op. cit.*, 1971, 1985a, 1985b.
15. Lumsden, *op. cit.*, 1982.
16. Wilson and Sober, *op. cit.*, 1989.
17. Seeley, *op. cit.*, 1989.
18. Wilson, *op. cit.*, 1985a, p. 1492.
19. Seeley *op. cit.*, 1989, p. 548.
20. William Morton Wheeler, "The Ant Colony As an Organism," *Journal of Morphology* **22** (1911), 307–325; *Social Life among the Insects* (New York: Harcourt, Brace, and Co., 1923); *The Social Insects: Their Origin and Evolution* (New York: Harcourt, Brace and Co., 1928).
21. See Garland Allen, *Life Science in the Twentieth Century* (Cambridge: Cambridge University Press, 1978).
22. See Michael Ghiselin, *The Economy of Nature and the Evolution of Sex* (Berkeley: University of California Press, 1974).
23. Charles Otis Whitman, "Prefatory Note," *Biological Lectures* **1894** (1895), pp. iii–vii. See also Jane Maienschein, "Shifting Assumptions in American Biology," *Journal of the History of Biology* **14** (1981), 89–113.

24. Wheeler, *op. cit.*, 1911.
25. Mary Alice Evans and Howard Ensign Evans, *William Morton Wheeler, Biologist* (Cambridge, Mass.: Harvard University Press, 1970).
26. Although Wheeler also pronounced on its possible extension to human societies, was interested in sociology, was the teacher of Alfred C. Kinsey, etc., it cannot be said that he developed any serious theoretical or experimental work in that area.
27. Wheeler *op. cit.*, 1923, p. 3.
28. Daniel Todes, *Darwin Without Malthus* (Oxford: Oxford University Press. 1989), p. 13.
29. Wheeler, *op. cit.*, 1911, p. 5.
30. Wheeler, while providing this "formal definition," admitted that it was impossible to adequately define an organism because it consists of a continuous, ever changing and never finished set of processes.
31. Charles Darwin, *On the Origin of Species. A Facsimile of the First Edition*, 1859 (Cambridge, Mass.: Harvard University Press. 1964), p. 236.
32. Ghiselin, *op. cit.*, 1974.
33. Ghiselin, *ibid.*, 1974, p. 30.
34. 1926, quoted in Evans and Evans, *op. cit.*, 1970, p. 224.
35. Quoted in Evans and Evans, *op. cit.*, p. 224.
36. Wheeler, *op. cit.*, 1911.
37. Wheeler, *op. cit.*, 1928.
38. Timothy Lenoir, *The Strategy of Life* (Chicago: University of Chicago Press, 1982).
39. Wilson and Sober, *op. cit.*, 1989.
40. Stephen J. Gould. "The Hardening of the Modern Synthesis" in Majorie Grene (ed.), *Dimensions of Darwinism* (Cambridge: Cambridge University Press, 1983), pp. 71–93.
41. See M. J. Wade, "A Critical Review of the Models of Group Selection," *Quarterly Review of Biology* 53 (1978), 101–114.
42. Sandra D. Mitchell and Robert E. Page, Jr. "Idiosyncratic Paradigms and the Revival of the Superorangism" (unpub. ms.).
43. Leo W. Buss, *The Evolution of Individuality* (Princeton: Princeton University Press. 1987), p. 20.
44. P. K. Visscher, "Kinship Discrimination in Queen Rearing by Honey Bees (*Apis mellifera*)," *Behavioural Ecology and Sociobiology* 18 (1986), 453–460; Robert E. Page and E. H. Erickson, "Selective Rearing of Queens by Worker Honey Bees: Kin or Nestmate Recognition?" *Ann. Entomol. Soc. Amer.* 77 (1984), 578–580; K. C. Noonan, "Recognition of Queen Larvae by Worker Honey Bees (*Apis mellifera*)," *Ethology* 73 (186), 295–306; R. E. Page and G. E. Robinson, "Nepotism in Bees." *Nature* 346 (1990), 708; R. E. Page, G. E. Robinson and M. K. Fondrk, "Genetic Specialist, Kin Recognition, and Nepotism in Honey-Bee Colonies," *Nature* 338 (1989), 576–579; Thomas Seeley, "The Honey Bee Colony as Superorganism," *American Scientist* 77 (1989), 546–553.
45. Wilson, *op. cit.*, 1971.
46. See Sandra D. Mitchell and Robert E. Page, Jr. *op. cit.* (unpub. ms.) for more of the biological details.
47. Wheeler included in a listing of the hierarchies of organisms those colonies with nutritional but not reproductive functional organization, *op. cit.*, 1911, p. 6.

DEFINING THE ORGANISM IN THE WELFARE STATE: THE POLITICS OF INDIVIDUALITY IN AMERICAN CULTURE, 1890–1950

GREGG MITMAN
University of Oklahoma

To speak of the "transfer of metaphors between biology and the social sciences" implies a sense of movement from one space into another. Implicit is an understanding of dichotomy, of boundaries to be transgressed. Framing the problem in this way legitimates the disciplinary boundaries central to the professional identity of the biological and social sciences and to their professional authority within the public sphere. For example, one might detail the preponderance of anthropomorphic terms such as "spite" and "slavery" in sociobiology, citing how such transgressions erode the disciplinary authority of the social sciences and breach the nature/nurture divide. Instead of reifying these boundaries, can we bring them into question? Can we alter these disciplinary topographies, shattering the linear movement of words from one space to another, and situate the metaphors within a general field of meaning? Once we begin to explore science *as* culture, the dichotomy erected between the biological and the social begins to break down. In the subject matter of this essay, biology is a human science.[1]

Early twentieth-century debates among biologists in the United States over the nature of individuality serve as a fruitful entry into the borderlands of the biological and social realms. For us, posed at the brink of the twenty-first century, genetic make-up has become an integral part of individual identity, especially with the recent media coverage on DNA typing as court evidence and the "miracles" to be expected from the human genome project. A decisive step in the emergence of this genetic definition of individuality came with the separation of the germ plasm from the soma, fostered by August Weismann's efforts to repudiate the inheritance of acquired characteristics in the latter part of the nineteenth century. If the soma merely serves as the differentation of hereditary substance, then the environment plays little, if any, defining role in the individual's constitution. The self is prior to and independent of the environment in which the fertilized egg develops and matures into an adult. Refracted through the political lens, the genetic self becomes an isolated being, solitary, without need of others for existence. It is not a social self; it is the-self of Hobbes. The inherent nature of the individual, its biology, is separate from the constraining external world imposed from without. As this genetic conception of individuality came to prevail, germ plasm signified nature as the domain of biology and everything beyond, and nurture as the province of the social sciences. Framing the biological self in genetic terms sets up a Hobbesian dichotomy between the individual self and society where individual interests are preserved through the establishment of a neutral framework of rights. The individual is foreground to the community, and the focus is on preservation of individual rights from the intrusive powers of the state, which threaten to erode individuality at every turn. Freedom in this context is understood in the negative sense as "freedom from interference by external forces" that limit an individual's actions.[2] Here a frontier is drawn, as Isaiah Berlin suggests, "between the area of private life and that of public authority," an area where the individual self is independent of social control.[3] It is a vision of self that harmonizes well with the rights-based liberalism of American political discourse in the post-World War II period.

Within the past ten years, alternative definitions of the self outside the framework of rights-based liberalism have received the most support within the field of communitarian democratic theory. For communitarians, the self is not independent of the external world but is "embedded in the story of those communities" from which identity is derived – "whether family or city, tribe or nation, party or cause."[4] Subjectivity, as William Cortlett writes, is "viewed as being located in the world-as-lived." The self is a situated self, a self whose identity emerges from the environmental complex in which it lives. In this instance, the self is a social self, ever dependent on its relations with the external world for meaning and definition. Whereas the part is prior to the whole in rights-based liberalism, the "group is a more fundamental consideration than the individual" in communitarian theory.[5]

These communitarian critiques reopen a window onto the past when a definition of the individual as situated self was a common topic of discussion among biologists in the United States. Throughout the late nineteenth and early decades of the twentieth century, individuality could mean everything from the unity and order in behavior of protoplasm, to hereditary constitution, to something in between. By the late 1940s and early 1950s, consonant with the entrenchment of the modern synthesis and molecular biology, a definition of the individual had narrowed to mean a "unique, genetically homogeneous, entity."[6] Intermingled with early-twentieth century definitions of individuality are a set of keywords such as community, cooperation, freedom, and the common good that were utilized by biologists, but whose etymological roots were derived from political discourse.[7] Exploring the precise etymology by which current definitions of the individual came to fruition and how a web of political, social, and scientific interrelations have bolstered and supported its meaning are the tasks of this essay. In analyzing political keywords used within biological debates on the nature of individuality, I am treating metaphors not as illustrative devices, but as heuristic tools used by scientists to gain the support of diverse audiences through appeals to language constitutive of early twentieth-century while, Anglo-Saxon, American culture. Yet metaphors are not purveyors of culture in any monolithic sense. It is important not to lose sight of the discrepancies and conflicts in meaning that arise when a shared political discourse is utilized by individuals who are members of different scientific subcultures, address different constituencies, and have different political allegiances.

By situating biological discussions of individuality within a changing discourse of American culture from positive to negative freedom, from the common good to individual rights, we can begin to discern how changes in the American political and cultural landscape helped create an environment conducive to a genetic definition of self, while exploring how this discourse became appropriated and rewritten in different cultural domains. To do this, we must look to the period when the genetic view was just one among many definitions of the individual vying for recognition. Reconstruction of the story begins with the problem of individuality in relation to late nineteenth-century American cultural anxieties over laissez-faire liberalism and the construction of selfhood in an emerging mass market society.

In the Interests of the Whole

By the 1890s, the nineteenth-century liberal ideal of American democratic society as an aggregate of atomistic individuals was under serious strain. In the transition from a rural, agrarian society to an urban, industrialized economy, the invisible hand of the marketplace appeared incapable of keeping competing interests in balance. The surge of riots and strikes, the most violent of which included the Haymarket bombing in Chicago in 1886, the Homestead strike of 1892, and the Pullman strike of 1894, coupled with the economic instability of the 1893–97 depression pointed to the need for the imposition of authority to restore a sense of community and order that the traditional Victorian bonds of church and family could no longer provide. During the 1890s, social scientists in the United States began to invert traditional liberal theory, placing much less emphasis on individual autonomy and more on social control. The ideological conflict between the individual and society that had marked the Gilded Age gave way as American sociologists came to recognize the urgent need to understand the processes of socialization that bound individual members into a harmonious whole.[8] Rhetoric of the common good and the public will became the rallying cry among reformers, politicians, and social scientists; individuals, they asserted, should subvert their own interests for the welfare of society. Idealism became the philosophical creed that held this fragmented society in order.[9]

Within the public sector, Progressives looked to the ethos of rational social planning and scientific management to reinstitute authority and check the spread of social decay.[10] But what authority held sway within the private sphere? Eroding faith in religion left God an ineffective cure to heal the moral degeneration of self and institute a sense of control. Caught within the web of an urban marketplace, the individual found a fractured existence, alienated within the factory and unconnected to the community life that had once marked the agrarian landscape of nineteenth-century America. By the turn of the century, the self had become fragmented, "a commodity," as Jackson Lears writes, "to be assembled and manipulated for private gain." It was through consumption, through the purchasing of things, that Americans increasingly came to define individual identity.[11]

Amid what Lears has described as the sense of weightlessness, where was the individual to seek solace, to find a renewal, a regeneration, of self? The problem was especially acute, considering the apparent need for greater social control would seriously threaten individual autonomy. One area to which the fractured self might turn in reestablishing a sense of identity, of wholeness, was biology. The biological self, defined either in the sense of genetic identity or in the innate, unconscious drives of Freud, increasingly became throughout the twentieth century the impenetrable core that remained resistant to the pervasive influences of society. The liberal sociologist Edward A. Ross, for example, looked to racial character and later to "Freud's stubbornly egocentric biological drives" as a place where individual identity was assured.[12]

In the critique of laissez-faire liberalism, which began in the 1890s and continued into the World War I, the concern with social integration and control seemed to outweigh any fears of loss of individual freedom and identity. Indeed, the appeal of idealism to American social scientists in the 1890s, with its neo-Gegelian emphasis on the state as an organism, brought the classic liberal understanding of individual freedom into question. John Dewey's attempts to forge an organic theory of democracy in the 1890s illustrate this shifting relation between the individual and society quite well. Dewey traversed the road from neo-Hegelianism to pragmatism while head of the department of philosophy at the University of Chicago from 1895 to 1905. Throughout, he remained committed to a communitarian view of democracy in which the individual self develops and grows, not in isolation from the external world, but through the interaction with its environment, through the association with other individuals of which it is a part. "It is through association that man has acquired his individuality," wrote Dewey, "and it is through association that he exercises it. The theory which sets the individual over against society, of necessity contradicts itself."[13]

Dewey put forth a theory of democracy that integrated the previous dualism between individuality and community. Although an idealist metaphysics lay at the foundation of his organicism in the 1880s, his immersion in functional psychology had by the turn of the century provided him with a naturalistic foundation psychology for his organicist views. Dewey's famous 1896 paper, "The Reflex Arc Concept," illustrated how such ontological divisions as mind/matter, subject/object, and stimulus/response could be unified without an appeal to some metaphysical absolute. Stimulus and response were not "separate and complete entities," Dewey insisted, but were instead "divisions of labor, functioning factors within the single concrete whole."[14] Adaptation, viewed as the equilibration between organism and environment, was the purposeful end through which stimulus and response were integrated into a functional whole. New environmental conditions called forth new responses, until the individual, through "conscious deliberation and experimentation," resolved the disconnected movements into a harmonious adjustment once more.[15] Thus the individual was continually in the process of self-realization through its activity in the world. Dewey believed that only through the participation in democratic life would the self be fully realized. Here was a vision of individual development congruent with the epigenetic outlook of many embryologists of the period. The community, furthermore, was the vehicle through which conflicts of individual interest would be resolved into the interest of the social whole. And it was the social scientist who could best find the means by which social integration and coordination was assured.

Freedom in this context differs sharply from the negative notion of freedom implicit in an atomistic view of democratic society where the role of government is to ensure the securement of individual right, to protect the private from the constraining forces of the public sphere. Dewey envisioned freedom, not as the absence of constraints imposed by others, but as the freedom of the individual to find his/her place in society. As Robert B. Westbrook writes, freedom in this sense is understood as "the opportunity to make the best of oneself as a social being."[16] In this instance, the social whole is seen as a positive force in the development of individual identity. Positive freedom recognizes individual liberty to be most fully realized when it is congruent with the common good.[17]

There is a strong resonance between these critiques of laissez-faire liberalism and biological discussions of the organism. The emphasis on the common good, the relationship between individual and community, and the conflicts between negative and positive freedom were not only part of the narrative structure of the professional discourse of politicians and social scientists, but of biologists as well. These very issues, I suggest, became written onto the organism and were at the center of debates begun in the 1890s around the inadequacy of the cell theory and the search for a biological definition of individuality, debates that continued in the United States into the period of World War II.[18]

The Lessons of *Volvox*

In an 1890 address before a gathering of fellow biologists at the Marine Biological Laboratory in Woods Hole, Massachusetts, Charles Otis Whitman (1842–1910) inverted the "society as organism" metaphor pervasive in American social science throughout the 1880s and 1890s. "On the same grounds that the sociologists affirms that a society is an organism," Whitman asserted, "the biologist declares that an organism is a society." The metaphor, Whitman insisted, was not based upon "superficial or fanciful resemblances, but upon analogies that lie at the very foundations of organic and social existence."[19] In equating the organism with society, Whitman made the biological social and thereby blurred the boundaries between biology and the social sciences. Moving beyond the parochial discourse of biological science, we can thus inquire into the significance that discussions of the organism played within American culture. If the organism was indeed a society, what particular place did the individual cell have in its constitution? Was the cell a signifier of the atomistic individual that had marked mainstream American liberalism in the Gilded Age?

Whitman's origin story of the evolution of the Metazoa offers suggestive insight into the ways in which late nineteenth-century critiques of laissez-faire liberalism redefined definitions of individuality. While the origins of the cell were, for Whitman, lost within an uncertain and probably unretrievable past, he had fewer doubts about the decisive factors in the appearance of multicellular organisms. In its early history of life on earth, the cell, according to Whitman, had an "independent nomadic life." Such examples could be found within the Protozoa, which had "persistently declined every temptation to part with individual freedom," and had thus excluded "all those larger possibilities of life which we see unfolded in the organic world." Division of labor and the "sacrifice of personal independence for the sake of physiological union" were the decisive steps in evolutionary advancement. The beginning of such "composite individuality" could be found in the *Volvox* colony, which Whitman viewed as the likely predecessor of all higher plants and animal life. Present within the colony were two types of cells: those involved in locomotion and those responsible for reproduction. Together, they comprised a functional whole. In the Volvox colony, the physiological unity of the organism was itself a defining aspect of the cell's individuality.[20]

Yet the lessons that the *Volvox* colony offered to the Protozoa had become lost in American society. Individual interests and needs were being pursued at the expense of social welfare. While "specialization had . . . forced its way to the front, organization . . . lags lamentably behind the needs of the times."[21] Whitman, who became the first chairman of the zoology department at the University of Chicago in 1892, perhaps knew only too well the centrifugal forces at work in the urban metropolis, forces that fragmented and fractured the individual into a thousand different selves. Looking around the city of Chicago in 1892 when the population had surpassed one million, the fifty-year-old biologist may have agreed with Jane Addams that the "social organism" had "broken down."[22] Whitman longed for "that kind of organic association which permits each unit to work for itself while making it the servant of all the rest." Implicit in his analysis was the assumption that individual freedom is best realized in "subordination to the entire welfare."[23] He envisioned freedom in the positive sense of Dewey, hoping to align the interests of the individual with those of the group – in this case, the emerging community of professional biologists whose support he desperately needed to ensure the success of the Marine Biological Laboratory.

Professional affiliation increasingly became an important therapeutic for the restoration of identity in late-nineteenth-century American culture, and the professionalization of biology in the United States was one of Whitman's central concerns. In the Marine Biological Laboratory, Whitman found an idyllic place for the establishment of professional community. Whitman offered a stern reminder that his fellow biologists not forget the "centripetal forces" that prevent science from "flying into disconnected atoms." Organization was an essential corollary to specialization. From the 1890s until his death in 1910, Whitman insisted on the importance of organization as a guiding force in biological and social evolution. By appealing to an idealist, teleological concept of organization, Whitman, like Dewey, worked with the inherent tensions between the professional interests of the individual and those of the group. The loss of freedom that would result in the sacrifice of individual welfare to the good of the group was trivial compared to the benefit that each individual would receive through "the power and influence of the whole."[24]

Whitman's idealist emphasis on the organism as a whole made him suspicious of biological theory that relied upon the cell as the fundamental unit of organization. In a paper read at the Zoological Congress of the World's Columbian Exposition in 1893, Whitman challenged the cell theory's preeminent role in embryological explanation. The cell doctrine of Schleiden and Schwann had focused attention on the cell as an independent life, a view that had become magnified with the cascade of cytological discoveries in the latter half of the nineteenth century. From the perspective of cell theory, "higher organization" began with "cell formation" and reached its "fullest expression in the mutuality of constituent cells." The cell had become the elementary building block; the organism was merely the result of a "building process" achieved through division and precise patterns of cleavage.[25] Wilhelm Roux's mosaic theory of development, and its alliance with Weismann determinism, was a perfect example of the cell doctrine redressed in new garb. Development, in Roux's theory, occurred by a qualitative parceling out of hereditary particles at cell division, so that each cell became a particular type. Thus, the pattern of the organism was determined by internal forces situated within the cells themselves. The organism, in this instance, was a "community of individualities, bound together by interaction and mutual dependence."[26]

Whitman objected to this particular rendition of the organism as a cell state. He was much more sympathetic to the positions of Hans Driesch and Oscar Hertwig, citing experimental evidence that indicated cleavage patterns alone did not determine organization. The organism was not built up from the interaction, struggle, and symbiotic associations of individualities such as cells or biophores; instead the organism existed as an individual from the onset of development, and "continued as an individuality through all stages of transformation and subdivision into cells." "At every turn," Whitman insisted, "the organism dominates cell-formation, using for the same purpose one, several, or many cells, massing its material and directing its movements, and shaping its organs, as if cells did not exist, or as if they existed only in complete subordination to its will."[27] The individual cell did not exist prior to, and independent of, the social organism; rather, the social whole was an essential part of individual identity.

It is significant that Whitman in describing the cell doctrine allied it with a Weismannian, neo-Darwinian interpretation of the organism as a product of the struggle of, and interaction between, individual cells. This interpretation would later prove particularly useful to American biologists in their attacks on German biology during World War I, even though a wide range of political interpretations of the cell-state theory within Germany could be found.[28] Whitman in his description chose to portray the cell doctrine as a form of laissez-faire liberalism, a view that, at least in the interests of professionalism, he clearly held suspect. He instead on treating the organism as a social whole and held the "personality of the cell" subordinate to the greater good. His appeal to the organism as a directing force in cell formation strikes a concordant note with political musings of the common good and public will. Unlike Dewey, however, who situated the self in the external world, Whitman looked not to a shared social environment for the primary bonds of community, but to a shared heredity. The secret of organization could be found, Whitman argued, in fundamental elements of living matter that he called idiosomes. These bearers of heredity were the "real builders of the organism," whose actions and control were "not limited by cell-boundaries," and whose continuity was not "broken by cell walls."[29] Each individual cell, by virtue of its membership in the social organism, shared a common heredity substance with its fellow cells that guided its development and controlled its place in the organic order. Just as Whitman subordinated the individuality of the cell to the shared hereditary properties of the organism as a whole, so eugenicists would later subordinate individuality of the organism to the shared hereditary foundation of the race. Before turning to the ways in which race upheld the ideal of the common good, however, we need to first consider alternatives to heredity as the unifying bond of community.

The Organism as Social Beast

Throughout the last decade of the nineteenth century and continuing into the first two decades of the twentieth, biologists pursued questions pertaining to the unity of the organism and individuality on a variety of fronts. Regeneration studies offered one experimental inroad into the question of whether organic form was the product of an inner, self-directed formative substance or was governed by external environmental conditions. Interest in regeneration could be seen as therapeutic to the fears of moral and racial degeneration being expressed within the medicalized and natural scientific language in the latter part of the nineteenth century.[30] If the organism's regenerative power was itself dependent on the environmental milieu, as some biologists argued, then the social environment would be an important element in rehabilitating the individual. Such a position would sanction the need for social control and the consolidation of authority into the hands of a scientifically-trained professional elite such as the biological or social science professions. It would also establish the authority of biologists as social engineers, since, in this view, the biological and social were one and the same. Charles Manning Child (1869–1954) was one defender of the individual as social self. A leader in the study of regeneration, especially planarian regeneration, from the time of his appointment as an associate in zoology at the University of Chicago in 1896 until his retirement as full professor in 1934, Child serves as a useful example of someone whose definition of the organism as social beast intersected with Progressive concerns with authority and social control.[31]

Although a generation apart, Child and his colleague Whitman each received their doctorates under Rudolph Leuckart at the University of Leipzig, and both put forth the "organism as society" metaphor. Unlike Whitman, who situated individual identity in a hereditary substance labeled idiosomes, Child immersed self-identity completely within the surrounding environment. The individual, Child insisted, was not "independent and self-determining in its origin," but instead arose from the "relations between living protoplasm and the external world."[32]

Child's firm commitment to the belief that the individual has "its *being* in an external world" placed him in close alliance with John Dewey's understanding of the individual as a situated self. It was Dewey, Child reminded his readers, who had "pointed out that biology forces us to maintain that 'living goes on in, and because of an environing medium.'"[33] Integration of the organism into a harmonious whole was ever dependent on the action reaction complex between environment and the organism, in short, on the organism's behavior. The individual was continually in the process of adjustment and equilibration to surrounding environmental conditions and was thus always in the process of redefinition. Morphollaxis, Child argued, offered a clear instance of such self-renewal. In a regenerating planarian, a small piece may recreate its missing part by utilizing already-present cells. In this process, some of the fragment's cells change both their shape and their function to reform as regenerated parts. Each cell thus responded to the changed environment by altering its individual constitution to meet the demands of the social organism. How steadfast Child was in his insistence on the individual as a dynamic, social self is evident in his criticisms of a theory of regeneration put forward by a former Whitman student, Samuel Jackson Holmes.[34]

In an attempt to explain the process of regeneration, Holmes suggested that the organism was best conceived as a symbiotic association of individual cells. Consider, for example, an organism made up of differentiated cells – say, A, B, C, D, E, F, and G – that are mutually dependent upon one another. Between these differentiated cells are undifferentiated cells that have yet to develop. If one removes cell A from the organism, the remaining differentiated cells will no longer receive the benefits of A necessary for their existence. Holmes argued that the "social pressure" of the remaining differentiated cells would thus cause the undifferentiated cell to develop and grow into a new A, regenerating the missing part.[35]

In Holmes explanation, the organism is an aggregate society of interdependent individuals. The development of the individual, in the case of the undifferentiated cell growing into form A, is dependent on an adjustment to surrounding environmental conditions, rather than on some inner, self-determining mechanism, as Weismann maintained. Thus, as Holmes argued, organic form was the "result of functional activity, or more specifically, functional equilibration."[36] Individuality was, in this instance, a product of the social environment. Child was sympathetic with Holmes' attempt to explain organic form solely as a product of functional activity, but he took issue with the "lack of plasticity" that each individual displayed once formed. In Holmes' explanation, once differentiation took place, individuality became fixed. Child suggested that if A was removed, the social relations between the remaining cells changed, and the differentiated cells would reorient their behavior along new lines. As the community changed, so did the individuals within it.[37]

The problem with corpuscular theories such as Weismann's germ plasm hypothesis, Child suggested, was that the organism was "merely a composite of . . . [the] orderly activities" of individuals, be they "gemmules, determinants, units, pangenes, specific accumulators, or whatever."[38] Priority in these theories was placed on individuals and their interactions and exchange, but the unity and order of the social organism, Child insisted, came not from exchange, but from "law and order," from "co-ordination and control of the activities of its constituent units; in short [from] some degree and kind of government." Order, be it in human societies or the social organism, was based "primarily on authority . . .

and its transmission."³⁹ While lower levels of organization were characterized by primitive autocracies that relied on coercive power rather than on cooperation and consent, Child suggested that the most advanced social organisms were physiological democracies, where a highly specialized and efficient type of leadership, namely, the brain, was responsive to the more subordinate parts. Experimental evidence for the existence of such authority could be found, Child believed, in the physiological gradients of organisms. He suggested that areas of increased metabolic activity would occur in the undifferentiated protoplasm in response to environmental stimuli. These areas of excitation would become imprinted in the protoplasm, establishing a higher metabolic rate than the region farthest from the source. According to Child, the major gradient of development originated along the anterior-posterior axis of the organism; thus, the apical region of the head dominated and controlled differentiation in the subordinate parts. He subjected regenerating *Planaria* fragments to metabolic poisons over a lifetime of research, accumulating evidence that *Planaria* are indeed metabolically graded.⁴⁰

In rejecting any theory of individuality that smacked of predeterminism, Child placed the welfare of the whole over that of the individual parts. If the individual only found definition and meaning through its interaction with the environment, then individual rights would themselves be partially congruent with the public interest. Conflicts would arise, for heterogeneous environmental conditions resulted in individuals with different behaviors, but all would be integrated and regulated into a functioning social order under some higher mechanism of coordination and control. The form of individual freedom implicit in this rendition of the cell state is not absence from governmental restraint, rather it is freedom from domestic disorder and instability. Cancer within the body physiologic or body politic was, Child argued, an instance where the individual constituents had become "complete anarchists," failing to respond to physiological control. Such a situation threatened individual opportunity as the social organism languished with disease.⁴¹ Yet, Child's resolution of individual conflict through integration into the social will could equally be seen to threaten individual privacy, an implication that would become only too apparent during World War II. "Total community," as Stanley Benn suggests, "is totally incompatible" with the ideal of individual autonomy embedded in classic liberal theory.⁴²

A Situated Self

In the 1910s, American political and social rhetoric was adorned less in the language of individual autonomy and rights and more in impassioned appeals to the public will and the common good. The social organism and the individual as situated self continued to strike a responsive chord, especially in the writings of American biologists during World War I. William Emerson Ritter (1856–1944), director of the Scripps Institution for Biological Research, for example, employed Child's dynamic conception of the individual and the Progressive rhetoric of community and social bonds to launch a sustained attack on all that was wrong with the materialist Weltanschauung at the heart of German biology and culture.

In *The Unity of the Organism*, published in 1919, Ritter, who became acquainted with Child during Child's summer sojourns to the West Coast environs of Scripps Institute and Friday Harbor, adopted a similar conception of the individual as social self. Although Ritter praised Whitman's attack on the cell theory and his attempts to place the organism as a whole primary to the cell in development, he criticized Whitman for ultimately retreating to an elementalist position by invoking those mysterious hypothetical hereditary substances Whitman called "idiosomes." "The very men who have admitted the rights of the organism as against its cells in development," Ritter cried, "are yet far from admitting those rights as a general proposition." The renowned cytologist E. B. Wilson was no different than Whitman in Ritter's eyes. While Wilson had been critical of the cell theory in the 1890s and pointed to the importance of the organism as a whole in development, he became more and more confident in the early 1900s "of finding the 'key to all ultimate biological problems' in the cells, or maybe in the chromosomes."[43]

Both Whitman and Wilson reflected, Ritter believed, biologists' inability to break away from a conception of the organism as an aggregate of individuals, a position that Ritter linked to the cell state doctrine of Virchow and Haeckel. Holmes's theory of the organism as a symbiotic aggregation was the most recent attempt to account for development by reference to "the principle of aggregation," and Ritter found it troubling enough to warrant a sustained critique. Although Holmes explained such formative processes as regeneration by discussing the functional equilibration, or adaptation, of the parts to the whole, Ritter expressed an uneasiness toward the rudiments of individuality in

Holmes's theory. Holmes seemed to depart little from Roux's mosaic theory which accounted for development on the basis of a struggle for existence among the parts of the organism. While Holmes placed less emphasis on struggle than Roux and more on interdependence and mutual benefit derived from interaction between the parts, individuality still came prior to association. Within all symbiotic associations, Ritter informed his readers, individuality never extended beyond the constituent parts. In the case of a lichen, for instance, the algae and fungus both have to reproduce themselves, so that each preserves its identity. Ritter even suggested that since in "all known cases of symbiosis one of the members to the partnership actually enters the other and lives upon it, . . . the 'living together' is really a sort of parasitism."[44]

Ritter's equation of symbiosis with parasitism and his criticism of the cell state theory and aggregate theories of the organism in general are indicative of Ritter's suspicion that interactions between atomistic individuals could never bind the organism, community, or state into a unified whole; only anarchy would ensue. Order was essential to liberty. When Jacques Loeb described the ability of a posterior fragment of *Planaria* to regenerate a new head as an "emancipation of the part from the tyranny of the whole," Ritter took issue. "Tyranny," Ritter exclaimed, "is not merely absolute power; it is such power exercised unjustly or 'in a manner contrary to law or justice.'" This was decidedly not the case in the instance of the organism as a whole directing the development of its parts. No justice was breached. Nowhere in nature, Ritter insisted, was the "law of the parts of the organism more just than the law of the whole." Those who defended the cell as the fundamental unit of organization in biology upheld the law of the atomistic individual; they were "anarchists" in Ritter's words "toward all the most common, most objective, structures and laws of nature."[45]

Ritter reminded his American colleagues of the source of the cell state theory and the "materialistic Weltanschauung" that it bred. German biologists had seized on the natural selection and gemmule-pangenesis hypotheses of Darwin and turned them into a scientific justification for "militaristic brutism." August Weismann's espousal of the struggle for existence between biophores as the perfecting force in the organism was particularly to blame. Ritter was not alone in implicating Weismann's determinant hypothesis and its neo-Darwinian underpinnings as a cause of World War I. But the response engendered by American biologists could equally serve nationalist ends. If nature's lesson was subordination of individual freedom to the common good and not the struggle for existence, as many American biologists maintained, this could easily be interpreted as a rallying cry for the American people to unite in making the world safe for democracy.[46]

Ritter's implicit appeal to the common good stemmed from a rather idiosyncratic conception of the self that rested on the presupposition of the organism as a chemical reaction system; each individual, Ritter suggested, has "the chemical value of an elementary substance."[47] In the organic world that elementary substance was the personality, which Ritter defined not just in terms of integration at a psychological level, but at the physico-chemico-biological level as well. Like Child and Dewey, Ritter believed the self was realized through lived experience. It did not develop, to use Dewey's words, "as a sort of unrolling push from within."[48] This was one reason why Ritter was so hostile to preformationist theories of development. For Dewey, the environment to which the individual responded was largely that of human social interactions. Ritter wanted a definition of self other than that offered by social psychology, one that extended to the rudimentary biological and physicochemical nature of the organism. While he saw his theory closely akin to Dewey's notion of the self as a situated being, Ritter suggested that his hypothesis departed from social psychology in that the roots of the individual were found "not only in the social relationship of the human species, but extend right on through these into sub-human relationships, even down into the very constitution of inorganic nature."[49]

Ritter's definition of the individual as an organic personality realized through interaction with the environment – be it at the physical, chemical, biological, or social level – meant that "persons are organically interdependent upon one another."[50] Other individuals are part of the social environment to which a given personality responds and develops, and are therefore integral to its individual being. Hence, the common good or social will is not antagonistic to the individual, but stands prior to and supportive of individual identity. Cooperation, mutual service, and the sacrifice of self to community were as natural to such a conception of individuality as competition and selfishness are to a genetic understanding of self.

The Social Bonds of Heredity

While the idealist overtones of the common good might keep the surface layer of the body politic intact, it could not sustain the more penetrating gaze of cultural critics in the 1920s. The influential journalist Walter Lippmann, who had helped unite public sentiment through his contributions to the *New Republic* during World War I, offered a closer look into the commonweal in the war's aftermath, only to find a sham. Lippmann's *Public Opinion* and *The Phantom Public* were just two of many exposés written during the 1920s that revealed the extent to which the common will was little more than a smokescreen, managed and manufactured through the propaganda and advertising campaigns of private interests to achieve their own economic and political ends. In the 1920s, democratic theory became shaped by the realism of political scientists as they turned their attention to empirical studies of voting behavior, only to find a largely uneducated public whose actions were shaped primarily by unconscious and irrational drives rather than rational and intelligent behavior. Such revelations did not bode well for communitarian theories of the individual as social self. If the common good was simply a media achievement of a select few, Ritter's understanding of the individual as an organic personality could become merely a self fabricated for the culture of consumption.[51]

The problem with a concept of individuality dependent on the realization of self through the environment was that nothing in the individual seemed capable of resisting the forces at work in a consumer society, which were ready at every moment to capitalize on the individual's longing for community and turn it into a commodity for mass culture. This is not to suggest that the common good was completely abandoned in the face of 1920s democratic realism. The rhetoric of Franklin Roosevelt's National Recovery Administration in 1933 was heavily laced with appeals to "the great whole," and American intellectuals such as Waldo Frank and Lewis Mumford crafted a radical social and political vision of organic community in the 1930s that owed much to Dewey's understanding of the self realized through participation in a democratic culture.[52] But the seeds of suspicion had already been sown: the idea of the public will was just a sugarcoating to make a democracy driven by interest-group politics more easy to swallow. By the 1950s, democratic theorists had rejected the placebo of public unity for a more realistic Madisonian picture of democracy in which competing interests and pluralist divisions preserved the public by preventing any single interest from dominating the rest.[53]

If Ritter's organic personality or Child's dynamic individual were too easily consumed by the crowd, heredity offered an impenetrable core for the establishment of identity, unmalleable, even in the midst of group life. The erosion of the common good in American political discourse did not threaten a genetic definition of individuality, because self-identity was not dependent on community. Throughout the first decades of this century, biologists like Whitman did appeal to the common good by pointing to the shared hereditary material that bound individuals together into an organic whole. But this social organism view was not a necessary condition for a genetic definition of self, a realization that became ever more apparent to biologists during the interwar years. Take away the external environment, and the hereditarian core of the individual remained. This was decidedly not the case with definitions of the individual as situated self. In a communitarian view, the social environment is an integral part of individual identity. Take away the environment, and essential aspects of the individual disappear. Here, individual and community are necessarily part of a functional whole.

The views of the Princeton biologist Edwin Grant Conklin (1863–1952) on individuality illustrate the commonalities between a hereditarian and social definition of self in the first part of this century and the advantages that a genetic view offered amid the climate of democratic realism in the 1920s. Conklin, trained under William Keith Brooks at Johns Hopkins University, was a frequent contributor to developmental studies, which occupied many members of the MBL community during the 1890s and early 1900s.[54] From his embryological research on Ascidians, Conklin was led to believe that some sort of organization existed in the cytoplasm of the unfertilized egg and was an important factor in controlling development. "All the early stages of development, including the polarity, symmetry, type of cleavage, and the pattern, or relative positions and proportions of future organs," Conklin argued, were "foreshadowed in the cytoplasm of the egg cell."[55] The characters that determined the general phylogenetic groups thus resided in the cytoplasm. Specific traits of the individual, the characters of Mendelian genetics, were determined by the chromosomes of the egg and the sperm. It is important to note, however, that the polarity in the cytoplasm did not emerge, in Conklin's view, from differential reaction of protoplasm to environment, as Child maintained. In Conklin's developmental views, polarity had a material basis that was passed on from generation to generation. He criticized Child's theory of individuality for its failure to provide a structural foundation for physiological gradients. "Disembodied functions," Conklin suggested, "are as unknown in biology as are disembodied spirits." Function was inseparable from structure, and it was structure that provided the basis for individuality in Conklin's view.[56]

By the 1910s, Conklin had become a firm believer in the power of eugenics to improve the direction of biological and social evolution. In a 1919 article on "Biology and Democracy" published in *Scribner's Magazine*, Conklin assured his readers that "recent work on development and evolution shows that the influence of environment is relatively slight, that of heredity overwhelming."[57] Since the environment played little defining role in the individual's constitution, Conklin could not escape the implication that a conflict between the freedom of the individual and the good of society would arise. Unlike Child and Ritter, who saw individual identity integrally dependent on the social will, Conklin faced a more difficult task in appealing to Progressive rhetoric of the common good. Such appeals he did make, however, profusely and with passion. "A change in the conception of liberty," Conklin observed, "has been coming over the nation. Democratic freedom is not the freedom of isolation or anarchy, . . . the lawlessness of Bolshevism and revolution . . . but the freedom of fellowship, common service, and mutual esteem; not freedom from social control but freedom from the tyranny of selfish individuals and classes."[58] Subordination of individual freedom to the common good was essential, when such actions were necessary for the welfare and continued existence of the race. "Race preservation, not self preservation," Conklin argued, was "the first law of nature." Evolution operated for the good of the species. Among humans, intellect and the wish for individual freedom interfered with the instinct that obeyed nature's first law. Social control and the enactment of eugenic measures were thus necessary to ensure the future development of the race.[59]

Conklin moved easily from a hereditarian definition of individuality to a social organismic framework in defending the interests of the common good. "Each individual is a minor unit in the great organism of mankind," he wrote.[60] His ability to make that shift is important because it indicates the extent to which biologists in the 1910s and 1920s understood race and, later, the population, in organicist terms.[61] Even though a hereditarian definition of the self might seem to insulate the individual from society, this was not the case in the early twentieth century, especially when illuminated through the prism of race. Not until the development of a nontypological definition of the population in the 1930s and 1940s, which replaced the ideal type of race, did a genetic conception of self that sheltered the individual from the powers of the state emerge.[62]

In his appeals to the social organism, Conklin, like other biologists of his generation, rejected the notion of negative freedom and sought a positive version where the individual was free to reach his/her potential within the bounds of society. Freedom, in Conklin's view, did not imply equality. All individuals were not equal because each individual possessed different ability and potential based on their hereditary constitution. Hereditary inequality did not, however, justify aristocracy. Aristocracy, Conklin argued, was founded upon the law of entail, which embraced an outdated understanding of heredity. Under aristocracy, privilege, wealth, and social position were conferred to offspring, not on the basis of their innate ability, but on the basis of their parent's social environment. Such a system condemned "low-born genius to the humblest work and put well-born blockheads in exalted places." But the law of Mendel was democratic. Because new combinations of genes occurred in each generation, every person was given a "new hereditary deal," regardless of the environment into which they were born. Not all individuals were equal in a democracy, Conklin recognized, but individuals were "free from purely artificial restraints in finding their own levels."[63]

Since individuals differed in their innate abilities, any hope of participatory democracy was doomed to fail. Conklin shared with other democratic realists of the twenties a similar dismay over the findings of military mental testing that revealed a largely illiterate and unintelligent American electorate. Democracy would only succeed if government was not *by* the people, but *for* the people. "Majority rule," Conklin believed, "would level society down to general mediocrity were it not for the instinct of the people to follow leaders."[64] Power was best placed in the hands of responsible elites whose decisions would be guided by the expert opinion of social science professionals to ensure that the interests of the individual and the commonweal were best served. Thus Conklin was able to bind individuals as genetic selves into the social organism "like that of ants or bees in a colony, where each individual is free to serve as best it can under the control of the colony as a whole."[65]

The Biology of Democratic Pluralism

In each case examined in the previous sections, individuals contributed to the reshaping of a shared political discourse of individuality, community, and the common good to articulate different concerns and to reach different constituencies. Whitman was preoccupied with the establishment of a professional discipline of biology and the need for individuals to subvert their own specialized interests for the good of the academic community. Child and Ritter cast their net wider; their interests centered on order within society and the cultural authority of biology within that same society. Conklin was also concerned with social order and the authority of the biologist as social expert, but his writings on individuality were oriented specifically to the members of the eugenics movement, of which he was a part.

Despite their different emphases, however, each biologist invoked the social organism to make individual rights compatible with the public interest. Each defended a position of positive freedom: namely, the freedom to find one's place under the social whole. This Progressive emphasis on positive freedom, which was intimately connected to rhetoric of the common good, eroded throughout the 1930s and 1940s in the United States, partly in response to the rise of totalitarianism abroad. In his analysis of changing perceptions of liberty in American culture, Michael Kammen argues that during the 1930s "liberty tended to be discussed in terms of external threats rather than in terms of . . . domestic disorder and instability."[66] Concerted efforts were made to reeducate the American public about the Bill of Rights, and a more pronounced rhetoric of individual rights and freedom occupied American political discourse during the war and immediate postwar period than was apparent in the Progressive era. The threat of communism and McCarthyism with the onslaught of the Cold War only heightened American intellectuals' defense of liberty in negative terms, "as freedom from rather than freedom to."[67] During the 1950s, this preoccupation with civil liberties was coupled with an attack on mass culture by David Riesman, Eric Fromm, and others that reasserted the value of individualism in America. Riesman's *The Lonely Crowd*, for example, revealed a society where cooperation and conformity had resulted in a population of automatons, individuals who had no independent or authentic existence. The greatest need in postwar America, Riesman insisted, was to encourage "people to develop their private selves – to escape from groupism – while realizing that, in many cases, they will use their freedom in unattractive or 'idle' ways."[68] Encroachment of the state on individual rights had become much more important than the sacrifice of individual identity to the public self. Whereas Conklin saw Bolshevism as a form of rampant individualism in the aftermath of World War I, communism became linked to totalitarianism and group conformity during the 1950s. In the revival of liberal individualism in postwar America, the communitarian ideals of Rousseau, Hegel, and others appeared particularly suspect.

Stripped of the common good, the social organism framework that held Conklin's society of genetically distinct individuals together collapsed, revealing not a unity, but an aggregation of competing genetic selves. In postwar America, a pluralist society of competing interests did not result in disorder and anarchy, as Conklin, Child, and others of their generation feared. Instead, it was precisely those diverse and competing interests between individuals and groups that sheltered democracy from the threat of dictatorship. Ridding the genetic conception of self from its interwar social organism baggage was a challenge that a younger generation of biologists would take on with enthusiasm. Natural selection operated, not for the good of the species, but in the interests of the individual. Under the banner of the modern synthesis, George Gaylord Simpson, Theodosius Dobzhansky, and others defended the genetic identity of the individual as the pillar of a biological democracy.[69]

Throughout the interwar years and into the early postwar period, ecology remained one area within the biological sciences most resistant to a genetic definition of the individual organism. The development of ecology at the turn of the century was heavily dependent on an explicity organismic, physiological, nonhereditarian perspective that was closely wedded to a communitarian definition of self. The community was defined as an assemblage of unrelated organisms that had gathered together as a consequence of similar physiological responses to environmental conditions. Yet the community, as Frederic Clements argued, despite the unrelatedness of individuals, was, in a physiological sense, an organism that underwent growth and development. Each individual received its identity from the environmental complex within which it lived and participated in the sustenance and evolution of the community – the common good. In the 1930s, Clements utilized the social organism ideal at the center of community ecology as a naturalistic foundation for land management policies on the Great Plains that placed the interests of the whole above those of the parts. Authority was to be placed in the hands of scientific experts within federal agencies such as the Soil Conservation Service, and the Division of Grazing to institute a regional planning scheme that included depopulation, resettlement, changes in farm size, and new agricultural practices. "Coordination of process and practice on farm or ranch must be reflected in the organization of community," Clements reasoned. "Within such a huge organism" as the United States, he continued, "the whole is much greater than the mere sum of its parts, and hence the need for coordination and correlation far transcends all other considerations whatsoever." As Ronald Tobey has persuasively shown, Clement's organicism clashed with the Madisonian view of American society that became the basis of interest group pluralism in 1950s democratic theory.[70]

The communitarian view implicit in ecology was challenged by a definition of the organism as shared genetic complex in the 1950s and 1960s. A transformation of community ecology took place in the postwar years, directing attention away from organicist tropes such as cooperation, integration, and the common good to a theory of community structure modeled on competitive interactions between closely related species. This is one reason Henry Allen Gleason's individualistic concept of the community, which received scant support during the 1920s, was vindicated in ecology during the 1950s. The community, in Gleason's theory, was not the ideal common good, but "a coincidence," a happenstance assemblage of genetically distinct organisms that were drawn from surrounding plant populations and flourished or died depending on the fluctuating and variable environment into which individual organisms migrated. The revival of Gleason's individualistic concept of the community is just one example among many pointing to the erosion of the social organism and the pervasiveness of a genetic conception of self within biological discourse in the years following World War II.[71]

While the individual as a genetic self has throughout the twentieth century served an important ideological function in establishing individual identity independent of social control and the fragmenting forces of a consumer culture, recent technological developments in molecular biology bring into question a genetic self as the guarantor of autonomy and individual rights. If in the twenty-first century, biotechnology enables governments, corporations, or individuals to alter genetic make-up at will, buying and trading genetic information, then a biological self defined in hereditarian terms offers no firm foundation for the construction of identity. We need not wait, however, for the miracles of biotechnology to fracture the genetic self. Lynn Margulis has recently challenged this "atomistic model of organismic identity." If, indeed, eukaryotic cells are derived from symbiotic associations of bacteria, we are all genetic chimeras.[72] Our individuality again opens to question. To where then might we turn?

Acknowledgments

My thanks to John Beatty, Hamilton Cravens, Jane Maienschein, participants of the ZIF conference on "The Transfer of Images and Metaphors Between Biology and the Social Sciences," and members of the 1992 MBL Seminar on "Individuality: Understanding Units and Levels of Organization in Biology" for their comments on an earlier version of this paper.

Notes

1. The treatment of biology as a human science in the twentieth century has been most systematically explored by Donna Haraway, *Primate Visions: Gender, Race, and Nature in the World of Modern Science* (London: Routledge, 1989); Stephen J. Cross and William R. Albury, "Walter B. Cannon. L. J. Henderson, and the Organic Analogy," *Osiris*, 2d. ser., **3** (1987), 165–192; and Robert M. Young, "The Naturalization of Value Systems in the Human Sciences," in *Science and Belief: Darwin to Einstein*, Block VI, *Problems in the Biological and Human Sciences* (Milton Keynes: Open University Press, 1981), pp. 63–110.
2. Sam B. Girgus, *The Law of the Heart: Individualism and the Modern Self in American Literature* (Austin: University of Texas Press, 1979), p. 12.
3. Isaiah Berlin, "Two Concepts of Liberty," in Michael Sandel (ed.), *Liberalism and Its Critics* (New York: New York University Press, 1984), p. 17.
4. Sandel (ed.), *Liberalism and Its Critics*, p. 6.
5. William Corlett, *Community Without Unity: A Politics of Derridian Extravagance* (Durham: Duke University Press, 1989), p. 23.
6. Leo W. Buss, *The Evolution of Individuality* (Princeton: Princeton University Press, 1987), p. 13.
7. The attention to keywords and their interrelations in unravelling cultural and social meaning owes its greatest debt to Raymond Williams, *Keywords: A Vocabulary of Culture and Society*, rev. edn. (New York: Oxford University Press, 1985). For a similar approach in analyzing the changing meanings of American political discourse,

see Daniel T. Rodgers, *Contested Truths: Keywords in American Politics Since Independence* (New York: Basic Books, 1987).
8. On this inversion of liberal in American social science, see Dorothy Ross, *The Origins of American Social Science* (New York: Cambridge University Press, 1991). The development of an interdependent market economy and its impact on the rise of professional American social science is treated at length in Thomas L. Haskell, *The Emergence of Professional Social Science: The American Social Association and the Nineteenth-Century Crisis of Authority* (Urbana: University of Illinois Press, 1977).
9. For a historical analysis of the "common good" in American political rhetoric, see Rodgers, *Contested Truths*.
10. Most American historians now agree that "progressivism" never constituted a coherent and unified movement. Numerous explanations have been put forth to characterize the reformist initiatives of the early twentieth century. For a representative sampling, see Robert M. Crunden, *Ministers of Reform: The Progressives' Achievement in American Civilization, 1889–1920* (New York: Basic Books, 1982); Samuel P. Hays, *Conservation and the Gospel of Efficiency: The Progressive Conservation Movement, 1890–1920* (Cambridge, Mass.: Harvard University Press, 1959); Martin J. Sklar, *The Corporate Reconstruction of American Capitalism, 1890–1916: The Market, the Law and Politics* (Cambridge: Cambridge University Press, 1988); James Weinstein, *The Corporate Ideal in the Liberal State, 1900–1918* (Boston: Beacon Press, 1968); and Robert H. Wiebe, *The Search for Order, 1877–1920* (New York: Hill & Wang, 1967).
11. T. J. Jackson Lears, *No Place of Grace: Antimodernism and the Transformation of American Culture* (New York: Pantheon Books, 1981), p. 37. See, also, T. J. Jackson Lears, "From Salvation to Self-Realization: Advertising and the Therapeutic Roots of the Consumer Culture, 1880–1930," in Richard Wightman Fox and T. J. Jackson Lears (eds.), *The Culture of Consumption: Critical Essays in American History, 1880–1980* (New York: Pantheon, 1983), pp. 1–38. For a somewhat different perspective on the reconstruction of self in relation to an emerging consumer culture in America, see Casey Nelson Blake, *Beloved Community: The Cultural Criticism of Randolph Bourne, Van Wyck Brooks, Waldo Frank & Lewis Mumford* (Chapel Hill: University of North Carolina Press, 1990).
12. Ross, *The Origins of American Social Science*, p. 237.
13. Quoted in Robert B. Westbrook, *John Dewey and American Democracy* (Ithaca: Cornell University Press, 1991), p. 44.
14. John Dewey, "The Reflex Arc Concept in Psychology" (1896) in *The Early Works of John Dewey, 1882–1895* (Carbondale, Ill.: Southern Illinois University Press, 1972), Vol. 5, p. 97.
15. John Dewey, "Evolution and Ethics," in *The Early Works of John Dewey* 5 (1898), 53.
16. Westbrook, *John Dewey*, p. 43.
17. For an analysis of positive freedom in late nineteenth-century political thought, see David Nichols, "Positive Liberty, 188–1914," *American Political Science Review* 56 (1962), 114–128.
18. Norton Wise provides a fascinating analysis of the emergence of ideas on statis-

tical causality in the context of individualism and its relationship to *Gemeinschaft* and *Gesellschaft* in central European culture from 1870 to 1920. His approach strikes a resonant chord with some of the themes raised in this essay. See M. Norton Wise, "How Do Sums Count? On the Cultural Origins of Statistical Causality," in Lorenz Krüger, Gerd Gigerenzer, and Mary S. Morgan (eds.), *The Probabilistic Revolution. Vol. 1: Ideas in History* (Cambridge: MIT Press, 1987).

19. Charles Otis Whitman, "Specialization and Organization, Companion Principles of All Progress – The Most Important Needs of American Biology," *Biological lectures Delivered at the Marine Biological Laboratory of Woods Hole* (1891), pp. 1–2. On Whitman, see Edward S. Morse, "Charles Otis Whitman," *National Academy of Sciences Biographical Memoirs* **7** (1912), 269–288; Frank R. Lillie, "Charles Otis Whitman," *Journal of Morphology* **22** (1911), xv–lxxvii; Jane Maienschein, introduction to *Defining Biology. Lectures from the 1890s* (Cambridge, Mass.: Harvard University Press, 1986); idem., "Whitman at Chicago: Establishing a Chicago Style of Biology?" in Ronald Rainger, Keith R. Benson, and Jane Maienschein (eds.), *The American Development of Biology* (Philadelphia: University of Pennsylvania Press, 1988), pp. 151–182.

20. Whitman, "Specialization and Organization," pp. 6–7.

21. *Ibid.*, p. 23.

22. Jane Addams, "The Subjective Necessity for Social Settlements," in Henry C. Adams (ed.), *Philanthropy and Social Progress* (New York, 1893), p. 4. On the search for community in late nineteenth-century America, see, for example, J. Ronald Engel, *Sacred Sands: The Struggle for Community in the Indiana Dunes* (Middletown, Conn.: Wesleyan University Press, 1983); Jean B. Quandt, *From the Small Town to the Great Community: The Social Thought of Progressive Intellectuals* (New Brunswick, N.J.: Rutgers University Press, 1970); Wiebe, *The Search for Order, 1877–1920*; R. Jackson Wilson, *In Quest of Community: Social Philosophy in the United State, 1860–1920* (New York: Oxford University Press, 1968).

23. Whitman, "Specialization and Organization," pp. 23, 9.

24. *Ibid.*, pp. 22, 25. Both Maienschein, "Whitman at Chicago," and Philip J. Pauly, "Summer Resort and Scientific Discipline: Woods Hole and the Structure of American Biology, 1882–1925," in *The American Development of Biology*, pp. 121–150 emphasize this organizational ideal in Whitman's work. Pauly has also emphasized the MBL as a place of community for a nascent tradition of American biology.

25. Charles Otis Whitman, "The Inadequacy of the Cell-Theory of Development," *Journal of Morphology* **8** (1893), 641.

26. *Ibid.*, p. 648.

27. *Ibid.*, pp. 646, 653.

28. For an examination of the organism as cell state doctrine in the culture context of Imperial Germany, see Paul Weindling, "Theories of the Cell State in Imperial Germany," in Charles Webster (ed.), *Biology, Medicine, and Society, 1840–1940* (Cambridge: Cambridge University Press, 1981), pp. 99–155.

29. Whitman, "Inadequacy of the Cell Theory," pp. 645, 657.

30. See, e.g., Daniel Pick, *Faces of Degeneration: A European Disorder, c. 1848–c. 1918* (Cambridge: Cambridge Univ. Press, 1989). For a historical account of regeneration studies in biology during this period, see Jane Maienschein, "T. H. Morgan's

Regeneration, Epigenesis and (W)holism," in Charles Dinsmore (ed.), *History of Regeneration Research* (Cambridge: Cambridge University Press, 1991) and Gregg Mitman and Anne Fausto-Sterling, "Whatever Happened to Planaria? C. M. Child and the Physiology of Inheritance," in Adele Clarke and Joan Fujimura (eds.), *The Right Tools for the Job: At Work in Twentieth-Century Life Sciences* (Princeton: Princeton University Press, 1992), pp. 172–197.

31. For biographical information on Child, see Libbie H. Hyman, "Charles Manning Child, 1869–1954," *National Academy of Sciences Biographical Memoirs* **30** (1957), 73–103.
32. C. M. Child, *Individuality in Organisms* (Chicago: University of Chicago Press, 1915), p. 41.
33. C. M. Child, "Behavior Origins From a Physiologic Point of View," *American Medical Association Archives of Neurology and Psychiatry* **15** (1926), 174.
34. For a slightly different contextual analysis of Child's theory of individuality in relation to liberalism and Chicago pragmatism, see Sharon Kingsland, "Toward a Natural History of the Human Psyche: Charles Manning Child, Charles Judson Herrick, and the Dynamic View of the Individual at the University of Chicago," in Keith R. Benson, Ronald Rainger, and Jane Maienschein (eds.), *The Expansion of American Biology* (New Brunswick, N.J.: Rutgers University Press, 1991), pp. 195–230.
35. Samuel Jackson Holmes, "The Problem of Form Regulation," *Archives für Entwicklungmechanik* **17** (1904), 265–304.
36. S. J. Holmes, "Regulation as Functional Adjustment," *Journal of Experimental Zoology* **4** (1904), 422.
37. C. M. Child, "The Physiological Basis of Restitution of Lost Parts," *Journal of Experimental Zoology* **5** (1908), 497.
38. Child, *Individuality in Organisms*, p. 22.
39. *Ibid.*, pp. 21, 27.
40. On Child's gradient theory in the context of organicism in embryology, see Donna J. Haraway, *Crystals, Fabrics, and Fields: Metaphors of Organicism in Twentieth-Century Developmental Biology* (New Haven: Yale University Press, 1976).
41. C. M. Child, "Biological Foundations of Social Integration," *American Sociological Society Publications* **22** (1928), 35.
42. S. I. Benn, "Individuality, Autonomy, and Community," in Eugene Kamenka (ed.), *Community as a Social Idea* (New York: St. Martin's Press, 1982), p. 55.
43. William E. Ritter, *The Unity of the Organism*, 2 vols. (Boston: Richard G. Badger, 1919), Vol. 1, pp. 14, 163.
44. Ritter, *Unity of the Organism* **1**: 185–186. For a history of the symbiosis controversy, see Jan Sapp, "Symbiosis in Evolution: An Origin Story," *Endocytobiosis and Cell Research* **7** (1990), 5–36.
45. Ritter, *Unity of the Organism* **2**: 158–160.
46. *Ibid.*, pp. 353–357. For an analysis of American biologists' response to World War I and their anti-war evolutionism, see Gregg Mitman, "Evolution as Gospel: William Patten, the Language of Democracy, and the Great War," *Isis* **81** (1990), 446–463; idem., *The State of Nature: Ecology, Community, and American Social Thought, 1900–1950* (Chicago: University of Chicago Press, 1992).

47. Ritter, *Unity of the Organism* **2**: 186.
48. John Dewey, "The Need for a Recovery of Philosophy," in *Creative Intelligence: Essays in the Pragmatic Attitude* (New York: Henry Holt & Co., 1917), p. 10.
49. Ritter, *Unity of the Organism* **2**: 306.
50. *Ibid.*, p. 353.
51. For a discussion of democratic theory in the 1920s, see Edward A. Purcell, Jr., *The Crisis of Democratic Theory: Scientific Naturalism and the Problem of Value* (Lexington: University of Kentucky Press, 1973); David M. Ricci, *The Tragedy of Political Science: Politics, Scholarship, and Democracy* (New Haven, Yale University Press, 1984). For an analysis of the growing suspicion of the common good among American intellectuals in the twenties, see Rodgers, *Contested Truths*, pp. 198–203; and Westbrook, *John Dewey*, pp. 275–318. On "personality" as the ideal of self in twentieth-century American culture, see Warren I. Sussman, "Personality and the Making of Twentieth-Century Culture Culture," in *Culture as History: The Transformation of American Society in the Twentieth Century* (New York: Pantheon Books, 1984), pp. 271–286.
52. Quoted in Rodgers, *Contested Truths*, p. 204. On community as a cultural ideal in the 1930s, see Blake, *Beloved Community*; Richard H. Pells, *Radical Visions and American Dreams: Culture and Social Thought in the Depression Years* (1973. reprint, Middletown, Conn.: Wesleyan University Press, 1984).
53. On the importance of interest group pluralism in democratic theory during the 1950s, see Robert Booth Fowler, *Believing Skeptics: American Political Intellectuals, 1945–1964* (Westport, Conn.: Greenwood Press, 1978); Purcell, *The Crisis of Democratic Theory*.
54. For biographical accounts of Conklin's life, see Garland E. Allen, "Edwin Grant Conklin, *Dictionary of Scientific Biography* **3**: 389–391; E. Newton Harvey, "Edwin Grant Conklin," *National Academy of Sciences Biographical Memoirs* **31** (1958), 54–91. For historical analyses of Conklin's biological research and his public science, see J. W. Atkinson, "E. G. Conklin on Evolution: The Popular Writings of an Embryologist," *Journal of the History of Biology* **18** (1985), 31–50; Jane Maienschein, *Transforming Traditions in American Biology, 1880–1915* (Baltimore: Johns Hopkins University Press, 1991).
55. Edwin Grant Conklin, *Heredity and Environment in the Development of Men* (Princeton: Princeton University Press, 1915), p. 176. On Conklin's belief in the importance of cytoplasm in heredity, see Maienschein, *Transforming Traditions*; Jan Sapp, *Beyond the Gene: Cytoplasmic Inheritance and the Struggle for Authority in Genetics* (New York: Oxford University Press, 1987).
56. E. G. Conklin, "The Basis of Individuality in Organisms from the Standpoint of Cytology and Embryology," *Science* **43** (1916), 526.
57. E. G. Conklin, "Biology and Democracy," *Scribner's Magazine* **65** (April, 1919), 410.
58. *Ibid.*, 407.
59. Conklin, *Heredity and Environment*, p. 484.
60. Conklin, "Biology and Democracy," p. 408.
61. *Ibid*, pp. 407, 408.
62. Hamilton Cravens, *The Triumph of Evolution: The Heredity-Environment*

Controversy, 1900–1941 (Baltimore: Johns Hopkins University Press, 1978); Haraway, *Primate Visions*, pp. 197–203.
63. E. G. Conklin, *The Direction of Human Evolution*, rev. ed. (New York: Charles Scribner's Sons, 1934), pp. 145–146.
64. *Ibid.*, p. 149.
65. Conklin, "Biology and Democracy," p. 407.
66. Michael Kammen, *Spheres of Liberty: Changing Perceptions of Liberty in American Culture* (Madison: University of Wisconsin Press, 1986), p. 151.
67. Herman Belz, "Changing Conceptions of Constitutionalism in the Era of World War II and the Cold War," *Journal of American History* **59** (1972), 657.
68. David Riesman, *Individualism Reconsidered and Other Essays* (Glencoe, Ill.: Free Press, 1954), p. 37. See, also David Riesman, Reuel Denney, and Nathan Glazer, *The Lonely Crowd: A Study in Changing American Character* (New Haven: Yale University Press, 1950). On the revival of individualism in 1950s American culture, see Booth, *Believing Skeptics*; Richard H. Pells, *The Liberal Mind in the Conservative Age: American Intellectuals in the 1940s & 1950s* (New York: Harper & Row, 1985).
69. For a more extended analysis of the collapse of the social organism ideal in the war and postwar years in ecology and evolutionary biology, see Mitman, *The State of Nature*. The importance of the individual in American biologist's defense of democracy during World War II, especially those biologists gathered around Columbia University who were influential in the development of the modern synthetic theory of evolution has yet to be fully explored. See, however, John Beatty, "Theodosius Dobzhanksy, and the Moral and Political Significance of Genetic Variation," Typescript; Gregg Mitman, "From the Population to Society: The Cooperative Metaphors of W. C. Allee and A. E. Emerson," *Journal of the History of Biology* **21** (1988), 173–194. Some of the important primary texts that bear on this issue include George Gaylord Simpson, "The Role of the Individual in Evolution," *Journal of the Washington Academy of Sciences* **31** (1941), 1–20; idem., *The Meaning of Evolution* (New Haven, Conn.: Yale University Press, 1949); Edmund W. Sinnott, "The Biological Basis of Democracy," *Yale Review* **35** (1945–46), 61–73.
70. Frederic E. Clements and Ralph W. Chaney, *Environment and Life in the Great Plains* (Washington, D. C.: Carnegie Institution of Washington, 1936), pp. 51–52. Ronald C. Tobey, *Saving the Prairies: The Life Cycle of the Founding School of American Plant Ecology, 1895–1955* (Berkeley: University of California Press, 1981). On Clement's importance in the history of plant ecology, see Joel B. Hagen, "Organism and Environment: Frederic Clements's Vision of a Unified Physiological Ecology," in Ronald Rainger, Keith R. Benson, and Jane Maienschein (eds.), *The American Development of Biology* (Philadelphia: University of Pennsylvania Press, 1988), pp. 257–280; Robert P. McIntosh, *The Background of Ecology: Concept and Theory* (Cambridge: Cambridge University Press, 1985); Donald Worster, *Nature's Economy: The Roots of Ecology* (San Francisco: Sierra Club Books, 1977). For a more detailed analysis of ecology's debt to a nonhereditarian view of the organism, see Mitman, *The State of Nature*.
71. Henry Allen Gleason, "The Individualistic Concept of the Plant Association," *Bulletin of the Torrey Botanical Club* **53** (1926), 16. On the revival of Gleason's views in postwar American ecology, see T. F. H. Allen, Gregg Mitman, and T. W. Hoekstra,

"Synthesis Mid-Century: J. T. Curtis and the Community Concept," *Journal of the Wisconsin Academy of Arts and Sciences* (forthcoming); Robert P. McIntosh, "H. A. Gleason – Individualistic Ecologist' 1882–1975: His Contributions to Ecological Theory," *Bulletin of the Torrey Botanical Club* **102** (1975), 253–273; Malcolm Nicolson, "Henry Allen Gleason and the Individualistic Hypothesis: The Structure of a Botanist's Carrer," *Botancial Review* **2** (1990), 91–161.

72. Dorian Sagan and Lynn Margulis, "Epilogue: The Uncut Self," in Alfred I. Tauber (ed.), *Organism and the Origins of Self* (Dordrecht: Kluwer Academic Publishers, 1991), p. 365.

IV: Economics

A PLAGUE UPON YOUR HOUSE: COMMERCIAL CRISIS AND EPIDEMIC DISEASE IN VICTORIAN ENGLAND

TIMOTHY L. ALBORN
Harvard University

Introduction

The transfer of metaphors does not only take place between well-defined sciences: it is also a commonplace of what another contributor to this volume has called "everyday" language.[1] Within that realm, however, the problem of transfer and the meaning of metaphors take on different dimensions. This paper, in discussing nineteenth-century uses of epidemic language to describe commercial crises, demonstrates some of these differences. For one thing, the strong claim that metaphor transfer plays a generative role in discovery is more difficult to sustain when the participants in a debate are interested in implementing policy (for example) rather than in forming new theories. The commercial reformers described below used metaphors to enrich the meaning of their new forms of action, but the actions themselves derived at least as much from professional interests and perceptions as from the intellectual content of the borrowed language. Related to this claim is the idea that everyday language, to a greater degree than self-consciously "scientific" discourse, is embedded in a more general moral discourse. As a result, the meaning of the metaphors is not as stable as in the more formal language of science, the purveyors of which at least make an effort to shelter themselves from wider moral concerns. Although instability of meaning is also present in the more restricted domain of "science," its significance becomes more apparent as one descends from more to less technical discourse.

Never is interplay among different realms of everyday language more evident than in times of crisis, when such language is marshaled first to depict events as anything but everyday and then to control those events. A crisis is an occasion for specialist problem-solvers to advertise their wares in front of a wider-than-usual audience. People stop and listen to what generals have to say during a war; they pay special attention to the advice of meteorologists during a hurricane. At such times, specialists find cause to slip out of their typically dry world of numbers and technical language into the unfamiliar territory of popular discourse – territory that they often sprinkle with an abundance of metaphors, rushed into service to suit the moment. Crises, in short, as occasions when specialists reach into bordering domains to find language capable of popular comprehension, are prime sites of metaphor transfer. When, as is often the case, more than one interest group claims to be best at solving the problem in question, crises are also occasions when the borrowing of language from outside a field's boundaries facilitates a redrawing of those boundaries: whichever interest group is best at manipulating metaphors has a better chance of being asked to attend to such crises in the future.[2] Occasions for using metaphors in this way abounded in early- to mid- Victorian England, where unprecedented prosperity was interrupted by recurring crises in both commercial relations and public health, and where no single interest group in either health or commerce ever had more than a tenuous hold on public esteem. When economists measured out ten-year intervals between each succeeding financial panic, and when health officials traced the destructive path of each new bout of epidemic disease, social and epidemiological language frequently crossed paths. Between 1825, when the *Morning Chronicle* called commercial crises the "grand disease of England," and 1875, when a lecturer on banking called panics "the Asiatic cholera of the commercial world, epidemic, most contagious and fatal," few observers of British commercial society could resist resorting to biological language to describe their unstable surroundings.[3]

As long as people who talked about commercial crises occupied the everyday world of politics and persuasion, they could succeed in implementing policy only by using words that were on everyone's lips. In England during the 1820s and 1830s, the most obvious storehouse for such words was that of natural theology, which provided, as Robert Young has suggested, a "common context" for moral questions that equally embraced the physical and the social.[4] As participants in that tradition, early Victorians who sought to understand and exhibit the moral ramifications of commercial crisis seldom thought twice about employing "natural," often malthusian, references to epidemics. In borrowing from contemporary views of epidemic disease, they were doing more than bolstering the social authority of a proposed commercial intervention by appealing to the higher judge of the life sciences: they were appealing to the moral authority of God, which was higher than either commerce or nature. Epidemic language came to the aid of commercial writers who constructed crises as the "natural" end of individuals who had produced too many goods in greedy anticipation of undue profits. It did so by reminding readers that God's retributive justice, which unfailingly delivered pestilence as a positive check to working-class population growth, was no less rigidly administered among the commercial classes.

In addition to providing a common linguistic denominator for discussions of commercial crisis, natural theological language was sufficiently ambiguous to allow competing specialists to move the debate in new directions, and ultimately to outstrip the original context altogether. Most contexts in Victorian England, common or not, were composed of a large array of competing interest groups, forever pressing at the limits of language to redefine the world in forms more suitable to their own designs. One way to accomplish this task of redefinition was to use the same words as one's contemporaries but with slightly different connotations. As a result, by the time of what Young has called the "fragmentation" of the common context of natural theology in the 1850s, what had appeared, for the most part, was not whole new vocabularies but rather the same old words with new meanings.[5] The language used to describe commercial crises in the 1850s is a case in point. Natural theology as a common moral context for all crises, whether economic or medical, had by this time given way to a more secular ordering of the world's affairs; but that did not stop economic writers from employing epidemic language. The old language, bereft of its natural theological implications, still played the important role of calling attention to the need for a concerted social response. All that had changed was that the new response, in both commerce and public health, was far less focused on individual responsibility than it had been in more Malthusian times. The new authorities viewed crises as side effects of a complicated environment, where blame was due to the failure of the state and its allied professions properly to direct the flow of natural resources. As economists edged away from the earlier passive response and towards a more "modern" interventionist approach, they gained from parallel changes in the connotation of "epidemics" suggested by public health officials. These changes allowed them to continue referring to crises figuratively in language everyone thought they understood – a fever was, at least apparently, still a fever – and lure support for their new approaches to commercial stability away from competing interest groups.

Crisis and Atonement: 1825–1845

Britain's unique economic circumstances in the early nineteenth century were accompanied by new conceptions of the dynamics of growth. The business cycle, although not articulated as a formal theory until 1900, already started to appear in economic discourse by the 1830s to account for observed waves of recovery and decline. The banker Samuel Jones Loyd (later Lord Overstone), who played a major role in commercial legislation throughout the middle third of the century, repeatedly claimed that trade "revolved apparently in an established cycle."[6] Such claims for periodicity of commercial crises had been missing from Adam Smith's account of economic growth, which in many other ways corresponded with Loyd's. While Smith had recognized widespread instability in eighteenth-century commerce, he assumed the possibility of stable growth in the future so long as people followed his prescriptions for economic statesmanship. The transition from Smith to Loyd on commercial stability reflected a significant shift in their respective conceptions of human nature. Smith assumed that the "effort of every man to better his condition" was sufficiently "uniform, constant, and uninterrupted" to counteract overwhelmingly the occasional bankrupt who partook in "injudicious and unsuccessful" undertakings.[7] This attitude stood in marked contrast to Loyd's position sixty years later, which located the periodicity of commercial fluctuations in the natural speculative tendencies of business. In 1840, for instance, he called it a matter of "human nature" that "speculations will occasionally occur, and bring with them their attendant train of alternate periods of excitement and depression."[8]

Images of nature figured prominently in such recastings of Adam Smith's economic man, of which Loyd's stands as only one example. One major "natural" corollary of periodical commercial crisis was the regular outbreak of mortality owing to epidemic disease; but Loyd and likeminded economists had a host of other natural metaphors at their disposal, to which they were just as quick to appeal. Boyd Hilton, who has itemized such references in *The Age of Atonement*, exhibits the use of meteorological, volcanic, sexual, and diluvian language, in addition to biological metaphors, among writers in the 1830s who tried to evoke the purgative effects of commercial crises.⁹ Such natural language served as shorthand for a shared moral construction of events, revealing common assumptions about the relation between God and nature. Hilton's work, which locates commercial and medical language in the common context of evangelical theology, goes far in identifying the relevant moral messages that encoded references to commercial crises between 1825–45. From there it is further possible to detect how this common moral context (as signified by the presence of the metaphor) also reinforced specific institutional formations and provided the basis for debates from which new constructions and outcomes of crisis evolved.

The Response to Epidemic Disease, 1825–1845: From Retribution to Registration

Since the meaning of "epidemic" was far from stable in the nineteenth century, it is first necessary to locate its significance in the early Victorian period before discussing the epidemic language used by Loyd and his contemporaries. A leading popular assumption about epidemic disease was that it signified divine retribution in some form. This idea, already present in 1825, especially gained ground in the wake of the cholera epidemic of 1832, which was variously described as punishment from on high for intemperance, infidelity, sectarianism, popery, and overconfidence in medical science. The connections drawn between God and cholera were not always very precise and they differed from sect to sect.¹⁰ But the underlying assumption, that epidemic disease was a sign of God's dissatisfaction with the community, provided the groundwork for a broad consensus concerning the proper choice of response. This response appeared under several guises, covering theories of disease causation and therapeutics.

The prevailing theory of disease associated with retribution has been broadly labelled "anticontagionist," in contrast to eighteenth-century theories attributing disease to person-to-person transmission. The earlier "contagionist" view increasingly came under fire in the nineteenth century on medical, economic, and moral grounds. Among a number of medical arguments marshaled by anticontagionists, the leading one was that most fevers appeared too suddenly to be communicated via individual contact. Historians have also posited economic and political contexts for the shift to anticontagionism, on the logic that those who opposed quarantines for commercial reasons also tended to oppose contagionist theories of disease. Most significant from the perspective of this paper, anticontagionism also signified a major shift in the moral economy of disease, in which medical officials came to construct epidemics as an aggregate consequence of irresponsible behavior on the part of individuals. This conception reversed the contagionist model, where agents external to the body infected groups of innocent people.[11] Within this general framework there was room for much etiological disagreement, so much so that some historians of medicine have recently called into question the utility of lining up early-nineteenth century theories of disease on a contagionist-anticontagionist axis.[12] Regardless of whether one accepts the standard version or sides, for instance, with Christopher Hamlin's proposed split between predisposing and exciting causes, however, it remains the case that the ascendent medical assumption in the early nineteenth century ascribed responsibility for disease to victims as much as or more than to their surroundings. People whose theories of disease transmission kept them in the "contagionist" camp often remained convinced that one's personal habits directly affected one's susceptibility to disease.[13]

Only extreme anticontagionists, in any case, rejected all contagionist explanations and accompanying cures. Most were satisfied to demonstrate that only some diseases, such as small pox, were spread via contagion or inoculation, and to offer evidence that contradicted contagionist explanations in the remaining illnesses, which received the label of "epidemics." A typical demarcation between contagious and epidemic disease, appearing in the *Quarterly Review* in 1825, described the former diseases as those where "the bodies of the sick give out a noxious material, which excites them in the bodies of the healthy," and the latter as those where the "causes are so diffused as to affect many persons in the same place at the same time."[14] For the commonly recognized contagious diseases, such responses as quarantine and (in the case of small pox) vaccination were generally accepted. But in the restricted space of epidemics, cures ascribing to the patient moral responsibility for the disease were assumed to exhaust the capacities of the clinician. Other health officials, such as the sanitationist Thomas Southwood Smith, expanded the class of diseases where "noxious material" was involved but still held individuals at least partly responsible even in those cases.

Locating responsibility for epidemics in individual behavior reified Thomas Malthus's story about divine retribution, in which the undue multiplication of individual acts of reproduction led directly to the positive checks of famine and pestilence. There were, however, different ways to tell this story, and competing storytellers soon split off into opposing therapeutic camps. In one camp were those who remained attached to the original Malthusian lesson on population growth, namely, that passive acceptance of one's plight was the only viable response to pestilence. These people defended more traditional cures like bloodletting: assuming, as Hilton has suggested, that since "the body must have generated its own infirmities," a well-placed leech appeared to offer the best chance of removing the disease from the body. The same doctors and public health officials blamed epidemics on inadequate diet, using the Malthusian logic that pestilence was a warning sign of population pressing against scarce food resources.[15] In another camp were those who focused on Malthus's "preventive" checks to population growth, most pointedly that of moral restraint. Pledging allegiance to the New Poor Law, these people argued that once the "less eligibility" principle was in full swing, people would never die of fever for want of food. Their proposed interventions, registration of deaths and sanitary reform, reflected their utilitarian ideology of prevention by means of rational social control. Registration was a means of locating with more precision the relation between population and pestilence, and sanitation was a way to clean up what was assumed to be the diseased environment of the poor. Although both these interventions would ultimately refocus attention away from individual responsibility, when they first appeared the primary emphasis was on teaching potential victims of epidemics to mend their ways.

In 1832, when cholera first struck in England, the most prominent response appealed to passive therapies that accorded with the Malthusian construction of disease as a positive check; only later did a more utilitarian preventive response to epidemics prevail. In a period where perception of social esteem outweighed therapeutic expertise as a professional drawing card, many doctors geared their response to the 1832 epidemic to correspond as closely as possible with the common beliefs of their patients. To avoid being perceived by victims' families as performing untried medical experiments on their patients, doctors tended to rely on conservative remedies like bleeding and vomiting. Although they were mainly drawing from a much a earlier classical medical tradition stressing the need to keep the body's constitution in a healthy balance, doctors who performed these purgative remedies were also repeating at the level of the individual patient the more general aim of what might be called Malthusian social medicine: the restoration of society to a balanced level of population and food. This practice helped improve doctors' esteem relative to quacks, whose exaggerated promises of cure received fatal disproof in high-mortality diseases like cholera and other fevers. The state, which in the 1830s had no organized means of prevention, could secure itself from the worst of public dissatisfaction over its response to cholera by repeating the common wisdom that located a nationwide illness in collective misbehavior. When governors and churchmen were not organizing national fasts and prayers, they offered medical advice that stayed firmly within conventional medical wisdom. A central board of health, which had been hurriedly established in 1832 to address the cholera crisis, urged victims to eat their beef roasted instead of boiled and to provide leeches with a daily diet of ten to twelve ounces of blood.[16]

In the decade after 1832 this passive medical response to epidemics gave way to the more interventionist policies of Edwin Chadwick and his fellow utilitarian social reformers, without entirely losing its emphasis on individual responsibility. The new interventionist approach gained credence in part because the medical crisis of epidemic disease was at this time incorporated into a more general construction of crisis in the realm of social administration. This political atmosphere produced a more polarized debate on epidemics in which both sides claimed to be capable of preventing such occurrences in the future, leaving little room for passive acceptance of God's will. The debate pitted anti-Malthusians like W. P. Alison, who argued that increased welfare expenditure would prevent epidemics, against Chadwick and his allies, whose rearguard defense of the New Poor Law indirectly led to new theories of disease. Since the latter group claimed that poor law reform would soon put a stop to lack of food as a cause of crisis, they instead looked for supplemental causes.

One proposal was that the mechanism of the New Poor Law, while ultimately capable of eliminating positive checks to population growth, needed time to do its job. In the interim, Chadwick argued, it could be supplemented by inculcating among the poor a better understanding of the dire consequences of overpopulation. He assumed that a well-publicized registration system, pairing population returns with mortality statistics, would accomplish the necessary education. By 1836, Chadwick was urging that the recently-established General Registrar's Office be used to track correlations between epidemics and the condition of paupers. He hoped that a better supply of mortality statistics would stimulate both the state and the public to take a greater interest in discerning the causes of disease. One possible educational benefit of registration, he suggested in a republication of his *Essay on the Means of Insurance*, was that the resulting data would give "safety to the immense mass of property insured, so as to enable every one to employ his money to the best advantage for his own behalf" – in other words, a general registry would promote prudent institutions like friendly societies and sickness clubs, which at the time were frequently seen as a preventive against undue mortality and disease. More directly, and more generally, Chadwick urged that another important use of a registry of sickness and death was "[t]he direction of the mind of the Government and of the people to the extent and effects of calamities and casualties."[17] As such, his medical rationale for registration did not stray very far from the ascription of moral responsibility typical of his time: only by participating with a more activist state could the poor ensure their wellbeing. At the same time, the very act of collecting information on sickness in the form of impersonal numbers meant that future health officials found it easier to eliminate individual responsibility as a primary focus of concern. Much to Chadwick's chagrin, such a transformation is exactly what his hand-picked appointment to the Registrar's Office, William Farr, accomplished ten years later.

Poor sanitation was the other supplemental cause of epidemics forwarded by Chadwick and his allies; it was similarly transitional between blaming victims for their disease and ascribing responsibility to the state. When Neil Arnott and J. P. Kay, under Chadwick's direction, reported on the causes of London fever in 1837, they included among those "originating in [victims'] habits" crowded lodgings, poor housekeeping, and pig-keeping. This classification suggested a sharing of responsibility for epidemics between the state, which needed to educate the poor to take better care of themselves, and the individual, who needed to comply with the state's directives. Southwood Smith added a wrinkle to this prescription in 1838 by differentiating between the "exciting" influence of "febrile poison," which was in itself sufficient to produce fever, and such predisposing individual habits as pig-keeping, which merely increased the likelihood of fever in the presence of the poison.[18] Both in his 1838 report and in an earlier textbook on fever, Smith took the important step of coupling external material causes of disease with a more traditional focus on the individual victim. Such a pairing was evident in his textbook, where he claimed that a fever victim's "small and heated apartment in London" was "perfectly analogous to a stagnant pool in Ethiopia, full of the bodies of dead locusts." As John Pickstone has argued, this portrayal of the disease victim "as merely a form of decomposition" blurred the boundaries between "the human body and external agents." Pickstone adroitly uses the example to draw attention to the "disappearance of the patient" often located much later in nineteenth-century therapeutics.[19] It is possible, however, also to find in Southwood Smith's comparison lingering traces of a traditional focus on individual responsibility for disease. By binding victims with "dead locusts" in the same metaphor, he implicated them in an overarching ascription of responsibility – as Chadwick had done when he crystallized the sin of imprudence in the objective grid of mortality statistics. Not until the 1840s, when second-generation sanitary reformers like Farr decoupled the individual from the external agent, did "epidemic disease" come to connote a radically new ascription of responsibility.

The Response to Commercial Crisis

We are now in a better position to judge what early Victorian economic writers meant when they used epidemic language to describe commercial crises. All the implications of epidemic disease outlined above – its attribution to divine retribution, the etiological and therapeutic response – appeared in analogous form following the commercial crises occurring in 1825 and 1837–38. The common appeal to retribution, while the most direct in its attribution to a single source, was too general to produce any specific parallels at the level of intervention. Epidemics and commercial panics were merely two items on a long list of disturbances that meshed with the millenarian prediction of judgment: others might include Catholic emancipation and the Reform Bill.[20] The etiological and therapeutic parallels, however, reveal underlying assumptions and bases for transitions to new responses. At these levels commercial reformers shared their public-health contemporaries' view that crises most often resulted from individual moral infirmities. They also derived competing therapies from this general ascription of responsibility, with some reformers favoring more passive appeals and others advocating a stronger role for the state in achieving the same ends. As with public health reformers, the latter group's political successes paved the way for further reforms in the 1850s which would move the debate even further from its original context.

It is not too far-fetched to describe the usual interpretation of disturbances in the British economy between 1825 and 1845 as essentially "anticontagionist." In commerce as in medicine, the tendency at this time was to describe national outbreaks of crisis as aggregates of individual excess, rather than attributing them to a source external to the individual. The two types of excess most frequently cited were overproduction, owing to a failure on the part of capitalists to anticipate levels of demand; and overspeculation, owing to investors' imprudence in forming new companies. As with their medical counterparts, commercial observers attached to this analysis the proviso that such an interpretation did not cover all cases of "disease." Some commercial crises, it turned out, were in fact "contagious," in the sense of being due to external factors, but most qualified as internally generated and therefore "epidemic" in nature. Hence the economist J. R. McCulloch, in 1826, distinguished between crises that "depend on political contingencies, and may therefore be considered as in some measure accidental, and those which arise from the miscalculation of individuals, or from some defect in the system under which the industry of any given country is conducted." He restricted his view to cases that were not due to "political contingencies": the latter, he observed, were "plainly beyond the sphere of the Economist."[21]

McCulloch's distinction between what might be called epidemic and contagious commercial crises was grounded in an emerging methodological distinction in political economy between "the sphere of the Economist" and what William Nassau Senior called in 1836 the "art of government," which was "more important, but far less definite" than the economist's science.[22] This distinction mirrored the anticontagionist response to disease, where the consequences of individual behavior took center stage as a subject of debate, but also where the precise boundaries separating individual and state responsibility were constantly under negotiation. Firmly outside the domain of individual behavior stood "contagious" causes of crisis such as foreign tariffs and wars, which most economists were content to leave to the diplomats to sort out. Within the "anticontagionist" domain of individual agency, their job was to discover the human limitations to economic growth and stability (McCulloch's "miscalculation of individuals"), and to determine where domestic policies exacerbated these "natural" limitations. The first part of this job followed the doctrine of retribution implicit in Malthus's original conception of disease as a divinely ordained check to population growth. When economists turned to the realm of domestic policy, though, it became more difficult always to declare God's will to be the last word. With more or less continuing emphasis on individual responsibility, all proposals for a state response to crisis implicitly tacked on various "environmental" causes and cures, just as public health officials had done once they started to focus less on the individual and more on external agents that the state could help remove. This commercial environment was populated by "removeable" ignorance on the part of investors, as well as more tangible external agents like corn, currency, and company shares.

In the wake of the crisis of 1825, the old-fashioned Malthusian construction attributing crises to individual miscalculations was still at its height and talk of government intervention was relatively scarce. The evangelical economist Thomas Chalmers diagnosed the crisis along strict Malthusian lines: "what disease does with the redundant population, bankruptcy does with the redundant capital of our land; relieving the overdone trade of its excess, and so reducing capital within those limits beyond which it cannot find any safe or profitable occupancy." When Chalmers moved from diagnosis to therapy, his proposals also did not stray far from Malthus: "we should not object to the moral preventive check of Malthus being extended from laborers to capitalists; and a higher style of enjoyment is the instrument, in both cases, of putting it into operation." Although Chalmers took his call for a "higher style of enjoyment" to more doctrinaire extremes than Malthus, he was correct in implying that Malthus had, at least later in life, argued that an increased consumption of luxury good would put off population by raising the income perceived to be necessary for raising a family.[23] At a more formal level, Chalmers's call for increased consumption on the part of the rich also echoed the specific cures proposed by doctors in the wake of the 1832 cholera. Just as the doctors had ascribed to the Galenic idea that a victim of fever needed to be bled in order to restore the body's proper balance of blood, Chalmers argued that merchants needed to perform a similar operation on their balance books, which showed too much in the savings column and too little in the spending.

Notwithstanding the claims of some twentieth-century economists that Chalmers was a proto-Keynesian, there was virtually no room for the state in his proposals for increased consumption. These were strictly aimed at individuals, with all the personal invective he could muster. Unfortunately for Chalmers and for economists of a similar stripe, however, the burgeoning forces of competition turned out to hold more attraction for most capitalists than even the loudest religious supplications. So when another bout of overspeculation led to crisis in 1837, economists turned to solutions that combined individual responsibility with some form of state action. They focused, in particular, on three "defects in the system": protectionism, monetary policy, and company laws. These defects signified different "environmental" causes of crisis: protectionism suggested disequilibrium in corn prices from year to year, monetary policy addressed an improper level of note issue, and company laws were concerned with undue supplies of share capital. Most economists in the 1830s agreed that protectionism exacerbated crises by hindering capitalists' ability to predict supply and demand, and assumed that corn would cease to be a nemesis once merchants were free to import it to supplement bad domestic harvests. Free trade would cause corn prices to stabilize, sending a ripple effect through the economy as the price of labor and other goods attached to corn followed suit.[24] Since corn law repeal was a thoroughly negative piece of legislation, it was an instance where most economists could agree on the need for state action despite holding widely different views about how to apportion responsibility for commercial crises. People like S. J. Loyd or Thomas Tooke, who assumed that crises were mainly due to the irremediable speculative penchant of merchants, still opposed protectionism as a contributing cause; while utopian economists like James Wilson and Richard Cobden assumed that free trade in corn would spell an end to crises forever. Even such attenuated concord was not possible in discussions about monetary and company policy, which called for the construction of new laws rather than simple agreement to abolish an old one.

In the debate over monetary reform following the 1837 crisis, the most successful strategy politically was employed by Loyd's "currency school." This strategy focused on overissue of bank notes as a significant contributing cause of crises, but, as in Loyd's advocacy of free trade, assumed that crises would continue to appear as retribution for immoral trade practices regardless of anything the state might do. Other proposals included allowing bankers to print notes entirely at liberty so that competition would regulate note issue, giving the Bank of England discretionary authority over the supply of money, and establishing a national bank. Loyd's plan, which attached note issue to the amount of gold bullion in England, sought only to prevent financiers from indulging what he assumed to be businessmen's inherent tendency to overtrade. Note restriction was not intended to make life easier on the trader; rather its task was to reinforce the "universal law in human affairs" that overall prosperity was possible only in exchange for such "occasional inconveniences and pressure" as were bound to arise when the Bank Charter Act forced down the supply of notes. Hard-won experience, not state management, was the only way the moral lessons of commercial crisis could be driven home.[25] As conservative doctors had done following the cholera of 1832, Loyd won approval for his pessimistic scheme by comparing his opponents' prescriptions to quack remedies. Joint-stock bankers and more radical anti-gold writers responded similarly in medical language, claiming that the currency school's misguided therapies would suck businessmen dry.[26]

One reason the different sides in the currency debate so easily slipped in and out of medical language was that they were repeating, at a formal level, many of the same arguments about disease causation then under discussion by doctors. Businessmen, like fever victims, exhibited bad habits that were related to crises. Bank notes, like Southwood Smith's "febrile poison," were agents external to the victim that affected commercial "health." The problem was to rank these various factors and (since political success frequently depended on being able to propose a simple solution) to single out one of them as the leading cause of the crisis. Loyd, by insisting that currency merely exacerbated the "natural" tendency of merchants to overtrade, was echoing the claim of traditionalist doctors who argued that a patient's "predisposition" was the only necessary factor in an epidemic. At the same time, by acknowledging that some influence was due to currency, he stole the thunder from other reformers who paralleled Southwood Smith's logic in claiming that traders were incapable of miscalculating unless in the presence of an ill-regulated currency. Much of Loyd's success in the debate stemmed from his ability to win the support of Manchester free traders, who suspected that currency was a problem but who worried that these other reformers would endow the state or the Bank of England with arbitrary control over note issue.[27]

Loyd's response to commercial crises was politically successful as long as people were convinced that free trade was not the sole cause of crisis and that currency was the only external "poison" worth considering. Currency and protectionism, however, did not exhaust the list of possible diagnoses mooted in the turbulent 1830s and 1840s. Another apparent cause of crisis was the growth in company formation, especially among railways and banks, that was already exhibiting destabilizing effects in the 1837–38 crisis and would produce even more convincing evidence of instability in the railway mania and subsequent crash of 1845–47. Many of the company failures during these crises had cast doubt on the assumption that capitalists were fully responsible for their misfortunes. When a bank or insurance company went under, the "capitalists" who appeared in bankruptcy court were frequently blind investors who had little if any control over the firm's management. To come to terms with this breakdown in agency, politicians and economists appealed to the premise behind Chadwick's call for better vital statistics. Just as epidemics appeared in part because the poor had been unaware of the connection between overpopulation and pestilence, so had the recent commercial crises struck because capitalists were insufficiently aware of the consequences of entrusting their funds to strangers. In both cases the most popular panacea for ignorance was registration: of births, deaths, and marriages for Chadwick, of company shares for politicians in the 1840s. Focusing on company shares in this manner suggested a further link between economic policy and public health reform: the share capital could be interpreted as a dangerous external agent, akin to Southwood Smith's poison, that investors could be taught to control more wisely. But it was not until after the 1847 crash that economists started to pursue the public health metaphor in that direction.

The man behind registering company shares was William Gladstone, who introduced the Joint-Stock Companies Act in 1844 while serving as Robert Peel's president of the Board of Trade. The act permitted projects applying for full registration with "provisional" status to allow prospective investors to watch over their proceedings; and required companies that had reached the stage of final registration to publicize their accounts as well as the names and addresses of their partners. As with Chadwick's proposal, Gladstone's registration bill enjoined individuals to participate with the state to ensure their well-being. The Joint Stock Registry, at public expense, identified with whom investors would be doing business, while investors needed to take it upon themselves to examine the register before purchasing shares and once they did so were fully liable for a company's losses. Although the specific functions of Gladstone's act did not correspond directly with the public health application of the General Registration Act, it is possible to observe less literal links with preventive medicine both in the circumstances of the legislation and in its effects. The hearings of the parliamentary committee on companies, chaired by Gladstone, was littered with comparisons between crises and disease. One lawyer who testified, for instance, referred to "those paroxysms of speculation, which would seem to recur periodically," called the 1825 crash a "national epidemic," and produced a table showing the "mortality" of companies formed between 1834 and 1837.[28]

The institutional fact of registration also suggested parallels with vital statistics, both to witnesses at Gladstone's committee and to observers after 1844. The lawyer John Duncan, anticipating registration in 1843, observed that "it is the duty of the legislature to keep some degree of control over the birth and course of life of joint-stock companies, having reference to the different objects for which they may be formed." In the years immediately following Gladstone's act, a number of projects appeared with the express intent of using the resulting company statistics in the same way that an actuary used vital statistics. One such firm, the Commercial Casualty Mutual Assurance and Indemnity Society, reasoned from the "scientific application of the principles of assurance" in the case of "the uncertainty of human life" to "a well-matured scheme for the alleviation of the evils inseparable from a state of bankruptcy."[29] More generally, the registration of joint-stock companies indicated a Malthusian regard for the social consequences of individual action. At the level of metaphor, companies could be described as resulting from "marriages" among partners, and improvident projects could be measured by levels of "infant mortality" among their offspring. In 1853 the commercial writer John Francis employed such language when he expressed confidence that company registration, like vital statistics, could produce empirical clarification about the victims of crisis: "Every six or seven years mercantile epidemics – analogous to the cholera, the influenza, or the typhus, of an unhealthy season – which seem to defy all calculation, and to level the lofty as well as the low – revolutionize our money system . . . there is now plenty of information on which to base some estimate of the [resulting] annual losses of special classes."[30]

Crisis Management: 1845–1870

The end of the 1840s, so notorious for its political convulsions, also witnessed major crises in commerce and public health. First, in 1847, three years of speculation in railway shares crashed to the ground when the major lines started making calls on their investors; combined with the Irish potato famine, the result was the most severe depression of the century. The following two years a new wave of cholera took 62,000 lives, twice as many as in the 1832 outbreak. Both crises assisted in effecting a new response to future commercial and medical disruptions that focused more on the environment and less on the individual. In the cholera epidemic of 1848–49, William Farr first used statistics generated from a revamped General Registrar's Office to support his "zymotic" theory which attributed cholera to disease-bearing miasmata in the atmosphere. When cholera returned five years later, he was able to use the resulting mortality statistics to confirm his theory, as well as amend it to account for John Snow's contention that the disease was actually water-borne.[31] The transition from individual to environment was less striking but equally evident in the case of commercial crises, where merchants and legislators soured on Loyd's restraint-minded solutions following the Bank Charter Act's exacerbation of the crash of 1847. Instead of blaming the crisis on imprudent business practices, they redefined the problem as one of excess capital flowing through a poorly-organized credit system. The trick, much like the new approach to epidemic disease, was to discover the unhealthy capital and reroute it in healthier directions.

Throughout this period of parallel transitions, people continued to talk about commercial crises using epidemic language, but the meaning of their language shifted to keep up with the changing connotations of each malady. These meanings produced new ascriptions of responsibility, new legislative responses, and new forms of professional organization, all of which were comparable between public health and commerce. By locating the source of crisis in aggregates of unhealthy objects (water-borne sewage or surplus capital) as opposed to aggregates of unhealthy people, the new responses to panics and epidemics shifted responsibility to very specific institutions that were easy to target. In the case of public health, the institutions were the water companies that served the various cities of England. In the case of commerce, they were the increasingly specialized joint-stock banks and brokerage firms that floated capital back and forth from firm to firm. Reconceiving responsibility in such ways produced new forms of legislative intervention: sewage regulation, for public health; and limited liability, for commerce. Finally, the parallel transitions had similar professional consequences. In both cases, the successful interest groups were those that reorganized themselves to meet the less personal conditions of crisis control. These were the managerial classes of the late nineteenth century: the sanitary engineers and accountants whose professional responsibility it was to see that day-to-day problems did not reach crisis proportions.

Medicine: From Atonement to Environment

Already by 1848, when cholera struck England the second time, the country was receding from its moral consensus that had once made retribution appear as the natural explanation. By the third strike in 1853–54, the notion of an angry God had become passé. Lord Aberdeen refused Archbishop Summer's request in 1854 to call a day of prayer in response to the epidemic, and a year earlier Lord Palmerston had similarly pronounced that the "sources of contagion . . . , if allowed to remain, will infallibly breed pestilence, and be fruitful in death, in spite of all the prayers and fastings of a united but inactive nation."[32] Action, not penitent submission, was the order of the day, and a specific sort of action: a search for Palmerston's "sources of contagion" wherever they might by found, whether in earth, air, or water. The only place people stopped looking for disease was in the realm of individual behavior, which affected proposals for cure as well as prevention. Bloodletting, by the 1850s, was on the way out as an orthodox medical practice, and the interventionist approaches to public health by Chadwick and Southwood Smith, when applied by later reformers, moved still further away from blaming epidemics on the victims.

The public health officials who carried out the active search for sources of disease in the 1850s did so by building on Chadwick's suggestion to keep better vital statistics and on Southwood Smith's decision to target external poisons. Chief among these officials was William Farr at the General Registrar's Office, who finished what Chadwick had started at that bureau. Instead of using vital statistics merely to trace the results of improvident marriages, Farr suggested that mortality rates could be used to isolate specific locales where epidemics were most fatal.

Following the cholera years of 1848 and 1849, Farr correlated these regional findings with elevation, which he had interpreted to be the only statistically significant variable. He assumed that the areas of higher mortality, which corresponded to lower elevations in London, were caused by the greater concentrations of miasma-producing organic matter in lowland soil. Later work by John Snow during the 1853–54 epidemic demonstrated that the relation between cholera and elevation was in fact coincidental: the actual circumstance of higher mortality was due to polluted water that was provided to the low-elevation, low-income houses along the Thames. Although there remained room for disagreement about exactly how mortality statistics could be used to prevent future epidemics, the new methodological focus was significant. Instead of using mortality statistics to link people's behavior directly to their demise, or to persuade the victims of disease to lead better lives, Farr and Snow linked mortality to factors over which the victims had little or no control.[33]

Farr and Snow also finished what earlier public health reformers had started by resolutely decoupling external disease-causing agents from individual behavior. By isolating strictly geographical variations in the occurrence of cholera, Farr directed attention away from personal habits even more than Southwood Smith had done, and directed attention towards the organic matter he assumed to exist in profusion at these lower levels. Snow took the isolation of external sources of disease a step further in his famous Broad Street Pump study of 1854, in which he discovered that contaminated water supplied by the Lambeth Water Company was responsible for nearly all the cases of cholera reported in the vicinity of the pump. This study appeared to rule out the personal habits of residents as even an important predisposing cause of the epidemic, since residents on the same block could be shown to survive while their neighbors died simply by virtue of being served by a different water company.[34] A common theme in this debate was that there was nothing intrinsically bad about such external agents, as long as they were disposed of properly. Christopher Hamlin has shown how the earlier providentialist perspective, which had once been used as an excuse for inaction, came to the service of the sanitary movement by suggesting that sewage, once properly treated, could be turned to good use as fertilizer – all to the greater glory of God.[35]

Both Farr and Snow translated their increased emphasis on external disease-causing agents into policy proposals that assumed a new sharing of responsibility for the prevention of cholera. Instead of expecting potential victims of disease to acquiesce to state-imposed sanitary reforms, they targeted the water companies whose practices appeared to be responsible for the correlations uncovered by their statistical studies. Farr's report on elevation led to the Metropolitan Water Act of 1852, requiring companies to provide water from higher altitudes which, according to Farr's zymotic theory, contained lower concentrations of organic matter. Snow's studies, and the zymotic theory of disease on which they were based, led to a more complicated call for water companies to clean up their act. As Hamlin has suggested, the zymotic theory made it hard for health officials and chemists to search for particular agents in the water, since it attributed disease to undetectable quantities of putrified matter that only became poisonous once they decomposed in the body. Unlike the more straightforward theories of Chadwick and Southwood Smith, which suggested that it would be possible to test different supplies of water to detect unhealthy levels of putrefaction, Snow and his allies provided no means to test water directly. The result was, ironically, a more sweeping call for water reform. If any epidemiological link could be established between disease and water supply, as Snow had done in 1854, suppliers had no higher appeal to chemical analysis as they might have had with Chadwick's theory. In later years, following the return of the cholera in 1866, the water analyst Edward Frankland continued to take sanitary reform in Snow's direction by emphasizing that a study of the possible past contamination of water supply, rather than on-the-spot chemical analysis, was the only way to determine impurity. To convince health officials that this strategy was to be trusted he needed to fend off methodological attacks from competing chemists, some of whom were in the employ of the water companies.[36]

Professionals and Prevention: Limited Liability as a Cure for Crises

Concurrent with the changing responses to epidemic disease in the 1850s, commercial participants were also reformulating their reactions to crises. Ultimately their response paralleled that of their medical peers, but only gradually, as evidenced by the opposition of many capitalist spokesmen in the late-1840s to centralized public health reforms. James Wilson and his colleagues at *The Economist*, for instance, while fully supporting the repudiation of fasts and prayers as responses to epidemics, thought Farr's "enlightened views" would work even better without central agencies like the General Board of Health in the way. In the words of one *Economist* columnist's message to the leaders of the public health movement: "Show the inhabitants of the towns that your recommendations will be advantageous, and they will as surely adopt them as the merchant will go to the place where he can carry on the best trade."[37] This logic, both in its continued emphasis on the need for individuals to take responsibility for reform and in its free-market analogy, informed the position of many mid-century economists on commercial crises. They were no longer interested in talking about crises as retribution for capitalists' sins; but they still thought capitalists could be taught to manage their money in such a way as to prevent crises in the future. At the same time, many of them – Wilson included – were now willing to accept, in keeping with Southwood Smith's public health pronouncements, that the merchant's education would be incomplete without a lesson on the potentially poisonous effects of the capital that flowed through the economic system.

Wilson decided to focus on capital as a potentially dangerous external agent in the context of a debate with Loyd over the allegedly inherent prudence of businessmen. The crisis of 1847, coming as it did a year after the repeal of the corn laws, seemed to suggest that something besides protectionism was necessary to account for merchants' miscalculations. When Loyd's allies reasserted that the new crisis signified a natural, divinely ordained punishment for imprudence, Wilson and *The Economist* felt compelled to defend the prudence of the British merchant. As an *Economist* writer argued in 1852, Loyd's theory that "commerce revolves apparently in a cycle" was bound to be incorrect, since the businessmen whose miscalculations were alleged to be the source of instability were in fact "the agents for equalising natural variations."[38] The repeal of the corn laws had put people like Wilson in much the same position public health officials like Chadwick and Southwood Smith had found themselves in after the New Poor Law, when epidemics were no longer admissible as evidence of positive checks to population growth. If, as Wilson had claimed before 1846, free trade would eliminate all causes of crisis, there was no point in harping about retribution. Instead it was necessary to search for supplemental causes, ones that would eventually disappear with a combination of improved knowledge among capitalists and sound advice from economists. In his *Capital, Currency, and Banking* (1847), Wilson attributed the recent crisis to an excess of fixed capital invested in railways, which had "absorbed the floating capital of the country to an extent beyond all proportion to our means." His lesson from this diagnosis was to call on capitalists to be more careful to exercise the prudence that the newly free market would, in any case, ultimately teach them: it was "incumbent upon us to see that none of our exchangeable commodities are unproductive and useless."[39]

Despite Wilson's standard recourse to individualism in this context, his actual analysis shifted the problem from one of individual behavior to one of resource allocation. Here surfaced an important parallel with emerging analyses of epidemics: the idea that disease was caused by an accumulation of unhealthy matter in one place. Wilson's oversupply of fixed capital was not unlike Southwood Smith's "febrile poison," in the sense that both suggested a redistributive solution. Polluted sewage could be turned into fertilizer; depreciating stocks could be turned into productive capital.[40] And while Wilson assumed that businessmen were the natural agents for reducing the toxic concentration of investment in railways, much as Southwood Smith had called on potential disease victims to help themselves, other onlookers soon took up the position that the problem required more serious structural changes in the economy. The leading change advocated by those who directed Wilson's analysis into interventionist directions was to legalize limited liability as a general form of incorporation. Until their efforts bore fruit in the form of the Companies Act of 1856, the only firms qualifying for limited liability had been railways and other specially-chartered companies. All other firms were required by law to hold their shareholders responsible for any losses they might suffer. Two different types of arguments appeared defending limited liability as a response to crisis. The first took up where Wilson left off, suggesting that the railways' special status as limited firms had attracted dangerous proportions of capital from its regular circuit. Limited liability, it was argued, would equilibrate investment and prevent too much concentration of capital in one place. The second defense of limited liability posited a different dangerous pool of capital, this time resting in the hands of the burgeoning credit institutions of mid-Victorian England. People taking this position hoped that limited liability, by spurring overall investment, would reduce the demand for credit and prevent the build-up of capital in the hands of bill-brokers and joint-stock banks.

Those who defended limited liability as a means of preventing over-investment in a single direction specifically criticized Wilson for not carrying his diagnosis of the 1847 railway crisis far enough. Matthew Begbie, in a pro-limited liability tract published in 1848, included a chapter on "Panics, – Their Cause and Effect" in which he took issue with Wilson's claim that free trade would "naturally" teach future investors to make better decisions regarding what to do with their money. He observed desire that their

> desire to obtain a higher rate of interest than the Three per Cents., allowed in the Funds was a very natural desire – that Railwayism promised to yield that higher rate of interest . . . and that the state of the law of partnership actually *forbade their employing their money in any other partnership*, public or private, without involving them in unlimited liabilities, no matter how limited the amount each person might have adventured.

In this case Begbie ironically sided with Loyd over Wilson in assuming that it was "natural" for investors to desire more than 3% in dividends; but he added the important twist that the occasional crises to which such natural behavior led could be averted by means of structural changes in the economy, instead of accepted as punishment from on high. Several witnesses at the Mercantile Law Commission, which met in 1854 to debate the merits of limited liability, repeated this wisdom with epidemic overtones. The Christian Socialist John Ludlow argued that the system of chartering railways opened "a single door, as it were, to limited liability"; he concluded that "the whole enterprise of the country rushing in by that door is likely to produce a speculation fever." This view was seconded by the Oxford economist G. K. Rickards, who claimed that general limited liability, by equalizing investment preference, would prevent capital from reaching dangerous levels. As things stood, whenever a new type of company enjoyed a monopoly over limited liability, "the tide of capital pours into it at a flood, enlarges it beyond all reasonable bounds, and converts what, if confined within moderate limits, might have been extremely beneficial, into a source of ruin."[41]

While some defended limited liability as a means of siphoning off unhealthy capital from chartered companies, others thought it would prevent crises by keeping money out of the hands of bankers and billbrokers. According to this logic, limited liability would encourage businessmen to pump their surplus finds directly into capital-hungry projects, instead of entrusting it to credit agents who were more likely to call in their loans at the first sign of a crisis. Charles Holland, a Liverpool merchant, linked the existing company laws with the popular practice of surplus income being deposited in banks and "loaned out profusely by them to their customers." he concluded that "the *credit system* of this country . . . must always be liable to those violent shocks while capital is thus diverted by legislation from its natural channels." This view paralleled that of the London warehouseman John Howell, who claimed that "money in the hands of agents is more mercurial than if invested and rooted as the capital of sleeping partners; the trader is consequently liable to be ensnared into undue speculation, by its cheapness at certain times, and greatly exposed when time of pressure comes; – hence panics, and from panics spring monopolies." As with public health reformers who newly defined "dangerous" classes of people more by their polluted environment than by their moral habits, proponents of this view reoriented the traditional blame for commercial crises. Howell warned in 1851 that "a young man beginning entirely with borrowed capital . . . is a dangerous customer if he borrows money which can be called from him at any time when the lender begins to be fearful."[42]

The combination of these arguments, as well as others having more to do with social than economic stability, was enough to provide Robert Lowe with the support he needed to pass a Companies Act in 1856 calling for general limited liability – to be followed in 1862 by even more comprehensive legislation. It is possible, as Boyd Hilton has suggested, to place these new laws, along with parallel shifts in public health reform, into a more general story about the decline of individualism and the rise of "collectivism."[43] A new social focus on insitutions and collections of people, in areas ranging from health to labor, helped to consolidate the efforts of commercial reformers who sought to distance themselves from the credo of individual responsibility that had fueled the anti-corn law campaign. And the same was true in reverse: many advocates of limited liability did what they could to move public health reform in an interventionist direction. Richard Slaney, who headed a parliamentary committee on limited liability in 1851, had chaired a similar committee in 1840 on sanitation. Christian Socialists like Ludlow, who played a leading role in bringing limited liability to public attention, toyed with the idea of forming a "Health League" in 1849 as a sort of vigilante sanitary corps. Lord Palmerston, who had encouraged people to seek out the sources of contagion in 1853, introduced the first partial limited liability legislation in 1855.[44]

These changes in the politics of commercial crisis forced defenders of the older natural theological stance into a corner. The rhetorical value of their previous ploy – referring crises to God's "natural" plans for restoring an imbalanced economy – had irretrievably declined. They were left with the option of continuing to use epidemic language in hopes of forcing the debate back to its original context; but this tactic was also unsuccessful, since epidemic language itself had shifted in meaning during the interim. Loyd, now elevated to Lord Overstone, displayed this problem when he used an epidemic metaphor to oppose limited liability in 1855 during the Crimean War. When the war ended and rates came down, he warned, limited companies "will germinate and fructify, and the commercial world, like Nature under a poisoned atmosphere, will teem with all monstrous things." But Overstone, once a master at turning language to his political advantage, was no longer in control of this metaphor. The "poisoned atmosphere" he worried about called to mind William Farr, not Thomas Malthus, and made his plea for increased responsibility among capitalists seem all the more outdated.[45] Again in the commercial crisis of 1857, when bank loans to American businesses made a bad situation worse, Overstone referred to the "malaria" imported from overseas which had entered England at "our ports – at Liverpool and Glasgow, where joint-stock banks soon proved the unsoundness of their position." Although this appeal to malaria allowed him to get a shot in at his favorite target, joint stock banks, its contagionist connotation also directed attention to the dangers of too much reliance on credit – the same external agent limited liability advocates had a plan for subduing.[46]

Parallels between the new approaches to public health and commerce did not only appear at the level of language and ideology. Just as the public health reforms of the 1850s created new alignments of professional responsibility, the turn towards limited liability in 1856 produced new directions for the commercial professions. One of the features of the new legislation was that it encouraged all applications for limited liability to abide by a standardized deed and to return uniform balance sheets on a regular basis. In one sense, this threatened the traditional business of conveyancers, who had long lived off the writing and interpreting of complicated deeds and accounts; but for enterprising company lawyers like John Duncan, Lowe's act offered a new lease on life. In a guide to the 1856 act, published out of his "Office for Establishing Joint-Stock Companies," he advised that the typical lawyer "cheerfully submit – he must hope to make up by quantity of the companies to be established for the heavy amount which the preparation of a single deed of settlement under the former act of Parliament yielded."[47] Duncan's prognosis proved accurate in all but one important particular: it turned out that accountants, not lawyers, would receive the bulk of the new business generated by limited liability and related changes in the bankruptcy code.[48]

Despite (or perhaps because of) the fact that accountants rather than lawyers succeeded as guardians over commerce, the diagnoses and solutions put forth in the 1850s ended up being effective in their primary goal of diminishing crises. Crashes ensued in 1857 and 1866, but these increasingly affected smaller sectors of the economy. Following the collapse of Overend, Gurney in 1866, British commerce for the rest of the century was unsettled only by the occasional bank failure. Much as the discovery of the water-borne transmission of cholera appeared to lead to the ultimate containment of epidemic disease, limited liability and the development of a responsible accounting profession seemed to have ended regular commercial crises. All of this was much to the better, especially as far as the accountants were concerned; but the years after 1866 also, of course, were even more famous for the Great Depression. This leaves us with a question that is not so relevant in the public health context, where the conquest of epidemic disease was straightforwardly a cause for celebration: did the diminution of commercial crises come at the expense of economic growth? Such has been the conclusion of economists like Joseph Schumpeter, who argued that an economy without a business cycle is destined to stagnate; and similar conclusions have been reached by studying the maturation of the banking and actuarial professions in the late nineteenth century.[49]

Regardless of the ensuing debate among historians over the relative virtues of growth and stability, accountants at the turn of the century had little doubt that the professional benefits of stability outweighed the demerits of decline. Additionally, they were quick to display the parallel between their own role in stabilizing the economy and the role of public health officials in reducing the toll of epidemic disease. In 1901 the accountant Victor Branford, looking back on the history of his profession, wrote that "as the medical practitioner tends to develop through the intermediate state of preventive medicine into the more positive and more socialised hygienist, so a parallel development is observable as incipient in the . . . work of the accountant." He defended this by observing that accountants were in a prime position to solve "the problem of equilibrating production and consumption," evidence of which was "the relative and growing diminution of intensity of the commercial crises" following from "the greatly-accelerated development of accountancy attributable to the stimulus of the Joint Stock Companies Act of 1862." Before such legislative stimuli, the accountant had been a mere bookkeeper, "whose operations too often suggested the barber-surgeon with his ubiquitous lancet." But since then, the status of the accountant had increased in tandem with that of the medical officer:

as medical science, chiefly by changing its point of view from the individual to the general or social level, passes from the curative to the preventive stage, ... so the accountant can only consider himself qualified for his functions when he has trained himself by strenuous economic study to treat each and every business as an integral part of a larger whole ... The upholding of this the social idea of wealth – the necessary and complementary correlative of the social idea of health – is, I take it, the supreme function of the accountant.[50]

As Branford was well aware, a crucial context in the formation of responses to crises in commerce and public health was the changing status and composition of professional groups who were in the business of ascribing responsibility for stability. Crises in both areas of life were perfect opportunities for emerging experts to draw attention to their relative abilities to bestow order on their surroundings. But such experts could not step in to save the day until public perceptions of crisis had moved away from the common natural theological strategy of blaming individuals for their woes, and toward a new rhetoric that located the cause of crisis outside the individual.

Notes

1. See Peter Weingart, "Struggle for Existence: Selection and Retention of a Metaphor" in this volume.
2. See Alborn, "Economic Man, Economic Machine: Images of Circulation in the Victorian Money Market," forthcoming in P. Mirowski (ed.), *Markets Red in Tooth and Claw* (Cambridge: Cambridge University Press, 1993). All this holds equally for intellectual crises in, for example, a scientific discipline undergoing a paradigm shift.
3. *Morning Chronicle*, February 9, 1825, cited in Bishop Carleton Hunt, *The Development of the Business Corporation in England 1800–1867* (Cambridge, Mass.: Harvard University Press, 1936), p. 38; Leone Levi, *The Gilbart Lectures on Banking* (London, 1875), p. 94.

4. Robert Young, *Darwin's Metaphor: Nature's Place in Victorian Culture* (Cambridge: Cambridge University Press, 1985).
5. *Ibid.*, Chap. 6. Young accomplishes this search for new meanings of old words in his analysis of Darwin's intended connotation of "natural selection" and he misinterpretations of that phrase by people still imbued with a "natural theology" worldview.
6. Samuel Jones Loyd, *Reflections . . . on the Causes and Consequences of the Pressure on the Money Market* (London, 1837), p. 44. Mary S. Morgan, *The History of Econometric Ideas* (Cambridge: Cambridge University Press, 1990), pp. 1–43 discusses the prehistory of business-cycle theory.
7. Adam Smith, *An Inquiry into the Nature and Causes of the Wealth of Nations* (Oxford: Clarendon Press, 1776), pp. 342–343.
8. Loyd, *Tracts and other Publications on Metallic and Paper Currency* (London, 1858), p. 131.
9. Boyd Hilton, *The Age of Atonement: the Influence of Evangelicalism on Social and Economic Thought, 1795–1865* (Oxford: Clarendon Press, 1988), pp. 147–149. Hilton's observations on the "natural" connotations of commercial crisis in Victorian England are illuminating, especially at pp. 117–125, but he does not go beyond a suggestion here and there about the functions of language in carrying this out: see pp. 131 and 161.
10. R. J. Morris, *Cholera 1832: The Social Response to an Epidemic* (New York: Holmes and Meier, 1976), pp. 129–158.
11. E. H. Ackerknecht, "Anti Contagionism between 1821–1867," *Bulletin of the History of Medicine* **22** (1948), 568–593; see also Hilton, *op. cit.*, 1988, Note 9, pp. 156–160.
12. See Christopher Hamlin, "Predisposing Causes and Public Health in Early Nineteenth-Century Medical Though," *Social History of Medicine* **5** (1992), 43–70, and John V. Pickstone, "Dearth, dirt and fever epidemics: rewriting the history of British 'public health,' 1780–1850," in T. Ranger and P. Slack (eds.), *Epidemics and Ideas: Essays on the Historical Perception of Pestilence* (Cambridge: Cambridge University Press, 1992), pp. 125–148. Both authors follow Margaret Pelling's original historiographical approach in *Cholera, Fever and English Medicine 1825–1865* (Oxford: Oxford University Press, 1978).
13. See, for instance, the "modified contagionist" views of Edmund Parkes, discussed in Pelling, *op. cit.*, 1978, Note 12, pp. 70–74.
14. "Plague, a Contagious Disease," *Quarterly Review* **33** (1825), p. 219. See also the sources cited in John Eyler, *Victorian Social Medicine: The Ideas and Influence of William Farr* (Baltimore: Johns Hopkins University Press, 1979), p. 225n3.
15. Hilton, *op. cit.*, 1988, Note 9, p. 156; Pickstone, *op. cit.*, 1992, Note 12, pp. 128–135.
16. Michael Dury, "Medical Elites, the General Practitioner and Patient Power in Britain during the Cholera Epidemic of 1831–32" in I. Inkster and J. Morrell (eds.), *Metropolis and Province: Science in British Culture 1780–1850* (Philadelphia: University of Pennsylvania Press, 1983), p. 272; Anthony Wohl, *Endangered Lives: Public Health in Victorian Britain* (Cambridge, Mass.: Harvard University Press, 1983), p. 123. On the continuity of the classical tradition at this time see Pickstone, *op. cit.*, 1992, Note 12, pp. 128–131.
17. Chadwick, *An Essay on the Means of Insurance against the Casualties of Sickness,*

Decrepitude, and Mortality, London, 1836; reprinted in S. E. Finer, *The Life and Times of Sir Edwin Chadwick* (London: Methuen, 1952), pp. 154–155.
18. Hamlin, *op. cit.*, 1992, Note 12, pp. 62–65.
19. Smith, *Philosophy of Health* (London, 1830), p. 364; Pickstone, *op. cit.*, 1992, Note 12, pp. 144–146.
20. See Hilton, *op. cit.*, 1988, Note 9, pp. 43, 85.
21. McCulloch, "Commercial Revulsions," *Edinburgh Review* **44** (1826), 70.
22. Senior, *Political Economy* (London, 1850), p. 50.
23. Chalmers, *The Christian and Civic Economy of Large Towns* (Glasgow, 1821–1826), Vol. III, pp. 326, 127. On Malthus's defense of consumption see Geoffrey Gilbert, "Economic Growth and the Poor in Malthus' *Essay on Population*," *History of Political Economy* **12** (1980), 84–96.
24. See, e.g., D. P. O'Brien, *J. R. McCulloch: A Study in Classical Economics* (London: Allen and Unwin, 1970).
25. Loyd, *op. cit.*, 1858, Note 8, pp. 140–141.
26. See, e.g., Loyd, *Thoughts on the Separation of the Departments of the Bank of England* (1844) in *op. cit.*, 1858, Note 8, pp. 280–281, and Alborn, "Commercial Therapeutics and the Banking Profession in Early Victorian England," *Perspectives on the History of Economics* **5** (1990), 105–116.
27. On Loyd's success among advocates of laissez-faire, see Frank Fetter, *Development of British Monetary Orthodoxy 1797–1875* (Cambridge, Mass.: Harvard University Press, 1965), pp. 175–176.
28. *Report from the Select Committee on Joint Stock Companies* [*Parliamentary Papers*, 1844. VII] vi; *Ibid.*, 1843 evidence, q. 2354.
29. *Ibid.*, 1843 evidence, q. 2256; Cornelius Walford, *Insurance Cyclopaedia* (London, 1871–80), Vol. II, p. 6. For an example of the continuing impact of the parallel forms of company and demographic registration, see Peter Payne's *The Early Scottish Limited Companies 1856–1895* (Edinburgh: Scottish Academic Press, 1980) in which he discusses the "standard of commercial mortality" in Scotland (p. 30) and includes a table showing the "Average Length of Life of Dissolved Scottish Companies by selected industrial classification" (p. 101).
30. Cited in Walford, *op. cit.*, 1871–80, Note 29, Vol. II, p. 13.
31. See John Eyler, "William Farr on the Cholera: The Sanitarian's Disease Theory and the Statistician's Method," *Journal of the History of Medicine and Allied Science* **28** (1973), 79–98.
32. Hilton, *op. cit.*, 1988, Note 9, p. 250; Wohl, *op. cit.*, 1983, Note 16, p. 122.
33. Eyler, *op. cit.*, 1979, Note 14, pp. 114–122. See also Charles Rosenberg's discussion of Florence Nightingale's views on hospital conditions in *Explaining Epidemics and Other Studies in the History of Medicine* (Cambridge: Cambridge University Press, 1992), pp. 90–108.
34. Pelling, *op. cit.*, 1978, Note 12, pp. 204–229.
35. Christopher Hamlin, "Providence and Putrefaction: Victorian Sanitarians and the Natural Theology of Health and Disease," *Victorian Studies* **28** (1985), 381–411.
36. Christopher Hamlin, *A Science of Impurity: Water Analysis in Nineteenth Century Britain* (Berkeley: University of California Press, 1990), Chs. 5–7.
37. *Economist* **11** (1853), 1417; *Economist* **5** (1847), 412. See also William C. Lubenow,

The Politics of Government Growth: Early Victorian Attitudes Toward State Intervention, 1833–1848 (Newton Abbott: David and Charles, 1971), pp. 94–95.
38. *Economist* **10** (1852), 796.
39. *Economist* **5** (1847), 517, 519; reprinted in Wilson, *Capital, Currency and Banking* (London, 1847). See H. M. Boot, "James Wilson and the Commercial Crisis of 1847," *History of Political Economy* **15** (1983), 567–583.
40. See John Francis's speech on "Commercial Crises" to the London Banking Institute in 1852, in which he referred to the redistributive power of crises as "a proof of the truth of Lord Bacon's wise aphorism, that money is like manure, and to do good requires to be spread": *Banker's Magazine* **12** (1852), 564. Francis, it should be noted, retained the traditional providential view of crises: he assumed that they could not be prevented, and that redistribution could only come at the retributive expense of failed businesses.
41. Matthew Begbie, *Partnership "en Commandite," or Partnership with Limited Liabilities* (London, 1848), pp. 85–86 (italics in original); *Mercantile Law Commission, First Report* [*Parliamentary Papers*, 1854. XXVII], pp. 152, 231. See also Vere H. Hobart, *Remarks on the Law of Partnership Liability* (London, 1853), p. 20.
42. *Mercantile Law Commission, op. cit.*, 1854, Note 41, pp. 114, 178; *Report from the Select Committee on the Law of Partnership* [*Parliamentary Papers*, 1851. XVIII], q. 186. In part this argument came in direct response to Loyd's currency theory, which was under fire following its role in exacerbating the 1847 crisis: see Charles Holland's evidence, *loc. cit.*
43. Hilton, *op. cit.*, 1988, Note 9, pp. 255–267.
44. See Lubenow, *op. cit.*, 1971, Note 37, p. 72; John Ludlow, *The Autobiography of a Christian Socialist* (London: Frank Cass, 1981), pp. 154–156.
45. *Hansards* **141** (1856), 140–141.
46. Cited in J. W. Gilbart, *The Logic of Banking* (London, 1859), pp. 548–549.
47. Duncan, *Practical Directions for Forming and Managing Joint-Stock Companies* (London, 1856), p. 48.
48. On the battle between lawyers and accountants following mid-Victorian changes in company and bankruptcy laws, see Edgar Jones, *Accountancy and the British Economy 1840–1980: The Evolution of Ernst & Whinney* (London: Batsford, 1981).
49. See Alborn, *The Other Economists: Science and Commercial Culture in Victorian England* (unpub. Ph.D. diss., Harvard University, 1991).
50. Victor V. Branford, *On the Correlation of Economics and Accountancy* (London, 1901), pp. 38–42.

EVOLUTIONARY METAPHORS IN EXPLANATIONS OF AMERICAN INDUSTRIAL COMPETITION

MARY S. MORGAN
London School of Economics and University of Amsterdam

The expressions "struggle for existence," "survival of the fittest" and "natural selection" are the catchphrases associated with the theories of evolution that swept through American society in the late nineteenth century. The particular focus of this paper[1] is the way these phrases gained currency and were used in contemporary explanations of American industrial competition. The context for their usage was the well-known growth of "Big Business." Large firms organized into cartels and oligopolies known generically as "trusts," and their ruthless methods of competition became disturbingly widespread in the American economy, and hence in the subject of economic commentary, from the 1880s onwards.[2] In order to make sense of the new phenomena and offer policy prescriptions, American economists used the metaphors both to explain the historical emergence of the trusts, and to categorize and interpret the changes in behavior and in industrial structure.[3]

It would be easy to see the import of these metaphors as just another example of the overwhelming influence of biological theories of evolution in the social sphere. But, as the massive literature on Social Darwinism in its various forms testifies, notions transferred from evolutionary biology to the social sciences invariably underwent some metamorphosis in the process.[4] What happened to these evolutionary ideas when they transferred into economics?

The Doctrine of Competition

Hofstadter[5] has claimed that evolutionary thinking did little to transform nineteenth-century economics because the latter already had natural law doctrines of competition, self-interest, and fitness. Although there were additions to the vocabulary of economics, there were no substantive changes as there were in other social sciences. At the general level, Hofstadter is correct: the notion of competition was so central to both fields (indeed, biology took the idea from economics in the first place) that the metaphors were easily translated into an existing framework. Competition was, after all, an essential element in both the older classical political economy and in the newer marginal economic science, and the metaphors reinforced this centrality. There was no clash of cultures, merely an enhancing of the importance of a notion.[6] This accounts both for the acceptability of evolutionary metaphors and their failure to make any transformation within economics.

When we look more deeply into the case of industrial competition, immediately it appears that Hofstadter's claim needs to be revised. Evolutionary thinking is portrayed, both by Hofstadter and by Fine,[7] as sympathetic towards two doctrines, competition and laissez-faire, which were widely accepted in mid-century economics as being both consistent and inseparable. But these commentators miss an essential point, for by 1900 the doctrines of competition and laissez-faire were no longer regarded as compatible by most American economists.

The laws of industrial competition in classical economics differed from those that came into American economics (along with the evolutionary metaphors) in this period, and the outcomes of the two theories differ. In classical economics, competition between firms worked more effectively as the numbers in an industry *increased* and the self-interested competition of individuals and firms created a benign outcome for the economy as a whole (the so-called "invisible hand" argument). Laissez-faire was consistent with the laws of competition both in their action and in their outcomes. In the later American economics that I will discuss, two things were different. First, competition got stronger as the number of firms *reduced* and then disappeared altogether when a trust took over. The laws of competition led to the destruction of competition; it was self-annihilating.[8] Second, as we will see, neither the action of the new competition nor the outcomes were seen to be benign, for increased competition got fiercer and nastier, and monopolistic behavior exploited the consumers and drove small "fit" firms out of the market. It was precisely because neither actions, nor outcomes, nor side effects of the new competition were necessarily benign, that the new analysis of competition was not necessarily consistent with a laissez-faire view, nor even with a procompetition view.

This is not just a logical point, most of the economists of the period argued in favor of government interference of one type or another, economists clearly did change their views. But, to what extent we can say that these changes in economists' beliefs about the character and workings of competition were the result of the metaphor transfer is much more problematic, and is what I will address here.

One of the difficulties with metaphors is how to assess their role. There are various propositions and schools of thought about the way metaphors work, discussed by Sabine Maasen elsewhere in this volume.[9] One possibility is that they simply remain metaphors when they transfer fields; another that they stimulate transformation in the importing field; another that they play a constitutive part in their new scientific home; or finally, that they might play a much more general supporting or enhancing role. Alternatively, and perhaps more usefully here, one could ask questions about the ability of metaphors to invade the importing science to different levels in its theory and practice: did the metaphor become essential to ideas, theories, models, or the entities in the theory/models? Did metaphors play important roles in descriptions, in policy prescriptions, or even in methods of analysis?

The Intellectual Context: The Metaphors' Sources

The evolutionary metaphors were not pristine biological terms transferring to an alien field. Rather, they were the result of an earlier cross-fertilization between the biological and social sciences. The "struggle for existence" originated in the dismal population economics of Malthus and, in a well-known metaphor transfer, was taken up by Darwin for his evolutionary theory. So, the phrase had an established home inside economics already. It is when we find the "struggle for existence" combined with the "survival of the fittest" that we have signs of a new god imported into economics, but even here, we need to be aware of the sources on which American economists drew. The "survival of the fittest" came from Herbert Spencer, whose intellectual dominance across a spectrum of fields from the 1860s – biology, sociology, psychology, and ethics – was widely acknowledged at the time, though hardly recognized today.[10] His thinking was, by all accounts, particularly influential in America where he was widely published and read.[11]

Although Spencer made no lasting contributions to what now constitutes the field of economics, he used political economy as a role model for sociology because of its appeal to impersonal forces (neither gods, nor natural objects) and for its scientific character. In his influential *Social Statics*,[12] he used economic metaphors, notions, and examples freely: metaphors from economics permeated his organicist analysis of both the body and society, while the economy and the way its elements interacted were portrayed in organic terms. The biological and the socio-economic were indefinably mixed. As the economist Ely later remarked:

"It is probably due to Herbert Spencer more than to any other one person that we have come to recognise the applicability of evolution to the various departments of the social life of man. We have an evolution of the body, and also evolution of the mind, and we have an evolution of society, which is the highest form of life."[13]

The appeal of Spencer for American social scientists lay across the political spectrum. For conservative thinkers, there was his strong ethical program that tried to graft onto his scientific analysis. This ethics involved an abhorrence of all government interference, which appealed to libertarian and conservative economists of they day, such as William Graham Sumner, a committed laissez-faire free trader and professor of political economy at Yale university who was much enamored of Spencer's writings.[14] Sumner was shunned by the bright young men who formed the American Economic Association in the 1880s on a platform of a new economics, which involved commitments both to an historical method and to reformist government action.[15] Yet even these young men, like Richard T. Ely, found Spencer worth reading. In his *Principles of Sociology*,[16] Spencer included an historical analysis of the economic evolution of society with some exceedingly pertinent coverage of the organization of industrial society. And, despite his views on government, he was predisposed towards organizations of individuals coming together within society to provide for their communal needs. All this fitted in with the historical, and even socialist, bent of the younger economists, though they jibbed at his laissez-faire views. Indeed, by virtue of his concern with human and social development, Spencer can be seen as fore-runner of those late-nineteenth-century American economic commentators who emphasised the increasingly cooperative nature of society.

The fact that evolutionary notions about the economy as a whole figured quite strongly in various aspects of American economic discourse reflects a variety of local causes and contexts, apart from the current of evolutionary thinking in biology represented in Ely's homage to Spencer. American economists were historically minded: Sumner was, by his own account, influenced by Buckle, while most of the younger American economists went to Germany for their postgraduate education, where they were taught by members of the German historical school of economics and came to share their belief that state action could reform society.[17] These experiences were reflected in their adoption of historical methods of analysis and their considerable interest in history of their economy and its institutions. In doing so, they also drew on a native tradition of optimistic writing about the social and economic evolution of their society.[18] It was in these intellectual contexts that American economists read Spencer, Darwin, and later evolutionary theorists, particularly Lester Frank Ward.

Evolutionary Metaphors at Work

It is important to emphasize that not all economists were persuaded by these evolutionary metaphors: some used them seriously, others used them in some circumstances and not in others, and other economists rejected them altogether. There is no easy generalization in this field. Nevertheless, my analysis suggests that the evolutionary metaphors were used in two different ways: in "laws of nature" explanations and in "institutional" descriptions. And they worked at various different levels of economic thinking and practice: in description and explanations of the emergence of large firms; in a more limited way in the theories and models of firm behavior, and finally in the policy prescriptions.

Descriptions/Explanations of the Emergence of Large Firms: Struggle, Survival and Predatory Behavior

The early literature on Social Darwinism claimed laissez-faire "Big Businessmen," such as Andrew Carnegie, as one of their main captives.[19] It was an obvious area for colonization, though recently it has become fashionable to challenge this influence of Darwin, suggesting that evolutionary rhetoric was a useful buttress for entrepreneurs' ruthless behavior, but nothing more.[20] Neither this literature, nor Spengler[21] have paid much attention to how specific evolutionary ideas were used by contemporary economists in their analysis of industrial competition, whether in the behavior of firms in the market place or in between-firm competition. In this social science arena, the two metaphors of "struggle for existence" and "survival of the fittest" played a role both in organizing the phenomena and transforming economists' descriptions.

Where evolutionary metaphors were used seriously, they functioned to explain how the new forms of competition and industrial organization came about; they described both the process: "the struggle for existence," and the outcome: "the survival of the fittest" firms. The struggle for existence had hitherto been applied in population economics at the level of the individual (or sometimes in Malthus' work, at the level of the "class"). Despite the influence of evolutionary thinking, this struggle was nevertheless often left somewhat vague – was it with nature or with fellow men? Sumner's economic analysis incorporated both the Malthusian "struggle for existence" against the elements and the Spencerian "survival of the fittest" in this struggle. Individuals are forced by circumstances to struggle for their existence and they compete with each other (and other groups) in the process. But not everyone makes it in this "competition of life"; those that are fit by natural advantages and apply themselves will wrest their living from a "niggardly" nature and survive. Those that are poorly equipped and lazy will not survive.[22] His observations suggested "a ceaseless war of interests" between individuals and groups throughout the economy.[23]

With the introduction of the idea into industrial economics, the struggle was explicitly with other firms (though I have not found any writer who went so far as to describe this as competition within a species). The "struggle" denoted new and violent forms of competition which had replaced the older style behavior of prices being determined amicably in the market place. For John Bates Clark, one of the younger group of economists, the outcome of the process was the survival of the fittest firm – fittest here denoting the strongest (rather than in evolutionary biology, the most fitted for a given environment – and the failure of the weaker firms.

The science adapted to such conditions was an economic Darwinism; it embodied the laws of a struggle for existence between competitors of the new and predatory type and those of the peaceable type which formerly possessed the field.

> Though the process was savage, the outlook which it afforded was not wholly evil. The survival of crude strength was, in the long run, desirable. . . .
> Predatory competition between unequal parties was the basis of the Ricardian system. This process was vaguely conceived and never fully analyzed; what was prominent in the thought of men in connection with it was the single element of struggle. Mere effort to survive, the Darwinian feature of the process, was all that, in some uses, the term competition was made to designate.[24]

By portraying competition as subject to a law of nature and locating it both within evolutionary theory ("economic Darwinism") and subject to the iron laws of economics ("The Ricardian system"), Clark scored a double hit. The resulting outcomes of the process could be described as inevitable: the growth of large-scale firms and monopolies[25] along with the collapse of small firms were all natural outcomes in the evolution of the economy.

The detail of the process of struggle was itself subject to an institutional analysis dealing with the mores of economic behavior: large firms indulged in "predatory" and "unfair" competitive practices. These competitive weapons, such as price cutting below costs, or refusal to supply, were designed to drive other efficient firms, particularly smaller, local firms, out of the market. The term "predatory" did not imply that the large firm always swallowed up the small ones (in the sense that they bought them up) but did suggest life-threatening behavior. Such behavior was also a feature of intertrust competition (i.e., large firms battling other large firms).

> The power of trusts to crush competitors is dependent upon three kinds of unfair dealing. The first is local discrimination in prices. The trust may sell goods for less than cost in a limited section of the country, where an independent producer is operating, while it sustains itself by charging high prices in the large remaining area. Even though the competitor may greatly excel the trust in the economy with which he makes goods, he may be forced out of the business by this predatory policy.[26]

Thus the biological meanings of "predatory" were not entirely paralleled in Clark's economics since the competition was mainly within the industry (the species) rather than between different industries. In addition, the metaphors of violence – "cut-throat," "destructive," and "extermination of rivals" – which were more commonly used than "predatory" to describe the new forms of competition, were in many ways more apt for the context of the descriptions being used, which often involved moral or ethical commentaries on firm behavior.[27]

Theories/Models of Behavior: Economic Competition as Natural Selection

A gap remained between the two levels of the natural law description of the emergence and success of the large firms suggested by the "struggle" and "survival" metaphors, and the institutional and moral analyses of the detailed behavior (including "predatory" practices). This gap left open the question of how it was decided which firms succeeded. Here our third evolutionary metaphor sometimes made its appearance – economic competition as the selective process that decides which firms survive and which die off. Naming economic competition as the "natural selection" mechanism did not provide an immediate explanation in economic terms of what was going on, but did suggest to economists where they should look for the new economic laws to fill the intervening layer.

For Sumner, there was no problem in interpreting the outcome of large firms as the result of "natural selection." "Natural selection" operates through competition. Competition, he argued, "develops all powers that exist according to their measure and degree."[28] So, a firm with capital assets and good managers has more power to invest and grow and increase capital assets than one with neither. The growth of the large capitalistic firm is due to "natural selection" operating at the level of individual entrepreneurs. As part of the competition of life, those individuals (and their firms) who have both natural merits and inherited aptitudes and are also subject to good fortune, will grow economically stronger at the expense of the others (with neither advantages nor good luck); thus millionaires and big business (monopolies and trusts), as well as bankrupt firms (and destitute paupers), are equally the outcomes of the process of "natural selection" or economic competition.[29] Here we have a very close translation of the metaphor into the economics sphere: competition "develops" existing advantages in Sumner while in Darwin's evolutionary biology, natural selection "preserves" advantages.

The reformist Richard T. Ely expected a similar outcome, not because the firms had been selected by unaided natural law, but because they had been selected through man's agency.³⁰

Competition is the chief selective process in modern economic society, and through it we have the survival of the fit. But what do we mean by "the fit"? . . . Modern society itself establishes, consciously or unconsciously, many of the conditions of the struggle for existence, and it is for society to create such economic conditions that only desirable social qualities shall constitute eminent fitness for survival.³¹

Ely was much influenced by Lester Frank Ward, who like Spencer, was a man of broad intellectual interests.³² The Ward/Ely position was to draw to distinction between the laws of evolution applied to animals and those applied to man. Both lead to the "survival of the fittest," but in the case of humankind, man may intervene in selection and society itself establishes the conditions for competition. Ely's "Man makes Nature do its perfect work" captures their views.³³ Not only does society regulate competition in its various forms, but, as they thought, the law of evolution follows the path of increasing cooperation, evident in the America of their day in combinations of labor (trade unions) and capital (trusts).³⁴

Both Ely and J. B. Clark presented the institutional workings of competition as subject to evolution; the new forms of competition between large and small firms and between the trusts were treated as a standard stage in the historical evolution of the economy. For both of them, competition worked in the economy to ensure the survival of the fittest firms. But, whereas Ely saw society as providing the regulator of this competition, Clark depicted industrial competition as a straightforward parallel to the laws of nature: "economic Darwinism."³⁵ Clark utilized Cairnes' notion of effective competition and his theory of non-competing groups³⁶ to describe how competition between individual firms within an industry gradually gives way to combinations and trusts in that industry, and then the trusts begin to compete with other trusts outside their industry.³⁷

Combinations are the product of social evolution, and can have no permanent existence until the Darwinian contest between the weak and the strong has completed its work. The surviving competition must be few, strong, and nearly equal. . . . Rivals do not combine so long as one is conscious of the power to exterminate the other. Moreover, strength for such a contest consists not merely in the size of a producing establishment, although that is an element to be considered; it consists primarily in advantages for economical production. Location is important, but the paramount influence is the mastery of cheap methods. Natural selection locates industries in the most favorable localities, and brings them to some equality in method; and until this is done there is no chance for an economic truce.[38]

Clark's co-author, Giddings, wrote,

Just as recurring waves of bankruptcy from time to time sweep off the competitors that are essentially weak, . . . so an industrial depression of unusual severity or duration forces one or another partly to unload his stocks at any prices that he can get, regardless of combination agreements, and consummates the extinction of those producers whose disadvantageous situation or antiquated methods make their cost of production relatively high.[39]

We see, in these two commentaries, competition as a natural process likened to the process of natural selection in which those firms with certain advantages of locality and those with the advantage of cheaper methods will survive; the others will become "extinct." But this was as far as the analysis went, there was no suggestion that the differences between firms formed the essential variation needed for economic competition to work.

Although these several authors used various different versions of evolutionary thinking to provide arguments to place between the institutional descriptions and the outcomes of selection, the evolutionary metaphors and their accompanying notions did not quite fully translate into economics by creating economic theories or laws of firm behavior. Sumner had Darwin's theory of "natural selection" working on individuals, according to their advantages and disadvantages but his description did not relate to any existing economic theories of the way firms behaved. Neither did Ely's discussions, in the context of his cooperative theory of evolution, provide an explanation of how the new competition worked to join up with the economic explanations of the growth of large firms. Clark and Giddings came closest to providing the link (as we can see in the quotes above), but even they did not provide a text that completely linked an analysis of the firm's activities with their analysis of the process of dynamic competition; they merely hinted at the "advantages" in economic production which made certain firms more fit for survival.

This is not to say that these authors were silent on how the new competition worked in economic terms, merely that they did not themselves link their economic theory building with the descriptions and explanations provided by the evolutionary metaphors. Ely for example, before his writing on evolutionary themes, had been one of a number of economists to use the explanation of increasing returns to scale in certain industries[40] as a way of explaining the growth of trusts.

> When it is stated that a business becomes relatively more profitable in proportion as the amount of capital invested increases, it is already granted that large concerns have the power to crush small ones, for if business is more profitable, it is because production is cheaper, and if the big man produces cheaper he will crush the little man.[41]

Clark on the other hand may well have experienced the evolutionary metaphors as a stimulus. In this earlier work he had decried the antisocial character of the modern competitive behavior and believed (or hoped) that competition would be replaced with a better distributive mechanism. As we have seen, his discussion of economic Darwinism in the late 1880s suggests a conversion to the notion of competition as a natural entity that could not be suppressed: it would always be there in reality or as a "potential competition," and it was an essential part of the dynamic industrial process. In his later writings, competition was put on a pedestal and revered as the power that made the static economy tend to its ideal state.[42]

Clark's final notion of competition and Darwin's notion of natural selection have many of the same qualities. Darwin's "natural selection" is a highly active principle which continually operates on everything.

> It may be said that natural selection is daily and hourly scrutinising, throughout the world, every variation, even the slightest; rejecting that which is bad, preserving and adding up all that is good; silently and insensibly working, whenever and wherever opportunity offers, at the improvement of each organic being in relation to its organic and inorganic conditions of life[43]

Clark portrayed competition as the all-powerful regulator, down to the last unit of capital and the last laborer.

> Competition . . . causes establishments that are so equipped to get out of their capital the utmost service that it is capable of yielding to survive, and incapable ones to fail. . . . Competition acts as a leveller, by reducing the earning power of the final increments of different men's capital to equality. This it does by putting out of the field the competitor whose last increment of capital . . . creates less than the standard amount of product.[44]

Clark's competition, like "natural selection," was an all-seeing, all-interfering God-like actor; competition was the final arbiter in the economic realm, just as natural selection was in the natural realm. Clark's use of evolutionary metaphors in his dynamic economics *may* have played an important role in his understanding the action of competition in the static case in a new light, but significantly he did not use the evolutionary metaphors in his marginal analysis of the firm, which propounded his new vision of competition. Why not? Perhaps because of an important difference: Clark's economic competition acted as a leveler – it created and maintained uniformity; "natural selection" also acted on variation, but without the same reductive leveling effect.

So far, the evolutionary metaphors have been used in a variety of positive ways. In one case of interest, they were rejected. Arthur Twining Hadley, who overlapped with Sumner at Yale, supposed that there was an evolution of economic institutions, with better institutional and habitual ways of doing things driving out poorer ones.[45] Hadley, like Sumner, saw the presence of powerful businessmen as another example of "natural selection" at work.[46] But, he explicitly rejected Sumner's claim that economic competition between firms was the same as the biological struggle for existence: "Competition is a totally different thing the Darwinian struggle for existence. Competitive ethics is not a mere glorification of force."[47] And, long before this, Hadley had used his observations on the railroad industry to develop a sophisticated economic theory of the new competition in which he had rejected the other catchphrase: "It is all very well to talk of free competition and survival of the fittest. But permanent competition is only possible when the unfittest can be physically removed – a thing which is impossible in the case of an unfit trunk line."[48] This suggests that the metaphors may have played a different role for Hadley in his theory of large firm competition. It was, perhaps, because he felt the metaphors did *not* sufficiently organize the phenomena, that he provided a theoretical account of large-firm competition that he felt did explain the phenomena.[49]

So, explanations of why some firms grew at the expense of others depended either upon economics without reference to evolutionary language (Adams or Ely, following Jevons; or Hadley) or invoked Darwinian competition without respect to economic advantages (Sumner), or used evolutionary ideas in dynamics but not in static analysis (Clark). Discussions of costs of production and other interfirm variations in efficiency remained, except in general terms for Clark, separate from evolutionary explanations. The only significant economic analysis of large-firm competition that linked both dynamic and static elements was undertaken by Hadley who rejected the metaphors in this context as misleading or plain wrong.

Policy Prescriptions

Policy questions came to a head when the growth of "trusts" accelerated in the period 1896–1904, known as the "great merger movement," and caused a mass of hand-wringing and analysis of their behavior.[50] Germany was experiencing the same growth in scale and dominance by large firms over the same period, but it did not seem to be such a problem for German commentators for there was both popular and official support for the trusts, and an interventionist state willing and able to intervene in the working of the market.[51] By contrast, American laissez-faire attitudes, combined with a history of trade legislation that granted tariff protection to industry, left the trusts pretty much a law unto themselves.[52] American economists faced with the full gamut of antisocial behavior by the trusts, agonized at length over the problem of how to separate out the virtues of large-scale firms from the attendant vices of monopoly behavior.

A comparison with the infamous ways evolutionary metaphors and arguments were used in eugenics might be useful here. Eugenic prescriptions for intervention in the human population to weed out the "unfit" and ensure the survival of the more "fit" could be presented as helping out on the side of natural selection, i.e., the actions were consistent with the theory of "natural selection." The same incentive for action could not be made in the case of interfirm competition, for the policy problem was not so simple. The large, strong firms which survived the competition were seen as good for the economy as a whole for a number of reasons. Economists of all hues believed that there were benefits of large-scale production and approved of the trusts for they were associated with greater productivity, increased output, and lower prices. They were generally regarded as one of the sources of the period's prosperity and of American dominance of the international economic community. But these large monopolistic firms were also believed to be the source of corruption of the political process and public and commercial life, and were accused of criminal and generally antisocial or vicious competitive behavior.[53]

These "bad" characteristics of viciousness and criminality became associated with the "robber barons" running these firms, for example Ross's "criminaloids" and Veblen's dubious "captains of industry" who knew how to manipulate finance and gain power but understood nothing of production.[54] At the same time, small businesses and businessmen were portrayed as virtuous and innocent victims in the economic rhetoric of the time; small did not necessarily equate with unfit, indeed these firms were often considered to be "fit" (in the sense that they were efficient producers, but unable to stand out against the predatory practices of the larger firms). This contrasts strongly with the eugenics arguments, where the poor and destitute individuals were portrayed as vicious and criminal.

Unlike the eugenics and population case, the policy problem for economists was to deal with the vices of the strong firms, not the weak. Ward's discussion of the problem of monopolies and trusts lead him to a comparison between competition in nature and in man which nicely captures this problem.

There (in Nature) competition is pure. It continues as long as the weaker can survive it, and when these at last go to the wall and the better adapted structures survive and triumph, it is the triumph of physical superiority, and the strong and the robust alone are left to replenish the earth. But when mind enters into the contest all genuine competition is crushed out, and while it is still, in a certain sense, the strong that succeed, it is strength which comes from superior cunning, necessarily coupled with stunted moral qualities, intense egoism, and underdeveloped sympathies, and always aided more or less by the mere accident of position. In no proper sense can it be said that this class is the fittest to survive the society. . . . Free competition as it exists in nature would be preferable to the present industrial state. . . .[55]

Because of these difficulties, many economists self-consciously argued against a simple transfer of notions from evolutionary biology to economics when it came to policy arguments.

In fact, some economists made the distinction between economic and biological competition the basis of their policy prescriptions. The institutional level of analysis turned out to be crucial in this respect. Economists such as Clark, Ely, and Hadley differentiated "true" economic competition, in which services and goods are rendered to the customer, from biological rivalry in which no useful by-products occur and in which survival/nonsurvival are the sole outcomes. Hadley expressed it thus:

[Competition] is a fight for existence organized in such a way that the outside world is benefited by it, instead of being injured by it. When two snakes or two tigers are struggling as to which shall get the same bird, the probability of an advantageous outcome for the bird is very slight. When two bosses are struggling as to which shall get the same workman, the probability of an advantageous outcome for the workman, the probability of an advantageous outcome for the workman is very large. . . . The more snakes there are, the worse for the bird; but the more bosses there are, the better for the workman. Competition is what its name implies – a simultaneous *petition* or offering of services under which the man who offers the best service is chosen.[56]

Economic competition, they argued, was a social process, subject to social institutions, and was differentiated from its biological or natural counterpart. Ward, and Ely following him, took a very similar line in interpreting the vicious competition between large firms as biological, for it demonstrated all the wastage of resources (such as in excessive marketing) they associated with the competition of nature. For them, this was merely a transformation stage along the social road to increased collaboration.[57]

This differentiation between the economic (or social) and the biological was matched by the distinction between "normal" or "fair" economic competition and "abnormal" or "unfair" industrial war. The latter was likened to biological rivalry; it involved predatory pricing, cutthroat or destructive competition, which destroyed both competition and fit firms and brought no value or benefit to the customer. Economists used these analyses to separate out the good (economic, or social) from the bad (biological, or war-like) forms of competition when it came to policy advice. Clark wrote:

Often enough is their [the trusts'] policy predatory. They do not literally kill men, but to a large extent they do kill competition. They often make property in the shape of rival plants very insecure. Indeed, one of the pressing questions is, whether the independent producers who have been crowded out of the field are unfortunate sufferers from natural progress, or whether they are the victims of a wrong against which society should protect them. Mere centralization means a crushing out of competitors by a process that, however hard it is for them, is in a way legitimate; for it is an incident of the process of the survival of the fittest. The large and economical establishment survives, and society gets a benefit from the fact. But centralization that goes to the length of quasi-monopoly takes a different color, for it may exterminate competitors in ways that do not benefit society. The employers who are forced out of the field are not then vicariously sacrificed for the good of the public as a whole. On the contrary, the sacrifice of them works exceedingly ill for the public, it must be stopped if society is to avoid graver evils than have recently come upon it from any economic cause.[58]

Fair competition was good in the dynamic economy because it ensured the survival of the fittest firms. Unfair or abnormal competition (certain pricing and supplying policies) were bad because they forced small "fit" firms out of existence. Economists' policy prescriptions aimed to get rid of the vices of monopoly power in order to enjoy the virtues of the large-scale fit corporations. As Ely wrote:

"Competition along with large scale production, brings its own problems, but they are easy of solution as compared with the problems of monopoly, because competition is compatible with private property in capital, and with private production. Where competition exists, the problem is its regulation in such a manner as to secure its benefits, and to remove, where it is possible, and where it is not possible, to mitigate, its evils."[59]

The advice to government to regulate the monopolies and protect the small firms and consumers appeared to go against the action of competition as the natural selector. However, such intervention was not inconsistent with all evolutionary thinking. For example, though Spencer lauded competition, his belief in moral development lead him to take a sympathetic view of the victims of industrial competition. The case of an infamous New York retailer of this period, who had driven a number of small retailers out of business, caused him to comment on the "commercial murder" involved.[60] For Spencer, moral restraint practiced by the individuals on their competitive instincts was the solution. For Ward, Adams, and Ely, healthy economic competition could only be ensured by government intervention and legal regulation (collaborative action) to remove the vicious behavior.[61] Clark struggled to retain his later belief in the efficacy of competition as an inextinguishable force either in reality or through the "natural" threat posed by potential competition,[62] while Sumner, true to his laissez-faire beliefs, desired only to remove legislation that protected the trusts.

Evolutionary Metaphors: The Baggage Argument

The problem for American economists of the period was that they faced new economic phenomena which the old theories and descriptions did not fit. The evolutionary theories provided a language to describe the new outcomes, and could organize the phenomena in the sense that they provided description and, by virtue of their place in the theoretical scheme of their original field, gave an interpretation to those descriptions. In a general way then, we might say the metaphors of the "struggle for existence" and the "survival of the fittest" hooked onto the problematic phenomena and organized the evidence. They could be used to interpret the large firms' survival, the small firms disappearance, the fierceness of the competition free of social constraints ("Nature, red in tooth and claw") and finally the role of economic competition could be interpreted as the "natural selector."

Nevertheless, the metaphors did not appear to have created a new economic theory of the firm. The metaphors were successful at the level of descriptions but apparently failed to work their way into new theories or models. Why? I argue that it was primarily because the baggage these metaphors carried with them from evolutionary thinking either did not fit into economics or was insufficiently developed to be of any use. This "baggage" argument works at a variety of levels: in the descriptions, theories, and models of economic behavior and in the more general methods and approach of economics.

Descriptions, Theories and Models

I have already remarked that the metaphors carried with them links to older economic theories of population. Was this an advantage? Their usage in economics was doubtless eased because they were part of a more general revival of Malthusian population economics, updated to deal with the late nineteenth-century American circumstances by economists such as Henry George.[63] But this same advantage may also have created barriers to the movement of the metaphor within economics, for Malthusian ideas were associated with theories of population and subsistence wages, rather than with interfirm competition. This is perhaps one reason why bridges to theories about firm competition proved difficult to build, and this may have limited the invasion by the metaphors.

Another reason why these metaphors did not fully translate into an economics of the firm may have been the difficulty of transporting the evolutionary theories that accompanied the metaphors. The Malthusian metaphor of population struggle had been transported by Darwin and applied wholesale to the natural world; in describing the struggle for existence Darwin wrote, "It is the doctrine of Malthus applied with manifold force to the whole animal and vegetable kingdoms; for in this case there can be no artificial increase of food, and no prudential restraint from marriage."[64] Darwin also transported the accompanying Malthusian theory of population by assuming the tendency to geometric growth rate applied across the board.

"That each organic being is striving to increase at a geometrical ratio; that each at some period of its life, during some season of the year, during each generation or at intervals, has to struggle for life, and to suffer great destruction."[65]

Natural selection provided the check on this tendency to increase geometrically, replacing the Malthusian checks in the human case, namely the ability to increase food output and to exercise restraint on marriage. But, the transport back into economics of evolutionary metaphors was hindered because neither the elements nor the microprocesses of Darwinian evolution were readily applicable to firms' competition (and the methods were partly antagonistic as we shall see).

First, the baggage of biological elements carried from Darwinian evolution, such as species, origin, reproduction, and extinction, and notions about varieties and sub-species, inheritability, and instincts, might travel to populations of people (maybe) but not obviously to firms. There is one clear exception: it would seem to have been perfectly feasible to have transferred the Darwinian notion of variation as the raw material for economic competition (i.e., "natural selection") to act upon it, but, this transfer did not take place. Even when an economists, such as Sumner, saw that the outcome of the struggle for existence was variation, he called it "inequality."[66] For other economists, such as Clark, the outcome of competition in the static case was uniformity, not variation. For the most part, there were no economic equivalents or parallels to the middle-level entities of Darwin's theory.

Second, Darwin did not tell economists how evolution happened. That is, there were no obvious microprocesses of models of how variation was created for natural selection to work upon. Even adopting Spencer's views on inheritance did not help the economists much for it was not clear that this would apply to firms since there were as yet only a rudimentary notion of genes and little understanding of how inheritance worked.[67] Even given the processes of "natural selection," the outcomes were not easy to understand. For example, Darwin, in discussing the difficulty of determining why a species was victorious over another blamed the complex interdependency of species.

> We can dimly see why the competition should be most severe between allied forms, which fill nearly the same place in the economy of nature; but probably in no one case could we precisely say why one species has been victorious over another in the great battle of life.[68]

Thus, the working parts of Darwinian and Lamarckian theories – the middle level elements and microprocesses – were either not obviously transportable or nonexistent. This severely limited the metaphors' ability to become constituted in economic theories and models. This is in marked contrast to the way Durkheim was able to use the concepts of the social division of labor and adaptation of groups to the changing circumstances.[69]

Methods and Approaches

Similar compatibility problems were evident in the methods of approach of evolutionary biology compared to economics. Darwin's and Spencer's theories were laid down in the natural history form, and depended on many individual descriptions at the specific level to motivate them. Darwin's difficulty (above) in pinpointing the relevant advantages that made one species/individual more fit for the circumstances enabled only a *post hoc* rationalization of outcomes. In Ward and Spencer, there were predictions, but of such a general sort (e.g., increasing cooperation, increasing heterogeneity) as to be fairly useless for describing why one firm in particular grew at the expense of another. The methods of natural history were to some extent inconsistent with those of classical economics, which relied on a priori deductive arguments and general rules.[70] Similarly, the fundamental focus of classical economics and the later marginal economics was concerned with laws that would tend to make all economic units the same, rather than with trying to explain variation between firms or variation over time.[71] As evolutionary biology dealt primarily with variation between individuals and over time, there was a potentially serious clash not only on methods but on approach.

Nevertheless, these arguments about clashing approaches and methods are not so persuasive as those about theories and models above, for it was precisely in the attempts by late nineteenth-century American economists to deal with these problems of changes over time and variations in outcomes where evolutionary metaphors operated. If we stand back a little with this in mind, we can begin to see that the metaphors helped define a specific American analysis of the problems of competition and monopolies which was not matched elsewhere.

The late nineteenth century was a period of innovation in theorizing about the firm. In particular, it was the period in which theories of monopoly and oligopoly were first properly developed.[72] For Americans, the problem was to explain how it was that some firms grow large in some industries and that large firms grow at the expense of small. For example, in the older theory, there was no particular reason why competition will cause some firms to increase in size and others to die off, provided there is sufficient demand. Under the stimulus of events and with the help of evolutionary thinking, American economists such as Hadley rejiggered the laws of competition so that there were built in reasons (independent of demand) why some firms will expand and some decline. Competition was reexplained in economic terms to pinpoint the sources of economic advantage and to incorporate new selective mechanisms, but not necessarily the advantages and mechanisms of evolutionary biology.

A comparison highlights what I have in mind here. Williams has pointed to an interesting national difference in approach to the problem of monopolies and competition.

The British were concerned to explore the impediments hindering the attainment of competitive equilibrium. The French were concerned to classify markets according to the number of participants and to state the conditions of equilibrium for each market category as elegantly as possible.[73]

I would add: American economists were concerned with the dynamics of monopoly and oligopoly – they wanted to understand why firms grew, why monopoly and oligopoly formed, and why the industrial structure in an industry switched between competition and various forms of monopoly. That is, American commentaries on monopoly and theories about large-firm competition were focused on evolutionary aspects, and, with the exception of Clark, not static equilibrium aspects.

Standing back even further, we can see that evolutionary thinking had seeped into the fabric of American economics of the period. Economists did not become evolutionary biologists (and we should hardly expect that they should), but they did, in some part, become evolutionary economists. Hofstadter was surely correct in pointing to an important general change in American economics which was connected to the methods and approaches of evolutionary thinking. The development of American institutionalism marked a new way of doing and seeing the enterprise of economics. It was inconsistent with mainstream economics (classical and neoclassical) in both subject matter and in manner of attack and in forms of explanation. The institutionalist economists provided descriptions of, and theories about, the process of cumulative change (the gradual small changes) which create the habits of thought and ways of behaving and the gradually changing structure that made up the economic world. There was widespread feeling among American economists of the period that economics needed to be made more of an evolutionary and historical science and they adapted the methods of natural history and anthropology in their institutional and evolutionary historical analyses. In the work of the leaders of the movement, such as Veblen and Commons, the institutional path was a pure one.[74] But, in the work of most economists that I have discussed here, these analyses went side by side with an economic analysis using the older methods and ideas of classical economics cautiously updated with the new marginalism and elements of German historicism.[75]

Evolutionary Metaphors: Success or Failure?

The propensity of these American economists to maintain different types of analyses for different levels of the problem of industrial competition has been one of the difficulties in locating the effect of our metaphors. Evolutionary ideas were important to the institutional analyses of competition, to descriptions of the changing forms of competition and sometimes permeated into the other layers, but not necessarily to all these layers at once. For example, Clark had maintained his marginal static analysis separately from his dynamic historical analysis involving the three evolutionary metaphors. Hadley had carried on an institutional evolutionary analysis of the forms of competition alongside a marginal analysis of large-firm competition. Sumner had simply updated the old Malthusian economics with the additional "survival of the fittest" and "natural selection" working on individual advantages to provide an explanatory model for monopoly capitalism based on evolutionary theories; but he did not connect it up with questions of costs and prices. Ely's version, too, had amounted to a straight application of Ward's evolutionary thinking to provide an historical explanation for monopoly capitalism, which, again, he did not link with his economic theory of economics of scale.

When we examine the economic commentaries explaining how competition works by the 1920s, we find very little left of the evolutionary ideas and their terminology (for a variety of reasons too complex to consider here). This is not evidence that they had been entirely passive and ineffective. In summarizing the case of industrial competition, I think we can say the metaphors transferred successfully, but they did not fully transform the economics of the firm because of their failure to carry enough useful baggage in the way of entities, theories, and models with them. On the other hand, they stimulated American economists to investigate and analyze the changes in industrial activity and to explain why economic competition was different from biological competition. They had a lasting effect in the idea of "predatory pricing" which remains to this day an important notion in antimonopoly legislation and judgments.

Notes

1. This paper draws on material collected during a research fellowship funded by the ACLS and Fulbright Commission during 1990–91. Further research was funded by a travel grant from STICERD in 1992. I am grateful to all three bodies. I also thank Roy Edwards for research assistance and the Lakatos project of the Centre for the Philosophy of the Natural and Social Sciences at the LSE who funded his assistance as part of the "Making Methods Fit" initiative.

    The paper was first given at a conference at ZiF, University of Bielefeld on "The Transfer of Images and Metaphors Between Biology and the Social Sciences" in June 1992. It was then given at a workshop on "Making Methods Fit: Transfers between Economics and Natural Science" sponsored by the LSE Centre for the Philosophy of the Natural and Social Sciences in October 1992. I thank Everett Mendelsohn, Peter Weingart and Sabine Maasen, the organizers of the ZiF conference (and their anonymous referee), participants at both conferences, and at a seminar at the University of Quebec at Montreal, as well as Margaret Schabas, John Beatty, Norton Wise and Geert Reuten, for their encouragement and comments on the paper at various stages.
2. See the classic work by A. D. Chandler, "The Beginnings of "Big Business" in American Industry," *Business History Review* **33** (1959), 1–31 and *The Visible Hand: the Managerial Revolution in American Business* (Cambridge, Mass: Harvard University Press, 1977), the general survey by Carl Degler, *The Age of the Economic*

Revolution (Glenview: Scott, Foresman and Co, 1977) and, more recently, N. R. Lamoreaux, *The Great Merger Movement in American Business, 1895–1904* (New York: Cambridge University Press, 1985).
3. Elsewhere, see M. S. Morgan, "Competing Notions of "Competition" in Late-Nineteenth Century American Economics," *History of Political Economy* (1993), forthcoming. I have given an account of the different notions of competition utilized by economists in their attempts to understand the "new competition" and its development. Their responses were surprisingly varied: competition was seen as an institution, a law of nature, an agent of natural law and an economic law. This paper extends my earlier analysis in a particular direction.
4. The classic work is R. Hofstadter, *Social Darwinism in American Thought* (Philadelphia: University of Pennsylvania Press, 1944); a more recent useful survey is D. C. Bellomy, "'Social Darwinism' Revisited," *Perspectives in American History* New Series **1** (1984), 1–129, and on evolution and social science, Carl Degler, *In Search of Human Nature* (Oxford: Oxford University Press, 1991). The only overall survey of evolutionary thinking in American economics is J. J. Spengler, "Evolutionism in American Economics, 1800–1946" in S. Persons (ed.), *Evolutionary Thought in America* (New Haven: Yale University Press, 1950). A more recent history of evolutionary ideas in economics (not specifically American) is by G. M. Hodgson, *Economics and Evolution* (forthcoming).
5. Hofstadter, *ibid.*, 1944, p. 144.
6. Sometime between the early classical economics of Adam Smith and the neoclassical economics of the twentieth century, the concept of competition did become more important and undergo some changes. No doubt, there was some influence from evolutionary biology. But this is boggy and treacherous ground, for such general changes are exceedingly difficult to pin down. For other influences at work, see K. G. Dennis, *'Competition' in the History of Economic Thought* (New York: Arno Press, 1977) for one of a few accounts of the changing notions of competition over the longer period.
7. S. Fine, *Laissez Faire and the General-Welfare State* (Ann Arbor: University of Michigan Press, 1957).
8. Some of the American economists arguments about the emergence of large firms bear similarities with those of Marx, but the outcomes differed, as we shall see.
9. S. Maasen, "Who is Afraid of Metaphors?" in this volume. By comparison with other fields, interest in the role of metaphors in economics is rather limited and recent, dating only from W. Henderson, "Metaphor in Economics," *Economics* (Winter, 1982).
10. Though see P. M. Hejl, "The Importance of the Concepts of 'Organism' and 'Evolution' in E. Durkheim's 'Division of Labor' and the influence of H. Spencer," and A. La Vergata, "Herbert Spencer: Biology, Sociology, and Cosmic Evolution," in this volume and Hodgson *op. cit.* (forthcoming). Note 4, Chap. 6.
11. See for example, Hofstadter, *op. cit.* 1944, Note 4, Chap. 2; Fine, *op. cit.*, 1957, Note 7, Chap. 2; and more recently D. Ross, *The Origins of American Social Science* (New York: Cambridge University Press, 1991), Part II.
12. H. Spencer, *Social Statics* (1850, American edition 1864).
13. R. T. Ely, *Studies in the Evolution of Industrial Society* (New York: MacMillan, 1903), pp. 6–7.

14. On Sumner, see Hofstadter, *op. cit.*, 1944, Note 4, Chap. 3; and B. Curtis, *William Graham Sumner* (Boston: Twayne, 1981).
15. See A. W. Coats, "The First Two Decades of the American Economic Association," *American Economic Review* **50** (1960), 555–574, for a discussion of the foundation of the association and J. Dorfman, *The Economic Mind in American Civilization 1865–1918* (New York: Arno Press, 1949), Vol. 3, Part II for a good general discussion of the economics of the period. See *Science Economic Discussion* (New York: The Science Co., 1886) for a contemporary debate between old and new style economists and R. T. Ely, "The Past and Present of Political Economy," *Johns Hopkins University Studies in History and Political Science*, Second Series, **III** (Baltimore, 1884) for a statement of the new school. Mary O. Furner, *Advocacy and Objectivity* (Lexington: University Press of Kentucky, 1975) gives an excellent account of the professionalization of economics at this time.
16. H. Spencer, *The Principles of Sociology* (New York: D. Appleton & Co., 1876–96).
17. See H. C. Adams, "The Relation of the State to Industrial Action," *Publications of the AEA* **16** (1987), 471–549, for a discussion of the German versus English influences on the economics of himself and his peers.
18. See Spengler, *op. cit.*, 1950, pp. 206–211, Note 4 and for early economics writings, Paul Conkin, *Prophets of Prosperity* (Bloomington: Indiana University Press, 1980).
19. See T. C. Cochran and W. Miller, *The Age of Enterprise* (New York: MacMillan, 1942), Chap. VI; Fine, *op. cit.*, 1957, Chap. 4, Note 7; and M. Lerner, "The Triumph of Laissez-faire" in A. M. Schlesinger Jr. and M. White (eds.), *Paths of American Thought* (Boston: Houghton Mifflin, 1963).
20. See R. C. Bannister, "'The Survival of the Fittest is our Doctrine': History or Histrionics?" *Journal of the History of Ideas* **31**(3) (1970), 377–398 and *Social Darwinism: Science and Myth in Anglo-American Social Thought* (Philadelphia: Temple University Press, 1979).
21. Spengler, *op. cit.*, 1950, Note 4.
22. See for example, W. G. Sumner, "Earth Hunger or the Philosophy of Land Grabbing," (1896) in A. G. Keller (ed.), *Earth Hunger and Other Essays* (New Haven: Yale University Press, 1913); "A Group of Natural Monopolies" (1888) in Keller (ed.), *op. cit.*, 1913; and "The Challenge of Facts," (undated) in A. G. Keller, (ed.) *The Challenge of Facts and Other Essays* (New Haven: Yale University Press, 1914). Sumner was a fan of Spencer, but also appreciated that Darwin represented the "best yet" theory of evolution in biology, see Curtis, *op. cit.*, 1981, Note 14, Chap. 5.
23. W. G. Sumner, "The Concentration of Wealth: Its Economic Justification," (1902) in Keller (ed.), *op. cit.*, 1914.
24. J. B. Clark, "The Limits of Competition" in *Modern Distributive Process*, with F. H. Giddings (Boston: Ginn & Co., 1888), pp. 2 and 3–4 [Originally published in *Political Science Quarterly* **2** (1887), 45–61].
25. We have to be careful here because "natural monopoly" already had a technical meaning in economics.
26. J. B. Clark, *The Control of Trusts* (New York, Macmillan, 1901; Second edition, with J. M. Clark, 1912), p. 33.
27. See for good examples J. B. Clark, *The Philosophy of Wealth* (New York, Macmillan,

1886), Chap. IX; A. T. Hadley, *Standards of Public Morality*, 1906 John S. Kennedy Lectures (New York: MacMillan, 1907), Chap. 2; and T. B. Veblen, *The Theory of Business Enterprise*, 1904, Chap. VIII (reprinted by Augustus Kelley, New York, 1965).
28. W. G. Sumner, "What Makes the Rich Richer and the Poor Poorer?" (1887), in Keller (ed.), *op. cit.*, 1914, p. 67.
29. See Sumner, *op. cit.*, 1902, Note 23.
30. Sumner has been portrayed in the literature on Social Darwinism as a conservative, which indeed he was. Although it is fashionable nowadays to deride the early literature, Goldman's distinction between conservative and reformist Social Darwinism still remains valid, at least in the field under discussion here, E. F. Goldman, *Rendezvous with Destiny* (New York: Knopf, 1953). On the reformist Ely, see J. R. Everett, *Religion in Economics* (New York: King's Crown Press, 1946), Chap. 3.
31. Ely, *op. cit.*, 1903, Note 13, pp. 139–140.
32. Ward was both a natural and social scientist. Though Ward himself was highly critical of Spencer, they shared many elements in their evolutionary thinking On Ward, see Hofstadter, *op. cit.*, 1944, Note 4, Chap. 4.
33. Ely, *op. cit.*, 1903, Note 12, p. 142. Ward went further than this and claimed that competition did not lead to the survival of the fittest, rather that the removal of competition in the natural sphere allowed the maximum development level to be quickly reached (as in the case of agricultural breeding) (see L. F. Ward, *Outlines of Sociology* (New York: MacMillan, 1897), pp. 257–258).
34. In the prediction of increasing interdependency and cooperation evident in the arrangements of industrial society, we might recognize the influence of Spencer at work here, as well as Ward.
35. See the above quote from Clark, *op. cit.*, 1888, Note 24, p. 2.
36. J. E. Cairnes, *Some Leading Principles of Political Economy* (London: MacMillan, 1874), Chap. 3, para. 5.
37. This might have been a further transfer from evolutionary biology, but there was no reference to within or between species competition, which might have indicated a metaphor transfer, nor was there any evolutionary language in Cairnes' original work.
38. Clark, *op. cit.*, 1888, Note 24, p. 15.
39. F. H. Giddings, "The Persistence of Competition" in *Modern Distributive Process*, with J. B. Clark (Boston: Ginn & Co., 1888), pp. 31–32. (Originally published in *Political Science Quarterly* **2** (1887), 62–78.)
40. Ely drew on Adams, *op. cit.*, Note 17, who in turn drew on Jevons' analysis.
41. R. T. Ely, *Problems of To-day* (New York: Thomas Y. Crowell, 1888), p. 124.
42. See J. B. Clark, *The Distribution of Wealth* (New York: MacMillan, 1899). In Clark's first book, *op. cit.*, 1886, Note 27, neither his description of population nor that of competition made use of the evolutionary metaphors which suddenly became evident, in a strong form, in his treatment of 1887/8. Thereafter they began to wane so there was again almost nothing left in his 1907 textbook, J. B. Clark, *Essentials of Economic Theory* (New York, MacMillan, 1907). On Clark's changing view of competition, see Morgan, *op. cit.*, 1993, Note 3; and Everett, *op. cit.*, 1946, Note 30, Chap. 2.

43. C. Darwin, *The Origin of Species* (1859; London: Penguin, 1968), p. 134.
44. Clark, *op. cit.*, 1899, Note 42, p. 254.
45. Thus those races or groups with better institutions had an advantage over those with poorer or more rigid laws and customs (see A. T. Hadley, *Economics: An Account of the Relations between Private Property and Public Welfare* (New York: G. P. Putnam's Sons, 1896), paras 20–26).
46. He distinguished between those who were truly competent in the organization of production, and thus deserved to succeed, from those who merely made it through financial manipulation, as Veblen was later to distinguish more clearly (see Hadley, *ibid.*, paras 133–135, and Veblen, *op. cit.*, 1904, Note 27).
47. Hadley, *op. cit.*, 1906, Note 27, pp. 59–60.
48. A. T. Hadley, *Railroad Transportation: Its History and its Laws* (New York: G. P. Putnam's Sons, 1885), p. 99.
49. See Morgan, *op. cit.*, 1993, Note 3; and M. Cross and R. B. Ekelund, "A. T. Hadley on Monopoly Theory and Railroad Regulation: An American Contribution to Economic Analysis and Policy," *HOPE* **12** (1980), 214–233.
50. For good examples, see *The Chicago Conference on Trusts* [held Sept 13–16, 1899] The Civic Federation of Chicago: Chicago, 1900 (republished, New York: Arno Press, 1973) and the *US Industrial Commission Reports on Trusts*. (1899 and 1900).
51. See P. Weingart, "Struggle for Existence – Selection and Retention of a Metaphor," in this volume, for the lack of influence of evolutionary metaphors in the commercial field in Germany. For a contemporary German commentary on American trusts of the period, see E. von Halle, *Trusts of Industrial Combinations and Coalitions in the United States* (New York: MacMillan, 1895).
52. Despite Sherman's antitrust act, little was done to control the evils of monopoly manifest in the trusts' behavior. Regulation of interstate commerce through the Interstate Commerce Commission turned out to be a more useful vehicle, but even this made scant difference.
53. See H. D. Lloyd, *Wealth against Commonwealth* (New York: Harper, 1894) and for a more restrained treatment, J. W. Jenks, *The Trust Problem* (New York: McClure Phillips and Co., 1903), Chap. X.
54. See E. A. Ross, *Sin and Society* (Boston: Houghton, Mifflin & Co., 1907) and Veblen, *op. cit.*, 1904, Note 27. The more well-known phrase, "the robber barons" was popularized by Josephson's 1934 book, M. Josephson, *The Robber Barons: The Great American Capitalists. 1861–1901* (New York: Harcourt, Brace and Co., 1934) which captures perfectly the late nineteenth century view of the chief entrepreneurs as medieval warlords who acted with force and outside the law. The description by Charles and Mary Beard, *The Rise of American Civilization* (New York: MacMillan, 1927), Vol II, Chap. XX, remains one of the most colorful of a large literature.
55. L. F. Ward, "The Psychologic Basis of Social Economics," *Annals of the American Academy of Political and Social Science* **3** (1893), 89.
56. Hadley, *op. cit.*, 1906, Note 27, p. 60.
57. See Ward, *op. cit.*, 1893, Note 55, pp. 4, 464–482 and Ely, *op. cit.*, 1903, Note 13, Part II, Chaps. I and II.
58. Clark, *op. cit.*, 1901, Note 26, p. 20.
59. Ely, *op. cit.*, 1903, Note 13, p. 97.

60. H. Spencer, *The Principles of Ethics* (New York: D. Appleton & Co., 1892–93), Book II, para 397.
61. See Ely, *op. cit.*, 1903, Note 13, Part II, Chap. IV; Adams, *op. cit.*, 1887, Note 17.
62. See J. B. Clark, *The Problem of Monopoly* (New York: MacMillan, 1904).
63. See for example, H. George, *Progress and Poverty* (London: Kegan Paul, Trench & Co., 1882).
64. Darwin, *op. cit.*, 1859, Note 43, p. 117.
65. *Ibid.*, p. 129.
66. For example, see Sumner, *op. cit.*, undated, Note 22, p. 25.
67. Perhaps this explains why, as Spengler, *op. cit.*, 1950, Note 3, p. 211, suggests, the theory of natural selection became almost synonymous with theory of evolution for most economists. More recent attempts to transfer evolutionary theories, post understanding of genes, etc. by, for example, R. R. Nelson and S. G. Winter, *An Evolutionary Theory of Economic Change* (Cambridge, Mass: Belknap Press, 1982) have proved problematic in some of the same ways.
68. Darwin, *op. cit.*, Note 43, p. 127.
69. See Hejl, *op. cit.*, 1992, Note 10.
70. An instructive comparison is with the examples of physics transfers into economics involving theories, entities, and models in nonhistorical forms (see M. Boumans, *A Case of Limited Physics Transfer* (unpub. Phd diss., University of Amsterdam 1992), and P. Mirowski, *More Heat than Light* (Cambridge: Cambridge University Press, 1989).
71. The importance of biology in changing our understanding of variation within populations and variation over time has been emphasized by T. M. Porter, *The Rise of Statistical Thinking* (Princeton: Princeton University Press, 1986).
72. See P. L. Williams, *The Emergence of the Theory of the Firm* (London, MacMillan, 1978).
73. *Ibid.*, p. 39.
74. Its most famous product was T. B. Veblen, *Theory of the Leisure Class* (New York: MacMillan, 1899) which in many respects is riven with evolutionary thinking. See also Veblen's important essay on the evolutionary approach in economics, "Why is Economics Not an Evolutionary Science?" *Quarterly Journal of Economics* **12** (1898), 373–397. It is notable that the modern disciples of American institutionalism call their organization the "Association of Evolutionary Economists."
75. See C. D. W. Goodwin, "Marginalism Moves to the New World" in R. D. Collison Black, A. W. Coats and C. D. W. Goodwin (eds.), *The Marginal Revolution in Economics* (Durham: Duke University Press, 1973) on the mixture of methods and gradual usage of marginal thinking in the period, and Dorfman, *op. cit.*, 1949, Note 15, on the German element.

BIOLOGICAL AND PHYSICAL METAPHORS IN ECONOMICS

GEOFFREY M. HODGSON
University of Cambridge

The aim of this essay[1] is to examine and compare the roles of both biological and mechanistic metaphor in economics.[2] Pioneers of modern economic theory often made reference to the mechanical analogy, and the metaphors of classical mechanics and pre-1860 physics still pervade modern economic science.

This essay begins with a brief discussion of the general role of metaphor in science. In accord with several modern philosophical studies, the treatment of metaphor as simply a literary device is deemed to be mistaken. Instead, metaphor is seen as being constitutive for science.

The essay follows this with a discussion of the limitations of the mechanistic metaphor in economics and offers several reasons why the biological metaphor is more appropriate for that subject. The mechanistic and biological metaphors are compared in terms of their affiliated analytical frameworks, including such issues as their conception of time, their treatment of information, their accommodation of phenomena such as reversibility, and their potential to address complex systems.

The concluding sections argue that because metaphors are indispensable in scientific discourse, the self-conscious evaluation of metaphor is preferable to an ultimately futile attempt to remove it entirely. It is suggested that the constitutive role of metaphor is such that the development of a subject like economics can proceed best by carefully replacing one set of metaphors by another of greater benefit, rather than putting fruitless and misconceived effort into wholly removing metaphor. The self-conscious evaluation of metaphor is preferable to the unattainable goal of a metaphor-free science.

In a short study of this kind the focus is inevitably narrow. First, we are confined to a limited number of brief examples, ignoring many other important instances of transfer of biological metaphor to economics.[3] Second, little attention can be given to the social and political context of the evolution of economic thought and to the sociology of the economics profession itself.[4]

The Role of Metaphor in Science

Practicing scientists often regard metaphors as mere literary ornaments. It is sometimes suggested that metaphors should be removed to reveal the essential theory below. Recourse to mathematical modes of expression is often motivated in part by a desire to remove all such "literary frills." However, modern philosophers of science take a very different view. For example, Mary Hesse complains, "It is still unfortunately necessary to argue that metaphor is more than a decorative literary device, and that it has cognitive implications whose nature is a proper subject of philosophic discussion."[5] Similarly, Max Black concludes in a prominent study of metaphor and analogy in science: "Metaphorical thought is a distinctive mode of achieving insight, not to be construed as an ornamental substitute for plain thought."[6]

Clearly, for a positivist, a formalist, or a nominalist the use of metaphor is a superficial matter, even a distraction: a confusing renaming of entities, which might have nothing to do with the entities' essence. But this kind of response is challenged by modern philosophers who argue that metaphor is constititutive and indispensable for all science.[7] As Friedrich Nietzsche wrote in the nineteenth century:

What, then, is truth? A mobile army of metaphors, metonyms, and anthropomorphisms – in short, a sum of human relations, which have been enhanced, transposed, and embellished poetically and rhetorically, and which after long use seem firm, canonical, and obligatory to all people.[8]

Metaphors may lead or mislead. By their nature, they are never complete, precise, or literal mappings. If they were precise representations, they would not be metaphors, and the juxtaposition of similar but different conceptual frameworks would be lost. This juxtaposition, involving a degree of similarity and dissimilarity, can have both creative and damaging effects.

Metaphors are more than similes. For example, to describe the economy as "evolving" is not simply to state that the economy develops like an organism or a species in the natural world. It also may prompt the investigator to consider the many meanings and ambiguities in the term "evolve" and the many extensions and facets of the implicit analogy between the natural and the social world.

In economics the triumph of formalism has done nothing to limit the extensive use of metaphor, with terms such as "human capital," "market forces," "consumer sovereignty," and "natural rates of unemployment." The use of metaphor affects not merely the phrasing, but the structure and substance of the discipline.[9] Yet the metaphorical references may be partially obscured by the progress of formalism in economics. As mathematical symbols replace words, the analogies may seem to disappear. This is an illusion. As Donald McCloskey puts it, "Noneconomists find it easier to see the metaphors than do economists, habituated as the economists are by daily use to the idea that of course production comes from a 'function' and of course business moves in 'cycles.'"[10]

A possible source of creativity in science is through the juxtaposition of two different frames of reference, so that already existing but previously separate ideas can cross-fertilize. Larry Laudan argues that the amalgamation of different research traditions may produce a sum greater than the constituent parts.[11] Arthur Koestler has coined the term "bisociation" to describe the kind of adjoining of different ideas that occurs in the act of scientific creation.[12] Providing such a bridge between different discourses and contexts, metaphor can thus be both creative and constitutive.

There is no guarantee that metaphor can play a creative role. Conceivably in a given context there is a large set of metaphors that would be employed to no positive benefit. However, the evidence suggests that the inspired choice of metaphor seems to be a major – if not the major – source of theoretical innovation in science.

The Role of the Mechanistic Metaphor in Economics

The use of metaphor in economics has often been explicit. Even at the foundation of modern economic science, Adam Smith appealed specifically to metaphor and Newtonian mechanics in his essay on *The Principles which Lead and Direct Philosophical Enquiries: Illustrated by the History of Astronomy*. As Brian Loasby points out, Smith used Newtonian astronomy as primarily a set of "connecting principles."[13] These were held to make sense of his experience and were deemed to be fitting for his own theoretical work.

It has been argued extensively, by Philip Mirowski, Richard Norgaard, and others that modern economics is still dominated by the metaphor of a mechanistic system.[14] Most commentators typify the kind of mechanistic ideas that permeate economics as essentially Newtonian, although Mirowski sees as crucial the additional influence of the energetics movement in physics in the latter half of the nineteenth century. Whatever the precise details of the account, the consensus is that economics is still heavily influenced by the kind of mechanistic thinking that dominated physics around the middle decades of the nineteenth century. In particular, the evidence for the substantial influence of physics on the architects of the "marginal revolution" is substantial. We shall now briefly consider a number of examples.[15]

Léon Walras (1834–1910) wrote that "the pure theory of economics is a science which resembles the physico-mathematical sciences in every respect."[16] The proposed unity of the methods of physics and economics is fully revealed in his article "Economique et Mécanique."[17]

For William Stanley Jevons (1835–82) the metaphor of physical science was all-pervasive.

Utility only exists when there is on the one side the person wanting, and on the other the thing wanted ... Just as the gravitating force of a material depends not alone on the mass of that body, but upon the masses and relative positions and distances of the surrounding material bodies, so utility is an attraction between a wanting being and what is wanted.[18]

Francis Edgeworth (1845–1926) was fond of similar analogies.

As electro-magnetic force tends to a maximum energy, so also pleasure force tends toward a maximum energy. The energy generated by pleasure force is the physical concomitant and measure of the conscious feeling of delight.[19]

Vilfredo Pareto (1848–1923) was likewise a consistent proponent of the mechanical metaphor. He saw "the equations which determine equilibrium" as "the equations of rational mechanics." That is why "pure economics is a sort of mechanics or akin to mechanics."[20]

The case of Alfred Marshall (1842–1924) is more complicated and requires a little more discussion. He is often quoted for his statement in his *Principles* that "the Mecca of the economist lies in economic biology rather than in economic dynamics"; but we are less often reminded of what immediately follows:

But biological conceptions are more complex than those of mechanics; a volume on Foundations must therefore give a relatively large place to mechanical analogies; and frequent use is made of the term "equilibrium," which suggests something of a statical analogy.[21]

Accordingly, it is argued elsewhere[22] that Marshall's actual invocation of biology was rather limited.

Importantly, even when turning to his biology, Marshall found more inspiration from Herbert Spencer than from Charles Darwin. In the nineteenth century, general scientific prestige belonged to the mechanistic kind of thought; as Alfred North Whitehead put it, biology aped "the manners of physics."[23] Spencer incorporated mechanistic thinking more completely than Darwin. For Spencer, evolution involved the continuous redistribution of matter and motion. Like the ancient atomists, he upheld the constancy of matter and the indestructibility of motion.[24] Accordingly, in this volume, Antonello La Vergata argues that Spencer's biology was thoroughly mechanistic.

The example of Marshall's appeal to biology is thus an ironic one. Much more than his contemporaries he saw the limitations of the mechanistic metaphor and perceptively saw biology as providing a more appropriate alternative. Yet his work remained almost entirely in a static and mechanistic mode, and even when he found inspiration in biology it was in the mechanistic ideas of Spencer.

In sum, the founders of modern mainstream economic theory all made explicit reference to the mechanistic metaphor, and indicated that their work was guided by such a vision of constitution and structure. However, the precise details of the transfer of metaphor need to be examined. Here some questions can be raised about Mirowski's account in his *More Heat Than Light*. He argues that the transfer was overt and generally self-conscious, supposing that the economists involved were continuously aping physics and its every twist and theoretical turn.

There are at least two major flaws in this version of the story. First, it underestimates the subtlety of the process of metaphorical transfer. While there are cases of the direct appropriation of concepts and mathematical formalisms – amply documented by Mirowski – the constitutive transfers are at a deeper and less conscious level, affecting the ontology, epistemology, and methodology of the subject. With such "deep level" transfers the contamination of economics by mechanistic thinking is even more profound than Mirowski suggests.

The second flaw is that Mirowski disregards the sociology of the economics profession. We may illustrate this by considering later developments in physics in the twentieth century and their influence on economics. For instance, Mirowski makes a case that quantum theory has had an effect on econometric practice from the 1930s.[25] Apart from this, however, the general effect of both biology and physics on economics since 1930 has been very minor.

If economics has been driven by physics in the direct and overt manner suggested by Mirowski, then he would have to account for the fact that the deeper implications of modern quantum theory have not been incorporated. Indeed, the organicist and indeterminist aspects of modern quantum theory are in profound contradiction to the Cartesianism, reductionism, and atomism at the root of both the earlier physics and of modern economics.

Consideration of these points suggests a slightly different story, although it cannot be discussed in detail here. Although Walras, Jevons, Edgeworth, Pareto, and Marshall did ape the physics of their time, the professionalization and institutionalization of economics since then has locked the subject into that particular genre. Once the new norms of scientific activity became established in the emergent economic journals and economics departments of the late nineteenth and early twentieth century, then they set down the self-reinforcing standards for subsequent research. Scholars seeking publication or promotion had a better chance of success if they conformed to these norms. As in other cases of institutional evolution, the phenomenon of locked-in behavior, which depends much on initial circumstances, is normal rather than exceptional.

Accordingly, the initial role of physics was crucial in contributing greatly to the initial intellectual environment and in legitimating the allegedly scientific credentials of the subject. But since the early decades of the twentieth century economics has been less driven by physics than by its own momentum. It is institutional practice, including a very outdated view of science, that drives economics along its present groove – not an overt and unabating desire to replicate everything going on in physics or other natural sciences. This does not mean that the role of the mechanistic metaphor in economics is diminished. Far from it: the metaphor is subtle and ubiquitous.

At least up to the World War I, economics was two-thirds a British affair. The development of economics in Continental Europe and in America followed a slightly different pattern. In both Germany and the United States, for instance, economics around 1900 was dominated more by often-bowdlerized biological metaphors taken from Charles Darwin, Albert Schäffle, Herbert Spencer, William Graham Sumner, and others. In the United States this led to the dominance of an "institutional economics" proclaiming "evolutionary" credentials in the 1920s and 1930s. The supremacy of neoclassical economics came later and with an internationally unprecedented degree of formalization, particularly with the publication of Paul Samuelson's *Foundations* in 1947. Like other neoclassical economists before him, Samuelson showed no hesitation in wrapping himself "in the banner of physics."[26] The locking-in of economics to a formalized, neoclassical theory emulating an earlier physics is very much a postwar phenomenon in the United States.

By comparison, the professionalization of other social and natural sciences has not always had the same kind of lock-in effect. Sociology, for example, has shown some capacity for drift and upheaval. In distinction, it is probable that the quite rigid lock-in exhibited in economics in the last hundred years is due in the main to the specific characteristics of its founding metaphor. Nineteenth-century physics did not only offer "scientific" credentials: it provided an armory of formal techniques, with differential calculus in pride of place. Accordingly, no other social science has used mathematical techniques so extensively.

Nevertheless, the cache of mathematical tools employed has typically been rather limited. It is only recently that the kind of mathematics bestowed in the last third of the nineteenth century is beginning to be overshadowed by other formal developments, particularly game theory. Although game theory has been used in economics since the 1940s, it did not become fashionable until the 1980s. Notably, it is now recognized that its application threatens core tenets such as the axioms of rationality.[27]

In overt terms, however, modern economics has almost forgotten the source of the crucial metaphor from which it gained so much formal and theoretical inspiration during the eighteenth and nineteenth centuries. The general and almost complete disregard for the history of economic thought by economists is now conspicuous by its absence from undergraduate syllabi and professional job descriptions. With the ever sharper focus on mathematical form rather than conceptual substance, attention has been shifted away from the nature and origin of core assumptions in economics.

This is one reason why the exposure of the role of metaphor is so important. It can reveal deeply embedded structures of thought and provide a clue to why the subject develops along one path rather than another. The reforming theoretician can then identify, and attempt to remove or alter, the malign features at the theoretical core. We now move on to examine the limitations of the mechanistic metaphor and briefly suggest how economics could be changed if this was replaced by something from modern biology.

Limitations of the Mechanistic Metaphor

The deficiencies of the metaphor taken from classical mechanics have been discussed at length elsewhere.[28] There are a number of perennial problems involved, all relating to the limitations of Cartesian philosophy and Newtonian principles. For instance, movement is reversible in the "conserved system" of Newtonian mechanics; there is no arrow of time. "Classical mechanics only knows motion, whereas at the same time the processes of motion are completely reversible and in no way give rise to any qualitative changes."[29] Although in some non-conserved, mechanistic theoretical systems the possibility of irreversibility emerges – such as with the addition of friction – the reasons are quite different from those in more complex systems. Above all, with mechanistic presuppositions, cause and effect can be mirrored by logical syllogism, and logical replaces historical time.[30]

It is easy to trace the derivation of the ideas of rationality and equilibria, the core concepts of neoclassical economics, from the inheritance of mechanistic thought:

> classical mechanics considers a system of material points upon which directional forces operate at a distance according to calculable laws of motion. The choice of paths is governed by the principle of least action, which may be termed the economic principle if we take the term in its widest sense as denoting a maximum-minimum principle.[31]

Hence, subject to a combination of forces, economic agents optimize to the point of equilibrium as if they were mere particles obeying mechanical laws.

In addition, there is a general difficulty of incorporating information, learning, and knowledge in a mechanistic scheme. In classical mechanics there is no place for thoughts and ideas: all is mere matter, subject to Newtonian laws. As Norbert Weiner remarked, "In nineteenth-century physics it seemed to cost nothing to get information."[32]

It is not proposed here that the use of mechanistic thinking in economics has been entirely without value. Nevertheless, the limitations are severe. In sum, the mechanistic metaphor excludes knowledge, qualitative change, and irreversibility through time. It entraps economics in equilibrium schema where there are no systematic errors and no cumulative development. Clearly much is missing here. The strength of the alternative, biological, metaphor is that a place can be found for these important features of economic life.

Biology as an Alternative Metaphor for Economics

Taking recourse to biology is not simply a tactic. It is held that real world economic phenomena have much in common with biological organisms and processes than with the mechanistic world of billiard balls and planets. After all, the economy involves living human beings, not merely particles, forces, and energy.

Nevertheless, there are risks involved in this trade; biology has been often abused by social scientists in the past, sometimes with horrendous social and political consequences. There was the episode of Social Darwinism and lamentable former associations of biological thought with sundry pro-aristocratic, racist, or sexist ideologies.

It is still widely assumed that evolutionary thinking involves the rejection of any kind of state subsidy or intervention, and the support for laissez-faire on the basis of the idea of "survival of the fittest." However, it is wrong to assume that evolutionary theorizing always points to the optimality of competitive outcomes, or to laws of evolutionary "progress," or to the sagacity of laissez-faire. According to modern theory, evolutionary processes do not necessarily lead – by any reasonable definition – to optimal consequences. Similar arguments apply in the economic as well as the biotic context.[33]

Moreover, biology has internal problems of its own; it is no panacea. Indeed, biology is not itself free of mechanistic metaphor and reductionist methods. A large number of biologists are committed to reductionism, even explaining biological phenomena in physical terms. However, there are also pronounced attempts to transcend such strains of thought.

The internal lack of consensus within biology is itself refreshing. Furthermore, a variety of forces and tensions point to an organicist ontology, a less rigid methodology, and the transcendence of mechanistic thinking. These indicators are discussed in the work of leading mainstream biologists such as Theodosius Dobzhansky and Ernst Mayr, as well as more heterodox scientists such as Niles Eldredge and Stephen Jay Gould, along with historians of biology such as Edward Manier.[34]

It is important to emphasize that all metaphors create difficulties as well as solutions. A preeminent problem with both the biological and the mechanical analogy is the conceptualization of the human agent. Several critical economists have argued despite the rhetoric, orthodox and mechanistic economics provides no room for real individual choice.[35] Biological natural selection invokes genetic replication and random variation or mutation, but seemingly affords no role for intentionality, purposefulness, or choice.

Clearly, as with many "hard core" metaphors, their transposition from one science to another may open up problems in their source as well as their destination. For instance, the problem of "vitalism," which involves choice, will, and purpose, has been persistent within biology. Although "vitalism" is now out of fashion, it raises real issues of importance, even if the notion is shunned by biologists who confine themselves to causal rather than intentional explanations. Dissenters to strict Darwinism, including Arthur Koestler, have tried to instate concepts of will and purpose in the science, but with limited effect. This whole problem should not be ignored or understimated; it is addressed at length elsewhere.[36]

The Value of the Biological Metaphor in Economics

Despite all the problems and dangers, modern biology provides a rich source of ideas and approaches from which a revitalized economics may draw. In all, the application of an evolutionary approach to economics seems to involve a number of advantages and improvements over the orthodox and mechanistic paradigm. For instance, it enhances a concern with irreversible and ongoing processes in time,[37] with long-run development rather than short-run marginal adjustments, with qualitative as well as quantitative change, with variation and diversity, with non-equilibrium as well as equilibrium situations, and with the possibility of persistent and systematic error-making and thereby non-optimising behavior.

In short, an evolutionary paradigm provides an alternative to the neoclassical "hard core" idea of mechanistic maximization under static constraints. The theory of rational choice at the core of mainstream economics relies on static assumptions, the notion of an eventually constant decision environment, and the idea of global rationality which are all challenged by evolutionary theory.[38]

Another extremely important reason why ideas from biology are of relevance to economics is that both economic and biotic systems are highly complex. They both encompass tangled structures and causalities, involve continuous change and embrace huge variety. Partly for this reason, there is the problem of levels of abstraction and appropriate units of analysis. This has been faced and debated by a number of prominent biologists, but far less attention has been given to this vital issue in economics. The more complete adoption of biological metaphors may help to redress the balance.

In biology there is not only a discourse concerning reductionism and the appropriate units of evolutionary selection, but also over the viability of further reduction from genetics down to molecular biology, and even below to chemistry and physics.[39] In contrast, confident in the Newtonian metaphor of the indivisible, "individual" particle, mainstream economics traditionally proscribes discussion of the psychological or social foundations of individual purposes and preferences as being beyond the bounds of the subject.

Partly because of the acknowledged complexity of the phenomena that it attempts to analyze, biological science exhibits a theoretical pluralism.[40] As David Hull points out, Darwin's methodology is not rigidly axiomatic. There is a rigorous deductive core, but it is deemed to prove little on its own and it is thus placed in the context of a mass of empirical material.[41]

Hence in biology there are deductive arguments combined with contingent empirical premises and conclusions. Typically, in biology a number of theories and explanations compete in their claims to identify the main, rather than the exclusive, cause in given real circumstances. Fortunately, biology does not present the near-monopoly of methods and approaches that threatens to stifle economics today.

There is another reason why the turn to biology is of value and significance. As Fritjof Capra and others have argued, the Cartesian and Newtonian worldviews have sanctioned habits of thought that involve an ultimately untenable conceptual division between humankind and the remainder of the natural world.[42] Yet humans live alongside others, and there are limits to the natural resources available and the tolerances of the ecosystems on the planet. The invocation of the biological metaphor surely helps remind us of these vital issues for the twenty-first century.[43]

The expression of the potential value of biological thinking for economics has been repeated at periodic intervals every since: importantly by János Kornai, with great finesse by Nicholas Georgescu-Roegen, rather one-sidedly by Gary Becker, Jack Hirshleifer and Gordon Tullock, with the clarity of a popularizer by Kenneth Boulding, and briefly and surprisingly by Frank Hahn.[44]

Notably the burgeoning recent literature on economic evolution and technical change has appealed to metaphors from biology.[10] However, relatively little attention has been paid to the deeper conceptual and methodological questions involved, e.g., units of selection, causality in evolution, reductionism. These issues go beyond a narrow focus on evolutionary selection mechanisms to include deeper ontological, epistemological, and methodological issues.

For example, there is an important discourse in biology that sustains units of selection and analysis above the most reductive elements – including, for example, groups or ecosystems – with important implications for social science.[45] Accordingly, there are currents of thought in biology that challenge the traditional reductionist approach to analysis involving the breaking down of phenomena to constituent parts. Such reductionism may be confounded, and higher levels of analysis legitimized.[46] One implication for economics is the reinstatement of macroeconomics as an autonomous level of analysis.[47] Finally, there are novel reconstructions of the concept of causality in biology linking it to the phenomenon of organization.[48]

Above all, taking these ideas on board involves greater reference to and deliberation upon biology in general and evolutionary theory in particular. In other words, the less casual appropriation of the biological metaphor can help to clarify and develop the emergent economic theory.

Conclusion

The main objective of this essay has been to examine the roles of the mechanical and biological metaphor in economic science. The view that economics is dominated by a mechanistic metaphor is now widely accepted, and the value of the alternative metaphor from biology has been suggested by a number of economists. Some specific indications of the value of the biological metaphor have been made here.

Mirowski has shown how the mechanistic metaphor was consciously appropriated and enshrined in the formalisms of post-1870 economic theory.[49] Yet it is a mistake to suggest that the emulation of pre-1860 physics has been a primary preoccupation of economists throughout the twentieth century. It is suggested in this essay that the function of the mechanistic metaphor is more subtle, and hence even more pervasive, than proposed by Mirowski.

Contrary to Mirowski, the problem for economics is not the general appropriation of metaphor from the natural sciences but more fundamentally the adoption of a positivist conception of science in which the role of metaphor is specifically denied. In this manner the source of the constitutive assumptions of modern economics has been ignored, despite their continuous transmission through the orthodox mathematical formalisms and their reinforcement through the sociological structure of the economics profession. For this reason, the ingrained mechanistic metaphor in economics is even more difficult to remove and to replace.

The argument that metaphors are indispensable in scientific discourse counters the mistaken and positivist view of a metaphor-free science. In addition, in regard to particular subjects such as economics, it would lead us to the self-conscious evaluation of metaphor, rather than futile attempt to remove it entirely.

As well as this appraisal of metaphor, and the examination of metaphorical transmission from one discipline to another, the adoption and replication of a particular metaphorical structure clearly needs to be examined in the wider context of the routinized practice and sociology of the academic discipline. In the case of economics this would suggest reform not simply of its theoretical content, but of its disciplinary structures and its evaluative procedures, in regard to university departments, academic journals, and professional associations. But these are issues that we have been unable to address here.

Notes

1. The author is grateful to participants at the conference on "The Problematic Connection – The Transfer of Images and Metaphors between Biology and the

Social Sciences," Zentrum für Interdisziplinäre Forschung, University of Bielefeld, Germany, June 22–24, 1992, and Richard Whitley in particular, for critical remarks on an earlier version of this essay.
2. Throughout this essay, the term "mechanistic" is taken to refer to classical mechanics or physics before 1860, at the very latest. Hence any reference to statistical mechanics and other more recent developments is excluded.
3. For a more extensive discussion of the past and potential use of biological metaphors in economics see Geoffrey M. Hodgson, *Economics and Evolution: Bringing Life Back into Economics* (Cambridge: Polity Press, 1993).
4. The sociology of the economics profession deserves more extensive study than it has received to date. Two of the relatively rare discussions by economists are found in E. Ray Canterbery and Robert J. Burkhardt, "What Do We Mean by Asking Whether Economics is a Science?" in Alfred S. Eichner (ed.), *Why Economics is Not Yet a Science* (London: Macmillan, 1983), and Robert J. Burkhardt and E. Ray Canterbery. "The Orthodoxy and Professional Legitimacy: Toward a Critical Sociology of Economics," *Research in the History of Economic Thought and Methodology* **4** (1986), 229–250. Historical studies of the process of academic establishment and professionalization of economics towards the end of the nineteenth century include Alon Kadish, *Historians, Economists, and Economic History* (London: Routledge, 1989); Gerard M. Koot, *English Historical Economics, 1870–1926* (Cambridge: Cambridge University Press, 1987); and John Maloney, *Marshall, Orthodoxy and the Professionalization of Economics* (Cambridge: Cambridge University Press). More general discussions of the sociology of science include Norbert Elias, Heninio Martins, and Richard Whitley (eds.), "Scientific Establishments and Hierarchies," Sociology of the Sciences **4** (1982), and Richard Whitley, *Social Processes of Scientific Development* (London: Routledge and Kegan Paul, 1974).
5. Mary B. Hesse, *Revolutions and Reconstructions in the Philosophy of Science* (Brighton: Harvester Press, 1980), p. 111.
6. Max Black, *Models and Metaphors: Studies in Language and Philosophy* (Ithaca: Cornell University Press, 1962), p. 237.
7. The number of works on metaphor in the philosophy of science is enormous. See the references and the excellent discussion of this literature in the essay by Sabine Maasen in this volume.
8. From W. Kaufman (ed.), *The Portable Nietzsche* (Harmondsworth: Penguin, 1982).
9. Warren J. Samuels (ed.), *Economics as Discourse: An Analysis of the Language of Economists* (Boston: Kluwer, 1990).
10. Donald N. McCloskey, *The Rhetoric of Economics* (Madison: University of Wisconsin Press, 1985), p. 74.
11. Larry Laudan, *Progress and its Problems: Towards a Theory of Scientific Growth* (London: Routledge and Kegan Paul, 1977), p. 103.
12. Arthur Koestler, *The Act of Creation* (London: Hutchinson, 1964).
13. Brian J. Loasby, *The Mind and Method of the Economist: A Critical Appraisal of Major Economists in the Twentieth Century* (Aldershot: Edward Elgar, 1989), pp. 1–5; Brian J. Loasby, *Equilibrium and Evolution: An Exploration of Connecting Principles in Economics* (Manchester: Manchester University Press, 1991), pp. 6–8.
14. See Philip Mirowski, *Against Mechanism: Protecting Economics from Science*

(Totowa, NJ: Rowman and Littlefield, 1988); Philip Mirowski, *More Heat Than Light: Economics as Social Physics, Physics as Nature's Economics* (Cambridge: Cambridge University Press, 1989); and Richard B. Norgaard, "Economics as Mechanics and the Demise of Biological Diversity," *Economic Modelling* (September 1987), 107–121.

15. Carl Menger (1840–1921) is omitted. It is sometimes held that Menger ranks with Jevons and Walras as one of the three main founders of neoclassical economics in the 1870s. However, the work of Menger is now appropriately regarded as outside the neoclassical mainstream (see William Jaffé, "Menger, Jevons and Walras De-homogenized," *Economic Inquiry* **14**(1) (1976), 11–24), and in any case it was not nearly so influential, especially in the Anglo-American world. Notably, Menger did not use mathematics and did not dress his theoretical work with the metaphors of physics.

16. Léon Walras, *Elements of Pure Economics, or The Theory of Social Wealth*, translated from the French edition of 1926 by W. Jaffé (1st edn., 1874; New York: Augustus Kelley, 1954), p. 71.

17. Léon Walras, "Economique et Mécanique" (1909), reprinted in *Metroeconomica* (1960), 1–13, and translated in Philip Mirowski and Pamela Cook, "Walras's Economics and Mechanics: Translation, Commentary, Context," in Samuels, *op. cit.*, 1990.

18. William Stanley Jevons, *The Papers and Correspondence of W. S. Jevons*, Vol. 7, ed. R. Black (London: Macmillan, 1981), p. 80.

19. Francis Y. Edgeworth, *Mathematical Psychics* (London: Kegan Paul, 1881; Reprinted New York: Kelley, 1961), p. 25.

20. Vilfredo Pareto, "On the Economic Phenomenon," *International Economic Papers* (1953), No. 3, 185.

21. Alfred Marshall, *The Principles of Economics*, 9th (variorum) ed. with annotations by C. W. Guillebaud (London: Macmillan, 1961), p. xii.

22. See Geoffrey M. Hodgson, "The Mecca of Alfred Marshall," *The Economic Journal* **103** (March 1993), 406–415; Geoffrey M. Hodgson, *Economics and Evolution: Bringing Life Back into Economics* (Cambridge: Polity Press, 1993); and Brinley Thomas, "Alfred Marshall on Economic Biology," *Review of Political Economy* **3** (January 1991), 1–14.

23. Alfred North Whitehead, *Science and the Modern World* (Cambridge: Cambridge University Press, 1926), p. 128.

24. See Milic Capek, *The Philosophical Impact of Contemporary Physics* (Princeton, N.J.: Van Nostrand, 1961), pp. 101–103.

25. See Philip Mirowski, "The Probabilistic Counter-Revolution, or How Stochastic Concepts Came to Neoclassical Theory," *Oxford Economic Papers* **41** (April 1989), 217–235.

26. Philip Mirowski, *More Heat Than Light, op. cit.*, 1989, p. 327.

27. See Robert Sugden, "Rational Choice: A Survey of Contributions from Economics and Philosophy," *Economic Journal* **101** (July 1991), 751–785.

28. See, for instance, Nicholas Georgescu-Roegen, *The Entropy Law and the Economic Process* (Cambridge, MA: Harvard University Press); John M. Gowdy, "Evolutionary Theory and Economic Theory: Some Methodological Issues," *Review of Social Economy* **43** (1985), 316–324; G. Sebba, "The Development of the Concepts of

Mechanism and Model in Physical Science and Economic Thought," *American Economic Review (Papers and Proceedings)* **43** (May 1953), 259–268; H. Thoben, "Mechanistic and Organistic Analogies in Economics Reconsidered," *Kyklos* **35** (1982), Fasc. 2, 292–306; and Thorstein B. Veblen, *The Place of Science in Modern Civilization and Other Essays* (New York: Huebsch, 1919; Reprinted 1990 with a new introduction by W. J. Samuels, New Brunswick: Transaction Publishers).

29. See H. Thoben, "Mechanistic and Organistic Analogies in Economics Reconsidered," *Kyklos* **35** (1982), Fasc. 2, 292–306, 293.
30. Joan Robinson, *History Versus Equilibrium* (London: Thames Papers in Political Economy, 1974).
31. Sebba, *op. cit.*, 1953, p. 269.
32. Norbert Weiner, *The Human use of Human Beings*, 2nd ed. (New York: Houghton Mifflin, 1954), p. 29.
33. For biology see Stephen Jay Gould and Richard C. Lewontin, "The Spandrels of San Marco and the Panglossian Paradigm: A Critique of the Adaptationist Programme," *Proceedings of the Royal Society of London*, Series B, **205** (1979), 581–598. For economics see Geoffrey M. Hodgson, "Economic Evolution: Intervention Contra Pangloss," *Journal of Economic Issues* **25** (June 1991), 519–533.
34. See Theodosius Dobzhansky, "On Some Fundamental Concepts of Darwinian Biology" in Theodosius Dobzhansky, M. K. Hecht, M. K. and W. C. Steere (eds.), *Evolutionary Biology* (Amsterdam: North Holland, 1968), pp. 1–34; Ernst Mayr, "The Nature of the Darwinian Revolution," *Science* (1972), No. 176, 981–989; Ernst Mayr, "How Biology Differs from the Physical Sciences," in D. J. Depew and B. H. Weber (eds.), *Evolution at the Crossroads: The New Biology and the New Philosophy of Science* (Cambridge, MA: MIT Press, 1985), pp. 43–63; Niles Eldredge, *Unfinished Synthesis: Biological Hierarchies and Modern Evolutionary Thought* (Oxford: Oxford University Press, 1985); Stephen Jay Gould, "The Meaning of Punctuated Equilibrium and its Role in Validating a Hierarchical Approach to Macroevolution," in R. Milkman (ed.), *Perspectives on Evolution* (Sunderland, MA: Sinauer Associates), pp. 83–104; and Edward Manier, *The Young Darwin and his Cultural Circle: A Study of Influences which helped Shape the Language and Logic of the First Drafts of the Theory of Natural Selection* (Dordrecht: Reidel, 1978).
35. Brian J. Loasby, *Choice, Complexity and Ignorance: An Enquiry into Economic Theory and the Practice of Decision Making* (Cambridge: Cambridge University Press, 1976).
36. See, for instance, Arthur Koestler, *The Ghost in the Machine* (London: Hutchinson, 1967), and Hodgson, *Economics and Evolution, op. cit.*, 1993, Chap. 14.
37. It should not be assumed, however, that all evolutionary processes in biology are irreversible: see G. S. Mani, "Is There a General Theory of Biological Evolution?" in P. Paolo Saviotti and J. Stanley Metcalfe (eds.), *Evolutionary Theories of Economic and Technological Change: Present Status and Future Prospects* (Reading: Harwood, 1991), pp. 31–57. An extended discussion of irreversibility in economics is given by Giovanni Dosi and J. Stanley Metcalfe, "On Some Notions of Irreversibility in Economics," in Saviotti and Metcalfe (eds.), *op. cit.*, 1991, pp. 133–159.
38. See W. S. Cooper, "How Evolutionary Biology Challenges the Classical Theory of Rational Choice," *Biology and Philosophy* **4** (October 1989), 457–481, and M. A.

Goldberg, "On the Inefficiency of Being Efficient," *Environment and Planning* **7** (1975), 921–939.
39. See Elliott Sober (ed.), *Conceptual Issues in Evolutionary Biology: An Anthology* (Cambridge, MA: MIT Press, 1984).
40. See Gould and Lewontin, *op. cit.*, 1979; and Ernst Mayr, "Darwin's Five Theories of Evolution," in D. Kohn (ed.), *The Darwinian Heritage* (Princeton: Princeton University Press, 1985), pp. 755–772.
41. See David L. Hull, *Darwin and His Critics: The Reception of Darwin's Theory of Evolution by the Scientific Community* (Cambridge, MA: Harvard University Press, 1973), pp. 3–36. Anthony flew has pointed out the similarity between the implicit scientific method of Darwin and that of Thomas Robert Malthus in this respect. See A. Flew, "The Structure of Darwinism," *New Biology* **28** (1959), 25–44 and A. Flew "The Structure of Malthus' Population Theory," in B. Baumrin (ed.), *Philosophy of Science: The Delaware Seminar* (New York: Interscience, 1963), pp. 283–307.
42. Fritjof Capra, *The Turning Point: Science, Society and the Rising Culture* (London: Wildwood House, 1982).
43. Georgescu-Roegen, *op. cit.*; Jeremy Rifkin, *Entropy: A New World View* (New York: Viking Press, 1980).
44. See János Kornai, *Anti-Equilibrium: On Economic Systems Theory and the Tasks of Research* (Amsterdam: North-Holland, 1971); Georgescu-Roegen, *op. cit.*; Gary S. Becker, "A Theory of Social Interactions," *Journal of Political Economy* **82** (November–December 1974), 1063–1093; Gary S. Becker, "Altruism, Egoism, and Genetic Fitness: Economics and Sociobiology," *Journal of Economic Literature* **14** (December 1976), 817–826; Jack Hirshleifer, "Economics from a Biological Viewpoint," *Journal of Law and Economics* **20** (April 1977), 1–52; Jack Hirshleifer, "Natural Economy versus Political Economy," *Journal of Social and Biological Structures* **1** (1978), 319–337; Gordon Tullock, "Sociobiology and Economics," *Atlantic Economic Journal* (September 1979), 1–10; Kenneth E. Boulding, *Evolutionary Economics* (Beverly Hills, CA: Sage Publications, 1981); Frank H. Hahn, "The Next Hundred Years," *Economic Journal* **101** (January 1991), 47–50.
45. For some of the many recent applications of evolutionary ideas to economics and to the theory of technical change, see Norman G. Clark and Calestous Juma, *Long-Run Economics: An Evolutionary Approach to Economic Growth* (London: Printer, 1987); Giovanni Dosi, Christopher Giovanni, Richard R. Nelson, Gerald Silverberg and Luc Soete (eds.), *Technical Change and Economic Theory* (London: Printer, 1988); Christopher Freeman (ed.), *The Economics of Innovation* (Aldershot: Edward Elgar, 1990); Friedrich A. Hayek, *The Fatal Conceit: The Errors of Socialism, Collected Works of F. A. Hayek*, Vol. I (London: Routledge, 1988); Joel Mokyr, *The Lever of Riches: Technological Creativity and Economic Progress* (Oxford: Oxford University Press, 1990); Joel Mokyr, "Evolutionary Biology, Technical Change and Economic History," *Bulletin of Economic Research* **43** (April 1991), 127–149; Richard R. Nelson and Sidney G. Winter, *An Evolutionary Theory of Economic Change* (Cambridge MA: Harvard University Press, 1982); Saviotti and Metcalfe (eds.), *op. cit.*, 1991; Ulrich Witt (ed.), *Evolutionary Economics* (Aldershot: Edward Elgar, 1993).

45. See Anthony J. Arnold and Kurt Fristrup, "The Theory of Evolution by Natural Selection: A Hierarchical Expansion," *Paleobiology* **8** (1982), 113–129.
46. See Geoffrey M. Hodgson, "Why the Problem of Reductionism in Biology has Implications for Economics," *World Futures*, (forthcoming 1993); and William C. Winsatt, "Reductionist Research Strategies and Their Biases in the Units of Selection Controversy," in T. Nickles, (ed.), *Scientific Discovery, Volume II, Historical and Scientific Case Studies* (Dordrecht, Holland: Reidel, 1980), pp. 213–259. Reprinted in Sober, *op. cit.*, 1984.
47. See Hodgson, *Economics and Evolution, op. cit.*, 1993, Chap. 16.
48. See John H. Campbell, "An Organizational Interpretation of Evolution," in D. J. Depew and B. H. Weber (eds.), *Evolution at the Crossrods: The New Biology and the New Philosophy of Science* (Cambridge, MA: MIT Press, 1985), pp. 133–167.
49. Mirowski, *More Heat Than Light, op. cit.*, 1989.

Sociology of the Sciences

1. E. Mendelsohn, P. Weingart and R. Whitley (eds.): *The Social Production of Scientific Knowledge.* 1977   ISBN Hb 90-277-0775-8; Pb 90-277-0776-6
2. W. Krohn, E.T. Layton, Jr. and P. Weingart (eds.): *The Dynamics of Science and Technology.* Social Values, Technical Norms and Scientific Criteria in the Development of Knowledge. 1978   ISBN Hb 90-277-0880-0; Pb 90-277-0881-9
3. H. Nowotny and H. Rose (eds.): *Counter-Movements in the Sciences.* The Sociology of the Alternatives to Big Science. 1979
   ISBN Hb 90-277-0971-8; Pb 90-277-0972-6
4. K.D. Knorr, R. Krohn and R. Whitley (eds.): *The Social Process of Scientific Investigation.* 1980 (1981)   ISBN Hb 90-277-1174-7; Pb 90-277-1175-5
5. E. Mendelsohn and Y. Elkana (eds.): *Sciences and Cultures.* Anthropological and Historical Studies of the Sciences. 1981
   ISBN Hb 90-277-1234-4; Pb 90-277-1235-2
6. N. Elias, H. Martins and R. Whitley (eds.): *Scientific Establishments and Hierarchies.* 1982   ISBN Hb 90-277-1322-7; Pb 90-277-1323-5
7. L. Graham, W. Lepenies and P. Weingart (eds.): *Functions and Uses of Disciplinary Histories.* 1983   ISBN Hb 90-277-1520-3; Pb 90-277-1521-1
8. E. Mendelsohn and H. Nowotny (eds.): *Nineteen Eighty Four: Science between Utopia and Dystopia.* 1984   ISBN Hb 90-277-1719-2; Pb 90-277-1721-4
9. T. Shinn and R. Whitley (eds.): *Expository Science.* Forms and Functions of Popularisation. 1985   ISBN Hb 90-277-1831-8; Pb 90-277-1832-6
10. G. Böhme and N. Stehr (eds.): *The Knowledge Society.* The Growing Impact of Scientific Knowledge on Social Relations. 1986
    ISBN Hb 90-277-2305-2; Pb 90-277-2306-0
11. S. Blume, J. Bunders, L. Leydesdorff and R. Whitley (eds.): *The Social Direction of the Public Sciences.* Causes and Consequences of Co-operation between Scientists and Non-scientific Groups. 1987
    ISBN Hb 90-277-2381-8; Pb 90-277-2382-6
12. E. Mendelsohn, M.R. Smith and P. Weingart (eds.): *Science, Technology and the Military.* 2 vols. 1988
    ISBN Vol, 12/1 90-277-2780-5; Vol. 12/2 90-277-2783-X
13. S. Fuller, M. de Mey, T. Shinn and S. Woolgar (eds.): *The Cognitive Turn.* Sociological and Psychological Perspectives on Science. 1989
    ISBN 0-7923-0306-7
14. W. Krohn, G. Küppers and H. Nowotny (eds.): *Selforganization.* Portrait of a Scientific Revolution. 1990   ISBN 0-7923-0830-1
15. P. Wagner, B. Wittrock and R. Whitley (eds.): *Discourses on Society.* The Shaping on the Social Science Disciplines. 1991   ISBN 0-7923-1001-2

Sociology of the Sciences

16. E. Crawford, T. Shinn and S. Sörlin (eds.): *Denationalizing Science. The Contexts of International Scientific Practice.* 1992 (1993) ISBN 0-7923-1855-2
17. Y. Ezrahi, E. Mendelsohn and H. Segal (eds.): *Technology, Pessimism, and Postmodernism.* 1993 (1994) ISBN 0-7923-2630-X
18. S. Maasen, E. Mendelsohn and P. Weingart (eds.): *Biology as Society, Society as Biology: Metaphors.* 1994 (1995) ISBN 0-7923-3174-5

KLUWER ACADEMIC PUBLISHERS – DORDRECHT / BOSTON / LONDON

MANAGERIAL DILEMMAS

Cases in Social, Legal, and Technological Change

ALAN F. WESTIN
Columbia University

and JOHN D. ARAM
Case Western Reserve University

BALLINGER PUBLISHING COMPANY
Cambridge, Massachusetts
A subsidiary of Harper & Row, Publishers, Inc.

Copyright © 1988 by Alan F. Westin and John D. Aram. All rights reserved. No part of this publication may be reproduced, stored in a retrieval system, or transmitted in any form or by any means, electronic, mechanical, photocopy, recording or otherwise, without the prior written consent of the publisher.

International Standard Book Number: 0-88730-181-9

Library of Congress Catalog Card Number: 87-22476

Printed in the United States of America

Library of Congress Cataloging-in-Publication Data

Westin, Alan F.
    Managerial dilemmas: cases in social, legal, and technological change / Alan F. Westin and John D. Aram.
       p. cm.
    Includes bibliographies.
    ISBN 0-88730-181-9 (pbk.)
    1. Management—Decision making—Case studies.
    I. Aram, John D.,
1942– . II. Title.
HD30.23.W47 1988
658.4'03—dc19   87-22476
                                                            CIP

Contents

Acknowledgments   vii

Introduction   1

CASE 1 **Medical Ethics and Business Decisions**
Weighing Professional Dissent within the Enterprise   27

CASE 2 **Sexual Discrimination and the Corporate Culture**
How to Deal with Business "Affairs"   35

CASE 3 **Reproductive Risk and EEO**
Balancing Enterprise Liability and Women's Rights   50

CASE 4 **Employee Protest vs. Employee Loyalty**
How Much Dissent Should a Corporation Tolerate?   61

CASE 5 **Testing Employees for Substance Abuse**
Individual Privacy vs. Organizational Responsibility   74

CASE 6 **Due Process in the Non-Union Firm**
Dispute Resolution and Human Resource Policy   93

CASE 7 **"Big Brother" in the Automated Office**
Balancing Employee Morale and Increased Productivity   104

CASE 8 **Building a New Information System at *The Call***
How to Combine Expertise with User Participation   115

CASE 9 **Creating a Computerized Medical Surveillance System**
Improving Health but Risking Lawsuits   127

CASE 10 Choosing Corporate Strategies in State Legislative Campaigns
Responding to a Sweeping "Right-to-Know" Proposal   143

CASE 11 Selling High Technology to Anti-Democratic Regimes
Are There Ethics in the International Marketplace?   157

CASE 12 Business and Political Action Committees
How to Define the Corporate Interest   165

Student Response Form   182

Acknowledgments

This book could not have been written without the assistance of a grant by the Exxon Education Foundation to the Educational Fund for Individual Rights. The grant underwrote extensive field work—visits to companies to interview managers and employees; interviews with union leaders, interest-group experts, and government officials; and collection of extensive primary materials about the issues treated in these cases. Exxon Education Foundation believed that this field work would make possible far richer, more realistic, and educationally more powerful case presentations than the usual casebooks supply, and that such a casebook could serve both business school curricula and management development programs within the corporate community. The authors are grateful to the foundation for recognizing the importance of addressing a casebook to the topic of managing change, and also for recognizing the desirability of developing novel formats (our in-basket approach) in case presentations.

At the Educational Fund, a gifted cadre of people worked on the case study project between 1982 and 1986. These included Michael A. Baker, the fund's research director; Luceil D. Sullivan, executive director and chief administrator; Alfred Feliu, who prepared legal memos and worked on first drafts of several cases; Raymond E. Smith and Moira Griffin, who helped with research and writing on the first case we developed, as our pilot effort; Mark Kozlowski and Lee Rifatierre, research assistants; Michele Ochsner, whose studies of right-to-know laws were of great value in the case we prepared on that topic; and Hope M. Campbell, who provided secretarial services to Professor Westin throughout the project and during book preparation. Once we decided on our in-basket format, Daniel Schiller was of prime service in working with Alan Westin on first drafts of many of the memos.

At Case Western Reserve University, Fred Freer ably researched the "Business and Political Action Committees" case, and his initial drafting of the case considerably influenced the final product. We also thank Michael Murphey for assisting in library research that contributed to the casebook introduction.

The authors also express their appreciation to the Weatherhead School of Management of Case Western Reserve University for providing the majority of secretarial and office support for the development of the casebook and the instructor's guide. The school generously made resources available for the project. In particular, Virginia Bailey faithfully and diligently prepared the bulk of the cases and instructor's materials and typed the introductions. Peggy Little and Claire Svet assisted in the considerable typing task at crucial phases in the project. We appreciate their patience as well as their skills with memory typewriters, word processors, and computers. We extend our gratitude to each of them.

After the cases began to be available for classroom testing, we were fortunate to have a diverse group of professors in business schools use the cases in their courses. We wish to thank the following for this classroom testing and valuable feedback; many suggestions they sent us were incorporated into later drafts:

> Herman Gadon, College of Business Administration, San Diego State University
> David R. Hampton, College of Business Administration, San Diego State University
> Sanford M. Jacoby, Graduate School of Management, UCLA
> Rosabeth M. Kanter, Graduate School of Business, Harvard University
> James W. Kuhn, Graduate School of Business, Columbia University
> David Lewin, Graduate School of Business, Columbia University
> Angus G.S. MacLeod, Institute of Industrial Relations, UCLA
> John F. Mahon, School of Management, Boston University
> Kossuth M. Mitchell, Alice Lloyd College
> Kathryn S. Rogers, Pitzer College
> Paul Salipante, Jr., Weatherhead School of Management, Case Western Reserve University
> Donna Thompson, Graduate School of Management, Rutgers University

We also received wise counsel from David Ewing, the executive editor of *Harvard Business Review*, and we benefited from interaction with personnel executives of Prudential Insurance and managers of Ohio Bell Communications who used some of the cases in management training programs during 1986.

Finally, we express our appreciation to the staff at Ballinger Publishing Company who have assisted in the development of this book and its accompanying instructor's guide. Carol Franco, Marjorie Richman, and Barbara Roth have been thoroughly supportive partners in this endeavor.

Introduction

General Motors was the country's largest corporation at year's end in 1986, as it was twenty-five years earlier at the close of 1961.[1] Although in both years the firm's core business was transportation equipment, differences, not similarities, stand out in comparing GM in the late 1980s with its former self in the early 1960s. Representative of many American businesses in the late 1980s, General Motors faced a threatening competitive and technological environment, demanding and occasionally conflicting social expectations, and unprecedented judicial, regulatory, and legislative challenges. Like most companies, it depended more than ever before on the commitment and creativity of its employees.

In 1961 GM held nearly 50 percent of the market for passenger cars in the United States and earned 14.8 percent on invested capital; at the close of 1986 the firm was struggling to maintain a 40 percent share of new auto registrations and earned less than 10 percent on invested capital. Although the firm was the largest industrial corporation in 1986, it ranked 356 as measured by ten-year average annual growth in total investor return.

GM's weakened position in the late 1980s was sufficiently noteworthy to receive cover-story attention by *Business Week* magazine in spring 1987, under the heading "General Motors: What Went Wrong."[2] The company's perceived afflictions represented common themes for much of American industry: lack of distinctive product design, neglect of manufacturing quality, overreliance on internally manufactured parts, inability to integrate computer-based technology into operations successfully, organizational turmoil, and loss of sensitivity to customer tastes.

At the beginning of 1962, GM managers looked forward to holding a preeminent position among the world's industrial companies. By 1987, GM managers were taking instruction from members of Asian companies—Isuzu,

Shinjin, Toyota, and Fujitsu—that had become welcome lifelines to GM's future. In 1961 General Motors operated a subsidiary in South Africa; in 1986 the company, under duress, announced its divestment of this 3,000-employee operation.

Finding a terminal on his desk in the late 1980s, a manager from the early 1960s would surely be puzzled about what to do with a microcomputer. Moreover, chances are much greater today that this manager would be a "she," not a "he." Earlier managers would also be required to master entirely new concepts of manufacturing, and they would find the extent of employee participation unfamiliar, and probaby uncomfortable. New worries, such as fuel economy, auto safety, environmental pollution, worker safety, consumer demands, nondiscriminatory employment practices, and organizational due process, would absorb much of these 1980s managers' time. Would 1960s managers even be able to comprehend, let alone meet, these demands?

A quarter of a century is an instructive yardstick by which to obtain a perspective on the organizational, social, legal, and technological forces facing today's managers. The early 1960s may ultimately be judged a period that separates a society defined by post–World War II social attitudes, industrial technologies, and institutional and international arrangements from a society transformed by dramatic shifts in technology, international competition, social consciousness, and public policy.

The following discussion scans several major changes that, having roots in far earlier times, appear to have come to full bloom during the past twenty to thirty years. The central objective of this introduction is to identify the implications of these changes for today's managers and to establish a context in which to study and discuss the cases that follow.

The Early 1960s: A Watershed for Society

A variety of social changes can be observed between the early 1960s and the mid-1980s. For example, while the U.S. population increased about 1 percent per year, from 180 million to 239 million people, the work force increased about 2.5 percent per year.[3] This means that the work force was considerably older in the mid-1980s than it was in 1960. It also means that a number of nontraditional workers, such as women, had entered the labor market. Specifically, the number of male employees increased less than 40 percent between 1980 and 1985, while the number of women employees increased nearly 120 percent.[4]

The year 1960 marked the beginning of a generally prosperous period for the U.S. economy. Median family income, as measured in 1984 constant dollars, increased from $19,711 to $26,433, about 34 percent, over the twenty-five years.[5] However, not all groups shared equally in this prosperity. For example, the absolute differences in the median black family income compared with the median white family income became greater. In 1960 the absolute difference between the two groups was about $9,000.[6] In 1984 median white family income was $27,686; median black family income was $15,432.

Introduction

In addition, even though the U.S. was more prosperous in the mid-1980s than it was in 1960, impoverished countries became more, not less, dominant over the world as a whole. For example, in 1960 the "less developed" countries represented less than 70 percent of the world's three billion people. With a population growth rate of about 3 percent per year, and too often having failing economies, the less developed countries in 1985 represented a full 75 percent of the world's population.[7]

These trends indicate a few ways in which the early 1960s serve as a point of reference for the current social and economic environment of U.S. firms. Numerous other economic, organizational, technological, and social changes affect today's managers. The following sections outline salient features of these changes, using the past quarter of a century as a convenient reference point.

The economic and competitive status of U.S. industry. The United States emerged from the Second World War as the unchallenged industrial power of the world. The transition to a peacetime economy was relatively smooth in the United States, and the postwar years witnessed rapid development of the American economy.

By the early 1960s, U.S. industry, at least in retrospect, began to experience major competitive difficulties. While productivity (output per hour) in the nonfarm business sector increased an average of 2.9 percent per year between 1947 and 1960, this same measure increased an average of only 1.3 percent per year between 1960 and 1986.[8] At the same time, the unit cost of labor, which had increased an average of only 1.1 percent in the earlier period, shot up more than 4.5 percent annually in the later period.[9]

America's economic status after 1960 appears inversely related to the success of the rebuilt economies of Europe and Japan. Since the early 1960s, Japan, Germany, and in some cases England have each outperformed the United States. For example, England increased output per hour in manufacturing 1.3 times faster than the United States between 1960 and 1983, and West Germany and Japan increased output per hour in manufacturing 1.9 and 3.5 times faster, respectively, than the U.S. over this span.[10]

These competitive issues would naturally be expressed in changing relationships among the developed nations in international trade. For example, in 1965, 22 percent of all exports of machinery and transportation equipment originated from the United States and only 5.7 percent originated from Japan. By the end of 1983, however, the United States share of world trade in these products had fallen nearly 25 percent and Japan's share had risen 185 percent.[11] By the end of 1986, the United States was running an annual trade deficit of more than $151 billion in manufactured goods.[12]

Growth in service industries constitutes another significant change in the structure of the economy over the past quarter of a century. Employment in the goods-producing sector of the economy increased at an annual rate of .4 percent between 1959 and 1984, while the annual rate of increase for the private service sector was 2.6 percent over these same years.[13] By the end of 1984, service businesses employed more than twice as many persons as goods-producing industries did.

A measure of significant change in the United States since 1960 is the fact that by 1985, public sector employment had increased from 8 million to almost 16 million, and employment in the goods-producing sector of the economy had increased from about 27 million to less than 30 million."[14] Of course, growth in public sector employment manifests the fact of increased government spending over this period. Between 1960 and 1985, federal government spending increased nearly ten and a half times.[15] In 1960 the federal government ran a slight surplus of $300 million and the national debt was about $291 billion. In 1985 the national government ran a deficit of more than $222 billion and the national debt had risen to $1,841 billion.[16] Surely a new set of assumptions and realities applies.

Employee-employer relationships and the internal environment of the firm.[17] The late fifties and the early sixties represent the height of what could, in retrospect, be called the "right to manage," a period of employer prerogative in which supervisory and administrative decisions were made relatively free from external controls. With the exception of union members, social mores gave uninhibited support to the use of managerial discretion in personnel matters. Private sector employees had no right of tenure in their jobs; they could be fired whenever it was considered necessary by the employer. In addition, a positive duty of loyalty to the firm—its operating rules and procedures, its reputation, and its commercial opportunities—was held by the courts to be an obligation of employment. Standards of personal morality, life-style, dress and grooming, associational activities, and political ideology could be (and were) set by companies as conditions for hiring and advancement. Discrimination in hiring and promotion on the bases of race, religion, sex, age, and handicap was not forbidden in these decades, and it was widely practiced. The law did not protect rights of expression for corporate employees on the job, or even off the job. Procedurally, there was no legal obligation for an employer to provide internal complaint and appeal mechanisms to deal with employee grievances. If employees were dismissed from their jobs and appealed to the courts, the general rule of common law was that no judicial relief was available—a private employer was free to fire an employee at will, for any reason, or even for no reason at all.

Only in rare circumstances could a woman be seen within management ranks; not only was presence of "the wife" socially required, but she was also expected to enhance the social position of the aspiring corporate manager, ensure family stability, and raise the children. One chapter of Vance Packard's popular book about corporate life, *The Pyramid Climbers*, published in 1962, was titled, "The Wife: Distraction, Detraction, or Asset?"[18]

The presence of a black person as manager was truly an exceptional occurrence among corporations. Moreover, little thought or consideration was given to the implications of this fact for the character and values of society.

Given impetus by the legitimization of collective bargaining with the enactment of the Wagner Act in 1935, the union movement may have reached a high point in the organization of the work force at the end of World War II, when 36 percent of manufacturing employees were covered by collective-bargaining agreements. By 1960, however, the Taft-Hartley Act, which outlawed certain union practices, such as secondary boycotts and sympathy and jurisdictional strikes, and the Landrum-Griffin Act, which strengthened federal regulation of internal union affairs, substantially reequilibrated the balance of power between union and management and set standards for the management of unions as organizations. Yet in 1960 unions continued to represent more than 32 percent of the private, nonagricultural work force[19] and established the major, and perhaps the only significant, institutional influence on the employment relationship.

The revolutionary impact of modern technology. The first fully electronic digital computer was built in 1946 by two engineers at the University of Pennsylvania using technology developed during World War II to break German military code. The first commercially available computer, called the UNIVAC 1, was produced in 1951, although few persons imagined the impact of this machine on industry and society. Observers forecast a very limited market for the new technology, estimating that a maximum of "ten" or a dozen very large corporations would be able to take profitable advantage of the computer."[20]

Subsequent developments in the electronics industry are well-known—the invention of the transistor, the development of solid-state technology, the miniaturization of components, and the advent of the microprocessor have stimulated the growth of an enormous industry. By the mid-1980s, the worldwide computing equipment industry alone accounted for more than $200 billion in revenues, and the software and services industry was estimated to be generating an additional $12 billion in sales.[21] The explosive growth of those industries from virtually zero in 1960 to these levels in 1984 constitutes a major economic phenomenon.

The impact of the computer revolution on existing businesses has been profound. Electronic technology in the form of automation has been used in manufacturing, publishing, communications, education, banking, and entertainment, and most services have been transformed as a result of electronic information processing. Decision making at all levels, in virtually all companies, and the shape of the corporation itself have been affected. It is not uncommon for a major company to be utilizing five, ten, or even twenty thousand microprocessors,[22] and for employees using personal computers to spend four or more hours a day working with this equipment.[23] Electronics has dramatically transformed the telecommunications industry, making possible more than 190 billion telephone calls per year and making new communication systems, such as teleconferencing, viable.[24] Advances in telecommunications are also having a significant impact on consumer services, finance, and the globalization of marketing.

Equally dramatic, yet still at an earlier stage, is the advent of genetic engineering and the biotechnology industry. As recently as 1962, James Watson and Francis Crick received a Nobel prize for building a model of the molecular structure of DNA, the substance that transmits genetic information between generations. Today, a large and promising industry is built on knowledge and technology deriving from this discovery—an industry that may result in cures for diseases and illnesses, such as cancer, that plague humankind. By the late 1980s, more than 400 U.S. firms were engaged in industrial biotechnology. By the early 1990s biotechnology products are expected to generate several billion dollars in revenues, with estimates ranging from $15 to $40 billion by the year 2000.[25] Genetic engineering technology may be capable of dramatically increasing the efficiency of food production, inventing new materials, and perhaps even creating new life forms. Just as the electronics revolution has fundamentally altered social consciousness, redefined management tasks, and posed new ethical dilemmas for society, biotechnology may similarly affect us, as managers and as citizens.

The visible impacts of information and genetic technologies in the past quarter-century should not be allowed to mask other significant technological developments that also affect corporations and managers. Major technological changes are also taking place in metallurgy and ceramics, in macromolecular chemistry, in bioengineering, and in the medical sciences, and discoveries often are drawn from more than one of these and other disciplines. In may respects, the rate of technological change today compared with that of a quarter of a century ago creates major threats and opportunities for the private enterprise sector and its managers. At the very least, technology presents unprecedented uncertainties and challenges to managers.

Changes in values and social ideals. The early 1960s mark a transition period in social attitudes and relationships. The decade of the 1950s may have been the high point of WASP supremacy and of corporatism. The key question of business managers, as reported by Vance Packard in his analysis of corporate life, was, "How does one please the bitch-goddess Success?" And Packard's analysis provided the answer in four rules: Be Dedicated. Be Loyal. Be Adaptable. Be Quietly Deferential.[26]

The late 1950s represented an age of people seeking social approval—wear a white collar and a gray flannel suit. Earlier that decade, David Riesman coined a phrase, "the other-directed person," to describe this aspect of the changing American character.[27] Riesman's book was aptly titled *The Lonely Crowd*. Another influential book of the 1950s, *The Organization Man*, by William H. Whyte, confirmed many of Riesman's themes for the business sector: the decline of the Protestant ethic, belongingness, togetherness, and "a generation of bureaucrats."[28]

By the turn of the decade, however, the seeds of social change were stirring. Rock 'n' roll music was incorporating itself into the American consciousness. In 1960, America elected its first Catholic President, John F. Kennedy. Disenchantment with the society's failure to comply with the Supreme Court's rejection of "separate but equal" facilities for blacks was gathering momentum and would lead to the historic Civil Rights Act of 1964. By the end of the 1960s, a new phrase, "the counterculture," celebrated the rise of a youth-oriented, antiestablishment individualism. Alternative life-styles were "in." Between 1960 and 1982, the number of marriages per 1,000 persons increased 1.25 times; the divorce rate per 1,000 persons, however, increased almost 2.3 times between those same years.[29]

Changes in social attitudes during the 1960s affected awareness of poverty and its effects in America. The civil rights movement increased in influence, and the women's movement became a social force of its own. The use of hallucinogenic drugs emerged as a significant issue for society, and a costly, unpopular, and unwinnable war in Vietnam opened latent cleavages among generations, races, and social classes in America.

In the early 1970s, the Watergate episode demonstrated to an alarmed public the extent to which secret corporate financial contributions were corrupting the American governmental and political process, as well as supporting a system of bribes and corruption in overseas business activities. By the late 1980s, massive insider-trading scandals on Wall Street and the unfolding of secretive and possibly illegal government actions in the Iran-Contra affair stimulated new questions about the role of ethics in private and public organizations.

These social currents were bound to alter the character of employee roles and relationships in corporations. As a formula for success, quiet deference took a quick exit in making way for a new spirit of independence and liberation. Corporate loyalties, traditional work roles, and rewards for longevity weakened. The concept of employee rights began to emerge from a new national value placed on personal growth, feelings, and individuality. Affected by changing social values and the shift toward services, union membership declined, and employees came to place individual respect above institutional loyalty. Participation, psychological ownership, decentralization, and entrepreneurship became key concepts for managers.

To these developments must be added the new sense of activism and personal moral commitment that grew out of the civil rights, antiwar, and student protest movements. Although this is often portrayed as a mood that affected youth, many middle-aged and older Americans also came to feel a new sense of personal responsibility for confronting unlawful or illegitimate authority with a moral protest. Many people who might not be willing to take such actions themselves came to admire those who did; they recognized such commitment as the vital force that brought about important changes in law and public policy that, in time, won overwhelming public approval. As millions of young people who had been activists or had supported personal activism during their college years moved into corporate employment in the 1970s, a major source of whistle-blowing also moved inside the corporate gates.

Along with personal activism came an increased concern among many professional employees in the corporation that they should not be forced to perform services that would clearly violate ethical standards. Such professionals now make up a major segment of the corporate work force—engineers, chemists, computer experts, accountants and auditors, lawyers, physicians, and behavioral scientists. Though their professional codes traditionally spelled out the ethical principles to be observed, the issue of whether professionals should refuse to continue working for an employer that directs a violation of professional ethics became a matter of growing concern among professionals and professional societies in the 1960s and 1970s.

Another trend has been the development of considerable ferment in the corporate world since the late 1970s in response to new demands for employee rights outside the unionized sector. Many companies with reputations for anticipative management have voluntarily instituted new policies that recognize rights of employee privacy and have adopted new procedures for hearing employee complaints. Some of these companies have also instituted codes of voluntary self-disclosure to make more information about corporate actions and policies available to employees and the general public, and have developed high-participation techniques of making decisions or organizing work.

Public policy and the firm: new rules. Government policy toward business had been relatively quiet since the decade of the 1930s, when unprecedented legislation in collective bargaining, farm price supports, securities regulation, minimum wage, and social security were passed. By 1960, these policies were a well-accepted part of the business system, wage and price controls from World War II were long discarded, and the airline, electrical power, petroleum, communications, and transportation industries were all regulated under one or another form of government control. The focus of public policy was along traditional lines, such as antitrust, as illustrated by the fact that a 1962 Supreme Court decision forcing General Motors to divest its holdings in the Du Pont Corporation represented significant government action at the time.[30]

The era before the 1960s was a time when American society and the law supported wide corporate autonomy. It was assumed that most corporate products were safe for consumers; that most substances used in production did not pose serious risks to workers or to the health of people living in nearby communities; that the environment was generally capable of absorbing industrial wastes; and that the manner in which corporations used their funds to influence the political process or to secure contracts abroad was not an important issue for public concern. When occasional problems arose, Americans looked to such remedies as lawsuits for individual damages, limited government regulation, and the "discipline of the market" to provide whatever control was needed over dangerous or improper actions. In this ethos, there was usually little public support or legal protection for an employee who blew the whistle on company practices that were alleged to be harming the public.

In 1962 Rachel Carson published her book *Silent Spring*, credited by many as having inspired modern environmentalism in the United States.[31] Carson convincingly demonstrated the growing strength and variety of lethal chemicals used by American industry since the Second World War, and she illustrated their devastating impact on animal and human life. In 1965 Ralph Nader wrote

Introduction

Unsafe at Any Speed,[32] an attack on GM's rear-engine Corvair, drawing public attention not only to that car model but to auto safety in general, and giving birth to the organized consumer movement typified by the battery of Nader public interest organizations.

The public awareness that corporate behavior could have serious effects on public health, the environment, and the political process, and the new demands for protection of equality and privacy rights in employment, led to major legal and organizational changes in the late 1960s and 1970s. First, federal and state laws were enacted to provide new rules for employee protection. Equal employment opportunity laws forbade discrimination in hiring, promotion, or firing on the basis of race, religion, nationality, sex, age, and physical handicap. Rules for occupational safety and health at the workplace were enacted, and new legislation protected employee pension rights, allowed corporate employees to see the contents of their own personnel records, and forbade employers to interfere with rights of political activity and association by their employees.

Many of these laws contained a provision forbidding an employer to punish an employee for claiming the rights protected under the legislation, or for reporting violations to public authorities. Complaints of employer reprisal could be filed with those government agencies set up to enforce the legislation (such as the Equal Employment Opportunity Commission, the Occupational Safety and Health Administration, and the Department of Labor). An employer found to have taken reprisals could be ordered to reinstate the employee or to pay damages for improper termination.

Second, a wave of similar federal and state legislation was enacted to protect consumers and the public from improper business practices. This legislation included the Truth in Lending laws, the Fair Credit Reporting Act and the Equal Credit Opportunity Act, the Environmental Protection Act, and the Foreign Corrupt Practices Act (pertaining to illegal overseas payments). In addition, there were amendments and new laws dealing with protection of the public in particular industries, such as registration of new drugs by the pharmaceutical industry or regulation of operations by the nuclear power industry. Many of these public protection laws also contained anti-reprisal provisions that forbade the firing of an employee for reporting violations to public authorities. The anti-reprisal rights were not well-known at first, and it took time for employees to learn how to use them. But by the end of the 1970s, they began to change supervisory relationships.

The cost of what many called the greatest wave of government regulation of business in American history was not lost on business. One economist, sympathetic to the plight of business under weighty new regulations, adopted a slogan from social protesters of the times and issued a call to "Free the Fortune Five Hundred."[33]

Common law legal doctrines were also changed in accordance with a more critical and demanding attitude toward business responsibilities and liabilities. The doctrine of negligence had generally been used by state courts to place the blame for product-related injuries and assess compensation for victims. Historically, injured parties had to show that the manufacturer was negligent in the production of a defective product. Later, due to the burden placed on the plaintiff by complex products and complicated distribution channels, the burden of proof shifted to the manufacturer to show that it was not negligent, that is, that it had taken reasonable actions to avoid injury to the consumer. By the mid-1960s, however, a new doctrine—that of strict liability—was being argued by injured parties and was being accepted by many judges and juries. Strict liability is a no-fault doctrine: if an injury occurred in the ordinary, intended use of the product, the manufacturer is in the best position to pay compensation and spread the cost of the injury among all users of the product. Thus, demonstration of due care and the absence of negligence became irrelevant—if someone is injured, the manufacturer must pay.

Such changes in common law not only have become anchored in product injury cases in the past quarter-century, but also represent a perspective or value that affirms the rights of the individual vis-à-vis corporations in other contexts. Expanded payments for worker injuries, as in the asbestos rulings in the 1980s, have been supported by the courts. Community members may be protected by concepts akin to strict liability when residents are harmed by exposure to hazardous chemicals or radiation. Some state and federal courts in the 1970s began to carve out a noteworthy exception to the common law doctrine supporting management's power to fire an employee at will. The courts had traditionally upheld management's absolute right to dismiss because of several judgments: the private employer's right to organize and administer the work force as it saw best to achieve efficiency; the concept that no one has a right of tenure in private employment unless there is an express contract; and the difficulties that judges would have in passing on disputed personnel decisions. Beginning in the mid-1970s, courts began to hold that, where important "public policy considerations" would be offended by refusing to question the employer's action, the courts could intervene.

Finally, the deregulation of significant sectors of domestic enterprise activity in the 1970s and 1980s has itself intensified competitive forces and complicated management decision-making for a number of companies. Ideologically and practically, it was concluded in the 1970s that the government could play a dysfunctional role for society in controlling entry and pricing competition in the airline-transport, trucking, railway, and long-distance telephone industries, and to some extent in the broadcast-communications and banking industries. In the late 1950s, those industries were secure in the accommodating hand of government regulation; by the mid-1980s, technological changes, inflation, and the high cost of governmnent involvement combined to free them of their regulatory bindings and set them loose in a demanding (and unsettling) world of market competition.

Implications for Managing in the Coming Decade

This brief survey of changes affecting the business sector implies that managerial work has simultaneously become more multifaceted and more risky over the past quarter of a century. The chance of failure is heightened by a fiercely competitive environment. Managers must grapple with the prospect of investing millions of dollars in uncertain technologies, and they must consider the risks to the company of failing to make such commitments. Creating a strong organizational culture in a diverse, individualistic society is no minor challenge. External groups, including the public sector, increasingly limit managerial discretion and often impose constraints on management decisions. The tasks of management have dramatically increased in complexity and challenge. Looking to the next decade, what competencies are required for managers?

Managing internal resources. Managers today are under enormous pressure to increase their firm's economic performance. Harsh international competition continues to erode many companies' markets. Existing competitors stand ready to capture market share from a company that falters, while the threat of new competition is ever present in most industries. Although major investments in new technologies, products, services, and markets are imperative today, they usually cannot be expected to yield returns for several years. Competition can be assumed to be unrelenting; in a real sense, managers must face tomorrow's competition today.

Simultaneously, managers of publicly held firms are forced to look over their shoulder on a short-term basis. Financial markets are unforgiving of weaknesses in quarterly earnings, and corporate raiders are poised to make a takeover bid for companies suffering depressed market values. Managers may experience the financial markets as expecting a promise of significant future earnings increases without giving companies the leeway to make the necessary short-term investments. Thus, managers of publicly held companies find no respite from extreme performance pressures. There are no hiding places.

What can managers do in the face of unrelenting competition and pressure for performance? Surely there is a premium on efficiency, and many functions of the firm are undergoing close examination for efficiency improvements. Improving the management of financial assets, for example, involves full sensitivity to the time value of money—maximizing the earnings of assets such as cash, and minimizing the cost of liabilities such as accounts payable. Increasingly, companies restructure financially, or participate in financial options and futures markets, in order to increase returns on financial assets and manage financial risk.

Under pressure, many sources of efficiency gains become more visible. Slack, for example, must be taken out of the management of physical assets; inventories need to be reduced, preferably to zero; and distribution and transportation systems require rationalization in order to cut the costs of carrying extra materials and moving goods unnecessarily. Also, companies use less expensive and more flexible part-time or temporary staff to accommodate swings in work-force demand. Financial management, physical resource management, information management, and personnel management are all impacted by the demand for efficiency. No corporate responsibility is left unexamined and unaffected by the response to a new economic and competitive environment.

Yet pursuing some of these economic policies will generate conflicts over many recently installed expectations about employee rights, job security, health and safety standards, quality of work life, and other important interests. And efficiency improvements alone are unlikely to allow corporations to meet the economic and competitive challenges facing them. Solutions to the dilemmas of American managers lie not only in doing the same things better or more efficiently. Successfully encountering the economic challenges of the modern world demands the ability to make step-function increases in performance, to achieve *radical* gains in productivity, quality, cost savings, and service. Significant improvements are unlikely to come from incremental efficiency improvements alone; managers may need to adopt a wholly new framework in which the implicit and often self-limiting "either/or" framework of many organizations is discarded.

In their best-selling book *In Search of Excellence*, Thomas Peters and Robert Waterman describe the final characteristic of excellent companies as "Simultaneous Loose-Tight Properties."[34] In some way, the high-performing companies included in that study—IBM, McDonald's, 3M, Marriott, and others—are able to combine a loose, entrepreneurial organization with tight control. Whereas many companies vacillate between various forms of centralization and decentralization, of control and autonomy, the excellent companies of the Peters and Waterman study were able to integrate both of these qualities at a high level of performance. The ability to combine apparent loose-tight qualities in practical operations expressed the essence of many other characteristics of excellent management.

Conventional wisdom among operations managers says that increases in manufacturing quality necessarily mean additional manufacturing cost; that is, product quality is improved only as higher-quality, hence more expensive, components or raw materials are used, as more costly machinery is purchased, and as more control operations are implemented. A legacy of the past in manufacturing management is that a linear relationship exists between cost and quality—increments of quality come only with higher manufacturing costs.

Yet some companies' experience with advanced manufacturing technologies and employee participation programs demonstrates the exact opposite—that it is possible to achieve major increases in product quality while actually *lowering* unit costs of production. In other words, a different set of assumptions about the creativity and resources of the work force, together with a reordering of organizational relationships and technology in the workplace, may turn former beliefs about costs and quality upside down. How often does a firm remain stuck in a low-performance routine by defining its management problem as "either/or," when defining the situation as "yes/and" would open new possibilities of achievement?

Organizations are prone to many "either/or" situations: The problem is seen as either the supervisor's or the subordinate's fault. Responsibility for product development will go either to Marketing or to R&D. Either the Vice President of Human Resources is right or the Legal Counsel is right. Either the union will give wage concessions or we will close the plant. Subordinates will either do what I say or wish that they had.

Introduction

"Either/or" organizational cultures limit alternatives, generate defensiveness and resistance to change, and expend precious organizational resources on debilitating struggles. Companies bogged down in "either/or" dynamics may face an increasingly difficult time in achieving the dramatic breakthroughs in innovation, efficiency, and effectiveness that are necessary to compete in the global economic environment of the present and the future.

Managing external relationships. Social changes of the past quarter-century mean that managers are increasingly expected, ethically and often legally, to respond to shareholders; to the communities in which their firm operates, to local and national media; and to a host of issue-oriented groups claiming to represent the public in consumer, environmental, health and safety, and employee concerns. Perhaps management at one time was largely a private, internal affair (though this portrait is sometimes overdrawn), but in any case, the past several decades demonstrate that a large number of constituencies have discovered the will and the ability to affect corporate actions. Managers in the late 1980s and early 1990s must realistically assume that a variety of "partners" will expect to participate—directly or indirectly—in corporate decision-making.

The term "stakeholders" is frequently used in reference to a wide range of parties claiming an interest in corporate decisions. Official stakeholders such as investors, as well as self-organized parties such as environmentalists, experience an important "stake" in corporate decisions. Greater social awareness of the consequences of corporate actions, coupled with the ability of interested groups to mobilize personal energies and public opinion, heightens managers' sensitivity to the values inherent in their decisions.

Several results flow from society's high level of consciousness, organization, and awareness of values. First, managers need to realize that, because our perceptions are real to us, perceptions and beliefs have real consequences. Stakeholder groups may or may not have a broad understanding of the firm and its multiple obligations. And they may or may not have complete or accurate information about the firm's actions and their effects. Yet perceptions lead to actions, and managers must take perceptions and the values that shape them as an influential context for their decisions.

Second, managers realize that to cope with and hopefully integrate a variety of internal and external pressures begins with a strong identity and sense of direction. Effective corporations must work hard to clarify their goals and objectives. Defining long-term goals and objectives is an unending and nearly continuous process that creates a framework necessary for solving short-term problems. Preparing a statement of goals and objectives is relatively easy; more difficult is the task of developing statements that are internally consistent, serve as criteria for planning and action, and are widely shared in a large organization.

A close look often reveals that management aspires to equally desirable yet conflicting values. Some officials may believe that product innovation is the key to corporate success, while others may argue convincingly that profitability is tied to being the industry's low-cost producer. Great advantages may be expected to derive from a major automation effort, but maintaining the stability of established organizational arrangements and individual roles is also important. Managers aspire to be socially responsive and take a high moral ground on social issues, but immediate financial and economic pressures absorb needed attention, time, and investments.

Objectives can never be assumed to be simple or stable; they require reconciliation of inherently conflicting impulses in the firm. Objectives also need to be evaluated based on new experience and learning, and they need to be affirmed or modified based on perceptions about a changing environment.

Finally, managers must be competent in establishing authentic communication, often within a setting of confusion, ambiguity, and conflict. All too frequently, ego-involvement, blaming, and stereotyping create damaging breakdowns in relationships among parties. Managers and leaders of other organizations can easily allow their immediate conflicting priorities to override their long-run common interests in the survival and prosperity of the firm. Personalities, rather than rational pursuit of long-term interests, readily dominate the situation.

Competent managers in a complex world treat information as preliminary and perceptions as provisional "truths." They also view negotiation as an important form of cooperation, and are able to be strong advocates without becoming adversaries. A confluence of interests—a common ground—that is needed for all parties to continue to be invested in the relationship is rarely possible without the patience and skill to create trusting communication.

In short, effective managers in the new social milieu recognize that their work inherently involves multiple and often conflicting wants. Managers of the future will need to recognize the constant necessity of conceptualizing organizational direction and mission; they will accept the legitimacy of diverse stakeholders; they will appreciate that, as managers, they enact institutional values; and they will seek to achieve mutual benefits in relationships among interested parties.

The Purpose and Design of the Cases

This book presents decision-making cases dealing with many of the issues, value dilemmas, and situations confronting contemporary managers. Twelve cases have been prepared to reflect modern organizational realities pertaining to multiple and conflicting values, the impact of advanced technologies, pressures from internal and external interest groups, legal and regulatory uncertainties, public relations considerations, and the definition of corporate goals in conflicted and ambiguous settings. The following questions, a sampling of issues arising in the cases, illustrate tasks demanding managers' attention today:

- How should the firm respond when a professional employee refuses to work on a company project because of that person's deeply felt but somewhat speculative concerns about product safety?
- Can a plant manager protect the health of the fetus of a female employee working near toxic chemicals, without risking employment discrimination against the woman?

Introduction

- Is it possible to avoid intrusive and dehumanizing testing procedures yet ensure that an organization is controlling drug and alcohol abuse?
- Can a company develop employee grievance procedures that both employees and supervisors feel are fair?
- What can management do to see that office automation produces both greater work efficiency and higher employee morale?
- To what extent should managers take ethical issues into account in deciding whether to sell computers used for internal security purposes to antidemocratic foreign governments?
- What should be the extent and type of corporate participation in U.S. congressional elections?

Business discussion cases allow a realistic portrayal of many realities of organizational life. Therefore, the case method, rather than, for example, issue-oriented essays, has been chosen as the medium for instruction of these topics. In developing this book, the authors have attempted to bring the issues alive and to present them as realistic slices of corporate life. The body of each case in the collection consists of a series of memos from line and staff managers to a central decision maker in a particular organization. This memo, or in-basket, format allows the expression of personal viewpoints, values, reasonings, and recommendations often encountered in corporations. Issues arise and are considered within organizations only through the opinions and experiences of people, and problems are resolved only through personal action. For these reasons, the cases in this volume focus on organizational members in particular settings and these persons' feelings about key organizational issues.

In line with the objective of maintaining a realistic focus, virtually all of the case material has been developed from actual events directly or indirectly known to the authors, or taken from publicly available documents such as court rulings. Although the identities of individual case participants are fictitious, and in most instances organizational situations are invented, the substantive issues of the cases are in every sense real. The goal of each presentation is to give readers an opportunity to deal with real-world issues and actual managerial dilemmas.

Three types of learning can result from studying and working with these cases. First, the issues orientation of these cases exposes readers to a considerable range of substantive questions faced by contemporary organizations. Understanding can be gained about the employment-at-will doctrine, the intangible but influential ways that organizational norms about sex and race affect corporate decision-making, work measurement as a management tool, the health risks of industrial chemicals, the technology of human engineering, worker and community right-to-know laws, the rules of elections and campaign financing, and a number of other subjects. Tomorrow's managers will require familiarity with the facets and subtleties of many of the issues around which these cases are constructed.

Second, the cases provide readers with opportunities to develop diagnostic skills, an important ability in managerial problem-solving. Different managers vary in their formulation of the central problem in these settings. Goals emphasized by one manager contradict the objectives of another manager, information and viewpoints presented appear incompatible, and priorities stressed by some persons vary substantially from the preferences of others.

Given the range and divergence of opinions in each of the situations, case discussants, acting in the role of the central decision maker, are led to ask: What is the factual basis for each viewpoint presented? What personal stake might each manager have in the outcome of the issue? What is the validity of each proponent's reasoning about causal relationships in the situations? What goals or values should be sought by the firm in the setting, and how do these bear upon each viewpoint?

As readers, you will assume a variety of roles in these cases. First and foremost, the cases emphasize a comprehensive management perspective. A general manager—a company president or CEO, a chief operating officer, or a plant manager—is the focal decision maker in eight of the twelve situations. Persons holding these roles carry responsibility for the organization in a broad sense, and their decisions often reflect the interaction of a number of specialized functions, disciplines, and viewpoints in the firm. General managers are responsible for the whole, and their actions frequently need to integrate opposing factions. In addition, Senior Vice Presidents for Operations are the focal decision makers in two cases, and a Vice President of Staff Relations assumes this role in another case. In one case, a younger manager, assuming an "assistant-to" role, must propose a course of action to his supervisor on a controversial issue and must synthesize conflicting recommendations from older, more experienced, and more powerful managers in the firm.

A variety of other positions and roles present themselves across the twelve cases: manufacturing managers, human resource staff members, legal advisers, medical experts, labor relations managers, information systems specialists, market managers, public affairs specialists, and others. All bring information, recommendations, and personal biases to the issues facing their companies. The cases allow readers to assess the viewpoints and to evaluate the performance of these "players." How well prepared are they? Do they understate or overstate the case at hand from a functional perspective? As an occupant of a particular role, how would you rewrite its script?

Third, the cases present opportunities to consider the management of change processes in large corporations. Each of the situations calls for organizational development—for the company to be more effective, more efficient, more adaptive, and more focused after the decision than it was before the decision. Turning conflict-ridden and "either/or" situations into development opportunities is a vital managerial skill, one that calls for an understanding of how to create, nurture, and reinforce constructive processes in the organization.

Introduction

These cases stress managerial process skills—authentic communication, clarification of values and assumptions, management team development, conflict resolution, expansion of the range of identified solutions, and the building of commitment throughout the organization.

The cases included in this book have been designed to put readers, at least vicariously, in situations that correspond to the realities of managing in the next decade. The memo format intends to encourage personal involvement and identification with the different "players" in the cases, to present opportunities that provide for the acquisition of substantive knowledge, and to permit practice of the diagnostic and process skills required of effective managers.

Perspectives on Decision Making for the New Management Context

This book is about decision making, even though it presents little theory and offers few research results on managerial decisions. Numerous theories of management and a plethora of organizational research are available to managers today; an important task is to translate current knowledge into practical guidelines for dealing with the social, economic, legal, and technological forces facing corporations. In addition to the development of new knowledge, educators need to stress the enhancement of decision-making abilities for business leaders.

Consequently, this book is about the *practice* of management decision-making. The chief aim is to present factually based situations that illustrate current organizational dilemmas; the book does not seek to refine existing theories about decision making or attempt to develop entirely new approaches to understanding management. Rather than seeking to generate new theories, this book stresses application of knowledge for managerial analysis and decision-making skills.

In order to further our goal of managing the dilemmas of contemporary corporations, it behooves us, as readers and authors, to recall several influential perspectives on decision making from the management literature. From the viewpoint of the cases in this volume, three influential writings have particular relevance for understanding the decision-making process.

First, James G. March and Herbert A. Simon, in their classic book *Organizations* (1958), describe the role of perception and personal identification in decision making.[35] Their concepts provide a context for assessing how decision makers come to see particular information as relevant yet fail to attend to other important information.

Second, the popular notion of "muddling through," first articulated by political scientist Charles E. Lindbloom in 1959, adds an organizational perspective to understanding the process of policy development and administrative problem-solving.[36]

Third, insight into how managers' assumptions about people affect subordinates' work outcomes, articulated by Douglas McGregor in his renowned book *The Human Side of Enterprise* (1960), opens the management process to the possibility of creating more committed, and potentially more productive, organizations.[37]

As noted, each of these seminal works was first published between 1958 and 1960, years that were identified earlier in this introduction as a watershed in modern American social and economic history. Perhaps the authors of these writings perceived the growing complexity of society and sensed its incipient changes. Judging from the acknowledged standing of their concepts over time and the significant amount of additional writing each inspired, these authors made valuable contributions to our storehouse of basic concepts for dealing with contemporary organizational issues. These works, each a landmark in the field of management thinking, continue to serve managers well, and their concepts are worth summarizing for readers of this casebook.

Why organizational members identify with subunit goals. March and Simon present organizations as decision-making systems that seek rationality but inevitably suffer from the cognitive limitations of the human's ability to collect, retain, recall, and process information. Because no single person can comprehend all information necessary to administer a complex organization, overall tasks need to be divided and subdivided into smaller units. Basic organizational structures — departments, positions, and other subunits — derive from the fundamental characteristics and limitations of human problem-solving.

Biases are invariably introduced into organizational decision-making by the differentiation of goals and tasks at the departmental, or subunit, level. Persons in subunits of the organization simplify their complex world by selectively paying attention to certain information and criteria for decisions and ignoring other information and criteria. Moreover, selective attention occurs by subunit and subgoal *regardless of the impact on other subgoals or the goals of the larger organization*.

March and Simon's observations have important implications for understanding the decision-making situations in the cases developed for this volume. Each case illustrates the division of a complex organization into functional specialties and departments. Memos prepared for the key decision maker in each case reflect the selective perceptions and subgoal identifications described in decision theory. The cases confront managers with the kinds of results of the organizing process that are so aptly described by March and Simon.

Moreover, organizational members' perceptions and frames of reference persevere. People see things that are consistent with their established conceptions, and they filter out inconsistent information. Irrefutable, disconfirming information is "rationalized" away. "The frame of reference," state March and Simon, "serves just as much to validate perceptions as the perceptions do to validate the frame of reference."[38]

Existing perceptions are reinforced by means other than the psychology of individuals. In-group communication is affected by established frames of reference, thereby selecting, screening, and filtering internal organizational information. In addition, the division of labor inherent in complex organizations influences the dimensions of the environment to which members attend, and it affects the information they receive. Individual predisposition toward consistency, in-group communication, and the perceived environment reinforce one another to create stability and tenacity of organizational arrangements. Adapting the firm to potentially threatening social and economic changes becomes a significant managerial challenge.

March and Simon's analysis of the tendency toward specialization in organizations and of participants' identification with subunit values and goals is a valuable point of view for understanding the recommendations of many of the managers in this casebook. This view also raises the question of how processes of perception and identification may affect a company's key decision maker, especially when the organization needs the general manager to span and possibly integrate a number of specialized views.

How policy is formulated in complex problem-solving situations. Writing in the same decision-making tradition as March and Simon, Lindbloom provides an analysis of why a logical/rational approach to problem solving is an unrealistic concept for organizational policy development. Real-world problems, according to Lindbloom, resist precise definition and quantification. In fact, the only problems amenable to exacting quantification either are based on a number of questionable assumptions or are basically trivial. Policy development is messy—managers inherently deal with incompatible values; must often take action when little, if any, valid data exist; and have to reckon with tightly interconnected issues. A comprehensive, well-informed, and systematic evaluation of goals, policy alternatives, and decision consequences is simply not possible in most policy situations.

Each of the cases in this book calls for the formulation of organizational policy where most of the characteristics of messy problem-solving are present. Managers in these cases face conflicting values: maintaining discipline in the work force vs. having the flexibility to accommodate special employee problems; protecting human health vs. maintaining employment security; maximizing revenues vs. being sensitive to human rights claims. The very title of this book, "managerial dilemmas," conveys its focus on value conflicts.

Decision makers in the cases are often presented with partial information of questionable reliability. As much as these managers may desire information that is complete, clear, accurate, and based on a consensus view, some decision has to be made and the organization has to try to move ahead. Issues rarely will wait for a comprehensive investigation, a thorough search for alternatives, a detailed analysis, and precise recommendations.

As readers, you will experience the interconnectedness of many of the issues in these cases. Technological changes affect organizational structures; legal doctrines affect management practices; social attitudes (for example, about the role of women in society) affect employment relationships; and managerial processes affect the resistance to technological changes. Moreover, it is equally true that organizational structures affect technological changes; that management practices affect the extent and type of litigation a firm experiences; that technology affects social attitudes; that managerial processes affect organizational structures; and on and on. Managers should disabuse themselves of any notions about their chances of facing simple, freestanding, linear problems.

Lindbloom states that actual policy development involves *successive limited comparisons* between one's present situation and the expected consequences of incremental changes. Based on their own prior experiences, decision makers typically consider a relatively narrow range of policy changes, mostly lying close to what currently exists. Where opposing factions permit change, new policy is tried and a new base of experience is created for further consideration. In other words, ideal goals involving dramatic departures from the current situation are rarely viable on a practical basis. More frequently, organizations change through numerous iterations involving smaller, incremental steps.

Lindbloom warns us to make preferences for action specific to each situation. Abstractions without regard to concrete circumstances offer little help to managers. In addition, few situations can be addressed in absolutes; choosing between more control or more decentralization will usually depend on how much and what type of control currently exist in the organization.

Finally, Lindbloom argues the virtues of diverse views and fragmented, iterative decision-making. Countervailing interests serve as watchdogs for each other, and while they may reduce the speed of decision making, they also increase its equity and effectiveness over time. Different administrative experiences and contrasting interests need to be encouraged. The necessity of mutual adaptation among policy contestants adapts policies to the multiple interests that are part of organizational life and, in the long run, permits wiser decisions. Confronted with seemingly unmanageable problems, managers may gain solace by occasionally remembering this viewpoint.

How to realize the potential of an organization's human resources. Experienced managers recognize that the commitment, effort, and creativity of people are essential to an organization's performance. Consequently, a critical decision for managers in all organizations is how to manage people. In the most straightforward way, Douglas McGregor's concepts of Theory X and Theory Y state that managers make important choices about supervisory style that have major consequences for both employees and the firm.

McGregor observed that supervisors invariably hold implicit assumptions about people in organizations. On the one hand, some managers view subordinates negatively — they believe that the average person has an inherent dislike for work; that most people have to be coerced, controlled, directed, and threatened with punishment in order to work; and that people generally wish to avoid responsibility, have little ambition, and value security above other rewards. McGregor labeled this set of assumptions Theory X.

On the other hand, a very different set of assumptions characterizes Theory Y. Managers following this perspective believe that work is as natural as play or rest; that people will exercise self-direction and self-control in the service of objectives to which they are committed; that commitment to objectives is a function of the rewards associated with their achievement; and that the average human being seeks responsibility and is capable of imagination, ingenuity, and creativity in the solution of organizational problems.

Since all managers need to predict behavior and direct their organizations, McGregor pointed to the choices available to managers in accomplishing this goal. He concluded that reliance on formal authority, the *only* means of influence assumed to be available and relied on by most managers at the time, would be a poor choice in many circumstances.

The importance of managerial assumptions arises from the fact that assumptions have consequences. That is, a manager's assumptions about people constitute a significant factor in the subordinate's work environment, affecting that person's opportunities to use skills, to receive tangible and intangible rewards, and to experience himself or herself positively. Pejorative managerial assumptions are more likely to result in apathetic or hostile employees; positive managerial assumptions are capable of producing constructive, responsible employee behavior. In this way, a manager's current problems and successes can be a function of his or her earlier assumptions and actions. Managerial assumptions and subordinate actions are interdependent. And organizational problems can often be traced to the self-fulfilling quality of management's assumptions.

McGregor was an avowed integrationist; he wanted managers to meet the needs of the individual *and* the requirements of the organization. The principle of integration holds that the manager's task is not to control people directly, but to create "conditions such that the members of the organization can achieve their own goals *best* by directing their efforts toward the success of the enterprise."[39]

The managers in the cases that follow are engaged in making a number of decisions concerning the validity of competing theories about people's motivations. Will employee participation in the design and implementation of a new management information system build commitment and facilitate use, or will participation increase expenses, create needless delays, and risk derailing the project? Does asking employees, especially union members, to help plan a health information surveillance system in the workplace mean that management may be forced to modify the concept and design of the system and produce an ineffective or degraded system? What are reasonable limits to the participation of external social activists in a corporation's decision about whether to sell police computers to totalitarian governments? The cases in this volume indicate that choosing among competing assumptions remains a critical management decision and that integration of the demands of a variety of constituents with the objectives of the firm is a continuing requirement of the practice of management.

Douglas McGregor's perspective is also useful in understanding that the decision makers in these cases are acting solely at one point in time within an ongoing organizational process. Readers should be sensitive to how current circumstances facing a decision maker in a particular case result from a confluence of forces with far earlier roots in the firm and its environment. Occasionally managers in the cases can even be seen as "victims" of their own previous decisions or of the decisions of their predecessors in the firm.

Likewise, readers should be aware of how present actions in a particular case will create a set of conditions that give rise to new behavior and perhaps to different problems in the future. In other words, the cases do not pretend that decision makers can ever resolve situations finally and completely. Rather, the best a manager can do is to shape existing forces and create conditions in which positive, integrative behavior is more likely to result.

Ideas developed by March and Simon, Lindbloom, and McGregor are presented here solely as useful perspectives for reviewing, evaluating, and discussing the cases that follow. Readers are encouraged to explore how other management writers and research results may contribute to a conceptual understanding of the dilemmas and decisions posed in the cases. In particular, concepts pertaining to the management of power and interdependence in organizations; to awareness of the significance of symbols, language, and meaning systems; and to the importance of entrepreneurship, internal venturing, and innovations management all represent rich areas of recent thought that may also provide valuable insights for understanding managerial behavior in the cases. The present discussion is intended only to suggest several influential starting points for making the important bridge between practice and theory and to reaffirm the validity of doing so.

Bon Voyage

The authors hope that this introduction provides a context and a rationale for the cases presented in this volume. Our intention in preparing the cases has been to focus in on a number of pressing management issues of the day in a way that encourages the development of analytical thinking and a holistic, general management perspective. We also hope that readers find the cases personally engaging and entertaining, as well as challenging and thought-provoking. Finally, a reader's questionnaire, provided at the end of the book, gives you the opportunity to respond to the material and communicate to the authors your experiences in working with the cases. Your time and interest in responding would be greatly appreciated.

Notes

1. Information about General Motors in 1986 is from *Fortune*, April 27, 1987, pp. 364–365, and *Moody's Industrial Manual* (New York: Moody's Investor Service, Inc.) 1986, pp. 1395–1406. Information about General Motors in 1961 is from *Fortune*, July 1962, pp. 172–173, and *Moody's Industrial Manual* (New York: Moody's Investor Service, Inc.) 1962, pp. 2767–2774.

2. "General Motors: What Went Wrong," *Business Week*, March 16, 1987, pp. 102–110.

3. U.S. Department of Commerce, *Statistical Abstract of the U.S. 1986* (Washington, D.C.: Bureau of the Census, 1985), p. 5.

4. Ibid., p. 391.

5. Ibid., p. 450.

6. Ibid.

7. Ibid., p. 834.

8. U.S. Bureau of Labor Statistics, *The Handbook of Basic Economic Statistics* 41, no. 2 (February 1987): 86.

9. Ibid.

10. Edwin Dean, Harry Boissevain, and James Thanes, "Productivity and Labor Cost Trends in Manufacturing, Twelve Countries," *Monthly Labor Review*, March 1986, p. 4.

11. United Nations, Department of International Economic and Social Affairs Statistical Office, *1980 International Trade Statistics Yearbook* (New York: United Nations Publishing Division, 1980), pp. 1196–1197, and *1984 International Trade Statistics Yearbook* (1984), pp. 154–155.

12. "Making Brawn Work with Brains," *Business Week*, April 20, 1987, p. 56.

13. Ronald E. Kutscher and Valerie A. Pesonick, "Deindustrialization and the Shift to Services," *Monthly Labor Review*, June 1986, pp. 3–13.

14. Ibid.

15. U.S. Department of Commerce, op. cit., p. 305.

16. Ibid.

17. Some material in this and the following two sections has been drawn directly from Alan F. Westin, *Whistleblowing! Loyalty and Dissent in the Corporation* (New York: McGraw-Hill Book Company, 1981), pp. 4–10.

18. Vance Packard, *The Pyramid Climbers* (New York: McGraw-Hill Book Company, 1962).

19. Harry P. Cohany, "Membership of American Trade Unions," 1960, *Monthly Labor Review*, December 1961, pp. 1299–1308.

20. Richard Braddock, "Coping in the New Environment Brought on by Explosive Technological Innovations," *Communications News*, April 1985, p. 126.

21. *Ward's Business Directory of Major International Companies: 15,000 Leading Worldwide Corporations* 3 (1986): B-99, B-171.

22. "High-Tech Workplaces," *Business and Society Review* 56 (winter 1986): 90.

23. "Hardware at Work," *Business and Society Review* 55 (fall 1985): 91.

24. U.S. Department of Commerce, *1987 U.S. Industrial Outlook* (Washington, D.C.: Department of Commerce, 1987), p. 31-1.

25. U.S. Department of Commerce, *1987 U.S. Industrial Outlook* (Washington, D.C.: Department of Commerce, 1987), p. 11-8.

26. Packard, op. cit., p. 4.

27. David Riesman, *The Lonely Crowd: A Study of the Changing American Character* (New Haven: Yale University Press, 1950).

28. William H. Whyte, Jr., *The Organization Man* (New York: Simon and Schuster, Inc., 1956).

29. U.S. Department of Commerce, op. cit., p. 79.

30. *Moody's Industrial Manual*, 1962, p. 2767.

31. Rachel Carson, *Silent Spring* (Boston: Houghton Mifflin Co., 1962).

32. Ralph Nader, *Unsafe at Any Speed: The Designed-In Dangers of the American Automobile* (New York: Grossman, 1965).

33. Murray L. Weidenbaum, "Free the Fortune 500: An Economist's Response to Big Business Day," *Vital Speeches of the Day* 46, no. 14 (May 1, 1980): 421–425.

34. Thomas J. Peters and Robert H. Waterman, Jr., *In Search of Excellence: Lessons from America's Best-Run Companies* (New York: Harper & Row, 1982), Chapter 12, pp. 318–326.

35. James G. March and Herbert A. Simon, *Organizations* (New York: John Wiley & Sons, Inc., 1958).

36. Charles E. Lindbloom, "The Science of 'Muddling Through,'" *Public Administration Review* 19, no. 2 (1959): 78–88.

37. Douglas McGregor, *The Human Side of Enterprise* (New York: McGraw-Hill Book Company, 1960).

38. March and Simon, op. cit., p. 152.

39. McGregor, op. cit., p. 61.

Managerial Dilemmas

CASE 1

Medical Ethics and Business Decisions

Weighing Professional Dissent within the Enterprise

American Pharmaceutical Company is a 75-year-old firm whose corporate headquarters and research laboratories are in New Caledonia. A pioneer in the ethical drug industry, American Pharmaceutical has been a highly profitable company, primarily by achieving "lead time" in bringing new prescription drugs to market ahead of competitors.

The engine for such competitive success is American Pharmaceutical's Advanced Research Facility in the Corporate Research Park complex outside Metropolis. American Pharmaceutical has attracted an outstanding group of biomedical researchers to this facility, which has been called the Bell Labs of the drug industry. Top scientists from universities, government health institutes, and health regulatory agencies have been attracted by the frontier research, tradition of scientific inquiry, and high prestige afforded to research scientists at American Pharmaceutical.

This attraction was symbolized by two events in 1988. Philip Macready, Chief Executive Officer of American Pharmaceutical, received the National Medical Association's Industry Excellence Award for advancing public health in the United States through the company's research and development achievements in combating disease. Also in 1988, American Pharmaceutical created an annual prize of $100,000, to be awarded to the research scientist in academia who has made the most important discovery in the field of geriatrics, thereby advancing the health of senior citizens.

However, a major problem has arisen at American Pharmaceutical, involving the development of a new drug that the company expects to market shortly. You are Michael Armstrong, Senior Vice President for Operations, and have just received three memos.

Key Persons Involved

Michael Armstrong, *Senior Vice President for Operations*
Harry T. Jennings, M.D., *Executive Director of Laboratory Research*
Marie Pawling, M.D., *Director of Medical Research/Therapeutics*
Ingrid J. Scalise, *Director, Public Affairs*
Jacob Lindt, M.D., *Director, Prolidamine Project Team*
McKinley Knight, Ph.D., *Chief Toxicologist*
T. Grayson Hurd, Jr., M.D., *Director of Clinical Studies*
Charles Rouse, *Assistant Director, Marketing*

Memorandum

Date: April 3
To: Michael Armstrong, Senior Vice President for Operations
From: Harry T. Jennings, M.D., Executive Director of Laboratory Research
Re: Clinical Trials of Prolidamine

I regret that I must bring to your attention a potentially serious obstacle in our development of prolidamine for introduction into the American market. Although progress up to this point has been rapid and the product is ready for clinical testing, we now find ourselves faced with an unprecedented problem. A key member of the development team feels that we should *not* move ahead with clinical trials, and has carried her opposition to the point of insubordination.

Prolidamine was originally developed by our Danish affiliate, Maarten, for treatment of acute and chronic diarrhea. Under the name Prolidam, it has been successfully marketed as a prescription medication in several European countries for the past two years, and no medical or legal problems have arisen in connection with its use. There is at present in this country no formulation that offers comparable effectiveness and ease of administration. Thus, it seems reasonable to expect that prolidamine could eventually command a substantial share of the market for antidiarrheal medications, estimated by Marketing at more than $150 million per annum. Accordingly, it is our aim to introduce prolidamine, under the trade name Damolid, to the domestic market as quickly as possible. Current plans call for initiation of clinical trials by May 1; conclusion of clinical trials by June 1; FDA approval by December 1; and test marketing next spring.

Our present difficulties with prolidamine stem, paradoxically, from one of its most advantageous characteristics. The wide acceptance of prolidamine can be attributed in large part to the fact that it is administered in liquid form, making it especially well suited to the needs of infants, children, the elderly, and any other patients who cannot conveniently take solid medication. As the active ingredients have a markedly bitter taste, the new artificial sweetener dematril is used to improve palatability. The American Pharmaceutical project team that has been working on prolidamine since May of last year initially questioned the advisability of using large doses of dematril in prolidamine, in view of recent (but highly controversial) allegations that this substance may have carcinogenic effects on laboratory animals. (Background memos are available from Dr. Jacob Lindt, director of the prolidamine project team, and subsequent memos from Dr. McKinley Knight, the team's chief toxicologist, and Dr. Marie Pawling, its clinician). It was suggested that, as there are unresolved questions about dematril, an alternative formula containing sugar and/or smaller amounts of dematril should be developed.

On February 15, a number of members of the prolidamine team, including Dr. Lindt, Dr. Pawling, and Dr. Knight, met with management representatives, including myself; Charles Rouse, Assistant Director of Marketing (Prescription Medications Division); and Dr. T. Grayson Hurd, Jr., Director of Clinical Studies. The concerns of the prolidamine team over the dematril level in the original Maarten version of prolidamine were discussed at some length. Dr. Knight submitted a written summary of the objections to the Maarten formula, including a review of the clinical literature relating to dematril as a possible carcinogen. At this meeting, Dr. Pawling, who as the team's chief clinician would be in charge of clinical trials of prolidamine, asserted that she could not in good conscience administer prolidamine to human subjects in its present formulation. She suggested that an alternative means of sweetening the product be developed—a process that, according to Dr. Lindt, would probably take no more than six months.

Dr. Hurd then made the following points:

- That to date there is *no* incontrovertible evidence that dematril is a carcinogen in humans
- That, while the quantity of dematril in a *bottle* of prolidamine is large (about 44 times the amount currently allowed by the FDA in a 12-ounce can of diet soda), the prescribed dosage of prolidamine is quite small, so that the amount of dematril actually consumed by a patient using the medication during a 24-hour period would be less than that found in a single can of diet soda
- That the Maarten formula has encountered neither clinical complications nor regulatory problems during its two years of use in Europe

Thus, a change in the formula of prolidamine for the American market did not seem warranted. Rouse added that the six months required for such a change, though it might seem negligible to the research personnel, was by no means insignificant. A delay of this length might allow competitors to establish themselves in what is currently a wide-open market, as well as adversely affecting the profitability of the company for the fourth quarter of the fiscal year.

It was agreed that Dr. Lindt should inform the rest of the team as to the points raised by management, and initiate further discussions of the dematril issue. As a result of these discussions, and a number of follow-up meetings, the prolidamine project team decided on March 12 to approve the original Maarten formula and begin field trials at once. At this point, however, Dr. Pawling indicated that she refused to be bound by the team's decision. She stated to me personally, and in a memo dated March 16, that as a physician she considered it unethical to undertake clinical trials with a product when there were serious doubts—in her mind at least—about its safety. Accordingly, she will not be a party to human testing of prolidamine. She views the matter as one involving professional responsibility, and thus transcending any consideration of loyalty to American Pharmaceutical.

In nearly thirty years of research work—the past seventeen of them with this company—I cannot recall a comparable issue having arisen. While I have no firm evidence to support this view, I have received the sense from various discussions that Dr. Pawling is working on an alternative to dematril, which, if successful, would obviously enhance her scientific status in the company and beyond. I suspect that her high moral stance on this issue may simply be a tactic to increase the likelihood of recognition for her own work. In any case, I cannot see how we can allow her purely personal beliefs or motives in this matter to prevail over the considered judgment of the entire team, which includes many highly competent scientists. I therefore see no alternative but to remove her from the prolidamine project. I would appreciate it, however, if you would review the facts in the case and let me know whether you concur, before we embark on an action that could have consequences that are difficult to foresee.

Memorandum

Date: April 14
To: Michael Armstrong, Senior Vice President for Operations
From: Marie Pawling, M.D., Director of Medical Research/Therapeutics
Re: Unwarranted Personnel Actions

I am writing to protest, in the most vehement terms, the recent actions taken against me by my superior, Dr. Harry Jennings, Executive Director of Laboratory Research. Dr. Jennings has informed me that I am to be demoted; that I will no longer be permitted to work on therapeutic drug development, which has been the focus of my activities with American Pharmaceutical up to the present; and that management considers me "unpromotable." I feel that these actions are completely unfair and unjustified. Moreover, I am convinced that the motivation underlying them is not that which Dr. Jennings has alleged in his conversations with me. Although Dr. Jennings has made broad and unsubstantiated accusations of irresponsibility against me, I feel that in reality I am being punished for being *too* responsible. I refer to my refusal to take part in clinical trials of prolidamine, a decision that I feel is at the root of these attacks.

Although Dr. Jennings has questioned my loyalty as well as impugned my judgment, I consider myself to be a loyal—indeed, dedicated—employee of American Pharmaceutical. I have been working at the New Caledonia labs for nearly four years now, and I have always attempted to perform my assignments in a conscientious and professional manner. I have had no reason to think that

Dr. Jennings or anyone else was dissatisfied with my work, for I have received no complaints. On the contrary, my evaluations have always been excellent, to which my promotion two years ago attests. It is certainly *not* my habit to arbitrarily defy the wishes of management or the directives of my supervisors. Nor am I accustomed to imposing my will on my colleagues and peers. Thus, my stand on the prolidamine matter was not taken without long reflection and soul-searching. Since I believe this to be the heart of the matter, let me restate—and try to explain—my position:

- Dematril has not been *proved* to be deleterious to humans. However, neither has it been proved safe. Since the issue is unresolved, it is inappropriate to market a medication containing dematril in high concentrations.
- The amount of dematril in the recommended 24-hour dosage of prolidamine is not excessive by current FDA standards. However, this medication is intended to be used by children and the elderly. How can we be sure that a user will not exceed the recommended dosage? What is to stop a child or an ill or senile adult from drinking an entire bottle? Moreover, the cumulative effects of substances such as dematril are not yet well understood. What might be the consequences of prolonged use of prolidamine? We do not know.
- The tumors with which dematril is thought to be associated have a latency period of some seventeen years. Thus, it would be many years before dematril-induced damage became apparent. In the meantime, countless people might be put at risk.
- Alternatives to the current, high-dematril formulation could easily be developed.

I should add that as a physician, I feel a special responsibility in this matter. It is true that the other members of the project team eventually approved prolidamine for clinical trials (after being subjected to heavy pressure from management, I should add). However, no one else on the team is a physician; no one else has taken the Hippocratic oath; no one else would have the responsibility of actually dispensing the drug to test subjects. I do not feel that I can be absolved of this responsibility by considerations of corporate loyalty, "team play," or any of the other notions that have been advanced to justify ignoring one's conscience.

I hope that the actions taken against me will be reconsidered. I have no desire for further conflict with the company, and I hope that this unpleasant episode can be put behind us as quickly as possible. However, I must state that if management continues to treat me in this way, I will have to consider legal recourse—not so much to protect my own livelihood and professional standing (though these are important to me) as to defend the principle that the first obligation of a physician is to his or her professional conscience.

Memorandum

Date: April 26
To: Michael Armstrong, Senior Vice President for Operations
From: Ingrid J. Scalise, Director, Public Affairs
Re: Dr. Marie Pawling

In response to your request for an analysis of the issues in the Pawling case, here is my evaluation of what is at stake for American Pharmaceutical.

Pawling has threatened to resign if she is not reinstated to her former position, and her memo of April 14 makes it clear that she is likely to undertake legal action against the company if her request is denied. Thus, a great deal of trouble would be saved if Pawling were to be reinstated. However, it is difficult to see how we could proceed with clinical trials of prolidamine with Pawling as Director of Medical Research/Therapeutics. Even if we went ahead under the supervision of another physician, there is no assurance that Pawling would not then make her objections public and/or take legal steps to block testing of prolidamine. While this is speculation, my feeling—based on my contacts with Pawling and the tone of her communications on this subject—is that she would do just that. It therefore seems that there is no way to avoid publicity in this case, unfortunate as that may be for AP.

It is my understanding that management is committed to the speedy introduction of prolidamine, and that this decision is based on three considerations:

- Since most of the development costs have already been borne by Maarten, the drug can be introduced at little additional expense to AP.
- Marketing assures us that sales in excess of $110 million a year are highly likely.
- Owing to the expiration of our patents in the tranquilizer field, and to the general erosion of profits in the area of contraceptive products as a result of competition and shifts in public practices, the company badly needs profitable new products—preferably sometime within the next two years.

Accordingly, our efforts should be directed to anticipating the problems that might arise if we move forward with prolidamine and Dr. Pawling resigns as expected. Three problems are potentially serious and should not be minimized.

1. It is the contention of our legal department that the demotion of Pawling—or even her firing, should that become necessary—probably cannot be successfully challenged in court. As Pawling is not a contract employee, she has no tenure and serves on an "at will" basis. The courts in New Caledonia have long held that management can fire a noncontract employee at its discretion, in the absence of a union collective-bargaining agreement or a violation of any employee protection statute (such as Equal Employment Opportunity or Occupational Safety and Health). However, we should note that in the past decade, almost half the states have established various exceptions to the employment-at-will doctrine. Some of these are based on implied contract (such as promises

made at hiring or written in an employee handbook), and others are based on wrongful-conduct (tort) grounds (such as intentional infliction of emotional distress or invasion of privacy). One area of such new judicial doctrines involves protection of an employee who refuses to follow directions of the employer for reasons that the law recognizes as a legitimate "public policy," such as protecting public health or safety, and where the employee is then fired or punished by the employer for having done so.

In our state, New Caledonia, the courts have not in recent years considered revision of the traditional employment-at-will doctrine, and have not yet endorsed the public policy doctrine. We do not know how they might rule. Nor does our state have a whistle-blower protection law, as Pennsylvania, Michigan, New York, and half a dozen other states do. Even if these conditions were otherwise, Dr. Pawling, as I understand the facts, cannot cite any law or regulation that AP has broken, and she has not reported—and could not report—the company for a legal violation. Her basic claim would have to be that her physician's oath and her "professional code" of ethics allow her to take the actions she has, and that it is wrong for the company to remove her from her post or give her an adverse performance evaluation or undesirable assignment.

2. From a public relations viewpoint, however, our situation is less attractive.

    a. With respect to the product, any hint of controversy concerning its safety is likely to prove damaging. This is true even if the FDA and other regulatory bodies ultimately back us completely. In today's climate of alarm about the safety of the environment generally, and drugs in particular, a new product must be considered "above suspicion" if it is to succeed.

    b. With respect to the public image of AP, we would certainly want to avoid being seen as a company that would take chances with the safety of a new product, would bring pressure to bear on an employee who had raised concerns over safety, or would take retaliatory action against a "whistle-blower." Our reputation has always been that of a company committed to sound and thorough clinical research, with the greatest regard for public safety. We naturaly do not wish to jeopardize this reputation, or our reputation as a fair employer.

    c. Bad publicity could also impair our ability to recruit top younger scientists. If various medical or scientific groups were to take up Dr. Pawling's case as a cause célèbre, we could be made to look like the classic "blackhat" company, and experience a significant drop in the quality of our research staff.

3. At the same time, we do not want to encourage disgruntled or eccentric employees to feel that they can single-handedly veto or sabotage legitimate projects and activities of the company. If we back down in this case, we run the risk of compromising discipline and losing control of our own operations. We will be sending the "wrong message" to our personnel, leading them to think that their personal judgments and values must take precedence over the policies of the company and the objective, scientific weighing of evidence on which all research is based. Ultimately, it is hard to see how corporate management could

continue to run responsible enterprises, especially when new scientific and technological developments are involved, if self-appointed "whistle-blowers" can set themselves up as independent decision makers.

Taking all these considerations into account, I feel that we should make one more attempt to work out some kind of modus vivendi with Dr. Pawling. We might agree to submit this case to an outside confidential review by a panel of top industry research scientists *not* in the drug field. Failing to work out a resolution, however, I think we should hold to our position, and prepare as well as we can for whatever legal or public relations steps Dr. Pawling may decide to take.

Discussion Questions

1. As Armstrong, would you approve Jennings' recommendation to remove Pawling from the prolidamine project? What factors would you want to take into consideration before making your decison?

2. How should Armstrong deal with Jennings' speculation that Pawling was seeking to advance her own career by blocking the use of dematril as a sweetener for prolidamine?

3. As Armstrong, what would be your response to Pawling's memo dated April 14?

4. Explain Pawling's rationale for acting on the basis of professional ethics. What other professions or occupations might have members who are inclined to claim ethical responsibilities in today's corporations, and what are the implications of this situation for managers?

5. What procedures for receiving and resolving employee complaints do you believe the firm might have used to handle the Pawling problem? Why might the chances of handling the issue be improved with other procedures?

6. Assume that Dr. Pawling is forced to leave American Pharmaceutical and that she sues the firm for wrongful discharge. What values or interests would need to be balanced in arriving at a decision on the merits of her claim? If you were the presiding judge, would you award damages to Pawling or uphold the company? If you found for Pawling, and she had asked for reinstatement as well as damages, how would you rule on that request?

CASE 2

Sexual Discrimination and the Corporate Culture

How to Deal with Business "Affairs"

Computer Inventory Systems (CIS) is an information services firm that designs, tests, installs, and sometimes administers large-scale computer-based inventory and control systems, primarily for manufacturing and transportation companies. Founded in 1970 by Carl ("Tex") Bridewaite, a former IBM sales executive, CIS now has 12,000 employees; maintains offices in 28 U.S. and 11 foreign cities; and boasts that "40 percent of America's biggest and most complex industrial companies" are its clients. It had revenues last year of $563 million.

CIS started off by developing automated inventory and control systems for use in the steel, oil, and utility industries. It moved into the railroad and airline industries in the late 1970s, and the farm-equipment and auto industries in the early 1980s. Among information services companies, CIS has a reputation for being a "hard-driving, hard-drinking, and hard-selling" company, in what was seen as a "fit" with traditional leadership in the kinds of client industries in which CIS specialized.

Women make up 43 percent of CIS's overall work force, and are now majorities in the Programming and Field-Testing divisions. All of CIS's top fifteen senior executives are men, two of whom are black and one of whom is Hispanic. A woman has just been made Director of Customer Services and holds the highest post of any female in the firm.

CIS, a prototypical fast-growth, high-tech company, remains extremely profitable. It is known as an excellent place to work for people "who flourish in a high-competition, pressure-cooker environment." Salaries, benefits, and profit-sharing plans exceed industry norms, and there is very low turnover among CIS employees.

CIS has just been found guilty of sex discrimination in a case involving a female project manager, and has had damages of $374,864 awarded against it, plus legal fees and court costs. You are Philip Atwell, Chief Operating Officer, and have to decide (a) whether to appeal the federal court decision and (b) whether to make any changes in policies or procedures because of the case.

Key Persons Involved

Carl Bridewaite, *President and Founder*
Philip Atwell, *Chief Operating Officer, one of the original management group at CIS*
Felicia Green, *Project Manager, Expert Systems Development*
Peter Foley, *Vice President, Advanced Expert Systems Department*
Harry Merchant, *Senior Vice President for Human Resources*
Paul Costello, *Director, New Products Division*
Laura Stocker, *EEO/Affirmative Action Officer, Personnel Department*
Quentin Harris, Esq., *Chief Counsel for Employment Law*
Sidney Levine, *Vice President, Advanced Expert Systems Department* (before Foley)
Roger Delano, *Wage and Compensation Specialist*
Susan Horvath, *Programmer*
Fred Croce, *Personnel Manager*
Julie Caspar, *Project Manager, Expert Systems Development* (after Green)

Memorandum

Date: June 10
To: Philip Atwell, Chief Operating Officer
From: Carl Bridewaite, President and Founder
Re: Appeal of *Green* Judgment

I have just received a copy of the judge's opinion in the *Felicia Green* case and am enclosing a photocopy for you to look over. In a nutshell, we have been held liable for sex discrimination for our handling of the matter, and Green has been awarded a substantial financial settlement—including front pay, of all things.

I'm sure you will be as disappointed as I am about this ruling; nevertheless, the judgment is not necessarily the last word on the case. Our Chief Counsel for Employment Law, Quentin Harris, feels that we have an excellent chance of having the ruling overturned on appeal (see his legal analysis, also enclosed); Harris is ready to file our appeal brief as soon as we give him the go-ahead. There is some question in my mind, however, as to whether an appeal is really the wisest course. I have solicited the views of Harry Merchant, who, as you know, was very close to the whole drama—a participant, in fact—and of our EEO/Affirmative Action Officer, Laura Stocker. I am forwarding their memos to you. You will see that they offer conflicting advice. Stocker, in particular, raises some troubling issues; she feels that the case has implications that go beyond our immediate legal problems.

My own feeling is that we're in a very sticky spot, and should consider our course of action carefully before we move. I'd like you to review the judge's opinion and the accompanying memos and let me know what you think:

- Should we accept the verdict and pay the $375,000 (plus court costs and attorneys' fees) quietly, or should we press the case on appeal?
- Do we need to consider new policies on fraternization and sexual harassment, and specific complaint procedures for sex discrimination/sexual harassment issues?

I'll need your input by early next week.

Attachment

FINDINGS OF FACT
Excerpts from the Opinion of Judge Jefferson G. Holmes, U.S. District Court, Boulder, Colorado, in the case of *Green vs. Computer Inventory Systems*

1. Plaintiff, Felicia Green, is a white female citizen of the state of Colorado. Defendant is Computer Inventory Systems, Inc., a corporation based in Denver.

2. Plaintiff holds a B.A. in English literature. Plaintiff was hired by defendant in 1975 as a part-time data-entry employee. After being promoted to full-time status, she held the positions of Programmer, Systems Analyst, Senior Systems Analyst, and Project Manager for Expert Systems Development.

3. Plaintiff and Peter Foley, a first-line manager, began a personal, intimate, and illicit relationship in 1975, shortly after both were first employed by defendant. However, neither's job had anything to do with the other's until late in 1979.

4. As Plaintiff progressed with defendant from a data-entry job through the very responsible position of Project Manager for Expert Systems Development, she always received satisfactory to outstanding performance evaluations. Throughout her career, she was commended for her devotion, loyalty, and performance. There was not a "black mark" in plaintiff's personnel record with defendant until after she was removed from the Advanced Expert Systems unit.

5. During 1978, plaintiff worked as Senior Systems Analyst under the supervision of Sidney Levine, Vice President of defendant's Advanced Expert Systems unit. During the latter part of her tenure in that position, plaintiff oversaw the work of programmers and analysts developing sophisticated new expert systems for inventory control. On February 1, 1979, plaintiff was promoted to the newly created position of Project Manager for Expert Systems Development. Plaintiff continued to work under the supervision of Sidney Levine. Although she had no formal education or training in expert systems,

plaintiff performed in an exemplary fashion. Her contribution was recognized by Sidney Levine, who felt that plaintiff had "God-given talent" that enabled her to do such an excellent job despite her lack of formal training.

6. According to Levine, plaintiff was not being compensated appropriately for the work she was doing. Levine made repeated attempts to get her salary upgraded. He also thought that one of the reasons that defendant refused to upgrade plaintiff's salary was that plaintiff is a woman. Although defendant's Wage and Compensation Specialist, Roger Delano, testified that company policy is to evaluate the job and not the individual, it is clear that defendant rated plaintiff's job based upon its perceptions of plaintiff rather than upon the duties and responsibilities of the job. Delano was obviously rating plaintiff and not the job, in contradiction to his testimony.

7. During her tenure as Senior Systems Analyst and Project Manager for Expert Systems Development, plaintiff made a unique contribution to defendant's Advanced Expert Systems program. She helped shape the direction that the program took, and guided the research that led to development of software for just-in-time delivery systems. According to Levine, she played an integral part in the program's success.

8. In November 1979, Sidney Levine left defendant's employ. In December 1979, Peter Foley, with whom defendant had continued to carry on an affair, was promoted to Levine's former position, Vice President of Advanced Expert Systems. Foley thus became plaintiff's immediate supervisor. Prior to assuming his new job, Foley advised Paul Costello, Director of defendant's New Products Division, that he had previously had an illicit sexual relationship with plaintiff; however, he dishonestly reported that the affair had ended. Even though Costello was aware of the relationship between Foley and plaintiff, he nonetheless promoted Foley into a position that made him plaintiff's immediate supervisor.

9. After Foley became plaintiff's supervisor, he undertook to bring another female, Susan Horvath, into the Advanced Expert Systems program. Foley was also carrying on a social and personal relationship with Horvath. When Foley told plaintiff that Horvath refused to come into the Advanced Expert Systems Department as long as plaintiff was there, plaintiff telephoned Horvath to see if she could "work things out" with Horvath. Plaintiff had a second conversation with Horvath, during which insults were exchanged. Following that conversation, Foley became extremely upset with plaintiff, and on February 1, 1980, fired her from her position as Project Manager for Expert Systems Development because he was unwilling to continue the relationship with plaintiff.

10. Plaintiff attempted to appeal Foley's decision to fire her by requesting a meeting with Costello. Plaintiff had a meeting on February 4, 1980, with Foley and Harry Merchant, Senior Vice President for Human Resources at Computer Inventory Systems. Prior to the meeting, Foley met with Merchant and told him that he could no longer work with plaintiff and that he wanted her out of his department. During the meeting with Foley and Merchant, plaintiff became emotional and got on her hands and knees and begged for her job, stating that she had done nothing to warrant being fired by Foley. Merchant,

who did not understand why plaintiff wanted to remain in Advanced Expert Systems when Foley, her supervisor, wanted her out, did not allow plaintiff to explain her side of the story. Merchant stated that the defendant's employee handbook and the grievance/review procedures contained in it did not apply to plaintiff's situation.

11. The next day, plaintiff and Foley were scheduled to meet with Costello. Foley had a preliminary meeting with Costello, during which time he informed Costello about the "problem" that had arisen.

There were three versions of what occurred during the meeting at which Foley, Costello, and plaintiff were present. Both Foley and plaintiff testified that although she tried to explain her side of the story, she was not allowed to do so. Costello, like Merchant, told plaintiff that the employee handbook did not apply to her. The handbook itself, however, clearly states that the defendant's Problem and Complaint Procedures may be invoked when a problem arises that "may be based on a personal problem." Costello made it clear to plaintiff that he felt Foley's role in Advanced Expert Systems was more important than hers. After plaintiff had been denied the opportunity to explain and after Costello had lauded Foley and emphasized the importance of having him in the department, it was clear to plaintiff that both Merchant and Costello were going to side with Foley in the dispute, that no one was going to listen to her side of the story, and that she was not going to be allowed to exercise her rights as specified in the employee handbook. Costello and Foley testified that plaintiff then "volunteered" to leave the Advanced Expert Systems Department; plaintiff testified that Costello told her that she was being removed from her job. Costello did promise plaintiff that he would find her a comparable position in defendant's organization, then concluded the meeting by sending Foley back to work and telling plaintiff to "relax" and "go shopping." Both Costello and Foley applied the age-old double standard, namely, that the male is forgiven for illicit sex and the female is not.

12. Foley has not been disciplined by defendant in any manner for his part in the affair, for the disruption in the workplace, or for lying to Costello about the status of the affair. His performance reviews since plaintiff was fired have been high. Today he still holds his position with defendant.

13. Plaintiff's replacement in the job of Project Manager for Expert Systems Development was Julie Caspar, who had come into the department as a Systems Analyst shortly before plaintiff was fired. Caspar testified that she got along well with plaintiff in the workplace prior to plaintiff's removal.

14. The court finds Foley's testimony that plaintiff was subjecting him to sexual blackmail to be so preposterous as to undermine the credibility of his other testimony.

15. After plaintiff's removal from the position of Project Manager for Expert Systems Development was ratified by Costello, plaintiff became a "displaced person," still employed by defendant but assigned no job duties other than to seek a comparable position in defendant's organization. She remained on call but was asked to perform no other job duties. As one member of defendant's

Personnel Department testified, plaintiff maintained close contact with the Personnel Department in an effort to find a comparable job. Plaintiff received full salary and benefits during this period.

16. Defendant offered plaintiff a position as Head Librarian of the company's tape library. However, the Court finds that this job was not comparable to the position of Project Manager for Expert Systems Development. It required retraining and had nothing whatever to do with expert systems. Delano testified that defendant uses the Hay system to rate its jobs. He testified that the three areas considered when ranking a job under the Hay system are (a) know-how, (b) problem solving, and (c) accountability. He further testified that under criteria (a) and (b), the Head Librarian job was not comparable to plaintiff's position as Project Manager for Expert Systems Development. The only similarities between the two jobs were grade level and pay, according to Delano. Furthermore, defendant acquiesced in plaintiff's rejection of the job on the basis of noncomparability, and continued to assist her in attempting to find a comparable job.

17. Plaintiff applied for the position of Vice President for Systems Development. Defendant requires applicants for that position to have a graduate degree, which plaintiff does not have. She was refused that job.

18. During the time when plaintiff was a displaced person, Foley went to Costello to seek approval for a new position, Supervisor of Program Maintenance. Foley succeeded in obtaining approval for this position ahead of the time it was originally scheduled to be approved. The job required whoever filled it to travel out of Denver four days a week, to various locations where the company's inventory systems were in use. Once the job was approved by Costello, it was offered to plaintiff, even though the position reported to Foley. Foley thought that by giving plaintiff this job, he could get her out of his life, as he had already succeeded in getting her out of the Advanced Expert Systems Department. The Court finds that the position of Supervisor of Program Maintenance was not comparable to plaintiff's position as Project Manager for Expert Systems Development, and that Foley was trying to "send her to Siberia" with this wicked scheme. And again Delano testified that based upon the first two criteria of the Hay system, the two jobs were not comparable; the only similarities were grade level and pay.

19. After plaintiff initially refused the job of Supervisor of Program Maintenance, she was presented with an ultimatum. She could either accept that job or be terminated. When plaintiff did not accept the job, Foley was allowed to terminate her employment with defendant effective June 13, 1980.

20. During the period when plaintiff was a displaced employee, Fred Croce, Personnel Manager for defendant, who knew none of the details surrounding plaintiff's removal from the Advanced Expert Systems unit, began to question why plaintiff had been removed. He tried to get defendant to abide by the employee handbook and grant plaintiff a Grievance Committee Review. When Croce made his inquiries, Foley attempted to justify his removing plaintiff from his department by denigrating her work there. Yet earlier Foley had signed blank job-change applications for plaintiff that extolled her virtues. In addition, Foley

refused to cooperate with Croce in the Personnel Manager's attempt to determine what had happened to plaintiff. Plaintiff was not granted a Grievance Committee Review.

21. After plaintiff was fired by defendant, she sought and obtained several jobs, each as a programmer. She applied, unsuccessfully, for a position as a systems analyst with SmartSoft, Inc., by speaking with the president of that company. She is now employed as a programmer at a salary of $2,900 per month. Since her termination from defendant's employ, she has earned $71,504 in wages.

CONCLUSIONS OF LAW

1. On the issue of plaintiff's discharge, the Court finds in plaintiff's favor. Contrary to defendant's assertions, the meetings between plaintiff, Foley, and Costello did not result in plaintiff's voluntary, uncoerced resignation from her position as Project Manager for Expert Systems Development. She had already been removed from this position by the Vice President of the Advanced Expert Systems Department (Foley), and the result of the meeting with Costello was a confirmation and ratification of that action. This is a classic case of disparate treatment based on sex. There are two individuals, plaintiff and Foley, who were involved in an illicit affair that resulted in a disruption in the workplace. While plaintiff, the woman, has lost the job she worked so hard to develop, Foley, the man, has not been disciplined at all for his part in the affair or for his misleading defendant about the status of the affair when he was originally considered for the position of plaintiff's immediate supervisor.

Furthermore, Foley was authorized to try to get rid of plaintiff by assigning her to a traveling job. Defendant—through Paul Costello, Director of the New Products Division; Harry Merchant, Senior Vice President for Human Resources; and Peter Foley, Vice President of the Advanced Expert Systems Department—has indulged in the age-old discrimination against women based on a double standard. Such discrimination is in direct contravention of the principles of Title VII. Plaintiff made out a prima facie case of sex discrimination.

The burden then shifted to defendant to articulate some legitimate nondiscriminatory reason for plaintiff's discharge. Defendant claimed that plaintiff was discharged for refusing to accept a comparable job, the job of Supervisor of Program Maintenance. The burden then shifted to plaintiff to show by a preponderance of the evidence that defendant's proffered reason was pretextual. The Court concludes that defendant's proferred reason for plaintiff's discharge constituted a pretext. Of course, plaintiff bears the burden at all times of convincing the Court by a preponderance of the evidence that she was discriminated against because she is female. The Court concludes that plaintiff has demonstrated such discrimination by an overwhelming preponderance of the evidence. The Court finds Foley's after-the-fact attempt to justify his removing plaintiff from her position to be as uncredible as his testimony. After his approval of a maximum merit increase for plaintiff in January 1980, Foley's later

Attachment

denigrations of plaintiff's work are clearly unworthy of credence. The jobs offered plaintiff by defendant were not comparable to her previous position. Thus, when defendant discharged plaintiff, allegedly for refusing a comparable job when the true goal was to send her to Siberia, defendant compounded and ratified its earlier discriminatory actions.

2. Because of defendant's unlawful discrimination, plaintiff is entitled to full back pay, less any amount she earned at other work, from the date of her discharge to the present. The Court finds that plaintiff is also entitled to interest on the back-pay award.

3. The Court finds that plaintiff has fully complied with her duty to mitigate by seeking and accepting interim employment. Defendant is entitled to set off from the award the sum of $71,504, the amount plaintiff has earned since her discharge. Plaintiff's back pay is not tolled as a result of the two job offers by defendant. The burden is on defendant to reduce plaintiff's back-pay award because of failure to mitigate, and defendant has failed to carry its burden. Defendant is not entitled to set off any amounts received by plaintiff as unemployment compensation.

4. Plaintiff has diligently and earnestly sought reinstatement throughout this litigation. However, because of the hostility of defendant toward plaintiff, the Court finds that reinstatement would be inappropriate in this case. Additionally, reinstatement would require the displacement of a nonculpable employee, Julie Caspar, who has held the position of Project Manager for Expert Systems Development since plaintiff's discharge. In lieu of reinstatement, the Court awards plaintiff five years' front pay. The amount of this award is based on the difference between plaintiff's present salary and the salary that she would now be making (including cost-of-living adjustments and merit increases) were she still employed by defendant.

5. The total award to plaintiff is $374,864, calculated as follows:

Gross back pay (salary plus cost-of-living and merit increases)	$258,207
Fringe benefits	49,444
Interest	36,510
Mitigation	(71,504)
Net back pay	$272,657
Front pay (5 years)	$102,207
TOTAL DUE PLAINTIFF	$374,864

As the prevailing party in this action, plaintiff is also entitled to an award of costs and reasonable attorneys' fees. The parties are directed to attempt to agree on the amount of such costs and fees due plaintiff. If no agreement is reached within ten days of the docketing of this opinion, plaintiff's attorneys are directed to file a motion for attorneys' fees and costs with appropriate supporting documentation.

Memorandum

Date: June 4
To: Carl Bridewaite, President and Founder
From: Harry Merchant, Senior Vice President for Human Resources
Re: The *Green* Case

In response to your request of May 23, I will try to give as complete and candid an account as I can of what really happened in the Green affair. As you know, I had a ringside seat for much of what went on. I also had been friendly with Pete Foley for some time before the blowup, and learned some things from him that didn't come out at the trial, or were just hinted at.

First of all, it's important to understand something about the Green woman. She's not stupid—in fact, she did some pretty good work for us before all hell broke loose—but she's very neurotic. I'm not a psychologist, but she certainly strikes me as a hysterical type. Of course, it's true that Peter wanted to get free of her, even before he was promoted to Advanced Expert Systems. I don't blame him, either—she was jealous, insecure, and possessive. I do blame him a bit for not telling Paul Costello that he was still seeing her at the time of the promotion, but realistically, that's pretty understandable. He didn't want to jeopardize his opportunity to make a big move up, and he of course didn't relish the idea of revealing so much about his private life. What he told Costello was what he anticipated and hoped would soon be the case. He just didn't realize that Felicia was going to be so uncooperative and unprofessional about the whole thing.

As you know, the immediate cause of all the trouble was Pete's intention of bringing Susan Horvath into the department. Whether or not Pete was interested in her as a woman, I think it's safe to say that the whole thing would have happened anyway, sooner or later. If that hadn't triggered it, something else would have. Of course, when Felicia started calling Susan up and demanding that they reach some sort of understanding about Pete—and then calling her names when Susan refused to disavow any designs on him—it was inevitable that Pete would learn of it and have to get rid of Felicia. I don't see what else he could have done. He couldn't allow her to disrupt the work of the department with her private problems and anxieties.

A lot of the judge's decision evidently hinged on some pretty subtle questions—whether Pete really fired Felicia that Friday, whether she *believed* she had been fired at that time, whether Costello simply confirmed Foley's decision to fire her or made the decision himself, whether she agreed voluntarily to leave or was somehow coerced into saying that she'd leave. I wasn't present at the meeting with Costello, so I don't really know whether Felicia was volunteering to leave or merely acquiescing to a decision that she viewed as inevitable, if not already taken. I have to admit, though, that when I met with Pete and Felicia, it did seem to me that she thought she *had* already been fired at that time. You recall that this was after her confrontation with Pete, but before her meeting with Pete and Costello. She was trying to get Pete's decision to dump her reversed—that's why she asked about the grievance review procedure. I didn't think that our procedures were designed to handle such a situation, and I still don't. I think it's stretching the notion of "personal problem" pretty far to apply it to a situation in which an employee is having an affair with her supervisor and making intimidating phone calls to another woman who might work with him. I told her that she couldn't have recourse to a Grievance Review Committee, and Costello later told her the same thing.

All in all, I can't see that we did anything wrong in the whole affair. I'm not a lawyer, but if it were up to me, I'd certainly appeal. I think that we're being victimized by an unstable woman who created problems for herself and is trying to blame them all on us. I don't believe that she was mistreated in any way—certainly not because of her sex. She was moved out of the Advanced Expert Systems unit because she was making trouble, not because she was a woman or even because she was having an affair. Costello was very generous in offering her the opportunity to find another job in the company; he didn't have to do that at all, under the circumstances. And it's obvious that she didn't want just another job comparable to her previous one. She was trying to exploit our fear that we might have wronged her—seeking to use this incident as a stepping stone to a much higher position than she deserved or would have any reason to expect. As I said, with her emotional problems, I don't know how long she would have lasted in her old job, even if all this had never happened.

You asked me to consider also what implications this episode might have for our policies and procedures. Frankly, I don't think it has any. Our grievance machinery is well designed and has worked satisfactorily in the past. Of course, no rulebook can anticipate every possible contingency. People are complex and unpredictable, so unforeseen situations turn up now and then. The *Green* case was unique. I don't expect a repetition of it, and I don't see why we should rewrite our employee handbook because of it. There's no point in trying to construct a rigid mechanism for handling such human situations. It's better to deal with them on an ad hoc basis—informally, and with discretion. That's what we tried to do with Felicia. It's only because she was so stubborn—and lucked out by drawing a sympathetic judge who was willing to buy her whole sob story, hook, line, and sinker—that our efforts to resolve the problem fairly didn't succeed. I personally think that a victory in the appellate court will do us more good than a whole handbookful of fancy new complaint and grievance procedures. Failure to press an appeal, by contrast, will set a terrible precedent for the company. It will encourage our employees to feel that in any dispute with the company, they can always count on the courts to back them up. Fostering such an attitude is the last thing we want to do.

Memorandum

Date: June 7
To: Carl Bridewaite, President
From: Laura Stocker, EEO/Affirmative Action Officer, Personnel Department
Re: Implications of the *Felicia Green* Case for Personnel Policies

In responding to your request for an analysis of the *Green* case, I have tried to put aside the details to address the larger issues that could bear on the future of the company. I didn't know Felicia Green very well, and I wasn't involved in the events that led to the lawsuit, so I can't comment in depth on the personalities and motivations involved, as Harry Merchant can. What I do know about the matter leaves me with little personal sympathy for Felicia. Clearly she is a difficult and not very stable person, and made herself a royal pain to those around her. While it's presumptuous for me to judge her, I'd certainly like to think that I would never behave as she did. Nevertheless, I must say that while I have little sympathy for Felicia as a person, I sympathize to some extent with her *situation*. So, evidently, did the judge, and I think many female employees, both at CIS and elsewhere, are likely to share this feeling. I hope that by setting forth my view of the affair—as an EEO specialist, and as a woman employee—I can help you to see why this is so.

First of all, I think management does not sufficiently appreciate that Pete Foley's behavior has been very bad for the company. To begin with, he got involved in a sexual liaison with a co-worker. I don't wish to be harsh about this; we all know that these things happen. Nevertheless, it can't be said to represent very good judgment. More seriously, he deceived his superior about the relationship, telling Costello that the affair was over when in fact it wasn't. This was irresponsible, to say the least, and it was one of the main causes of the trouble that followed. Finally, instead of just breaking off with Felicia, he apparently tried to provoke her by telling her of his developing friendship with Susan Horvath and his intention of bringing Horvath into the department. Felicia, of course, responded hysterically, and I'm not trying to justify that in any way, but I think Foley must have known that she'd react that way. So in a sense it was Foley who first disrupted the workplace by using it as an arena for his personal relationships.

Second, I believe that the attempt to handle the situation informally was a big mistake. Everyone may have had the best of intentions, but look at what happened, and how the whole thing might be perceived by an outsider—or even another woman employee at CIS.

- All the male executives seem to have easy access to one another. Before the meeting with Merchant, Foley sees him alone to explain the situation from his point of view. Felicia doesn't get such an opportunity. Similarly, when the ex-lovers meet with Costello, Foley has a preliminary meeting to put his case; Felicia waits outside.

- Even in the meetings, Felicia feels that she isn't given a chance to explain her side of the story. (And the men involved don't dispute this.)
- Instead of clarifying the situation, the meeting with Costello confuses things more. Felicia claims that she was forced out, while Foley and Costello contend that she volunteered to leave. There is no formal record of what went on, no transcript, so the truth is almost impossible to arrive at.
- The informal attempt to find Felicia another job also fizzles. The company claims that she was interested only in jobs that were way above her head; she claims that the company offered her only jobs that were beneath her, or otherwise unsuitable.

Whether or not our handling of the situation was fair (and I am not in a position to make that judgment), there's no denying that there was at least the *appearance* of unfairness. I think that the very fact that the matter ended up in court (and worse yet, that we lost the case) shows that in such situations an informal approach can't be relied on. There should be specific appeal mechanisms available to women who feel that they have been subject to sex discrimination or sexual harassment. The fact that such cases involve highly personal, emotionally charged situations does not mean that they should be handled informally or "discretely" (which all too often just means paternalistically). On the contrary, that is precisely why a clearly formulated, objective grievance procedure is essential, as the Green fiasco demonstrates. It is the lack of such a procedure, more than Foley's unwise actions, that has led to our present legal vulnerability and exposed us to the displeasure of an unsympathetic judge.

As for appealing the judgment in the *Green* case, I think this would be a grave mistake. For one thing, it would send a very negative message to all women working for the company. For another, as the case goes up on appeal, it will inevitably attract more national publicity, which can only hurt the company. As I suggested above, the facts of the case don't present the company in a very favorable light. As the case unfolds, in the courts and in the press, we will be assuring ourselves of a bad image for another two to three years.

On a more general level, I think that we can learn some important things from the painful experience that this case has been for everyone concerned. We all know that in the past CIS has been an essentially male-dominated—even a "macho"—environment. This is inevitably changing, as the company expands and diversifies, acquiring more women employees in the process and promoting women into the higher professional and executive ranks. But the old ethos survives in many subtle and not so subtle ways. For example, note the judge's comments on Green's being underpaid. That situation, though it didn't figure directly in the legal arguments, may nevertheless have had some effect on the judge's sympathies. Unfortunately, it is far from unique; women in general tend to be less well compensated at CIS than at other firms. This fact, however, is only one symptom of a pervasive atmosphere in which women are not taken quite seriously. I think that the Green case provides us with an opportunity to recognize the need for a new ethos and new standards of behavior toward women, more in keeping with today's business environment and more appropriate for the kind of company that CIS must become if it is to thrive. Such a change can take root only if it is institutionalized. That is why I feel so strongly about the need for new appeal and grievance machinery.

In years to come, I hope that we can look back on the *Green* case as an important opportunity that we grasped, rather than as a setback for CIS.

Memorandum

Date: June 1
To: Carl Bridewaite, President
From: Quentin Harris, Esq., Chief Counsel for Employment Law
Re: *Green v. Computer Inventory Systems*

I have reviewed the judge's opinion in the *Green* case, and I strongly urge that we pursue an appeal as soon as possible. In my view, we have excellent prospects of success. The judge in the case has done a very slipshod job and committed numerous reversible errors. To wit:

- The judge conflated and confused two issues that should have been treated separately: (a) Green's *transfer* from the Advanced Expert Systems Department, and (b) her later *termination* for failure to accept any of the other jobs offered her. Once this distinction is clearly made, the finding of discrimination falls apart. Costello agreed to transfer Green out of the department because he believed that she had volunteered to be the one to leave, and sincerely, believed that Foley was more valuable to the Advanced Expert Systems program. The court simply refused to consider this legitimate, nondiscriminatory reason for defendant's action. As for Green's subsequent discharge, the absence of any comparably situated male employee who was treated differently prevents even a *prima facie* case of discrimination from being made. Instead of making this essential distinction, the judge treated the transfer and the dismissal as if they were a single act: Green was fired while Foley was retained. In fact, these were two separate actions that took place four months apart, and each was justifiable.
- With respect to the facts of the case, the judge accepted uncritically Green's contention that she was removed from the Advanced Expert Systems Department or coerced into leaving. The preponderance of the testimony, however (including Green's own), shows that she volunteered to leave. Similarly, the judge found that the two jobs subsequently offered Green were not comparable to her previous one. This finding is based on a distortion of the testimony of Roger Delano and on an excessively narrow definition of "comparable jobs." I do not think that either of these findings would hold up on appeal.

- Finally, the award of five years' front pay is arbitrary, excessive, and a gross abuse of judicial discretion. It rests on a misunderstanding of the proper purpose and scope of such awards, for it would put the plaintiff in a better position than she would have occupied in the absence of the alleged discrimination.

I think it is important to note that we had very bad luck in drawing Judge Holmes for this case. He is far and away the most liberal jurist on this particular bench and has a well-earned reputation for propounding far-out interpretations in civil liberties cases. He is almost always sympathetic to individuals in any sort of confrontation with organizations, institutions, or authorities—especially if the individual is a member of a minority group. Another judge might not have bought all of Green's "facts" so uncritically. You will note that Holmes' antibusiness slant is clearly evident in the language of his opinion. The appellate court will probably not take a very favorable view of such obvious bias. Luckily, as a result of the changing climate of opinion in the country, and of recent appointments to the bench, the judiciary as a whole is becoming more balanced in its approach to business and less willing to embrace questionable doctrines in civil liberties cases. I think that we can expect a more sympathetic hearing before the appellate bench than we received in the initial trial.

In conclusion, let me say that this could be a vital case, not only for CIS, but for the protection of employer interests everywhere. It could help shape EEO doctrine—which is already entering a period of rapid change—for years to come. So I urge that we press the appeal.

Discussion Questions

1. Reviewing the chain of events in this case from the perspective of Chief Operating Officer Atwell, what does this incident tell you about your organization? What management decisions or actions allowed the problem between Foley and Green to take an increasingly hostile course? What would you want your organization to learn from this experience?

2. Public attention to sexual harassment is a relatively recent focus for employer-employee relationships. Why might sexual harassment become an even more dominant issue in the coming years? What difficulties are posed by the problem of determining voluntary vs. coercive sexual conduct in employment relations?

3. Comment on the following legal questions about sex discrimination claims involving charges of sexual harassment:

   A. Should a charge of sex discrimination be rejected if the defendant can show that the plaintiff manifested provocative speech and dress in the workplace?

   B. Should the existence of sex discrimination in a company depend on finding that a plaintiff's employment status was affected?

   C. Should a company be able to defend itself on the basis that an employee's misconduct was unknown to officials?

4. Respond specifically to Merchant's conclusions. As Atwell, what is your reply to him?

5. What, if any, changes would you consider in company policies and programs? Do you agree with Stocker that the issue of sex discrimination at CIS calls for complaint procedures separate from the firm's regular grievance system?

6. Weigh the pros and cons of appeal as presented by Harris and Stocker. As the final decision maker, would you elect to appeal the decision? Why or why not?

CASE 3

Reproductive Risk and EEO:

Balancing Enterprise Liability and Women's Rights

Longlife Rubber Company, a tire manufacturer, holds a sizable portion of the tire market for farm vehicles and mobile heavy construction equipment. Longlife was recently acquired by Webster International, a conglomerate that has a variety of manufacturing firms within its ranks. Webster's CEO, Alexander Vague, has called on all units within Webster to "hunker down and meet the overseas competition that is our severest threat in the 1990s." "High technology," said Vague, "will be the key to that effort."

Longlife's Littlefield plant, one of ten Longlife plants in the United States, was built in the early 1950s and employs 850 production workers. Having begun developing and implementing new manufacturing processes to enhance productivity, Longlife recently decided to introduce a new technique for using vinyl chloride in the manufacturing processes at Littlefield. Currently the company is in the midst of making the necessary plant modifications.

Benjamin F. Jones, Plant Manager at Littlefield, is concerned that worker exposure to vinyl chloride may present a serious threat to the well-being of a fetus, and wants to take appropriate measures to avoid or limit this potentiality. Jones has requested that a number of corporate officials respond to the proposed "Fetal Protection Policy" that he is considering for implementation at the Littlefield plant. In response, Jones has received the following memoranda, and is ready to make his decision.

Key Persons Involved
Benjamin F. Jones, *Plant Manager, Littlefield*
Debra Weaver, *Director, Office of Human Resources*
Samuel Getz, M.D., *Medical Director*
Michael Murphey, *Vice President, Equal Employment Opportunity*
Peter Durfee, Esq., *General Counsel*
Peter Van Arsdale, *Marketing Manager*

Memorandum

Date: July 16
To: Benjamin F. Jones, Plant Manager, Littlefield
From: Debra Weaver, Director, Office of Human Resources
Re: Proposed "Fetal Protection Policy"

I have given careful consideration to the question you posed to me—"Would it be in the best interests of Longlife Rubber Company to exclude women of childbearing age from work that may expose them to vinyl chloride, a chemical known to adversely affect the development of a fetus?"

Perhaps I can be of greatest assistance by describing a wide spectrum of alternative plans that may be available to you. These alternative approaches grow out of my experience with similar (but nowhere as dramatic) situations in which new personnel policies were needed to counter known or suspected hazards. Please note, however, that while I am somewhat sophisticated in these areas, I am certainly not an expert on the legal and EEO implications of what you propose to do, and our staff experts on those matters will need to be consulted.

My direct (and short) response to your question is no, I do not believe a policy of excluding women from positions that might endanger the health of the fetus is in the best interests of the company. However, I urge you to keep in mind that my focus and primary concern are with the impact that such a policy would have on employee morale in your plant and in the rest of the organization, and with how our customers and the general public would view us as a result of that policy. With this in mind, permit me to summarize the options I see available to you, short of an exclusionary policy affecting all women of childbearing capacity.

1. *Alter the work setting so as to limit exposure.* There must be a way of containing the vinyl chloride in the manufacturing process so as to prevent its release, thereby reducing the exposure of *all* workers. If the cost of doing so is not prohibitive, this would seem to be the optimal choice. If not, then those employees who work with the chemical or who may be exposed to it might be fitted with protective gear—for example, insulated clothing or respirators.

2. *Use a rotational work plan.* Limit employee exposure by rotating workers; in this way, you will be restricting the time and amount of exposure of any worker.

3. *Substitute another chemical for vinyl chloride.* You many want to search for an alternative to vinyl chloride. If one exists, a cost-benefit analysis should reveal if substitution is feasible.

4. *Exclude women unless they can prove that they are not at risk of pregnancy.* Sterilized women and women beyond childbearing age would *not* be barred from working with vinyl chloride, under this "Fetal Protection Policy." We must be careful, however, not to make it appear as if we are requiring women to get sterilized as a condition of employment. I've attached a recent news clipping that details Chem-America's problem in this regard.

5. *Transfer pregnant employees.* You could allow all women of childbearing capacity to work at the site, but then (a) inform them of the dangers to a fetus of a mother's exposure to vinyl chloride, (b) require them to inform the company of when they are pregnant or are trying to get pregnant, and (c) transfer them at that time. The obvious problems with this plan are, first, how do we police this policy and, second, since women very often do not know they are pregnant until months into gestation, will the plan indeed protect the fetus as desired?

6. *Utilize disclosure and waiver.* Finally, you may want to open the positions up to all employees, but inform women of childbearing capacity of the dangers presented to the fetus by their working with vinyl chloride—that is, inform them of dangers to the best of our ability, given current scientific knowledge. If they chose to work with vinyl chloride (despite our warning), we would then ask them to sign a form waiving any future claims they or their offspring might have against Longlife.

I would also like to share with you the experience of a friend of mine, who was confronted with a similar dilemma at Gorman Chemical Company. The company reached the conclusion that exposure to a certain chemical used in its production process presented a danger to the viability of the fetus. Gorman proceeded to classify the jobs in the plant using the following three categories: "restricted," "open," and "controlled" jobs. The company barred women of childbearing capacity from "restricted" jobs, made "open" jobs available to all employees without restriction, and permitted women of childbearing capacity to be employed in "controlled" jobs (where exposure to the harmful chemical was limited) only upon completing a special counseling program conducted on company time. Further, the women employed in "controlled" jobs were required to notify their supervisors as soon as they knew they were pregnant. If I remember correctly, a lower federal court upheld Gorman's policy in the face of a claim of sex discrimination; however, the verdict was overturned on appeal and the case returned to the district court. As I understand it, the appellate court found that Gorman had not clearly established the necessity for such a policy. I believe the matter is still under litigation. I mention the Gorman situation just to show that apparently even a carefully thought-out and seemingly justified fetal protection policy is subject to challenge and may have to be defended in court.

Attachment **"CHEM-AMERICA'S FETAL PROTECTION POLICY SUBJECT OF LAWSUIT"**
Daily Blade. March 12

The National Civil Liberties Association (NCLA) today filed a sex discrimination lawsuit against Chem-America's policy of excluding women of childbearing capacity from jobs in the Pigments Department of the company's Harbor Island, New Caledonia, plant. The company, fearing injury to a fetus from the mother's exposure to lead, barred women from work in that department unless they could prove they were sterile. The NCLA suit alleges that Chem-America is guilty, under state law, of fraudulently misleading its employees, and asks the court to declare the exclusionary practice illegal; to require that sex-neutral policy be instituted and that the reproductive status of workers remain confidential; and to award back pay, seniority rights, and damages to the injured women.

Chem-America's fetal protection policy was instituted at the Harbor Island plant in October 1986. A company official told women employees at that time that the policy would soon be required by forthcoming regulations of the Occupational Safety and Health Administration. Out of fear of losing their jobs, five women at the plant submitted to sterilization procedures. OSHA later determined that exposure to lead affects the reproductive capacities of both sexes. Chem-America closed down its Harbor Island plant soon after.

The company argued that the policy was based on reputable medical studies demonstrating adverse effects on the fetus of exposure to lead. The danger to the fetus is aggravated by the fact that exposure tends to occur, if at all, in the first trimester, when the fetus is most vulnerable and the mother is least likely to know she is pregnant. The company contends that it neither conditioned further employment on the women being sterilized nor encouraged such actions on their part. The company claims it offered to transfer the women to other jobs within the plant.

The NCLA alleges that this practice is just another example of "romantic paternalism" and company "protectivism" that have served to deny women equal opportunity to employment throughout American history. The group contends that, in reaching its decision to exclude only fertile women workers from the Pigment Department, Chem-America ignored ample evidence that exposure to lead affects the reproductive capacities and genetic makeups of both men and women.

The Petroleum Workers Union is also critical of Chem-America's policy of exclusion. A union official commented, "If Chem-America and others can get away with removing women of childbearing age from these jobs, we will have established the principle of altering the worker to the configuration of the workplace instead of altering the configuration of the workplace to protect the worker." The solution, the union insists, is to make the workplace safe for all workers, not to ban particular classes of workers. Corporate medical directors agree, but counter that ensuring a clean and healthful workplace at all times for all workers is not always technologically or economically feasible. This problem is compounded by the difficulty in determining with any certainty what precisely is a "safe" level of exposure.

The lawsuit alleges that the emotional impact on the sterilized women has been devastating. Several of the women have required ongoing medical and psychiatric treatment. The women complained that they were ridiculed at the plant following their sterilization. One woman recalled a male colleague referring to her as "one of the boys" following her sterilization. While each of the five women had children prior to her decision to become sterilized, a number of the women have remarried and now face the prospect of not being able to have a child by their new husband. The lawsuit seeks to recover damages for the emotional distress suffered by the women.

Memorandum

Date: July 18
To: Benjamin F. Jones, Plant Manager, Littlefield
From: Samuel Getz, M.D., Medical Director
Re: Teratogenic/Fetotoxic Effect of Vinyl Chloride

As you requested, I have reviewed the medical literature on the issue of reproductive hazards, and my findings are summarized below. However, the medical community is only now beginning to give this issue the attention it deserves. What is clearly lacking is good, reliable basic research.

Types of Reproductive Hazards Reproductive hazards are generally broken down into three categories: mutagens, gametotoxins, and teratogens/fetotoxins. Mutagens alter the genetic makeup of living cells; gametotoxins make an individual infertile; and teratogens or fetotoxins harm the fetus *in utero* but have no effect on the reproductive processes of either males or females. Your concern is obviously with the last category, teratogens/fetotoxins.

Teratogens/fetotoxins have their most devastating effect on the fetus during organogenesis (days 18–60 of gestation). In particular, an embryo is most vulnerable during the period of early organ differentiation (days 18–30). High-level exposure will most likely result in the spontaneous abortion of the fertilized ovum, while low-level exposure will most likely produce defects that are apparent either at birth or in the child's early years. For your purposes, to avoid harmful exposure to the fetus, pregnancy must be ascertained soon after the second week of gestation.

Vinyl Chloride Vinyl chloride is generally viewed as being a mutagen—that is, intense exposure to the chemical may result in chromosomal abnormalities—and there are clear indications that it is a gametotoxin as well (impacts on fertility). While reliable data on the teratogenic/fetotoxin qualities of vinyl chloride are more nascent and uncertain, in my professional opinion vinyl chloride is dangerously teratogenic as well. As a result, I believe that your proposed "Fetal Protection Policy" is both scientifically justifiable and, from a corporate policy standpoint, desirable. In support of this view, may I just point out that in 1979 B.F. Goodrich Company instituted a similar policy involving all women of childbearing capacity assigned to vinyl chloride polymerization areas.

On the surface, a simple and efficient solution would be to utilize manufacturing processes and practices that minimize or eliminate all workers' exposure to the chemical. Of course, that is what we should be striving for. Unfortunately, I know of no industrial user of vinyl chloride who has been able to successfully institute such a practice, though efforts in the industry continue. This work, however, is further complicated by the fact that we have yet to determine with full confidence acceptable levels of exposure of the fetus to vinyl chloride, or any other teratogen/fetotoxin, to my satisfaction.

In my opinion, your "Fetal Protection Policy" is both scientifically justifiable and based on sound business reasoning, given the information currently available to the medical community.

Memorandum

Date: July 19
To: Benjamin F. Jones, Plant Manager, Littlefield
From: Michael Murphey, Vice President, Equal Employment Opportunity
Re: Implications of Proposed "Fetal Protection Policy"

I have received your inquiry about the EEO implications of excluding women of childbearing age from jobs in which they may be exposed to vinyl chloride, a chemical injurious to the well-being of a fetus. My answer to your question is that the implementation at Longlife of a fetal protection policy such as you have proposed for the Littlefield plant would present problems for the company's EEO practices, and could do damage to the company's image and legal standing far beyond what we can anticipate now.

The EEO issue your policy presents can be succinctly stated as follows: Is protection of the fetus from harmful exposure sufficient to justify the exclusion of all women of childbearing age from a certain class of jobs? My comments will be addressed to this issue.

We are confronted here with one of those rare situations in which concern for the health and safety of our workers conflicts with other legal obligations, namely, not to discriminate on the basis of sex. I do not want to minimize the safety and health dangers inherent in this situation; nor do I want what I say to be interpreted as suggesting that we endanger the well-being of our female workers or their potential offspring so as to avoid EEO liability. I will urge, however, that, if you decide that some sort of exclusionary policy is necessary (after you have explored all less drastic alternatives), you tailor that policy to your specific needs and seek to minimize the displacement in every way feasible. One sure way to incur EEO liability is to implement classwide policies without an unassailably legitimate and nondiscriminatory basis for doing so.

Please bear in mind that women have traditionally been excluded from production and line positions—often for the very reason we would be tendering here—"for their own protection." Men have always been able to choose whether or not to accept hazardous work; why not give women the same opportunity? Just as women have begun to move into jobs that were formerly denied them, we are asking them to confront another "biology is destiny" situation. In addition, you know as well as I how few jobs are available to women in general in a small town like Littlefield.

I have not forgotten that the policy is designed specifically to protect the fetus, and the woman, but isn't the woman and not Longlife the one with the responsibility to care for her child and look out for its well-being? We do not prohibit a pregnant worker from smoking or drinking alcohol, especially at home, yet those activities may be just as damaging to the fetus as working with vinyl chloride. I can already hear the charge of "romantic paternalism" from the women's groups. There will undoubtedly be a presumption against us that will be hard to overcome, no matter how noble our true intentions. It will also be important to show that we have investigated the effects of vinyl chloride on the male reproductive system and determined that they are not sufficient to warrant job exclusion.

Assuming that removal is necessary to protect the fetus, may I suggest that you institute a voluntary transfer system rather than a mandatory policy of exclusion. I think we can assume that all women who are pregnant or are planning pregnancies will choose not to endanger the fetus, particularly if a reasonable option is available.

You must understand how your proposed policy fundamentally undercuts so much of EEO law. Laws that protect against employment discrimination first and foremost seek to ensure equal opportunity and access to jobs. In effect, we would be asking a class of workers—potentially pregnant women—to bear the weight of our failure to provide a safe workplace. And in general, the idea of a "susceptible" class of workers is highly offensive to EEO principles. All identifiable groups have unique characteristics upon which basis a company may be able to justify disparate treatment. Carried to the extreme, we would be allowing genetic makeup to determine job availability.

Memorandum

Date: July 20
To: Benjamin F. Jones, Plant Manager, Littlefield
From: Peter Durfee, Esq., General Counsel
Re: Legal Liability under Proposed "Fetal Protection Policy"

I have reviewed your proposed "Fetal Protection Policy" to be instituted at Longlife's Littlefield plant, and have concluded that while we may become subject to suit under federal and state employment discrimination laws, that option is preferable to the potentially boundless liability that we may incur should the children of our women workers be born with birth defects resulting from exposure of the fetus to vinyl chloride at one of our plants. In short, if asked to choose between liabilities, I believe it is in Longlife's best interest to opt for the bounded danger of liability that would result from your proposed policy.

I will summarize existing employment discrimination law as it applies to this situation and will suggest some actions that may limit our potential legal liability.

Title VII Background Title VII of the Civil Rights Act of 1964 and Executive Order 11246, which applies to federal contractors, require that government enforcement agencies carefully review policies and practices that serve to exclude classes of individuals from equal employment opportunities. A company may try to offer a legitimate, nondiscriminatory reason for the policy or practice, which will then be reviewed by the enforcement agency to determine if it is merely a pretext for discrimination. Facially neutral policies that have a disparate impact on a protected class of individuals without nonpretextual justification will be found to be in violation of the law.

There are two recognized defenses to charges of employment discrimination: (a) the bona fide occupational qualification (BFOQ) exemption and (b) the "business necessity" defense.

The BFOQ exception applies to a suspect employment practice that "is a bona fide occupational qualification necessary to the normal operation of a particular business or enterprise" (42 U.S.C. section 2000[e]–2[e]. Courts have taken a notably narrow view of this defense. To come within this exception, all or almost all of the protected class must be unable to perform the particular job safely and efficiently. *Analysis under this defense is focused on the worker's ability to perform the job safely and efficiently, not on the health and safety of a third party*, here, the fetus. While it is possible that courts in the future may take a broader view of this defense, particularly when the causal connection between harm to the fetus and exposure to the toxic substance is clearly substantiated, in my opinion this is not a viable defense in our situation, and we should not rely on it.

The courts have created a second defense to employment discrimination lawsuits, namely, the "business necessity" defense. Here the focus is on the employer's ability to demonstrate that the policy or job criteria have "a demonstrable relationship to successful performance of jobs for which [they were] used." To be sustained, the business necessity defense requires a legitimate and compelling business purpose, a policy that clearly effectuates that purpose, and the absence of reasonable alternatives.

There are a number of considerations to note in asserting the business necessity defense. Congress amended Title VII in 1978 to explicitly bar discrimination on the basis of pregnancy. As a result, any fetal protection policy that applies only to pregnant women or to women of childbearing capacity will certainly be considered "facially discriminatory"—that is, discriminatory on its face because it overtly and intentionally singles out one class of employees for special treatment. Generally, the only defense permissible in such a case is the BFOQ defense, which, as I indicated above, is not likely to be applicable to a situation such as ours. However, in some of the recently decided cases of this sort, the court has held that the business necessity defense may be acceptable under certain circumstances.

In one case, for example, the court maintained that a fetal protection policy, though seemingly discriminatory on its face, might actually be considered "neutral in the sense that it effectively and equally protects the offspring of all employees" if the employer could show that the policy was designed to counter a genuine hazard that is confined to the unborn children of female workers (that is, the unborn children of male workers are not at risk). Such a policy could still be shown to adversely affect the employment opportunities for women, and thus be subject to challenge under Title VII. However, in such "disparate impact" cases, the business necessity defense can be used. And in the case under discussion, the court in fact "recognize[d] fetal protection as a legitimate area of employer concern to which the business necessity defense extends." In another recent case, the court held that the business necessity defense could legitimately be invoked because of the possibility that a huge tort suit brought by the congenitally malformed child of an exposed worker might have devastating financial consequences, and so disrupt the employer's ability to carry on its business.

Lest I make our prospects appear too rosy, let me state that it is not certain that all circuits courts would necessarily adopt either of these approaches. (Indeed, one circuit court has already rejected the idea that the possibility of exposure to tort action justifies a business necessity defense, saying that the potential liability is "too contingent and too broad a factor.") Moreover, even when the possibility of a business necessity defense exists, the burden on the employer is still great. It is necessary to prove that:

- Objective, scientific evidence suggests that there is a significant and substantial risk of harm to the unborn children or potential future offspring of pregnant or fertile women workers, but not of men, from exposure to a particular substance of condition.
- Such exposure is likely to occur in the absence of the employer's fetal protection policy.
- The policy effectively accomplishes its aims (that is, protects women from such exposure).
- The policy has not been adopted with intent to discriminate or as a pretext for discrimination.
- There is no alternative policy that would provide better protection or comparable protection with less discrimination effect (that is, less of a disparate impact on male and female workers).

Thus, the circumstances under which a fetal protection policy can be held valid are quite limited. Nevertheless, though such policies are hard to justify, the courts have so far declined to prohibit them altogether. I believe, therefore, that our legal position, though difficult, may be tenable if we are willing to defend our policy with energy and determination.

Tort Liability for Injury to Fetus Perhaps the most persuasive factor arguing for the proposed "Fetal Protection Policy" from a legal standpoint is the emerging right of the fetus to recover for injuries suffered during gestation. It is fair to say that most state jurisdictions allow recovery in one form or another in lawsuits brought on behalf of the fetus. Most cases have arisen in the product liability area, although fetal recovery has also been allowed under insurance policies and in actions for wrongful death of a stillborn child. What is crucial for our purposes is that *it is not clear whether the mother can waive the rights of the fetus*.

In other words, we cannot be sure that we are protecting ourselves against recovery in the name of the fetus by having the mother waive any claims her offspring may later have against us. If a child were to be born with birth defects resulting from exposure at one of our work sites, we may be liable for continuing damages for the length of that child's life. Clearly the proposed "Fetal Protection Policy" is the lesser evil from a liability point of view.

Suggestions for Limiting Title VII Damages   Assuming that the proposed "Fetal Protection Policy" is implemented, may I suggest that the following actions be considered so as to minimize our prospective employment discrimination liability:

1. Encourage the corporate medical director to document the danger to the fetus of exposure to vinyl chloride (if you have not done so already), and have him or her submit a memo to that effect for the files.

2. Ensure that the women being transferred do not lose job benefits or seniority as a result of the move, and try diligently to place them in equivalent or better positions.

3. Limit the breadth of the exclusion policy—for example, by applying it only to those women who are exposed to the chemical at certain levels or over a certain period of time.

4. Make it clear that proof of sterilization will not exempt women workers, in order to avoid what happened at Chem-America's Harbor Island plant and to get us out of the business of cataloging workers' reproductive capacities.

5. Monitor closely developments in scientific research on the effects of vinyl chloride on the fetus and on the male and female reproductive systems, and amend the policy in accordance with new developments.

Please bear in mind that none of these actions will permit us to avoid potential liability, only limit actual damages, if awarded.

If I or my staff can be of further assistance, please let me know.

Memorandum

Date: July 23
To: Ben Jones
From: Peter Van Arsdale, Marketing Manager
Re: The Vinyl Chloride Problem

Ben—I know you are reviewing our approach to vinyl chloride use in Longlife's new tire production process. You have lots of legal, employee relations, and public relations issues to sort out, but I think I need to put one more element into your in-box. (Sorry . . .)

We have learned that Korean Rubber Company, our fastest rising competition in both the U.S. and the overseas markets, will also be going to the new tire-manufacturing process using vinyl chloride. An Argentine firm, Nueva Epoca Tire Company, has also begun using this process. Neither company has any government occupational health or safety regulation to contend with; nor have the unions representing their women workers made any issue of possible fetal damage. Civil rights litigation over such issues is not present in either country.

If we have to use an expensive substitute for vinyl chloride, or if the air-purification and protective-gear expenses substantially increase our production costs, we could be left at a major competitive disadvantage in the upcoming marketing struggle with Korean Rubber and Nueva Epoca over our largest and most important product. In a sense, then, the jobs of hundreds or even thousands of Longlife workers could be affected by this decision.

Discussion Questions

1. Benjamin Jones is seeking to enhance the "best interests of the company." How would you describe the company's interests in this case? As Jones, would you take further action on any of Debra Weaver's suggestions?

2. Assess the adequacy of Dr. Getz's responses to Jones' request. What additional information might reasonably be expected from him? Does Longlife have a social responsibility either to conduct research or to suggest university-conducted research into the health effects of chemicals used in its manufacturing processes?

3. The National Civil Liberties Association and Michael Murphey both refer to the concept of "romantic paternalism." What do you understand this concept to mean? What actions might be taken to ensure that romantic paternalism has no role in a firm's policies and practices?

4. Murphey proposes a voluntary, rather than mandatory, transfer system for women in order to protect fetuses. What is his reasoning, and what are the limitations of such a policy?

5. Evaluate the ability of the company to defend a fetal protection policy on the basis of (a) the BFOQ principle and (b) the business necessity defense. Do Durfee's suggestions for limiting Title VII damages exhaust the company's sources of evidence for demonstrating that its exclusionary policy is a business necessity, that is, is not discriminatory?

6. Are there any other actions—for example, disability leaves of absence for pregnant employees—that the company might undertake to protect the fetus, yet not discriminate against women? How would you evaluate the degree of expense that should be incurred to implement these alternatives?

7. What is your response to Van Arsdale's memo about the threat of low-cost foreign competition? What other considerations are raised for Jones and other company officials by viewing the situation in a competitive framework?

CASE 4

Employee Protest vs. Employee Loyalty

How Much Dissent Should a Corporation Tolerate?

Great Rainbow Furniture Company was founded in Michigan in 1924. The corporation, now headquartered in Los Angeles, had sales of $870 million last year. It employs about 24,000 people in plants and offices located throughout the United States and Canada.

Great Rainbow's California Office Furnishings Division has its manufacturing facility in Urbana, California, outside Los Angeles. The work force is blue-collar, divided between manufacturing and shipping. Employees are represented by the Amalgamated Furniture Workers Union, Local 27.

Over the years, Great Rainbow has implemented EEO requirements and has achieved a proportion of black and Hispanic workers in its Urbana facility roughly equal to the proportion of those minorities in the metropolitan area population. However, advancement of black workers into supervisory and line management positions has been slow. Several charges of discrimination against blacks in promotion policies have been filed by black employees with the Equal Employment Opportunity Commission, and there has been picketing outside the Urbana plant by civil rights groups.

Six black employees of Great Rainbow have just sent a letter to the Urbana Board of Education—a major customer of Great Rainbow for desks and furniture—attacking an "affirmative action" award the school board voted to give the Urbana facility's Director of Personnel, Hunter Greene.

You are Perry Grayson, Great Rainbow's CEO, and have received the following four memos from corporate staff members. You must decide what action to take in response to the public protest of the six employees.

Key Persons Involved

Perry Grayson, *Chief Executive Officer*
Hunter Greene, *Director of Personnel*
Violet Smith, *Employee Relations Manager*
Harold Bluestein, Esq., *Counsel for Employment Law*
Tim Browner, *Vice President for Corporate Social Responsibility*
Members of the Black Rainbow Employees' Group
    Oliver Jones
    Louis Goldstone
    Blanche Ellison
    Norton Black
    Sterling V. Reade
    Rodney Johnson

Memorandum

Date: August 1
To: Perry Grayson, Chief Executive Officer
From: Hunter Greene, Director of Personnel
Re: Black Rainbow Employee Group

I'm sorry to say that we're having trouble with the Black Rainbow group again. You may already have received your copy of the letter they sent yesterday to the Urbana Board of Education—one of our major customers—accusing Great Rainbow of being antiblack. (I've attached a copy of their letter, which they sent to us at the same time that it went to the school board.)

    You've got to hand it to these people—they sure are persistent. Last year they picketed for a week, carrying big signs up and down the block in front of the plant: "Rainbow Is a Racist Employer." They've sent letters to the governor and the city council describing our hiring and promotion practices in very unflattering terms (to say nothing of what they call *me* in this most recent letter: "the standard-bearer of Rainbow's racism"). In addition, they've repeatedly filed complaints with the Office of Federal Contract Compliance, the National Labor Relations Board, and, of course, the Equal Employment Opportunity Commission.

    The EEOC, as you will recall, substantiated some of their claims; it accepted their allegations that Great Rainbow's employment practices were discriminatory in some respects (though the actual violations of EEOC guidelines were rather minor and limited). We negotiated a conciliation agreement with the EEOC, stipulating that we would correct certain "deficiencies" in our practices. Nevertheless, on the basis of the EEOC findings, the Black Rainbow group last month filed an action against us in California District Court, this time on our promotion policies. We are confident of winning this suit. It is not of great concern, but it is damned annoying.

Employee Protest vs. Employee Loyalty

What burns me up about this particular incident is that these people are trying to pressure us through one of our best customers. And ironically, they are complaining about a prize Great Rainbow received for an outstanding community relations program. The award-winning program is called Successways. As you know, each year the personnel department takes a bunch of sixth-grade kids from the Baldwin Street Elementary School through the plant, showing them all the interesting things they'll get to do if they come here to work when they grow up. The program is meant to inspire the kids—who are mostly Chicanos—to become productive members of society. And to show them that we, at Great Rainbow, welcome them as employees. (I've attached a copy of our newsletter on the Successways project.)

One would think that an antidiscrimination group like the Black Rainbow would be pleased to learn that this program is successful. Not at all. Because if Rainbow gets prizes for affirmative action, then the Black Rainbow employees can't characterize the company as racist any longer and will have nothing to complain about. We can't satisfy them. Last year, after the picketing, we set up a meeting between Black Rainbow members, some personnel management people, and Violet Smith, the Employee Relations Manager, who happens to be black and who came from the San Francisco plant just to meet with them. But not one of the Black Rainbow showed up.

There are three courses of action we can now consider:

1. *Tolerance*. As in the past, we can choose to do nothing. Up to now, we have always viewed Black Rainbow as a small, unsophisticated group of malcontents posing little threat to the smooth operation of business here. Their complaints about racial discrimination may have been unreasonable and unjustified, their actions irritating and sometimes embarrassing, but they have been relatively harmless. We have therefore believed that we would risk more by censuring them than by allowing them to gripe in peace. We have felt that by trying to accommodate the Black Rainbow employees, we would be broadcasting our "goodwill" to our other (600) black employees—who show very little support for Black Rainbow activities, by the way.

Alternatively, if we disciplined these letter-writing cranks, we would run the risk of inciting our other black employees. In addition, counsel has advised us that most of their protest activities are protected by law, even if they unfairly put Great Rainbow in a bad light. Taking action against them might only serve to get them media attention—locally or even nationally—which they would no doubt welcome.

I have serious reservations about continuing with this tolerance under the present circumstances, however. It seems to me that the Black Rainbow employees have become more than a nuisance; they may have seriously damaged our reputation with an important customer. Their letter could have a very negative effect on our relationship with the school board. The board was considering using our Successways program as a model for other companies in this area—something they will surely be reluctant to do if they think we discriminate. They might even stop doing business with us—a blow we cannot be expected to take lying down.

Moreover, if we don't respond to Black Rainbow's letter—which seems to me clearly libelous (though you'd have to check with legal counsel for an authoritative opinion on that)—we could be encouraging our employees to have disrespect for company management. What concerns me most is that by not taking some kind of significant punitive action now and establishing a precedent and a clear policy, we could be inviting trouble down the road. If some other employees attempt to interfere with a customer in the future and we want to punish them, our hands will be tied because in this instance we tolerated such behavior. Whatever we do or don't do now could have serious consequences in court in any subsequent case.

2. *Disciplinary hearing*. A disciplinary hearing, well publicized in the shop, might be the answer. It would let employees know that management will not simply wink at behavior that could damage business. Perhaps the Black Rainbow employees who signed the letter should be suspended without pay for a while.

3. *Termination*. There is a third option, which I think is preferable: firing the people who signed the letter. Surely this is wholly justifiable in view of the seriousness, recklessness, and irresponsibility of their actions. (See the attached memo that I wrote on the subject of our termination policy.) As I mentioned, the group has little support from the majority of our black workers. I also doubt that the union would be willing to make Black Rainbow a cause célèbre. I think that the morale of *all* workers—black and white—would be improved by getting rid of these troublemakers. In short, I feel that the Black Rainbow employees have really stepped over the line this time, and should be dealt with accordingly.

Attachment   "SUCCESSWAYS PROGRAM RECEIVES AWARD"
Article from the *Great Rainbow Employee Newsletter*

Great Rainbow's outstanding Successways program has won an award from the Urbana Board of Education.

Successways is an extension of Great Rainbow Furniture Company's Affirmative Action Programs. It is designed to produce better-qualified people for the work force of the future.

The program was started in 1982 at the Urbana plant, in cooperation with the nearby Baldwin Street Elementary School. Each year since then, speakers from various functions within Great Rainbow have visited the school to talk with sixth-grade students. Audiovisual materials and other items are supplied throughout the year to help the students learn about the work world. And to highlight the program, students are taken on a tour of the Great Rainbow facility to see the workers in action.

The program is very successful. On June 9, Hunter Greene, Director of Personnel at the Urbana plant, will receive the award on behalf of the company at a special meeting of the school board.

Attachment

134 Central City Avenue
Urbana, CA

July 31

Urbana Board of Education
100 Center Street
Urbana, CA

Dear Members of the Urbana Board of Education:

Members of the Black Rainbow employee group of Great Rainbow Furniture Company were shocked and outraged when we heard of the affirmative action award your board gave to Hunter Greene.

Hunter Greene is the standard-bearer of bigoted racism at Great Rainbow. He is the Director of Personnel, so you don't get hired without him.

We filed the first charges of discrimination against Great Rainbow ten years ago, and we have filed twice more in the past four years.

You should know that many of our charges of racism were substantiated by the Equal Employment Opportunity Commission and one case is pending litigation at this moment.

We take offense with the honors being given to these bigots. Most of us in the Black Rainbow group have been fighting racism and discrimination at Great Rainbow for the past ten years and some even longer.

We would like an immediate reply from you explaining why you didn't take into account Great Rainbow's total affirmative action situation. Please send your reply c/o Oliver Jones at the above address.

Yours truly,

The Black Rainbow:

Oliver Jones	Norton Black
Louis Goldstone	Sterling V. Reade
Blanche Ellison	Rodney (Red) Johnson

MEMO ON COMPANY POLICY ON TERMINATION   *Attachment*

Great Rainbow's policy on terminating employees has always been flexible. Most terminations are the result of poor job performance and are the responsibility of the employee's immediate supervisor.

We have, however, fired workers who cause disciplinary problems. Two months ago, we fired four men (truck loaders) who were drinking and roughhousing on Great Rainbow property. Drinking on company property is a dismissible offense as outlined in the *House Rules* issued to all employees when hired, and agreed to by our union. Although these men had not yet failed to get their loading done on time, they were disturbing other workers. Thus, those terminations were actually job related.

I can find no record of an employee being fired for interfering with a customer. And I can find no written statement describing company policy toward employees who interfere with customers. The *House Rules*—included in the copy of the Collective Bargaining Agreement that all Bargaining Unit Employees receive—and the posted rules—appearing on bulletin boards around the plant—do not address this situation.

Just as there is no written policy concerning impermissible contact with customers, there is no written policy concerning appropriate or inappropriate forums for the bringing of complaints—about discrimination or other matters. Nevertheless, the Black Rainbow employees knew that the school board is an important customer of ours. Two of the group's members work in Shipping and must sign the order forms when products are to be delivered, and so have some idea of the amount of purchases by the school board. It is *clearly* inappropriate to complain about company policy to a customer. I don't believe we need a written statement to that effect—we don't have written statements forbidding rape and murder either. Some wrongdoings are too obvious to require written policy statements.

Memorandum

Date: August 2
To: Perry Grayson, Chief Executive Officer
From: Violet Smith, Employee Relations Manager
Re: Proposed Terminations of Black Rainbow Employees

I have read the letter to the school board, and I understand your concern. However, we may do more damage to our image by firing the protesters than by keeping them in our midst.

It is fair to assume that the Black Rainbow employees sincerely believe that they are the victims of racial discrimination at their workplace. They are, however, only six employees; Great Rainbow employs more than 600 blacks in a variety of positions. The group has existed for almost ten years; at its peak, its membership was approximately twenty. It would seem that the group's influence has diminished, rather than increased, over the years. I believe that this is because Great Rainbow's record on affirmative action has been very good, and that this is recognized by the great majority of employees.

However, if we fire the Black Rainbow members, we may be misunderstood by our other black workers. The action may give credibility to the Black Rainbow's claims. As a result, we may face discipline and morale problems—a situation that the Black Rainbow, however hard it tried, was unable to achieve while all members were employed here. As we have seen at other companies, discipline and morale problems can result in costly delays and interruptions in business.

I think it would be a grave mistake to underestimate the resentment that such an episode might engender among our black workers. Nor can we automatically assume that because our response is directed against blacks, our white work force will be entirely unconcerned. No doubt there are some white workers who feel that the blacks have been "coming on too strong," "wanting too much," or "demanding special treatment." Such employees will no doubt welcome disciplinary action against the Black Rainbow people. But for other white employees, the issue may cut deeper than color. They may see it as an infringement upon their rights of free speech and protest, and some of them may feel just as alienated by our actions as black workers will.

The effects may be felt more keenly outside the company. Even a small group of vocal and organized ex-workers—with plenty of time on their hands—could damage our reputation in the black community and make recruiting difficult. These employees have contacts in a number of civil rights groups—among them, the NAACP, the National Community Affirmative Action Center, COACH (Crusade to Originate and Accommodate Community Harmony), and the ACLU. We can expect these organizations to rally community support for the Black Rainbow people. They may also help the Black Rainbow employees with legal support in challenging their dismissals, which we can expect them to do in every possible way. And since they already have a suit pending against us in district court, we will be open to charges of reprisal, as well as racial discrimination. It seems to me that we should consult with the legal department before doing anything hasty that might involve us in fresh difficulties in the courts and before the regulatory agencies.

Concerning the union: union support for the Black Rainbow employees is unlikely. Although they are all members of the union, none of them is or has been an officer. In fact, none of them has engaged in union activities other than paying union dues. In addition, the Black Rainbow employees' demand for "equality" is too vague to be attractive to union leadership here. Frank Fernandez, head of our Local 107, concerns himself only with practical specific demands—wages, overtime, time off, and so on—the gamut of traditional union concerns. "Discrimination charges are best left to civil rights groups," he said to me recently.

However, we should not overlook three facts:

1. If we fire the Black Rainbow employees, and if they challenge their terminations, the union may feel obliged to support them in order to avoid charges of unequal representation. Each of the six has worked here seven years or longer; one fellow has been in Shipping for twelve years. Their work records are satisfactory.

2. Black employees—like the Black Rainbow members—have been apathetic about union involvement. This could be a good issue on which to galvanize black membership in the union.

3. If the Black Rainbow employees claim they were fired because of racial discrimination, it might make an attractive story for the media. Frank might enjoy a good embarrassing fight with us in the papers.

Therefore, although chances are that the union will not strongly support employees fired for disloyalty, we can't rule that out. If our decision is to terminate them, we must be prepared for a prolonged battle that includes arbitration, and possibly the courts.

The school board is indeed an important customer—but do we have any reason to think that they will take the Black Rainbow letter seriously? I doubt that the letter has damaged our relationship with the board to the extent that a quick explanatory phone call couldn't set things right again. As Employee Relations Manager, one of the chief architects of the Successways program, and a black, I am an appropriate spokesperson for Great Rainbow in this delicate matter. I would be happy to contact Regina Plumrose, the board president, on the company's behalf. I have also developed a good rapport with board members Ethel Silverstein and Joe Whiteside, and I am sure that I could persuade them to look at this episode in proper perspective.

I think a severe warning to the Black Rainbow employees is called for, and should be sufficient. Our response to this situation should be cautious and discreet; overreaction is likely to be counterproductive. Please let me know what your decision is.

Memorandum

Date: August 3
To: Perry Grayson, Chief Executive Officer
From: Harold Bluestein, Esq., Counsel for Employment Law
Re: Proposed Termination for Disloyalty

I have received your letter of August 1 regarding the six employees who signed a letter to a major Great Rainbow customer charging the company with racism. You ask whether Great Rainbow can terminate the employees who made the unproven charge, and whether the terminations are supported by the Collective Bargaining Agreement between the company and the Amalgamated Furniture Workers Union, Local 107.

According to Article XII, Paragraph 1 of the agreement, employees can be terminated only for "just and sufficient cause." In our opinion, the employees' conduct in writing such a letter was an act of disloyalty to their employer, Great Rainbow Furniture Company. In *NLRB v. Local Union No. 1229, IBEW* (346 U.S. 464), the Supreme Court stated: "There is no more elemental cause for discharge of an employee than disloyalty to his employer." Therefore, we feel that the employees' behavior is ample justification for their termination under the provisions of the agreement, and we feel reasonably confident that the company's position would be upheld if the terminations were challenged in arbitration or some other legal proceedings. However, the issues are somewhat complex, and we cannot be absolutely certain of prevailing.

The case involves a number of factors that must be considered in light of recent arbitration decisions. We have outlined the key issues below and noted how Great Rainbow seems to stand in relation to each.

1. Was the act or conduct expressed orally or in writing? If in writing, it appears to have a greater impact on the outcome of the case. Clearly this factor strengthens Great Rainbow's argument.

2. Was the act directed toward persons within the organization or outside it? If the latter, the seriousness of the act or conduct is accentuated. In addition, the recipient of the act or conduct is also an important factor. If the act or conduct was directed toward a governmental or regulatory enforcement agency, before a public hearing, to a competitor, to a customer, or in another public forum, arbitrators appear to take a more serious look at the propriety of the employee's act.

The six employees sent their letter to an elected body, which was also a major customer. This factor also strengthens Great Rainbow's case.

3. If the act or conduct was directed toward a customer or competitor, did it directly or indirectly cause damage to the business (competitive business position or company "image"), or interfere with the advantageous relations between the organization and the customer or competitor? If "harm" to business is found, arbitrators will generally support some degree of discipline.

This factor is unclear in Great Rainbow's case. Thus far, no loss of business is attributed to the employees' letter and none is expected. According to Great Rainbow's sales department, the school board buys from Great Rainbow because the company has traditionally offered it the best price for supplies, and the board is required by law to give its business to the lowest "responsible" bidder. Even if the board were to take the employees' letter seriously (and there is no indication that it has done so), and want to "punish" Great Rainbow for its alleged racism, the board could not stop business with the company for that reason.

However, charges of racism are serious and pose a threat to the good image of Great Rainbow Furniture Company—both with customers and within the community. We would expect an arbitrator to consider a reputation for racism to be "indirect" damage to business.

4. Did the employees exhaust internal avenues for redress of their grievances? Were those avenues reasonably accessible and usable?

Clearly the six employees did not exhaust internal avenues for redress—they did not even attempt to make their complaint known internally before writing their letter. If internal avenues are not used, arbitrators find this factor highly persuasive in support of the company position. However, there is some question whether those avenues are clear at Great Rainbow. Is there a prescribed method for making complaints at the company? No such method is posted or printed in an employee handbook. Although the absence of such instructions might count against the company somewhat, in our opinion an arbitrator will recognize the fact that an employee should take a complaint to an immediate supervisor or to the personnel director before complaining to a customer.

5. Were the statements known by the employees to be "true" or "false" or "undetermined" at the time they were made? "False" statements are a highly persuasive factor in supporting the company's disciplinary action.

There is at least one clear error in the employees' letter. The award given by the school board was not for "affirmative action" in general, but for one specific program, Successways. Evidently the employees did not know this, and believed the contents of their letter to be true at the time they wrote it. However, if the employees making the statements fail to conduct some investigation before acting, arbitrators find this a highly persuasive factor in justifying disciplinary action. Clearly the Black Rainbow employees made no such inquiry.

We think this negligence on the employees' part is a strong factor in Rainbow's favor.

In light of the factors enumerated above, we think an arbitrator would find that the employees who signed the letter acted disloyally. Therefore, in our opinion, the probable result of a legal challenge to the dismissals would be in favor of the company.

In addition, we think the union may refuse to support this case. The six employees have never participated in union activities, although they are union members. They have actively protested alleged racial discrimination at Great Rainbow for ten years, yet they have never brought their complaints to the union. And while last year they filed a lawsuit charging the company with discrimination, they did not request union support for the action.

Moreover, Local 107 has not shown an interest in espousing broad social issues. The union has always confined itself to specific, negotiable demands. Affirmative action programs, for example, have not yet appeared as a fringe benefit in the union's proposed contracts.

Another factor that might discourage union support is the strong case against the employees.

If the union chooses not to support the employees, the Black Rainbow group might still elicit the support of the ACLU or a black community group. However, the employees are already involved in a suit; this fact may deter support for another suit by them.

If they do not go to arbitration, there are three other forums in which they might challenge their terminations:

1. The Equal Employment Opportunity Commission. The Black Rainbow Employees could say that their termination was a reprisal for having filed a charge with the EEOC. The company would have to show that its grounds were entirely independent, which *ought* to be possible in this case.

2. The National Labor Relations Board. "Disloyalty" cases brought before the NLRB have been treated much the same as similar cases that were arbitrated.

3. California District Courts. The California courts are well-known for being highly sympathetic to employee rights. Conceivably, the Black Rainbow employees might claim that their first amendment rights to free speech and their constitutional rights to petition the government (school board) for redress of grievances were violated. Such a claim would take us into a rather unclear area

of the law. First amendment principles are not directly applicable to private employment relationships. Public sector employees have gained considerable court protection from dismissal for commenting on social and political issues in a way that appears unfavorable to their employers. Litigation resulting from this incident would represent one of the few clear tests of whether this principle will be extended to private employment relations.

However, the employees are already involved in one case in the district courts; it's unlikely that they will find an organization to support them in a second lawsuit. The cases take years to resolve, which may also discourage the employees from filing another suit.

In sum, we feel that Great Rainbow management is in a strong position if it should decide to discharge the employees on grounds of disloyalty.

Memorandum

Date: August 3
To: Perry Grayson, Chief Executive Officer
From: Tim Browner, Vice President for Corporate Social Responsibility
Re: The Black Rainbow Employee Group

In the past few months I have been talking with several black employees — both Black Rainbow members and others not associated with the group — and I feel that it is important that we try to understand their perspective before we decide how to deal with this latest incident. We see these people solely as troublemakers, but how many employees really seek to make trouble for its own sake? What have they to gain from doing so? I know that some management personnel view the Black Rainbow people as irresponsible and disloyal, but we should consider the fact that most of them have worked here for considerable periods of time. The signers of the recent letter that is causing all the trouble have worked, among them, some fifty-two years at Great Rainbow. As one of them said to me, "If that ain't loyalty, what is?" Another remarked: "We helped make Rainbow a big successful company just like management did, maybe more. We're Rainbow too."

I am suggesting, therefore, that we should consider taking them more seriously when they say they are motivated by a sincere belief that racism exists at Great Rainbow and that they feel they must oppose it — for their own sakes, on principle, and for the good of Great Rainbow. Can we in all honesty claim that there is no racism at all here? If so, then we must be unique in this country. That is not to say that there is an official policy of racism, or to fault any individual. I feel that the Black Rainbow charges against Hunter Greene, for example, are quite unjust. Yet we cannot deny what they have experienced. One of them recounted to me how once, when two or three black workers went on disability with back injuries, one of the foremen asked him, "How come so many blacks get these back injuries?" We tend to dismiss such incidents as trivial, but to the people involved they can be telling—indicative of a subtle but pervasive climate that they have experienced all their lives and that affects them every day. Another black employee told me how in 1978 he was arrested and accused of assaulting laborers trying to cross a picket line, even though several eyewitnesses denied that he had done anything of the sort. He was not surprised at being arrested. As he put it: "I was the only black on the picket line that day. And I had been making noises to the EEOC recently, into the bargain." We should bear in mind, too, that the EEOC *did* find that our affirmative action programs left something to be desired.

Perhaps, therefore, we should believe the Black Rainbow spokesman who said: "We're trying to make this place a leader in affirmative action, not a haven for racists. What we've done in the past by filing grievances and lawsuits was a high form of loyalty." If this is so, then we should consider the implications of taking severe disciplinary action against these people. Legally, we may have justification for firing them—but oughtn't we consider the larger issues that their actions raise? I think that positions on both sides are becoming frozen in a way that is unnecessary and very unfortunate—and the present episode threatens to make the situation much worse, unless something is done to defuse it. I think that instead of overreacting, we should use this opportunity to try to change the present climate of distrust. At the very least, if we hold some sort of open hearing on the matter of the letter and establish clearly what our position and policy are, we will be preparing the way for a fully defensible discharge of these employees at a later date, should efforts at resolving the problem fail. But those efforts are worth making, it seems to me. I am particularly concerned about the attitude that the EEOC is likely to take toward our dismissing workers while they have a complaint pending. Whatever our actual motivation, the incident will look very much like reprisal.

In an interview with a local black community newspaper, Oliver Jones was quoted as saying: "My dad worked on the railroad, and he helped organize the Pullman porters. He told me to stand up for what's right, and pay the price. That's what we did. We'd do it again, but maybe we'd know how to do it better." I agree that they should have known how to "do it better"—but I think that we at Great Rainbow should try to get that kind of spirit on our side rather than directed against us.

Discussion Questions

1. As Perry Grayson, what does Hunter Greene's memo indicate implicitly to you about the culture and policies of the Urbana facility? What "script" would you, as Perry Grayson, write for your personnel director?

2. Violet Smith argues that a decision to terminate the six protesters may do considerable damage to the company's relationships with other employees and with the public. Do you agree? Would you support her proposal to give a disciplinary warning to the signers of the letter?

3. Staff attorney Bluestein identifies five criteria that an arbitrator might use to evaluate the reasonableness of disciplinary action toward the protesting employees. Evaluate these criteria and assess the facts of the case in light of them. In the role of an arbitrator, would you sustain or mitigate discharge actions if Grayson were to take this approach?

4. Bluestein outlines other avenues of recourse available to these employees if they are fired. How strong would you perceive each of the following two actions to be:

    A. Filing a complaint with the National Labor Relations Board

    B. Claiming that the discharge actions were in retaliation for filing a discrimination suit against the company based on findings from an EEOC investigation.

5. Bluestein also gives brief consideration to the possibility that the Black Rainbow employees, if discharged, could sue the company in the state courts alleging company interference with free speech rights given to citizens by both the federal and California constitutions. How would you weigh the company's need to ensure employee loyalty against an individual's freedom of speech?

6. After meeting with and talking to a number of Black Rainbow employees and other company workers, Tim Browner has a different perspective and draws a different set of conclusions than Grayson's other advisers. As Grayson, what is your response to Browner's memo?

7. Considering all of these issues, what action, if any, would you take toward the persons signing the protest letter to the Urbana Board of Education?

CASE 5

Testing Employees for Substance Abuse

Individual Privacy vs. Organizational Responsibility

United Companies is a conglomerate with two major divisions: (a) a group of oil refinery companies located in four states and employing 7,000 persons, and (b) a multistate financial services division employing 4,500 persons in banking and brokerage operations. United thus has a diverse work force of executives, middle and line management, professionals, salespersons, white-collar clerical and customer service workers, and blue-collar production workers.

Like many firms, United has had incidents of alcohol abuse among its managerial and employee work forces. In 1984 a ring selling cocaine in the printing department of a branch bank was detected by local police and drew national publicity. In the refinery in Baton Rouge, Louisiana, the security department has reported several dozen accidents that security investigators believe to be drug related. And in one of United's banks the controller who embezzled $2 million was found to have been a heavy drug user who said he stole to keep up his habit.

Two years ago, United instituted a "Fitness for Duty" policy, dealing with the detection of alcohol- or drug-impaired behavior on the job. Now United's CEO, Charles Marston, is concerned that this policy may not be sufficient. He has called on various staff groups to reexamine the magnitude of the substance-abuse problem and to advise him about possible changes in the company's policy.

You are Charles Marston. After writing a memo on January 4, you have received the replies presented here. What action do you now decide to take?

Key Persons Involved Charles Marston, *Chief Executive Officer*
Howard R. Porter, *Senior Vice President, Administration*
Paula F. Astor, *Head of Employment and Labor Relations Group, Law Department*
Kevin Holloran, *Vice President for Employee Relations*

Memorandum

Date: January 4
To: All Department Heads
From: Charles Marston, Chief Executive Officer
Re: Employee Screening for Substance Abuse

I have just returned from last week's National CEO's Association meeting in Denver. One of the more interesting reports presented there concerned the problem of substance abuse in the workplace, which is evidently on the rise almost everywhere and starting to worry a great many executives. Some of the figures cited in the report were truly disturbing. Last year, drug and alcohol abuse resulted in productivity losses estimated at close to $40 billion. When the additional costs of increased medical claims, higher insurance premiums, absenteeism, theft, and accidents associated with substance abuse are figured in, the total drain on American business may have exceeded $100 billion. An equally alarming statistic is that by some estimates, nearly one worker in five uses controlled substances. But the problem isn't limited to these substances—abuse of over-the-counter drugs, prescription drugs, and alcohol is even more common than the use of illegal drugs. (I am enclosing two recent newspaper and magazine articles detailing the scope and magnitude of the situation.)

As you know, we haven't completely escaped these problems ourselves. You will recall that two years ago, a number of workers at our Texas refinery were found to have fairly severe substance-abuse problems—an incident that unfortunately was reported in the local press. It was this episode, coupled with our awareness of earlier problems and incidents, that led us to formulate and issue our Fitness for Duty policy. Despite some initial difficulty with the union, the policy has generally been well received by our employees. (In fact, it has become a model that several other companies have copied.) As you know, this policy provides for breathalizer or drug testing only when there is some reason to believe that an employee is impaired by alcohol or drugs while on duty. Our strong emphasis—both in our handling of the Texas episode and in the Fitness for Duty policy—is on offering help to our personnel under the Employee Assistance Program; this, rather than more punitive approaches, has certainly been a factor in the acceptance of our policies up to this point.

The question now arises, however, whether these policies are still adequate, and whether they will prove sufficient for dealing with a problem that seems likely to become increasingly severe in the years ahead. At Denver, the focus of the discussions was almost entirely on the issue of drug testing for employees. Estimates of the number of Fortune 500 companies that are now screening applicants or employees for substance abuse, or are seriously studying the implementation of screening programs, range from 30 to 50 percent, depending on whose figures you accept. Many medium-sized companies are following suit. It is widely felt that this approach is the wave of the future. Thus, we at United Companies had best confront the issue head-on:

- Do we need to institute a random screening program for our blue-collar and white-collar work force?
- If so, should we also test management personnel?

I would like your advice on these questions. Please send me your views within the next two weeks, as I intend to make a recommendation to the Management Committee on this subject at its meeting of February 5.

Attachment

UNITED COMPANIES FITNESS FOR DUTY POLICY
Adopted July 1, 1986

The unauthorized possession, distribution, sale, knowing transportation, or use of drugs or controlled substances, where admitted or verified, is prohibited. The unauthorized possession or use of alcohol or being under the influence of alcohol on Company property is also prohibited. These provisions apply to all Company employees, and to all persons entering Company premises or facilities. Employees required to meet federal safety qualifications and regulations may be subject to additional restrictions concerning the use of drugs — including over-the-counter and prescription drugs — and alcohol during nonworking hours.

United Companies reserves the right to require an employee to submit to a drug test or breathalizer test whenever the employee's observed behavior or other convincing evidence provides reason to believe that the employee's use of drugs or alcohol is likely to impair job performance or in any way jeopardize the safety of persons or property. If a drug screen indicates drugs in an employee's system or if a breathalizer test indicates that the employee is under the influence of alcohol, the employee will be terminated. (First offenders will have the option of seeking treatment and/or counseling and/or other rehabilitation services under the provisions of the Employee Assistance Program — see Publication 2-303E.)

Attachment

Drug, Alcohol Abuse Costs Firms $100 Billion A Year, Report Says

WASHINGTON (UPI) — Drug and alcohol abuse on the job is costing businesses an estimated $100 billion a year, but increased efforts to combat abuse are sometimes hindered by poor information and uncertainties, a business reporting firm said.

In a recent report, the Bureau of National Affairs Inc. said drug and alcohol abuse on the job is receiving "unprecedented attention from employers, unions and public policy leaders, but solutions to the multibillion-dollar problem are proving to be elusive and loaded with controversy."

"Financial costs range from medical bills and insurance premiums to productivity losses and business failures," the report said. "The bottom line losses, according to some estimates, may exceed $100 billion a year."

Lost productivity due to alcohol and drug abuse costs the U.S. $39.1 billion a year, the report said, adding that of the amount, $30.8 billion is attributed to alcohol-related productivity losses and $8.3 billion to drug-related losses.

Testing Employees for Substance Abuse

U.S. industry lost $81 billion in 1984 due to accidents "and people using drugs and alcohol on the job have three to four times the number of accidents as those who don't," the report said.

As a result, the report said testing for drugs and alcohol has increased.

"Surveys cited by the National Institute on Drug Abuse show that the percentage of Fortune 500 companies screening employees or job applicants for drug use rose from 3% to almost 30% between 1982 and 1985," the report said.

However, drug and alcohol screening has its problems.

According to the survey, critics say drug tests produce erroneous results, create employee and union relations difficulties, invade employee privacy and fail to indicate whether an employee is actually impaired by a drug.

Attention has focused on illegal drugs in recent years, but the report finds that legal drugs—alcohol, prescribed medicines and over-the-counter drugs—pose problems for far more employees than do illegal drugs.

"The National Institute for Drug Abuse has estimated that abuse of prescription drugs causes 60% of hospital emergency room admissions for drug overdose and 70% of all drug-related deaths," the report said.

In addition, the report notes that a 1981 study of drug testing labs conducted by the U.S. Centers for Disease Control—the latest such survey conducted—found error rates as high as 100% at some testing labs.

The study warns that employers "should be careful to examine the fairness of drug screens and searches. Innocent victims of errors may challenge managerial intransigence in court. There are clearly grounds for litigation seeking substantial damages in the doctrines of invasion of privacy and defamation."

Other findings in the report:

- Absenteeism among alcoholics or problem drinkers is 3.8 to 8.3 times greater than normal.
- Alcoholics have a two to three times greater risk of being involved in an industrial accident.
- Up to 40% of industrial fatalities and 47% of industrial injuries can be linked to alcohol abuse and alcoholism.
- Grievance procedures by workers appealing drug or alcohol-related firings cost employers an average of $1,050 each.
- Non-alcoholic members of alcoholics' families used 10 times as much sick leave as normal, according to one study.
- Average monthly health care costs in families with an alcoholic member were $207 per person, compared with $107 per person for families with no known alcoholic members.
- Alcoholics' average health care costs increased from $150 a month two years before treatment to $450 during each of the six months prior to treatment, and to $1,370 in the final pre-treatment month.

The Bureau of National Affairs is a Washington-based private company that has 60 publication covering business and economics, law, taxation, labor relations and environmental protection and other topics.

Source: Used with permission of *Investor's Daily* April 3, 1986

Attachment   USING DRUGS? YOU MAY NOT GET HIRED
by Ted Gest

Scores of applicants today are facing more than job-competency exams. They need to pass urine tests as well.

The problem is illicit drugs. Companies are cracking down to prevent accidents, absenteeism and low productivity that they blame on wide use of marijuana, cocaine and other substances ranging from illegal "angel dust" to prescription medications.

A new survey shows that 1 in 5 of the nation's biggest companies now give tests, and an additional 19 percent may join the trend within two years. Employers are beginning to test workers already on the payroll, and a few school systems may test both teachers and students.

The practice is spreading rapidly even though critics say tests are often inaccurate and that the drugs they do detect may have no effect on job performance.

GI Tradition. Testing got its foothold in the armed services, which long have insisted on strict prohibition against drug use by their personnel. In recent years, business owners and government officials have turned up more and more cases of civilian mishaps linked to drugs. Although no one has compiled a national toll of drug-caused accidents, several cases have been documented in the transportation and utility industries.

Drugs were a factor in a Burlington Northern rail crash in Wyoming last year in which two crewmen were killed. The railroad dismissed an engineer who had been smoking marijuana. Tests of crewmen on a train parked nearby turned up several more drug users, who also were fired.

The crash of a small plane last year in northern New Jersey that killed the pilot and three passengers also was the result of drug use, reports the National Transportation Safety Board. The board suspects that drug use was involved in train collisions in Arkansas, Atlanta and Miami. The Federal Aviation Administration fired three Miami air-traffic controllers found with marijuana and cocaine at work.

To weed out drug abusers, many firms require job applicants to submit urine for analysis that can detect a half-dozen or more drugs. The exam can find cocaine traces for two days after use and marijuana for several weeks.

Critics complain that the test often is inaccurate. Some samples are handled improperly, causing "false positives"—persons tagged as drug users when they are not. A study of 13 laboratories by the federal Centers for Disease Control showed an error rate of up to 66 percent. "Some testing is done by untrained people," notes Richard Hawks of the National Institute on Drug Abuse.

Others say initial indications of drug use should be rechecked on more-sophisticated equipment. When that is done, "results are almost 100 percent accurate," contends Claude Buller of North Carolina-based CompuChem Laboratories. Preliminary tests cost only a few dollars, but many firms don't pursue follow-up tests that often cost $50 or more.

When companies use the tests as a screening device, there is little that applicants can do to protest. As long as exams are required of everyone, discrimination claims are likely to fall flat. It may be a different story for tests of those already on the job, and some of these employees are charging that their privacy is being violated.

What may be the first drug-testing dispute to enter the courts erupted last summer in San Francisco at Southern Pacific Transportation Company, which forced nearly 500 workers without warning to submit urine samples. Computer programmer Barbara Luck was dismissed after she refused to participate on privacy grounds. Luck has a lawsuit pending against the company, as does an office manager whose test showed that he had used cocaine—a charge he denies. The man says he was sent to rehabilitation classes even though a follow-up test showed no evidence of drug abuse.

After reports of such cases, San Francisco last month became the first major U.S. city to bar employers from ordering tests unless there is clear evidence that a worker's drug use endangers others. The measure permits tests only for police, firefighters and rescue units.

Even so, more public and private employers whose work involves public safety are likely to continue drug checks. Burlington Northern intensified testing after last year's crash. Unions unsuccessfully challenged a plan by the Federal Railroad Administration to require all railroad job seekers to submit to tests.

Urinalysis may spread in sports. Big-league baseball was embarrassed last summer when the Pittsburgh trial of a drug dealer incuded extensive testimony on cocaine use by major-league players. Commissioner Peter Ueberroth has asked all 650 players to take tests. But the players' union has balked, arguing that all players shouldn't be made to comply if only a few have a drug problem.

School barriers. Moves to test teachers and students have run into roadblocks. A New York judge barred urinalysis for teachers on Long Island, and a New Jersey judge ruled December 10 against a plan to test high-school students. The courts said the procedure violated constitutional protections against unreasonable searches.

No court has stopped testing in a private firm, although the question of whether privacy rights are violated by such procedures may eventually end up in the Supreme Court.

In the meantime, the question facing businesses is whether testing is worth the costs. Some companies insist that it is. After Pacific Gas & Electric Company began to screen applicants, injuries among newly hired construction workers fell 40 percent. But North Carolina toxicologist Arthur McBay says he has seen no scientific evidence that the "millions of dollars being spent on handy-dandy screening programs" have reduced drug problems on the job.

Still, growing public pressure to do something about substance abuse will make screening routine in many employment offices.

Source: U.S. News & World Report Dec. 23, 1985
Copyright 1985, *U.S. News & World Report*.

Memorandum

Date: January 12
To: Charles Marston, Chief Executive Officer
From: Howard R. Porter, Senior Vice President, Administration
Re: Mandatory Employee Drug Screening

In response to your memo of January 4, I would like to say that I enthusiastically endorse your resolve to face the issue of substance abuse head-on, while it is still a manageable problem rather than a crisis. So far we seem to have escaped the worst, and it is my hope that decisive and farsighted action now will minimize our difficulties in the future.

It is hard to say just how severe the overall problem is at United Companies, but here is what we know about the situation at our various facilities.

1. *Refinery Division*. Our four oil refinery companies in Maine, Texas, Louisiana, and North Carolina currently employ some 7,000 people. About 60 percent of these employees are blue-collar workers, represented by the Petroleum Workers Union (PWU). Our Security Department reports that last year there were ten terminations of employees who were found to be under the influence of alcohol or drugs while on the job. Most of these employees were refinery workers at the Texas and Louisiana facilities, but a computer programmer and an officer worker were also fired. In addition, thirty-seven employees who admitted to substance-abuse problems were placed in counseling or treatment programs, at company expense, under the provisions of the Employee Assistance Program.

These figures do not necessarily define the full extent of substance abuse in our refinery operations, however. While no one so far has been caught selling drugs, there were rumors of drugs being readily available for a time at the Texas refinery. (This situation seems to have cleared up with the discharge of several workers there and the rehabilitation of a number of others, as mentioned above.) Moreover — and in some ways more alarmingly — there were a number of highly suspicious accidents at the North Carolina facility, one of which actually endangered the lives of several workers, as well as equipment worth millions of dollars. We have not been able to pinpoint the cause of these incidents, but most of the obvious possible factors, such as inadequacies in training, have been ruled out. Our Security people, though they have no hard evidence, think it quite possible that substance abuse may have been to blame.

2. *Financial Services Division*. Our bank in California and our brokerage firm in New York City together employ nearly 4,500 people, none of them unionized at present. Last year, six employees were terminated for drug use or intoxication at work. The most serious incident involved a small drug-selling ring that was uncovered in the printing office of the New Caledonia State Bank. Three employees were fired, and two of them were subsequently indicted; the cases have not yet come to trial. (In this instance, as in the Texas case of some years ago that you mentioned in your memo, the story made the papers and we

received some rather unfavorable publicity.) Here again, we cannot be sure that we know the full extent of the problem; the six cases that resulted in termination may represent just the tip of the iceberg. Twenty-three employees also sought help for substance-abuse or related problems and were referred for treatment under the Employee Assistance Program. The controller at our San Antonio midtown branch bank, as you will recall, said that continuing his drug supply was what led him to embezzle $2 million from the vault reserves.

I think it is important for us to recognize the full extent of the safety and security problems that substance abuse can entail. While some aspects of this are obvious, others tend to be overlooked. Of course, an intoxicated refinery worker could cause an accident that destroys expensive equipment, injures or kills plant personnel, and even endangers people in the surrounding community. (We all remember what happened in the Bhopal disaster.) But this is not just a blue-collar problem. What about the clerical worker who steals company property—or worse, sells confidential information—to support a drug habit? What about the computer operator who embezzles millions of dollars for the same reason, or, while stoned, accidently misdirects huge sums? Such things have happened elsewhere, and they could happen here. We are vulnerable to many sorts of harm that might result from the actions of our workers, and could be held legally accountable for the harm their actions might inflict on others.

It is for these reasons that I feel strongly about the following recommendations:

1. We should immediately begin routine screening of all job applicants. Many companies now do this, and to my knowledge there are no legal impediments. Clearly it is better to head off potential problems than to worry about how to deal with them when they have become deeply rooted. I look upon this approach as "an ounce of prevention."

2. We should institute a program of periodic screening for all Financial Services Division employees, perhaps with special emphasis on the testing of personnel who are likely to be in the possession of sensitive financial information, or who handle large sums of money, or who are involved in electronic fund transfers, and so forth. Such a program could help us eliminate unreliable people whose actions might someday prove very costly to us. In the absence of union representation, and given the publicity that substance abuse has received recently, I do not anticipate that our employees could successfully resist the implementation of a screening program, or would even attempt to do so. I see no need to extend such testing to management personnel at the present time. Apart from the one case of drug-related embezzlement by a high executive in San Antonio, we have no evidence of management involvement in any of the substance-abuse incidents that have come to our attention. Moreover, many managers would probably regard testing as indicative of a lack of trust. At best, this might lower morale; at worst, it could touch off a serious revolt among the very people whose enthusiastic leadership we most rely on.

3. The situation in our Refinery Division is somewhat different, in that the union strongly opposes random screening. I am enclosing a recent article from *Lifelines*, the health and safety newsletter of the Oil, Chemical, and Atomic

Workers Union. This is the "sister union" of our PWU, and the two usually have similar attitudes. The article spells out the union's position on this issue. Briefly, unions feel that any screening program should be an issue for collective bargaining — that is, we cannot implement such a program without their approval. They contend that random screening programs are inherently an invasion of privacy, in that tests can detect evidence of activities that the union feels are not a valid concern of the company (for example, the use of drugs off company property). So long as such activities do not affect an employee's job performance, the unions maintain that they are none of the employer's business. Thus, their position is that any screening should be only for "probable cause" — that is, on the basis of some objective evidence of impaired performance — and that employees found to have a substance-abuse problem should be given an opportunity to obtain treatment under an employee assistance program. (I should point out that what the union endorses is essentially our present policy.) In addition, the union has raised concerns about the accuracy of current testing procedures and the confidentiality of test results.

In view of this opposition, I think we should, as a first step, attempt to negotiate a drug-screening policy that will meet our needs while still being acceptable to the union. The union has expressed a great deal of concern over job-security issues lately. Perhaps we can arrange a trade-off during this summer's collective bargaining: some concessions on our part regarding job security if they will soften their stand against random screening. It may turn out, however, that the PWU feels that it cannot live with *any* program that involves random testing. If this proves to be the case, I think we should still go ahead and institute such a program. Our position should be that it is simply a management prerogative. It is our right to hire and retain the people we think can do the best job for us, and we do *not* feel that the people we want are those who use drugs — legal or illegal, on the job or at home. We must, of course, be prepared for the filing of grievances over the issue, and perhaps even a lawsuit. Nevertheless, other managements have stuck to their guns in the face of similar opposition, and we should do the same. The issue is too important to back away from; it is a test of our resolve.

4. There remains the issue of the test to be used. I am not prepared to make a recommendation on this issue just now: I propose that it be turned over to a committee for study, perhaps under the chairmanship of our Chief Medical Officer, Dr. Thomas Crippen. It should be noted, however, in view of the questions that have been raised about accuracy and confidentiality, that the test selected and the precise testing procedures adopted may be critical to the effectiveness and acceptance of the entire program. The tests currently available vary widely in their reliability, cost, and sensitivity, and in the nature of the substances they can detect. All of these factors should be taken into account before we commit ourselves to a particular procedure.

Testing Employees for Substance Abuse

DRUG TESTING PROGRAMS
Random Screening Is Not the Answer
by Sylvia Krekel, Occupational Health Specialist, OCAW

Attachment

On April 9, the Conoco Refinery in Denver was invaded by "rent-a-cops." Both Conoco and contractor employees were herded into lunchrooms and trailers for several hours while the hired private security police, assisted by drug-sniffing dogs, searched lockers and cars for drugs, paraphernalia, and firearms. Employees weren't even allowed to go the bathroom.

This is just another example of employers overreacting to perceived drug problems in the workplace.

Why are employers rushing to impose drug screening programs and what is OCAW's response?

Why the Rapid Growth

Substance abuse, including alcohol, is undoubtedly a problem in some workplaces. It is also probably true that many employers are blaming substance abuse for conditions that arise out of the nature of the workplace itself.

A major reason drug testing has become so popular today is that an entire new industry has grown up in the last 4 or 5 years marketing detection kits for drugs in the urine. These drug kits can cost as little as $10 or less per sample, and can be very unreliable.

Another reason for the rapid growth of drug testing is the wide publicity that has been given the issue of testing and its use in professional athletics. Any issue given a lot of publicity will have a band-wagon effect.

Finally a big impetus to drug testing was given by President Reagan's Commission on Organized Crime. The Commission recommended in March, 1986, that all Federal agencies immediately implement drug testing programs, and that both the public and private sector make it clear that any and all use of drugs is unacceptable.

OCAW Policy

More than a dozen OCAW-represented employers, many with multiple locations, have imposed drug screening programs, most in the past few months.

First, we should make it clear that OCAW does not condone the abuse of drugs and alcohol, and are just as concerned as the employer about employees using drugs or alcohol on the job or reporting to work under the influence. However, we also believe that what a member does on his or her own time, away from the job, is none of the company's business, as long as it does not affect the employee's job performance.

OCAW considers any drug screening program to be an issue for collective bargaining under the National Labor Relations Act (NLRA).

As a mandatory issue for bargaining, the company must engage in meaningful, good-faith negotiations with the Union regarding their proposed policy and its implementation prior to putting it in place. Merely advising the Union of the policy does not constitute bargaining and this should be made clear to the company, through NLRB charges if necessary.

OCAW Vice-President Robert Wages and the Health and Safety Dept. have developed a letter that should be presented to the company asking for further information on various aspects of the drug screening program. The letter is specifically designed to provide information on the "reasonableness" of the policy. Some of the areas of drug screening programs about which OCAW is concerned are:

- Random screening. As far as OCAW is concerned, random screening of employees makes any proposed drug screening program unreasonable and objectionable. Our position is that screening should always be for "probable cause" such as deteriorating job performance, excessive absenteeism, impairment on the job or other objective considerations. Employees found to be "positive" by a valid confirmed drug test should be given the opportunity to go into treatment through an Employee Assistance Program (EAP), without any discipline for a first offense. OCAW is fully supportive of EAP's.

  Union intervention accomplished the revision of the drug testing programs of two major oil companies; their programs were changed to exclude random testing in favor of "probable cause" testing. One was a result of a pending court order, the other the result of a stipulation at an arbitration hearing.

- Reliability. Many of the new laboratories that offer drug testing are a cause of concern because there are currently no licensing or certification requirements by the government. In some studies, test results from some of these labs have 100% error rates.

  OCAW is also concerned about the "chain of custody" of samples and the confidentiality of testing and reporting results.

  Any OCAW group that is faced with a drug testing program should immediately contact their assigned representative and Dan Edwards, Health and Safety Director.

Source: *OCAW Reporter* May-June, 1986

Memorandum

Date: January 16
To: Charles Marston, Chief Executive Officer
From: Paula F. Astor, Head of Employment and Labor Relations Group, Law Department
Re: Employee Drug-Testing Programs

I hope you will forgive me if I respond to your memo of January 4 at some length. Many people do not realize how complex the whole issue of drug testing really is. While screening for substance abuse has obvious attractions, it also has many legal ramifications that the Management Committee and CEO must take into account before making a decision.

First of all, it should be obvious that substance abuse is a problem of our society as a whole. United Companies did not create this problem, nor can we eliminate it single-handed. Nevertheless, we must make an effort to do our part in controlling it. This is a matter not just of corporate responsibility, but of corporate self-preservation. Our potential legal exposures, should we fail to do everything in our power to curb drug abuse by our employees, are enormous and extremely serious. Consider, by way of illustration, the following hypothetical scenarios:

- A refinery employee, under the influence of drugs or alcohol, ignites a fire that causes the explosion of a petroleum storage tank. A fellow worker is killed; in addition, several nearby houses are destroyed and their occupants injured. The injured parties, as well as the family of the deceased worker, sue United Companies.
- A computer programmer who has an expensive cocaine habit and is employed by our brokerage house diverts a million dollars in client funds to his personal account and disappears, surfacing a month later in a country with which the United States has no extradition treaty. The defrauded customer sues United Companies for losses incurred.

In our defense, we deny responsibility by demonstrating that in both cases all equipment was properly maintained, employees were well trained, adequate safety and security regulations had been formulated and were in force, sound management procedures had been followed, and so on. We contend that what occurred resulted from the actions of individual employees over which United had no control.

In vain! Under the doctrines of negligent hiring or negligent administration, we could still be found liable if the actions of the employees were the result of their drug use, and United, knowing the potential dangers that an impaired employee might present, had not done everything in its power to detect drug-abusing employees and exclude them from its work force. Thus, we have a substantial responsibility for the actions of our employees, and it is up to us to make sure we do not hire or retain people who might do serious harm to others.

How can we exclude such people from our work force? Some employers, faced with the threat of lawsuits and other challenges to drug testing from unions or employees, have seen the situation as essentially "damned if you do, damned if you don't." The situation, however, is not that bleak. The law is fairly clear: As a private corporation, we are bound only by our collective-bargaining agreements, statutes, and the common law as it pertains to employer-employee relations. We are not bound by the constitutional guarantees of the fourth, fifth, and fourteenth amendments relating to due process, searches and seizures, and privacy, as a governmental agency would be. Thus, our management *can* legally screen employees for substance abuse, even in a random-testing program, in order to ensure a safer workplace and limit our legal exposure.

Prudence dictates, however, that in setting up a random-screening program we attempt to follow certain guidelines and satisfy certain criteria that are not currently mandated by law. There are three good reasons for this:

1. By doing more to ensure proper procedures and safeguards than the law actually requires of us, we provide ourselves with an extra margin of protection against potential lawsuits. Litigation is so expensive and inconvenient that it is better to err on the side of safety, and thereby discourage any temptation on the part of employees or unions to sue, than to rely on prevailing in court.

2. It seems likely that in the coming years, many states will pass laws regulating programs such as we plan to establish and setting stringent standards for the protection of employee rights in matters such as privacy and due process. We would do well to anticipate such statutory changes now. That way, we can avoid the unpleasant experience of waking up one day to find that parts of our program are being successfully challenged in the courts of this or that state.

3. By making sure that our program is as fair as possible, we can blunt potential opposition and minimize the resentment that such programs often arouse. Good employee relations are a valuable asset for any company, and it is important that we do whatever we can to retain the trust and goodwill of our employees, as well as of the general public.

I feel that a successful drug-screening program should have the following characteristics:

- There should be a clearly defined need for such testing. This means that testing should be confined to employees whose activities could pose a danger to the health, safety, or security of their fellow workers, the public, or the company. Thus, many of our refinery workers might legitimately be tested, because if one of them make a mistake while impaired by drugs or alcohol, the consequences could be disastrous. Similarly, we could justify testing bank tellers and other employees who handle large sums of money. But we would not need to test our sales personnel or clerical workers. The risks posed by substance abuse on the part of a few such workers are not great enough to justify the invasion of individual privacy and abuse to dignity involved in wholesale testing. Although the constitutional protection of the right to privacy is narrow and does not apply to private corporations, such protection could easily be extended to private sector workers by state laws.

Note that it is not considered discriminatory to screen only certain workers or classes of workers, if the decision is clearly based on the nature of the hazards associated with their work. A decision to screen all blue-collar workers but not white-collar workers, by contrast, or to screen only black workers, might be held to be discriminatory.

- The test used and the associated procedures must meet high scientific standards of accuracy and reliability. (We would probably use urinanalysis.) This is necessary to forestall challenges based on the allegation that the tests violate requirements for due process. (Here again, the constitutional guarantee of due process is legally inapplicable to us, but could in the future be granted to workers by state statutes.) For the same reason, positive test results should be double-checked and confirmed by a second test before action is taken against an employee, and employees should be given an opportunity to contest positive results. (It is my understanding that Kevin Holloran intends to address some of the problems associated with tests and testing procedures in detail in a separate memo.)
- The company's policies with respect to drug use and its drug-testing program should be clearly explained in advance to all employees and prospective employees. Ideally, the drug-screening program should be set up with the cooperation of the union at unionized facilities. The drug policies of the company should be included in all employment contracts, and consent to testing should be specified as a condition of employment or continuation of employment once hired.
- The test should be administered with maximum regard for employee privacy, and the results should remain confidential. Test results should not be released to *anyone* outside the company, and should be available to management only on a need-to-know basis. Positive test results should not be shown to anyone until they are confirmed by a second test.
- Employees shown by a *confirmed* test to have a substance-abuse problem should be offered the opportunity to get help through our Employee Assistance Program rather than be discharged. If the employee refuses to use the program, or does not pass it, or is later detected to be using drugs again, we would have no alternative but to discharge him or her.

It should not be too difficult to devise a drug-screening program that satisfies all of these criteria. I have no qualms about recommending the establishment of such a program if the precautions outlined above are taken.

Memorandum

Date: January 17
To: Charles Marston, Chief Executive Officer
From: Kevin Holloran, Vice President for Employee Relations
Re: Drug Testing of Employees

I have seen Paula Astor's memo on the legal aspects of employee drug screening. She has certainly done an excellent job of clarifying some very complex issues. Nevertheless, I feel that before we can be in a position to embark confidently on a course of action, a number of difficult problems need to be considered. These problems involve not only the law but also employee relations, public relations, and our relationship with the union.

One particularly thorny problem is the question of whether to test management personnel. On the one hand, if we do test management personnel, we run the risk of alienating a great many of our executives, who would regard their inclusion in the program as a sign that they are not trusted. We certainly need to attract the best possible people to the executive ranks, and we cannot expect to get or keep such people if we impugn their integrity, even if only by implication. There would also be no easy way to draw lines between different levels of management—we would have to include everyone, from the Chairman of the Board on down. Such a program would be sure to generate a lot of resentment, along with passive if not overt resistance, which could not fail to undermine its acceptance among the rest of the work force. I foresee a great deal of pressure being exerted to exempt management from any screening.

On the other hand, the difficulties created by excluding management could be even worse. At present, the courts have held that there is no legal requirement that a screening program must include the entire work force of a company. This could conceivably change in the future, however, either by the passage of new statutes or simply by a change in interpretation. The courts might well rule that requiring only certain categories of employees to submit to drug testing violates the requirements for equal protection. And even today, some courts have upheld the position that a testing program cannot be limited to certain parts of the work force unless the employer can demonstrate a reason for such selectivity. Given the enormous publicity about cocaine use among professionals and executives, and the potential harm that an addicted executive could to the company (for example, by embezzlement or by misuse of privileged information), excluding professionals and executives might be very hard to do.

Quite aside from the strictly legal questions, there are psychological and public relations considerations involved in this decison. Many people might consider that testing workers but not managers is unfair, un-American, and so forth. Issues of class and even of race could easily arise; our managers are middle-class and predominantly white, while much of our work force, especially in our refinery operations, is blue-collar and includes a large minority contingent. I would not be surprised to hear claims that a testing program that has a disproportionate effect on minorities represents a violation of EEO requirements.

Another problem is the effect that a drug-screening program might have on our relationship with the union. Our past conflicts with the PWU have largely given way to cooperation in the past year or two, in part owing to the success of our Quality Team Program. It would certainly not be in our interest to jeopardize this greatly improved climate, which has made it possible to resolve many issues (such as those involving quality control, work rules, and the like) that previously gave us a lot of trouble. The union tends to see drug screening as a pivotal issue for this decade, and is likely to dig in its heels.

It should also be noted that our choice of testing methodology could have crucial implications for both the legal standing of the program and our employee relations. At present, the most widely used screening tests require the collection of urine specimens for analysis—a procedure that, since it must be done under observation to prevent fraud, many employees look upon as an embarrassing invasion of their privacy. The use of such specimens means that stringent precautions must be taken at all points in the testing process to prevent mislabeling, contamination, or tampering, as well as misreporting of results. Moreover, the specimens probably need to be saved, at least for a time, in case subsequent tests are required.

There are several different urine screening tests currently available. The most widely used is EMIT, an enzyme immunoassay that can be performed either in a lab or in the workplace. Such tests have the advantage of being quite inexpensive (about $15 per person); however, they are not always very accurate, even when done by a commercial laboratory (as quite a few reported incidents have documented). False negatives—which means that the test fails to detect a drug that is in fact present—are quite common. More seriously, false positives are also common. This means that the test erroneously indicates the presence of a drug that is not present, or confuses a legal substance (such as an over-the-counter cold or cough medication) with a restricted one (such as an amphetamine). I emphasize the danger of false positives because they can cause an employee to be unjustly accused of substance abuse, stigmatized, terminated, and the like. If the test results are contested and then shown to be incorrect, the company must then contend with a very unhappy employee who might well file a grievance or even a lawsuit for libel or slander. The courts have held that if a company acts recklessly or maliciously in disseminating incorrect information about an employee's drug tests, it can be held liable for damages. (In fact, revealing any information about an employee's test results is probably an illegal invasion of privacy, as the confidentiality of medical records is protected under the law.)

The inaccuracy of screening tests means that a positive test should always be confirmed by a more sophisticated (and expensive) test before the employee is confronted with the results or management is informed. Reliable confirmation can be obtained by means of gas chromatography and/or mass spectrometry, which are much more sensitive and accurate and can identify the unique chemical "fingerprints" of hundreds of different substances. Unfortunately, these techniques can be used only be trained laboratory personnel working with complex equipment, and so may cost ten times as much as the screening assays. For this reason, some companies and laboratories deal with the problem of confirmation by simply repeating the original screening test, or performing another screening test of a slightly different type. This approach is dangerous and should be avoided.

There is another problem associated with urine testing, however. Traces of many drugs remain in the body for long periods of time—hours, days, or in some cases even weeks. The length of time that a substance remains detectable in the urine varies with the nature of the drug, and to some extent with the individual involved—his or her weight, diet, body chemistry, and so forth. Thus, when a urine screening test gives a positive result, it tells us only that the individual has used a particular drug at some time in the recent past. It does not tell us that the drug was taken on the job, nor does it tell us whether the user is presently impaired by this substance. The fact that these tests do not measure impairment is a serious drawback, since the main purpose of such screening is to eliminate workers whose job performance is compromised in some way by substance abuse. Workers may contend that it is not the business of United Companies whether they smoked marijuana on Saturday night, so long as they come to work on Monday morning ready and able to do their job. Our only responsibility, they might say, is to ensure that United gets a day's work for a day's pay—not to monitor our employee's life-styles, impose our moral values, or even enforce the drug laws. The fact that our screening tests can detect lingering traces of restricted substances that may have been used off company property, on the employee's own time, puts us in the position of intruding into the private lives of our employees. Although this may not now be illegal for private employers to do, it would appear to be a violation of our own Employee Privacy Policy, instituted in 1981. Let me quote a few of the relevant passages:

- "United Companies will collect only such information about employees as is necessary and useful for hiring and personnel purposes. All such information will be treated as confidential, and will not be released to outsiders unless there is a legal requirement to do so (as in the case of IRS or Social Security records) or such release is consented to by employees (as in the case of information needed to obtain mortgages, etc.)."
- "Only management personnel with a demonstrable need to know can inspect an employee's personnel files."
- "The off-the-job activities of our employees—including, but not limited to, family, community, religious, civic, and social activities—are not legitimate concerns of United Companies, and United Companies will not collect information about such activities."

Fortunately, there is a possible solution to most of the technical problems I have discussed. Screening devices that measure brain electrical activity, rather than the presence of chemicals in urine or other body fluids, will be available shortly. Such noninvasive devices (the first of which, called the MONITOR 1000 system, will be on the market later this year) avoid the embarrassment associated with having to provide urine specimens, as well as the need for concern over the labeling and custody of the specimens. They are fast, cheap, and easy to use, requiring no laboratory work, and they can accurately detect and distinguish among the number of different substances, including alcohol. Most important, they *measure the extent to which the employee is actually impaired at the time of testing* — which is what the company needs to know, and (in the view of some people) all the company needs to know. Thus, the invasion of the employee's privacy is minimized.

It is difficult to weigh so many complex factors and decide on a clear course of action. Nevertheless, since you have requested a recommendation, here is what I suggest:

- I believe that we should institute a screening program for new applicants. Once hired, however, employees should be subject to drug testing only for "probable cause." This is our present policy, and it has worked well; I think the harm involved in expanding the scope of our drug testing would outweigh any potential benefits at this time. This is also what almost all state and federal court rulings in 1986–87 required *public* employers to have as their basis for testing, even when narcotics detectives, prison guards, and U.S. Customs Officers were involved.
- If we do decide to adopt a policy of random or periodic screening for our work force, we should definitely include management personnel in the program.
- Any screening program that we adopt should be based on the MONITOR or similar noninvasive brain-wave test. All positive tests should then be confirmed with a gas chromatography/mass spectrometry urine test.

Discussion Questions

1. Evaluate the Fitness for Duty policy of United Companies in terms of the company's interests and the interests of its employees. Do you feel that the company may be trying to act like a law enforcement agency on this troublesome social issue? Do corporations have a social responsibility to be involved in policing drug use, or are these problems best left to the proper government agencies?

2. Porter and Holloran advocate screening all job applicants for drug use. Do you support this proposal? Why or why not?

3. Porter recommends mandatory random testing of all nonmanagement employees. Do you agree with this policy? If mandatory random testing were adopted by the company, would you support Holloran's suggestion to include management?

4. Astor proposes restricting the drug-testing program to persons in jobs carrying significant risk to the corporation. What would be the advantages and disadvantages of this policy? Holloran would again argue that even a selective testing policy should apply to management. Do you agree?

5. What are the cost implications of broader employee-testing programs and/or of a vastly expanded employee assistance program? In your mind, are these additional costs justified?

6. Holloran holds out the possibility of brain-wave testing. If this technological advancement became feasible, would your response to proposals for either universal or selective random testing be affected?

7. To summarize, the major policy choices facing Marston are the following:
   a. Retain the Fitness for Duty policy (probable-cause testing)
      i. without preemployment screening
      ii. with preemployment screening
   b. Institute universal random testing
      i. not including management
      ii. including management
   c. Institute selective random testing
      i. not including management
      ii. including management

As Marston, what would be your recommendation to the Management Committee?

CASE 6

Due Process in the Non-Union Firm

Dispute Resolution and Human Resource Policy

All State Bank is a multistate banking company with 22,000 employees. It has one hundred branches operating in seventeen states, and maintains four regional data-processing centers. None of its employees are represented by unions, although there is a current organizing campaign directed at white-collar and service industries in several of the states in which the bank operates.

Since 1978, All State has had an internal complaint system, called the Fair Treatment Process (FTP). Employees at all levels can use the FTP to raise any grievance they have with management policies that have been applied to them; their grievances are then considered in a multistep process that culminates in a final decision being made by a bank executive vice president.

In recent years, the FTP has produced an average of 500 complaints per year, of which about 40 percent are decided in favor of the employee, 40 percent in favor of management, and 20 percent a compromise between the two. Last year, the Annual Employee Survey found that 89 percent of All State's employees knew about the FTP, and 70 percent considered it either an "excellent" or a "good" system. However, 24 percent of the employees said that the FTP "needs significant improving," and 34 percent said that they "would be concerned about possible reprisals" if they were to use the FTP.

In additon, several members of All State's personnel staff attended a recent Human Resources Conference at which the national organizer of the Financial Employees International Union said in his speech that he regarded "the existing Mickey Mouse complaint systems used by many banks and insurance companies as the union organizer's best friend. They promise fair treatment but they don't deliver." Finally, complaints by All State employees to the Equal Employment Opportunity Commission have gone up by 26 percent and 24 percent, respectively, in the past two years.

You are Samantha Sullivan, Vice President of Staff Relations for All State Bank. A personnel specialist, Laura Champion, has just proposed several changes in the FTP for your consideration. After reading the various memos that follow, what would your response be?

Key Persons Involved

Samantha Sullivan, *Vice President of Staff Relations*
Peter R. Gregg, *Director, Fair Treatment Process*
Laura Champion, *Personnel Specialist*
Harry S. Pierson, *Legal Services Department*
Marvin Brenner, *Senior Partner, Fosdick, McGlatchey, Smith and Brenner, Attorneys at Law (outside counsel to All State Bank)*

Memorandum

Date: May 22
To: Samantha Sullivan, Vice President of Staff Relations
From: Peter R. Gregg, Director, Fair Treatment Process
Re: Possible Modifications in FTP

I am forwarding to you a copy of a memo I recently received from Laura Champion in Personnel. It contains some very interesting suggestions about the FTP that I believe warrant our serious consideration. Champion's work has led me to reconsider some potential problems in the present system that could become increasingly troublesome in the future. (You'll recall the results of our latest Annual Employee Survey.) The remedies she is proposing, however, are not without their own difficulties. Before commenting further on her analysis, let me give you my own view of how our present system is working and my concerns about some trends that are starting to emerge

As you know, All State Bank inaugurated its present complaint resolution system, the Fair Treatment Process, in 1978. One of our principal concerns at that time was to create a work environment in which our employees would be unlikely to feel the need for a union to represent their interests, at least insofar as issues of fair treatment, grievance resolution, and relations with management were concerned. We also wished to discourage resort to outside agencies, such as the Fair Employment Practices Commission or the Equal Employment Opportunity Commission, by employees who had job-related complaints or concerns. Finally, we hoped to avoid or minimize productivity losses stemming from employees' disaffection or dissatisfaction with their work situation.

To achieve these aims, the company designed the FTP as essentially a system of two-way communication that allows appeal to several higher levels of management whenever an employee has a problem that cannot be resolved by means of review with his or her immediate supervisor. He or she is assisted in this by a personnel officer, whose job it is to help clarify the complaint, outlining it in three parts: the Situation, the Problem, and the Expected Outcome. If the grievance is not resolved in discussions with the personnel officer, it goes to Step 1 (a meeting with a regional vice president or department/division manager); then, if necessary, to Step 2 (a meeting with a division administrator or department head); and finally, if no resolution has been achieved, to Step 3 (a meeting with the appropriate executive vice president of the bank). The entire process is completely confidential, since the filing of a grievance is not recorded in an employee's personnel file. Indeed, employees are assured that their standing with the Bank cannot be adversely affected by their use of the FTP machinery.

I can illustrate the workings of this system with some figures for last year. Of a total of 500 grievances, 45 were filed by bank officers and the remainder by non-officers, which is about the percentage of these two groups in our work force. Nearly three-quarters of the grievances were filed by women employees, again tracking our female work force percentage. The largest number of complaints fell into three categories: "situation counseling" (which accounted for some 33 percent of all grievances), "training and/or advancement" (25 percent), and "supervision," which includes complaints concerning such matters as performance appraisal ratings, communication, and personality conflicts (16 percent). About 5 percent of the complaints were classified as related in some way to affirmative action.

As to the disposition of these grievances, 84 of them were withdrawn after the complainant had discussed the problem with personnel officers. Of the grievances that were carried further, a large majority—more than 80 percent—were resolved in Step 1. Nine percent were resolved in Step 2, and another 9 percent in Step 3. In 40 percent of the cases that reached Step 3, the issues in question were resolved in favor of the employee; in 20 percent, by a compromise result. In the remaining 40 percent of the cases, the employee's grievance was denied.

The grievance procedure outlined above is the heart of the FTP, but it is complemented by two additional mechanisms added during the 1970s. The QUERY program was designed to provide a confidential, two-way channel of communication for employees who have questions or problems that require the attention of and a response from management but who do not want to identify themselves or file a formal grievance unless it becomes absolutely necessary to do so. Employees send in their questions on prepared forms that are handled by a QUERY coordinator, who detaches the employee's name and sends the inquiry to the bank official best able to respond to it. A written answer is then returned to the employee. Anonymity is guaranteed—an employee's supervisor is not informed when he or she uses QUERY. Questions and answers of sufficiently general interest may be published in the employee newsletter. More than 1,000 questions were handled by the QUERY program last year.

The written QUERY program is supplemented by the more recently established QUERY-Line, a confidential telephone service that is available at certain hours of the day specifically to provide information and assistance to any employee who has an urgent problem. Here, too, anonymity is guaranteed; employees can obtain guidance or counseling about a particular situation, get a quick answer about a question of Bank policy, or be directed to the person who can assist them further. Staff from Personnel will also investigate the problem if that seems appropriate. QUERY-Line received nearly 2,000 calls last year.

On the whole, I think that the FTP and its supplementary programs can be considered a success. The system has certainly demonstrated to our employees a genuine interest on the part of management in their concerns and a willingness to work together to solve problems. I feel that this willingness is one of the major factors in the failure of unions to make inroads into any of our operations, even though two banks in our state did lose union elections last year, one for its data-processing employees and the other in its VDT customer service operations. In addition, the FTP has provided management with a means of monitoring the nature and level of employee dissatisfaction, as well as the application of personnel policies and the effectiveness of supervision. Up until two years ago, complaints to outside regulatory agencies had also been reduced, and until very recently, the amount of wrongful-discharge litigation in which we were involved had been reduced. (Even today, this statistic is lower than the industry average, and below that experienced by companies of comparable size.)

Despite these successes, I feel that we may be facing difficulties with the FTP in the years ahead. I base this conclusion on the following analysis:

1. We have conducted confidential surveys among our employees to gauge their attitudes toward the FTP: their willingness to use it, their opinion of its fairness and effectiveness, and so on. We have found that many employees (about 24 percent, in our last survey) do not believe that the system can produce fair decisions when complaints are dealt with solely by management personnel. The people holding this view are in the minority, but it is a significant minority, and one that appears to be growing. Perhaps this finding reflects a more general shift in worker attitudes, a shift toward increased consciousness and heightened distrust of management. In any case, we will have to contend with its consequences in the future.

2. In each of the past two years, the number of EEO charges filed and the number of lawsuits brought against the Bank for wrongful discharge have risen by about 25 percent. The greatest increases have been at our units in California and Michigan, where virtually all companies have experienced comparable increases. One of the great advantages of having a grievance machinery that is not dominated entirely by management is that it has been shown to reduce such burdensome litigation. This is especially relevant given our success in recruiting and retaining greater proportions of minorities and women. Having greater numbers of persons in protected classes means that we will have more people seeking external recourse to fight for their positions; such fights can prove very costly to the company, in addition to exposing it to bad publicity. And the continuing high rate of unemployment in this state can be expected to aggravate the tendency for individuals to take every step possible to retain their jobs.

3. Further strains on our employee relations appear imminent. It is now clear that some staff reductions will be necessary in the near future, as a result of the general economic climate and the increasing impact of automation on many

phases of our operations. This means that layoffs will be occurring at a time when jobs for the affected personnel are likely to be scarce. During such a period of job shortages, employees are inclined to contest terminations or layoffs. No matter how carefully we handle the issue of staff cutbacks, we are likely to find an increasing number of complaints from the personnel involved — age discrimination complaints, sex discrimination complaints, racial discrimination complaints.

4. There is another consideration that is perhaps the most important of all. In recent years, the unions, realizing that the jobs of the future will be in the office and service areas rather than in the traditional manufacturing fields, have mounted organizing drives aimed at the banks and insurance companies. Clerical workers and computer operators — who make up a large part of our work force — have been the particular targets of these efforts. A major theme has been the contention that only a union can secure unbiased grievance resolution for employees. Complaint systems run solely by management always back management, they contend — there is no one to stand up and speak for the employee. As many studies have pointed out, justice and fair treatment are very important to American workers today, so this appeal is a potent one. One of the best ways to counter it may be to strengthen and reorganize our complaint resolution machinery, so that it cannot possibly be viewed as a "management-dominated" system. This may, in the long run, be the best insurance we have against union sentiment gaining ground among our employees.

More than ever before, today's worker cares about the conditions under which he or she works and the way he or she is treated. People expect to enjoy their jobs, not merely endure them. Fair treatment in the workplace is a concern of employees, and a legitimate one. We can hardly expect to attract the best job applicants and retain the most productive workers if we are perceived as a company in which justice is sacrificed to corporate expediency and management always has the last word. (I need hardly remind you of the enormous expense associated with high turnover and low productivity rates resulting from employee disaffection.) In view of these factors, which are likely to play an increasingly significant role in employee-management relations, I think this is an appropriate time to address the question of whether the FTP needs to be modified, and to explore various ways in which this might be done. The options that Champion has proposed are bound to be controversial, but they seem to me to hold considerable promise. If properly implemented, they might result in an FTP that is stronger and better able to deal with the stresses and strains that I anticipate in the coming years.

Attachment **Memorandum**

Date: May 12
To: Peter R. Gregg, Director, Fair Treatment Process
From: Laura Champion, Personnel Specialist
Re: Alternative Approaches to Employee Grievance Mechanisms

I have recently been investigating some approaches taken by other companies to the resolution of employee complaints and problems, and I thought you might like to know more about their experiences. Many of these organizations employ approaches quite different from that of All State's FTP, and I believe that we might profit from adopting certain of their policies and mechanisms. In particular, I would like to direct your attention to two promising methods: employee peer participation in fact-finding or appeal boards, and the use of outside voluntary arbitration as a final step in grievance resolution.

1. A number of firms—such as Control Data and Citibank—have grievance resolution procedures that provide for employee participation, either on fact-finding boards or appeal boards. At Citibank, for example, disputes are resolved through the use of a Problem Review Board—an impartial panel that makes recommendations to senior management for the final resolution of an employee's grievance. The board is convened as a third step in the grievance procedure if the employee lodging the complaint is not satisfied with the results of the first two steps, which are essentially reviews by managers, not unlike those called for in our own FTP. If these appeals fail, the personnel officer selected to assist the employee through Steps 1 and 2 contact a staff relations specialist, who then oversees the selection of the Problem Review Board. The only permanent member of this board is the vice president of staff relations; the other four members are selected by the grievant. The only limitations on the selection of members are that:

- They must be from groups other than the grievant's
- They must be of rank equal to or higher than the grievant's
- One of them must be at the vice president level

The staff relations specialist also sits on the board in an advisory capacity, without a vote. He or she gathers all relevant evidence and presents it in written form to the board, which must meet within ten days to evaluate the record and make a written recommendation to the grievant's department head. This recommendation need not be unanimous, and can and does reflect any minority opinions. The executive receiving this report then has five days to review the recommendation and the entire record leading up to it and to reach a final decision.

Although it expects most problems to be resolved by staff relations personnel at earlier steps, Citibank considers the Problem Review Board procedure to have been very effective as a "court of last resort." Its existence tends to deter possible inequitable actions and encourages the expeditious resolution of disputes at lower levels. Moreover, the board has established an excellent reputation for fairness and is viewed as an effective appeal mechanism. This is reflected in the fact that nearly every board recommendation is accepted by the executive to whom it is submitted, although these recommendations have often reversed management decision. The board's credibility among Citibank employees is attested to by the steadily rising number of cases that are being brought before it, and the widening range of issues represented by those cases. The success of this mechanism can be attributed in large measure to a clear commitment on the part of senior management to the board and the approach that it embodies.

2. A number of firms — such as TWA, Polaroid, Kraft, and Northrop — now provide for binding outside arbitration as the final step in the resolution of employee grievances. Northrop Corporation, for example, has what is probably the oldest formal complaint and appeal system for non-union employees in this country. The basis for resort to this system is very broad — anyone with a work-related complaint has access to the grievance machinery. The procedure, which is similar to the blue-collar systems under typical union contracts, involves four steps:

- Discussion with the employee's supervisor.
- The filing of a written Grievance Notice, with the help of an employee relations representative, for adjudication by a management representative.
- Appeal to a committee of higher-management personnel.
- Final and binding arbitration by an outside agency, selected by mutual agreement of the grievant and the company. The expenses of the arbitration hearing are borne by Northrop, but if the grievant chooses for counsel to represent him or her, the grievant must pay those expenses.

Northrop believes that it is the possibility of binding outside arbitration that ultimately lends credibility to its grievance system. It is felt that the existence of such arbitration has the beneficial effects of forcing the establishment of written personnel policies and regulations, which makes for clarity and consistency, and of encouraging supervisors to exercise good judgment. The system works because of the firm commitment to it at all levels of management and the excellence of the employee relations staff.

Can we learn anything from these examples? I personally think we can. Of course, any system must be designed to harmonize with the character, aims, and philosophy of the corporation instituting it. We cannot simply import, unaltered, the grievance machinery of some other company and expect it to work perfectly in a totally different environment. But the approach to the problem of employee complaints that these companies have taken may have valuable lessons for us. What works for them *may* work for us, if we take care to adapt their solutions to our own needs. I think it is worth a try.

Memorandum

Date: May 28
To: Samantha Sullivan, Vice President of Staff Relations
From: Harry S. Pierson, legal Services Department
Re: Proposed Changes in the FTP

In response to Peter Gregg's memo of May 22, I would like to offer my comments on the changes in the FTP machinery proposed by Laura Champion on May 12, and evidently endorsed by Gregg. I must say that I find his approval of these far-reaching and (to my mind) very disruptive initiatives suprising and disturbing. In view of his central role in the development, implementation, and oversight of our fine existing FTP, I cannot understand how he can contemplate with equanimity a drastic overall change—"modifications" is much too weak a term—that would leave the current system virtually unrecognizable and destroy much that we have achieved in this area to date. I am sorry that I must respond so negatively, but I feel that these proposals are fatally flawed and could do much harm to All State. I will attempt to set forth my objections as succinctly as possible.

First of all, we must consider the impact of such changes on managerial authority. Do we want to take the ultimate authority and decision-making power out of the hands of company management and vest it in some other group or body? Whatever the makeup of such a body—whether it be composed of employees or outside mediators—and however well-intentioned it may be, we cannot expect its members to have the same understanding of company needs, aims, and philosophy, or the same sense of accountability, that characterizes the good corporate manager. Any loss of managerial control and dilution of managerial responsibility is always detrimental to an organization. Involving employees or outsiders in making personnel decisions that are properly the domain of management will inevitably lead to an erosion in both line and operational authority. It will become much harder for managers to make difficult personnel decisions with courage and conviction when they must worry about being second-guessed at a later time by people who may not understand the situation or have the company's best interests at heart.

There is another dangerous trend evident in these proposals: turning the corporation into a quasi-judicial body. Champion's suggestions involve much greater formalization of our complaint procedures. We are suddenly becoming involved in gathering evidence, making records of proceedings, writing minority opinions, and hiring arbitrators. These are activities for people professionally trained in the law; they are not the work of companies. Our firm works effectively because informal relations can be maintained. Legal experts such as myself are employed to know when documentation needs to be created on an issue and how best to do it. Believe me, we do not want this corporation trying to function as a court.

Worse, such an approach to employee grievances may encourage the courts to encroach even further on the right of companies to hire, fire, and generally handle personnel matters as they see fit. The time-honored concept of employment at management's sole discretion has been under attack of late, with courts in many areas recognizing an increasing number of qualifications and exceptions. The notion of procedural rights for non-union employees has been gaining ground, and every time a company hires an ombudsman, accepts employees on fact-finding or appeal boards that deal with personnel matters, or makes provision for outside arbitrators in grievance procedures, it implicitly recognizes the validity of this highly questionable doctrine and provides its proponents with a more solid foundation on which to base their claims. Surely the courts should not be given any excuse to usurp still more authority in this area, having already gone so far (perhaps too far) in their EEO and antidiscrimination rulings.

Moreover, in certain areas, such as EEO, I do not think the courts will allow binding arbitration anyway. On sensitive issues like this, an individual grievant may be able to sue us even after the claim has been denied by an arbitrator. Having the company pay for the expenses of an arbitrator creates an incentive for our employees to carry their complaint to the proposed final step. Why not? What will they have to lose? (This is especially true of discharged employees, who now make up 42 percent of our FTP cases.) And if the arbitrator's decision is not respected by the courts on certain cases, what will the company have gained?

Finally, I wish to make an obvious point that so far seems to have been overlooked: *We do not have to make these changes*, at least not at this time. The law does not require it—not yet, at any rate. Nor is there any real evidence that our present FTP is not working well, only speculation that it *might* not work so well in the future. Furthermore, the approval of 70 percent of our employees is a very positive and strong sign, as I read the Annual Employee Survey results. Why, then, go out of our way to borrow trouble? We should also keep in mind some facts about the history and philosophy of All State over the years. We are, and have always tried to be, a mainstream institution. We are not pioneers; we do not see our role as being in the forefront of experimentation or social engineering. Traditionally, we have adopted policies that have been tested and proved effective. Personally, I am quite content to let Citibank or Bank of America explore the frontiers of innovation in corporate personnel policy. Should there come a day when their approach to these matters has become universally accepted—or is mandated by the courts—then we will adopt it. Until it is common practice in the banking industry, however, I propose that we stick to our well-tested and very effective FTP machinery.

To provide another perspective on these matters, I have asked our outside legal counsel, Marvin Brenner, for his views on this question. His advisory opinion is attached, and I think it merits serious attention.

Attachment **Memorandum**

Date: May 24
To: Harry S. Pierson, Legal Services Department, All State Bank
From: Marvin Brenner, Senior Partner, Fosdick, McGlatchey, Smith and Brenner, Attorneys at Law.
Re: Changes in All State's Fair Treatment Process

You have asked for our advice about changes to the FTP that are under consideration at All State, based on our service to All State as its outside counsel in EEO, wrongful discharge, and other employment-related litigaton. We are pleased to respond, and to draw upon our representation of more than 110 other companies in their employee-related affairs.

I believe that All State should *not* make the changes being proposed. There could be two quite harmful consequences in the handling of litigation or in the support of management if such modifications were made:

1. Courts in those states which have read company procedures and handbook statements as creating contractual obligations between employer and employee might well say that the FTP, as modified, created an employee *right* to have peer representation and outside arbitration. They could say that employees joining the company had relied on the presence of those mechanisms in deciding to work at All State, and that existing employees have remained at the company in the same reliance. That would mean that the company's freedom to abandon such mechanisms if they did not work well might be seriously limited. It could prove very hard to go back to a less formal and "due process oriented" complaint system in the future.

2. It may sound funny coming from a lawyer, and one whose firm receives substantial fees for utilizing our legal system, but I believe that these proposals would substantially and unwisely "legalize" the handling of complaints at All State. Peer-based review boards and arbitration are highly legalistic procedures, and they invite thinking about employee complaints in lawyerly, courtroom-oriented terms. We already have a lot of this through the impact of television programs like *Divorce Court*, *People's Court*, *L.A. Law*, and all the *Perry Mason–*clone shows about lawyer-heroes (and - heroines) championing the causes of individuals against big and heartless institutions. As someone serving as the managing partner of a large, nationwide law firm, with 450 partners and 2,300 supporting staff, I have had to learn something about administration and employee relations over the past ten years. We don't have anything like the proposed changes to the FTP, nor would we consider them. We rely on a management-administered complaint procedure, with an ombudsmanlike alternative if someone wants to seek resolution of a serious problem through less formal means. The danger of creating a "shadow courtroom" system is that it

can transform what are personnel and interpersonal problems into legalized, "rights" and "wrongs" thinking. And you can never satisfy the civil liberties advocates once you embark down their road. Next they will be demanding that an employee have the option to have a lawyer to represent him or her at each stage, a written transcript of all proceedings, and the right to submit formal briefs. Does All State Bank really want to open up that kind of Pandora's box, in the most litigious society ever known to human history?

In short, I believe that with the present FTP, All State has already gone as far as good business, good human resources policy, and prudent management should take it. "Be not the first by whom the new is tried," Alexander Pope warned us. Damn good advice for this situation.

Discussion Questions

1. From a general management point of view, what is the rationale for an employee complaint system? What purposes or functions might it serve, and what drawbacks might it have? Should the same system be expected to work equally well for all employees and all types of issues?
2. Evaluate the reasons given by Gregg in favor of changing the FTP. What is your response to his general orientation?
3. What are your reactions to Champion's proposals for (a) a peer review board as the third step in the complaint procedure and (b) binding arbitration as the final step? What might be the advantages and limitations of each?
4. Respond to Pierson's and Brenner's objections to changing the FTP system. Which of their points carry weight with you and which do not? Do you agree with Brenner that All State should avoid being an innovator in the area of employee policy?
5. As Sullivan, how might you decide about making changes to the complaint procedure? What process would you envision to gain closure on Champion's proposal?

CASE 7

"Big Brother" in the Automated Office

Balancing Employee Morale and Increased Productivity

Acme Office Products is a large company with annual sales of $850 million and 11,000 employees. Among its chief products are typewriters, copiers, calculators, and office furniture. The firm maintains six service centers that retailers and individual customers can reach dialing a toll-free number. At the service centers, each of which employs about 200 persons, operators working at video display terminals (VDTs) use a sophisticated on-line system to take orders from customers, provide product information, trace the progress of customer shipments, and respond to complaints.

Acme is known as an advanced user as well as a producer of new information technology. It sees itself as continually seeking a "competitive edge" through applications of computers and telecommunications. The company is non-union, and has enjoyed good employer-employee relations throughout its 35-year history.

An important issue has arisen involving work standards and productivity at the service centers. The issue has been presented to Justin Acme, the CEO, in the following set of memos. You are Justin Acme's Administrative Assistant, Martha V. Lindsay, and have been asked to prepare a "decision paper" for him on this issue.

Key Persons Involved

Justin Acme, *Chief Executive Officer*
Harry S. Andrews, *Senior Vice President for Human Resources*
Martha V. Lindsay, *Administrative Assistant to Justin Acme*
B.E. Fisher, *Senior Vice President, Operations*
Helen Counter, *Director, Happy Valley Service Center*
Paula M. Washington, *Staff Specialist, Communications*

Memorandum

Date: September 19
To: All Customer Service Center Directors
From: B. E. Fisher, Senior Vice President, Operations
Re: Increased Productivity from Service Center Operators

I know you are all aware of the extent to which mounting costs have squeezed profit margins in our industry generally and at Acme in particular. As a result, competition has lately become more intense than ever, with each company seeking the crucial "competitive edge." A key element in our own drive to remain competitive has been the recent upgrading of our Service Center capabilities—most notably, the design and installation of our on-line computerized order-processing system, which is one of the most modern and sophisticated in the industry. This system has proved to be a great technological success and has enabled us to make dramatic improvements in the quality of service that we can offer our customers. The efficiency of our order processing has increased, while complaints have decreased and customer satisfaction is at an all-time high.

Nevertheless, such progress has carried with it a high price. The system's implementation required a major investment of capital, and its operation continues to represent a substantial item in our annual budget.

You will recall that Chief Executive Officer Justin Acme, in his memo of August 1, stressed the necessity for every business unit to find ways of reducing its operating expenses by 5 percent.

I believe that this is not an unreasonable goal. Our operators are now using VDTs of a very advanced design, which should enable them to work more efficiently than ever before.

Our number of transactions per day is growing at a rate that will reach about 18 percent by the end of this quarter, and is then expected, reflecting increased sales, to level off. On that basis, we would have to hire approximately 200 additional customer service operators to handle the increased call traffic. If we can avoid that increase in headcount, we could save $4.6 million during this next fiscal year (in payroll, benefits, training costs, additional terminals and workstations, additional supervisors, and additional line traffic). This would produce just about the 5 percent saving in operating expenses that the Customer Service Centers have been directed to achieve.

Currently, our customer service operators spend an average time per call of 48 seconds. If we can reduce that average time to about 40 seconds per call, our present work force could handle the 18 percent increase in number of transactions, and no increase in headcount would be needed.

You will recall, too, that when we installed the VDTs we also purchased some new, ergonomically designed office furniture for our Texas and North Carolina facilities, the oldest of the Service Centers. These improvements, together with the increased dedication that we must now expect from our employees, should enable us to meet or even surpass our goal.

I cannot overemphasize the importance of the challenge facing us today. The business climate is difficult and the competition is tough. In particular, it is increasingly obvious that we must be prepared to go "head to head" with Mammoth Business Products. Mammoth is the biggest of our rivals, and they obviously don't intend to cede any of their market share to us without a fight. I am sure that you are familiar with their advertising claims—that they are "lean and mean" and run "the tightest ship in the office supplies business." Unfortunately, these are not just empty boasts. Mammoth is extremely efficient—in part because of their stringent productivity requirements. They demand high productivity from their work force, and they get it. We cannot afford to settle for less from our people if we hope to remain profitable in today's environment.

Memorandum

Date: October 1
To: All Customer Service Operators
From: Helen Counter, Director, Happy Valley Service Center
Re: New Guidelines for Operator Transaction Time

I am sure all of you are aware of the role that our Service Centers play in maintaining Acme in its position as an industry leader. Management is very proud of the contribution of our call operators, and I have been asked to thank all of you for a job well done. However, our Service Centers, with their advanced VDT system, also represent a large expense for the company. To be economically justifiable, they must operate with a high degree of efficiency. Though you may not be thinking of this when handling calls, every second that you spend on the phone with a customer costs the company about 6 cents, which includes salaries, the cost of telephone time, the expense for running and maintaining the computer system, and other overhead factors that are taken into consideration.

In order to remain competitive, Acme must keep its costs under tight control. It is therefore essential that we reduce the average time that each operator spends in handling a customer call. At present, that average is 48 seconds. We believe that, with the excellent new equipment that is now in use in the Service Centers and with a bit of effort from our operators, about an 18 percent decrease—to 40 seconds—can be achieved without any sacrifice in the quality of service that we provide.

Our regular call-monitoring system, which has been employed successfully for the past six years, will continue to be used to compile statistical information about Service Center operations. In addition, regular, periodic listening-in on calls will continue to be done on a random basis by supervisory personnel. If an operator is found to be consistently exceeding the 40-second average time per call, he or she may be required to undergo retraining, or be reassigned if that is deemed necessary. Productivity—as measured by average operator transaction time per call—will in the future account for 50 percent of each operator's annual performance evaluation.

I am confident that all employees will understand the need for this policy and do their best to make it a success. If you have any questions concerning this memo, or other aspects of your job responsibilities, please feel free to speak to your supervisor. He or she will be happy to clarify company regulations and discuss any problems that may arise.

Memorandum

Date: November 2
To: Harry S. Andrews, Senior Vice President, Human Resources
From: Paula M. Washington, Staff Specialist, Communications
Re: Service Center Operator Reactions to New OTT Guidelines

As part of our regular program of annual "feedback meetings" with Acme employees, I have held three focus groups with Service Center call operators in the past month. In each session, the discussion turned almost immediately to the recently promulgated guidelines for operator transaction time (OTT). Indeed, the operators talked about almost nothing else. I must say that I was surprised and distressed at the tone of the discussion and the intensity of the feelings expressed. Generally speaking, we have enjoyed a well-earned reputation as a "people-oriented" company. Our employee relations have been excellent, on the whole, and we have come to expect a high degree of loyalty and job satisfaction from our personnel. Thus, it was a shock to encounter the degree of resentment and even bitterness that our new OTT policies have apparently evoked.

The theme that emerged again and again in these discussions was anger at the imposition of the new 40-second average OTT for Service Center operators. This was not the reaction of a few unusually irate or exceptionally vocal employees—there was total unanimity on the issue. Most of the operators contributed actively to the discussions. Even those who did not take the lead in complaining would chime in with comments to reinforce the speakers' points, or nod their heads vehemently to indicate that they fully agreed with what was being said.

The anger centered on three related issues:

- The time limit itself—its impact on the quality of service they are able to provide to customers, and on the quality of their work life
- The mechanism by which it is to be enforced (supervisor monitoring of calls, and performance appraisals based largely on OTT)
- The fact that it was imposed abruptly and unilaterally by management, without any consultation with the operators or studies of its feasibility, and the unresponsiveness of management to operator complaints

I will try to summarize the points that the operators made on each of these issues in turn.

1. One of the most frequently heard complaints is that the time limit of 40 seconds per call is making it impossible for operators to provide the kind of service that Acme is known for and that our customers have come to expect. As one operator put it, "Making the 40 seconds is now the main thing in everybody's mind, from plug-in to plug-out." Another operator compared the work to "a kind of beat-the-clock game," in which the 40-second target is much more important than what actually goes on during those 40 seconds. Operators gave numerous examples of the consequences of having to stay under the time limit at all costs. Some operators have resorted to cutting callers off; others routinely pass them on to other departments whenever possible, even when they know that the other department is not well equipped to handle their questions or complaints. Operators complain that they often must "duck" problems that they once would have undertaken to solve. Moreover, they no longer have the time to try to "push" other Acme products and services to potentially interested customers, as they are encouraged to do by management, and used to do routinely.

The operators are convinced that the lower quality of service that they now provide will in the long run alienate regular customers and cause Acme to lose business to its competitors. Said one operator, "This company is selling out its most precious asset—the way we treat our customers." The operators cited many examples of customers already asking them what has happened to "the old Acme courtesy and helpfulness." Another operator commented: "Managers don't allow us to be personable anymore. The customers hear the stress in our voice, and they feel we're rushing them off, just like the phone company operators."

According to the employees, the 40-second limit has adversely affected not only the quality of the service they provide, but also the quality of their work life and their job satisfaction. They complain that consciousness of the clock has filled their jobs with tension and anxiety. The pleasure that they used to get from helping people has been greatly reduced. As one operator put it, "Satisfying customers is what makes the pressure of our job bearable, and management has taken that away from us. Another operator summed it up this: "Management is taking all the pride out of our work."

At the same time, they feel that the constant pressure to beat the clock, added to the normal strains of the job (for example, dealing with angry or frustrated customers), is taking a toll on their physical and mental well-being. Some typical comments:

- "Stress has become a big problem around here lately. We take nerve pills if our doctor will prescribe them."
- "This job drains us daily. We are leaving with tension headaches. You go home and just want to fall into bed."

The problems are not just psychological. Many operators complained about the furniture at their VDT stations, stating that it is uncomfortable and in some cases the cause of physical problems. The chairs came in for particularly harsh criticism, with operators maintaining that they cause back pain and muscle tension. One employee claims to have developed severe neck cramps, and has been undergoing physical therapy in order to stay at work. Chairs with adjustable backs and better lumbar support, as well as armrests, are evidently needed; some of the shorter employees may need footrests as well.

Interestingly, these problems have only become acute since the new OTT standards were instituted. Apparently, the increase in job tension resulting from the time limits has exacerbated a situation that was previously tolerable, if less than ideal. Operators contend that new chairs were promised to them two years ago, but nothing has been done: "You have to bring in a doctor's note saying you are in medical trouble before you can get a proper chair." One employee summed up a widespread feeling:

- "Our job is bad for your health. I'm really serious. The strain and stress is taking its toll. And it's bad for your physical as well as mental health."

2. The operators were particularly resentful that management seems not to trust them. They believe that the timing of calls and the 40-second limit imply that the company does not think the operators are working hard. A few quotations from my notes make their feelings very clear:

- "They are treating us like children to be watched and punished."
- "It's absolutely enraging to be called in by our managers and asked to explain why our OTT is higher than 40 seconds one week, and be told we may need retraining. We are real veterans, and to be treated like children by some manager who has never been an operator and couldn't handle the pressure for even one day sends us up the wall."
- "Managers should trust operators. Almost every one of us is here to work, and we do care about the company and profitability. . . . But we get treated like children who have to coaxed or punished into working 'harder.' "

Clearly the operators perceive the time limit as not only difficult to achieve, but demeaning: "It's just Big Brother monitoring all the time." One operator put it this way:

- "It's simple—we've had our dignity taken away from us."

The idea that half their performance evaluation should be based solely on quantitative measures such as average OTT seemed absurd to most of the operators. Said one:

- "We didn't mind occasional monitoring in the past, but now, between the monitored calls and the work profile that the system generates, we feel that we're being treated like part of the machines. And the worst thing is that what they are collecting doesn't give a fair picture of what we are doing."

Another operator commented:

- "Our department is headed toward a production line. It's like building a truck, so much time per operation."

3. Finally, all the operators were critical of the decision to introduce the new OTT standards without prior study of their feasibility or their impact on the operators, and without involving the operators themselves in the decision.

- "The Operations people just set 40 seconds as the [OTT] standard, with no tests or employee involvement. They just dropped it on the [Service Center] managers and then they dropped it on us. There's no reality in this. We could have told them it wouldn't work out. But—no one asks us about anything."

Managers I spoke with in effect conceded that this was true. They acknowledged that there had not been good communication to employees about the reasons underlying such policies as the new OTT standards, and that employees were usually not consulted in such matters. One Service Center executive—unknowingly echoing the language of the operator quoted above—admitted that "things like this just get dropped on employees, with no warning, preparation, or participation."

Operators claim that they have raised all of their objections to the new standards at meetings with management personnel, but with no success. They perceive middle managers as powerless to change directives coming down from senior management. Said one operator:

- "I don't even go to the meetings anymore. It doesn't help; you just get a lot of evasive talk and nothing changes."

Needless to say, this apparent unresponsiveness has only increased the frustration and disaffection of the Service Center employees. They mentioned repeatedly that their morale is at an all-time low.

Since jobs are scarce and Acme is still generally perceived as a desirable place to work, they are reluctant to "rock the boat"; however, I personally am alarmed at the situation, which is unparalleled in my experience at Acme. I feel that top management ought to be deeply concerned, and give immediate attention to this problem.

"Big Brother" in the Automated Office

Memorandum

Date: November 12
To: Justin Acme, Chief Executive Officer
From: Harry S. Andrews, Senior Vice President for Human Resources
Re: Employee Attitudes toward Productivity Standards

I am forwarding for your consideration a disturbing memo from one of our Communications staff, Paula Washington, who has been studying the reactions of our Service Center call operators to recently instituted productivity standards governing the average time in which incoming calls must be handled.

As a human resource administrator, I find the situation described by Washington to be deeply troubling. On the one hand, I am well aware of the competitive pressures to which Acme is now being subjected and the need for Operations to find ways of controlling our costs. I do not doubt that promulgating OTT standards for Service Center call operators represents a bold and serious-minded attempt to comply with your August 1 memo mandating an across-the-board 5 percent reduction in operating expenses. I doubt that the decision to try this particular approach was made lightly, and I am reluctant to try to second-guess those who made it. On the other hand, I have grave doubts as to whether, in light of Washington's findings, we should persevere with this policy. The situation developing at our Service Centers is a serious one that could have far-reaching consequences for the whole company.

To begin with, I think we must consider whether what we are doing in the Service Centers is consistent with the people-oriented philosophy that has been such an important element in Acme's past success. There are two aspects of this question to be addressed. First, are we in danger of undermining our customer relations, as many of the call operators evidently fear? The operators, of course, have an axe to grind; nevertheless, I am inclined to think that they are sincere about this issue, and I don't believe we should dismiss their perceptions lightly. They are, after all, the people who actually deal with our customers on a day-to-day basis. They should be free to make the best possible impression, to present Acme in the best possible light. If our policies are making it impossible for them to do this, then we should consider changing those policies.

Second, we should ask ourselves what being "people-oriented" means in terms of our employee relations. Up to this point, our attitude toward new technology has always been that "people come first," as you yourself have emphasized on many occasions. Although much of our recent growth has been based on the use of advanced technology, we have never been a technology-driven company. Technology has been introduced in such as way as to make work easier, more interesting, and/or more creative—to help people do a better job. We have never used technology to make people work harder or to subject them to greater stress. This approach not only has paid off handsomely in building the image of Acme in the public mind, but also has earned us the loyalty of our employees and the reputation of being an organization that people want to work for. What is happening in the Service Centers now threatens to destroy public and employee goodwill that took us years to build.

Incidentally, I assume that most of us have read articles or seen TV programs recently on the theme of "Big Brother monitoring" in the high-tech office. Even though work settings are not private places, and supervisors have always watched employees and counted their output, the media attacks on employer use of computer software as "counting every call made and every claim processed, every minute, all day . . ." is the kind of anti-Orwell theme that may trouble many Americans. How *we* use this capacity needs to be considered in light of the fact that civil liberties and employee rights groups in 1987 made this a highly visible issue. And two states even had bills introduced to forbid the monitoring of individual employees by office systems and to allow only "group" data to be collected. I mention this just to remind us that our present legal freedom to set standards and monitor performance is not a God-given permanent right of employers, and that our employees are exposed to the same mass-media attacks on "Big Brother in the office" as we have seen.

In this context, I think that we must also consider the danger of unionization to which we may now be exposing ourselves. You will recall that our first Service Centers were opened in North Carolina and Texas, where there was little threat of union penetration. Now, however, we have centers in New York, Illinois, Colorado, and California—all states in which union organizing efforts are vigorous. In the past, we have not worried greatly about the unions, since they seemed to have so little to offer our employees. With our reputation as a progressive organization that treated its people well, and with satisfied, loyal personnel, we thought we had nothing to fear. But the spread of disaffection and discontent such as Washington has found in our Service Centers could quickly change all that. We must therefore ask ourselves whether it is worth jeopardizing our admirable labor relations record to achieve these productivity increases.

One approach that we might want to explore is adopting a set of systematic standards for all VDT monitoring by supervisory personnel. We could discuss with employee representatives the issues of just what kind of information is really needed to ensure both productivity *and* quality. We could explain to them just what quantitative data the system itself is capable of collecting, and try to reach a mutually satisfactory agreement on how such data should be used. Some companies have arrived at policies under which employees have access to their own records and can contest the accuracy or relevance of the data used to evaluate their performance. I feel that such an approach might go far toward neutralizing any charges of 'Big Brotherism" and sweatshop practices. In view of the attitudes that surfaced in Washington's sessions, I believe that *any* attempt on our part to open a dialogue with employees on these matters and solicit their input would greatly improve the general climate. More than anything else, our operators seem to feel insulted by our disregard for them—their opinions and their concerns. I feel that we have ignored them for too long.

I realize that this a complex question, and I hesitate to offer simple solutions. I should stress, too, that we are not *legally* obligated to change our current policies in any way. The law allows the sort of call monitoring and data collection that we now practice. Whether or not these practices are wise, however, is something that you will have to decide. The issue is important enough to require resolution at the highest level.

Memorandum

Date: November 25
To: Justin Acme, President
From: B.E. Fisher, Senior Vice President, Operations
Re: Service Center Operator Productivity Standards

As requested, I have reviewed the memos from Washington and Andrews about the reactions of our Service Center operators to the OTT guidelines, and frankly, I think the entire business is a tempest in a teapot. While I admire their commitment to good employee relations, I feel that Washington and Andrews are overreacting to the situation—I am not at all convinced that we have a serious problem here. I grant that there are a lot of ruffled feathers, and maybe we should do something to help people calm down. Nevertheless, as far as the substantive issues are concerned, I think we are in the right, and the whole thing will eventually blow over without our having to back down.

In retrospect, it is obvious that the introduction of the OTT standards wasn't handled as well as it might have been, and I'm willing to take some of the blame for that. We acted so quickly for two reasons. First, we felt that the installation of the new telecommunications package for use with the VDTs would allow us an increase in productivity. In fact, such an increase was part of the economic justification for the new package. We thought it would be a good idea to institute the new standards right at the start, so that people could know just what was expected of them and the system could start paying for itself, as it were. Second, the entire Operations department has been functioning under a great deal of pressure because of the impending introduction of our new line of color copiers. These machines will play a large part in determining our future profitability, and their successful launching naturally had top priority.

I think it is important to do so something to mollify the operators, who evidently feel unappreciated. Perhaps we could program some social events for these people to make them understand that we value their contribution to Acme. With regard to the more substantive matters, we should look into expediting the purchase of some better office furniture as soon as budgetary constraints permit. Perhaps new chairs could be considered as an item for next year's budget, if other costs can be kept down.

As for the 40-second OTT benchmark, however, I think we must keep it. Mammoth has similar (if not more stringent) productivity standards. They give each of their call operators a weekly printout with a breakdown of the operator's transaction times and other statistical measures of efficiency. If the operator's productivity lags below company requirements, they are reprimanded. If they consistently fail to meet their guidelines, they eventually receive a poor performance rating and forfeit any pay raises, and so on. If we can't get as much out of our people as Mammoth does, it will be impossible to meet the substantial discounts that Mammoth has recently been offering in an attempt to make inroads into our market share. Worse still, if our profitability slips, we run the risk of finding ourselves the target of a hostile takeover attempt. I am sure you have heard the rumors that have been circulating lately and are as disturbed by them as I am. Staying lean and efficient is probably our best defense.

I am all for a people-oriented philosophy. Nevertheless, these are difficult times in our business. The 40-second OTT standard is reasonable; our operators, with few exceptions, have met it, and have been able to live with it. I see no reasons why we shouldn't stick with it.

Discussion Questions

1. Based on how Fisher and Counter handled the OTT policy, can you speculate about the management style of this corporation? What are the advantages and disadvantages of the management style as implied by the OTT incident?

2. What level of validity would you place on the data collected by Paula Washington? Can methods such as focus groups improve managment's understanding of employee problems, or are focus groups likely to induce and magnify negative feelings and create a distorted picture of employee attitudes? How representative are the results of these efforts?

3. The positions of Fisher and Andrews represent sharply opposing views on how new technology should be implemented in the firm. Must Justin Acme choose between these views, or is there a common ground between them?

4. Andrews opens the possibility of ways other than OTT to measure employee productivity. What performance measures might be consistent with employee needs and at the same time attain company objectives? Would you support Andrews' suggestion to raise this question directly with employee representatives, or is this step likely to lead to unnecessary complications?

5. Andrews also invokes the spectre of unionization to argue against the OTT policy. Do you believe that this is a realistic argument, or is Andrews simply capitalizing on managerial anxiety about unions in order to strengthen the influence of his human resources department?

6. As Martha V. Lindsay, Administrative Assistant to Justin Acme, you need to prepare a memo outlining a decision for him on the OTT issue. What approach would you propose in your memo?

Building a New Information System at *The Call*

CASE 8

How to Combine Expertise with User Participation

The Call is a suburban newspaper west of the Hudson, in northern New Jersey. It is a profitable paper, part of a family-owned newspaper and television group, with a daily circulation of 150,000 and 1,200 employees. While it has been a leader in acquiring advanced offset presses (for color printing) and new computerized production systems, *The Call* has not—until now—been at the forefront of its industry in using video display terminals (VDTs) and other forms of office technology for its editorial and circulation operations.

The company has also been fairly traditional in its approach to employee involvement. A good deal of informal communication takes place between managers and employees, especially in the editorial and news departments, where employees are college educated and norms of common professionalism prevail. *The Call* has not used any quality circle or formal employee participation programs.

You are Jackson W. Olcott, Senior Vice President of Operations. A proposal has been made to create and use an employee-participative process in implementing a new office technology system in *The Call's* editorial and circulation departments. Olcott has sought and received several memos commenting on this proposal and must now decide how to respond. Would you support or oppose this proposal?

Key Persons Involved

Julian DeRobertis, *President and Chief Executive Officer*
Jackson W. Olcott, *Senior Vice President of Operations*
Dr. Eleanor Millspaugh, *Senior Vice President for Human Resources*
Edward D. Sobieski, *Director of Information Systems*
I.G. Aaronson, *Editor-in-Chief*

Memorandum

Date: August 8
To: Jackson W. Olcott, Senior Vice President of Operations
From: All Department Heads
Re: Design and Implementation of Computerized Systems

As you are no doubt aware, the Executive Committee meeting of July 25 was the first such session presided over by our new President and Chief Executive Officer, Julian DeRobertis, who last month succeeded President Stanton on his retirement. At the July 25 meeting, the Executive Committee approved a proposal by Mr. DeRobertis that *The Call* proceed as expeditiously as possible to implement the Master Plan for the introduction of modern information technology into many phases of its operations. It was agreed that the first step should be the development of automated systems for the editorial and circulation departments, as detailed in Section III-B of the Plan.

The feasibility studies undertaken last year as part of Phase I of the Plan clearly established the following general points:

- Competition from electronic media is increasing and industry profit margins are being squeezed. All newspapers must increase their efficiency to survive in this climate.
- In the specific case of *The Call*, competition from the larger New York City dailies, as well as the national *USA Today*, presents a formidable threat to our profitability. If we do not make effective use of modern information technology, we will find ourselves at a severe and increasing disadvantage with respect to these rivals.
- Our present office technology is seriously outmoded.
- Hardware and software costs are dropping rapidly, putting powerful automated systems within our current budgetary reach.
- A wide range of software packages and even complete newspaper systems designed specifically for the needs of many of our departments is presently available. Thus, if we prefer, we need not build our own system from scratch or develop our own software.

Specifically with regard to our own operations, the Plan proposes that we begin upgrading our capabilities with the implementation of two systems:

1. A computer-based system for the Editorial Department. This system would allow reporters and feature writers to create and revise pieces on VDT terminals. Text could be sent directly to other terminals for rewriting and copyediting, and (if desired) returned to the original creator's terminal with all changes highlighted for editorial discussion. Finished text would be transmitted directly from the Editorial Department to Composition. In addition, writers would have access, while composing their stories, to material in their own files and interview notes; to reference materials in the paper's own database and those of on-line research and information services; and, via an electronic library (an

"automated morgue"), to all past stories published in *The Call*. This system would include an "electronic bulletin board" for exchange of news and messages among Editorial personnel at any hour of the day or night. Ultimately, the system could be expanded to incorporate portable lap-top computers for use by reporters at home or in the field.

2. An automated system for the Circulation Department. This system would consist of a large number of on-line terminals linked to a central database containing information about each subscriber and his or her account history. Operators would be able to take orders or handle customer complaints on the telephone, keying in any action to be taken and updating the customer's file. The system would be tied directly to a computerized billing system. It would also be able to generate management information reports, such as demographic breakdowns of subscribers and the like.

Systems such as those described above, as well as similar ones that we hope to implement for other departments in the future, have already been developed and brought into use by a number of the more enterprising American newspapers. It is our intention to appoint a management team to inspect several of these pioneering systems at the earliest opportunity.

Clearly, upgrading of this magnitude will be a major undertaking, requiring the commitment of substantial human as well as financial resources. I am therefore requesting that each Department Head submit to me by August 30 an analysis of the key issues in the design of such systems as they bear on his or her department. Specific points to be addressed include:

- The goals that should be kept in mind in designing the system
- The existing constraints (fiscal, operational, organizational, human, and so on) that limit our options
- Problems that can be anticipated

I am particularly eager to obtain your response to an interesting suggestion made by Dr. Eleanor Millspaugh, Senior Vice President for Human Resources. Dr. Millspaugh has proposed that we adopt an employee-participative decision-making process in the design and implementation of these systems. I am enclosing her memo, which addresses both the general advantages and drawbacks of this approach, and the specific considerations that bear on our present situation at *The Call*.

Attachment

Memorandum

Date: August 6
To: Dr. Eleanor Millspaugh, Senior Vice President for Human Resources
From: Jackson W. Olcott, Senior Vice President of Operations
Re: End-User Participation in the Design and Implementation of New Information Systems

It is my understanding that *The Call* is now about to proceed with the development of automated systems for the editorial and circulation departments. It is the strong conviction of the Human Resources Department that serious consideration be given to according the end users of these systems a major role in their design and implementation. Our reasons for taking this position are spelled out below.

Ever since work on the preparation of the Master Plan began, early in 1985, Human Resources has been projecting the likely results of introducing information technology at *The Call* and studying ways of maximizing the benefits while minimizing the ensuing problems. You will recall that the Master Plan gives considerable emphasis to considering ergonomic and personnel factors in devising strategies for office automation. In the words of the Plan, "Human factors will ultimately determine the success or failure of implementing office automation systems."(p. 22)

Indeed, it was a suggestion in the Master Plan (see Section II-A) that inspired us to commission a well-known firm of management consultants specializing in information technology—the Socio-Technical Group—to study the feasibility of end-user participation in the design of automated systems for *The Call*. We are most impressed with the quality of this study and feel that the conclusions reached should be given very serious consideration. At the risk of repeating some of the points made in the report, let me set forth the view of Human Resources on this subject, together with some of the theoretical framework that underlies it.

It is generally stated that the most important resource of any organization is its people. This is certainly true of the Editorial Department at *The Call*, where the energy, imagination, creativity, dedication, and initiative of the staff are the foundation on which all achievements rest. It is only slightly less true in other departments, such as Circulation. While the work in such departments may not require the same level of creativity, it is nonetheless true that low morale, job dissatisfaction, and lack of dedication among such personnel can have a highly negative impact on efficiency and effectiveness. In this era of skilled workers and rising training costs, high turnover rates resulting from job dissatisfaction represent a major economic drain and are a substantial problem for many companies. Thus, *finding* good people, *keeping* them, and *helping* them to work at their highest levels of productivity are generally the keys to success in today's competitive environment.

It is equally well accepted that technology has a powerful effect on the people using it. There is no question that modern technological resources can help people to reach new heights of productivity and creativity. Unfortunately, when technology is introduced in a haphazard way without suffficient advance planning or *the right kind of planning*, it can have profoundly negative consequences that can reduce the potential gains for the company hoping to benefit from its use. The introduction of any new technology is potentially stressful; without proper planning, the introduction of automated systems can increase the stress on employees to unacceptable levels. This occurs for a number of reasons, including the following:

- Fears by employees that they lack sufficient training to handle the new system, and will be unable to perform satisfactorily
- Anxiety among employees that their job skills will become outdated, or even that their jobs may be entirely replaced by the machines
- Anxiety that even if their jobs survive, the jobs will be downgraded ("deskilled"), or restructured in ways that are uncongenial to them and over which they have no control
- Fear that their work will become more boring and routinized as the system performs the more complex functions, leaving them with only the less challenging chores and depriving them of the opportunity for creativity, initiative, or control over their own work—"mere extensions of the machine"
- Fear that the new technology will lead to increased social isolation, thereby reducing opportunities for interaction among peers
- Fear that the machines will permit management to monitor their work more closely and without their knowledge, set unrealistic productivity standards, and evaluate workers solely on the bases of relative output (the "Big Brother" syndrome)

In addition, many employees are concerned over health and safety issues that have repeatedly been raised in connection with VDTs. Although all scientific evidence to date suggests that VDTs present no special health hazards, employees have heard rumors and read scare stories in the press about radiation from VDTs and alleged cases of cancer, birth defects, and cataracts linked to intensive VDT use. Even people who discount these unfounded and sensational accounts may worry about potentially real problems such as eyestrain, muscular problems (backache, shoulder pains, and so forth), and harmful stress that may indeed accompany protracted VDT use if insufficient attention is given to ergonomic factors in system design. Anxieties of this sort—*whether they are well founded or not*—can produce serious mental and physical health problems.

What are the results of such problems? They are various: low morale, low levels of job satisfaction, high rates of absenteeism, low productivity, and high rates of job turnover (with the concomitant expenses of recruitment and training). Clearly, all of this is likely to be very costly to the organization, and may actually outweigh the benefits derived from the new system itself. Moreover, the system itself may never function smoothly or live up to its full potential if it is not embraced enthusiastically by the people for whom it was designed. Rather than using the new technology creatively, seeking ways to exploit its powers, and taking full advantage of its capacity to enhance human capabilities, employees may be driven by resentment and fear to adopt (consciously or unconsciously) an attitude of "passive resistance" that subtly but surely impairs the performance of both the people and the machines.

In addition to these human factors, certain organizational and operational aspects of technological innovation are often overlooked. When new automated systems are introduced, they are usually designed to do the same jobs as the system they replace, in basically the same ways. Thus, just as electric typewriters replaced manual typewriters, word processors are expected to replace electric typewriters—without any major change in what they are expected to do, or what role the person using them is expected to play in the organization. More often than not, this approach leads merely to accepting the limitations that were built into the old system and automating its inefficiencies rather than eliminating them. The full potential of automated systems can only be reached when system designers are willing to reexamine at a fundamental level *what the organization is attempting to do*, and then ask how modern technology can help in doing those things. Often this involves the restructuring of jobs, and even making changes in organizational structure.

Unfortunately, the people who design automated systems may not be well prepared to conduct an analysis at this level. Their area of expertise lies with what the equipment can do. They may not know a great deal about what the people in an organization actually do, how they do it, and how they feel about it. More important, they are unlikely to understand what people might *like* to do, or *could* do, if given the opportunity.

In light of the above considerations, it is easy to see why many people who have studied these problems believe that when automated systems are introduced, *optimum results are most likely to be achieved when the people who will use the system are involved as early and as fully as possible in its design and implementation*. Such involvement has four major components:

1. Participation in system design (both hardware and software components)
2. Participation in the reorganization of work tasks and the redesign of jobs
3. Participation in the design of the work environment
4. Participation in setting performance evaluation standards and measurement procedures

Such involvement often results in a system that is superior to what might otherwise have been produced—in flexibility, in adaptation to the specific needs of the organization and capabilities of the employees, and frequently in efficiency as well. Equally important, though less tangible, are the psychological benefits: increased job satisfaction, greater loyalty and dedication, and the alleviation of many of the anxieties spelled out earlier. Workers who have been consulted in the design of their tools and work environment have a greater sense of pride and of being appreciated by their organization. They have heightened feelings of responsibility and control of their own destinies, and thus are far less likely to blame management for everything that goes wrong. Most significant, they tend to feel that people are the center and the driving force of the organization and automated systems merely a tool that exists to serve people—not vice versa.

Obviously, securing employee participation in a project of this magnitude would be a complex undertaking, but it is one for which the Human Resources Department is well prepared, having devoted considerable research to the subject over the past three years. We can rely for guidance, moreover, on the experience of a number of other organizations, including some newspapers. The St. Paul *Pioneer Press* and the Minneapolis *Star and Tribune*, for example, have used quality circles and other employee participation methods for some time now. In addition, we have the specific recommendations of the Socio-Technical Group staff.

I am well aware that this is an approach that has not previously been adopted at *The Call*. However, there is a substantial body of literature suggesting that it is the approach most likely to lead to success, and one that will be increasingly adopted by sophisticated organizations in the future. The presence of new and innovation-minded top management at *The Call* encourages me to believe that now is an opportune time for an initiative that holds so much promise. If we are successful, we can expect to reap substantial benefits: a high degree of staff commitment and job satisfaction, reduced turnover, and increased productivity. I feel that these benefits greatly outweigh the risks.

Memorandum

Date: August 15
To: Edward D. Sobieski, Director of Information Systems
From: Jackson W. Olcott, Senior Vice President of Operations
Re: Employee Participation in Automated System Design

In response to your memo of August 8, I would like to comment on Dr. Millspaugh's recent suggestion that our employees play a major role in the design and implementation of the new systems we will develop for the editorial and circulation departments. As you are no doubt aware, I was not consulted to any significant extent by the Socio-Technical Group staff that prepared the report on this matter for Dr. Millspaugh. I do not know whether their recommendations would have been any different if I had been. Nevertheless, employee input in system design is a subject that concerns me closely, and that I have studied carefully. I am familiar with some of the literature on the subject, and I have had some firsthand experience—not always happy—with it at other organizations. I am sorry to have to say that I don't think it is a very practical idea for *The Call*.

To begin with, let me point out some of the constraints under which my department must function.

- First, we are understaffed. There is a shortage of good systems analysts and programmers, and we have not always been able to attract the kind of people we need.
- Second, we labor under the constraint of having to patch up our existing information systems, which, as everyone admits, are woefully outdated. This is a job that consumes many person-hours each month.
- Third, these problems will not, unfortunately, disappear just because we are developing new systems. In fact, they will get worse. The old systems will somehow have to be kept running while costly manpower is diverted to designing and implementing the new ones. In order for the new systems to be adequately tested and debugged, and to ensure a smooth transition, *both* systems will have to run simultaneously for several months prior to and during the changeover period. Moreover, since our upgrading is to take place in stages—department by department and function by function—it will be necessary to ensure a high degree of compatability between each new system and the remaining components of the older setup. I need not spell out here the technical difficulties that all this presents; suffice it to say that they are enormous.
- Fourth, developing and testing new systems is extremely time-consuming. Numerous vendor demonstrations must be scheduled, product specifications must be studied, hardware and software must be evaluated, lengthy tests must be conducted. Unfortunately, the time constraints in designing and implementing the systems we need are going to be acute.

To put it bluntly, therefore, we can't waste time fooling around with people who don't know the first thing about the capabilities and limitations of the equipment. Providing users with enough technical training for them to be able to participate meaningfully in the design process would alone consume most of the time and funds at our disposal. The goals set forth in the Master Plan for these systems are very ambitious, while the time and resources available for achieving them are limited. Under these conditions, essentials must take precedence over niceties. Of course, everyone wants a system that is "user-friendly" (to use the favorite buzzword). Many people don't realize, however, that there's a great deal more to an information system than whether it can be used by your 10-year-old daughter. Speed, cost, efficiency, reliability, ease of repair, compatability, expandability, and many other factors must be considered. Often these are more important than having VDT screens that are color-coordinated to match the chairs in the employee lounge. My experience has convinced me that *management goals for sophisticated systems can be met only if the design of the systems is in the hands of people who understand the equipment and what it can do.*

I hope these remarks do not give the impression that I am opposed to employee participation per se. On the contrary, I have high regard for Deming's work in popularizing the concept of quality circles. That approach obviously works very well in some situations, when it can directly affect product quality, improve problem solving, and help to reduce grievances. In fact, my own department makes limited use of quality circles in our programming function. This does not mean, however, that employee participation is a panacea or that it is appropriate in all contexts. I am familiar with much of the recent discussion in the information systems literature about people-oriented systems and the desirability of end-users in the system design. What these rather vague accounts do not make clear is that *up to this point, no major system has, to my knowledge, actually been constructed in accordance with these principles*. Thus, I feel that any attempt to include end-users in the system design process on a large scale represents a highly expensive, time-consuming, and speculative gamble. This is an experiment that we cannot, given our current constraints, afford to undertake. Input for system design and implementation from outside the Information Systems Department should be limited to (a) any outside *technical* consultants whose assistance might be needed, and (b) the *managers* of the units being automated. The latter are the people who should best understand the purpose of the system and the needs of its ultimate users.

Memorandum

Date: August 22
To: I.G. Aaronson, Editor-in-Chief
From: Jackson W. Olcott, Senior Vice President of Operations
Re: Employee Participation in Automated System Design

I would like to respond to your memo of August 8 concerning the possibility that staff members in the editorial and circulation departments may be invited to participate in designing the information systems that they will eventually use. Let me say at the start that my immediate gut reaction to this idea was highly favorable—as a matter of fact, I threw my hat in the air. (Old newspapermen like myself still wear hats, even in the newsroom.) Then, of course, I started to worry about the details; reservations, qualifications, nagging problems began to nibble away at my initial enthusiasm. I guess you could say that the purpose of this memo is to catalog some of those reservations, analyze them, and perhaps put them to rest by translating them into positive suggestions. I'd hate to see them spoil such a great idea.

    1. I know reporters pretty well, being one myself—or at least I like to think of myself as still a reporter at heart. I think I know feature writers, editorial writers, reviewers, etc., etc., too, having tried my hand at all these jobs at one time or another. And of course, I've been a copy editor too. I think I understand all these people, from the inside out. I know what makes them tick. Now, I know that I *don't* know the people who work in Circulation nearly so well. I do know, however, that they're probably rather different from reporters, writers, and editors. A lot more of them are women, for one thing. (Perhaps it shouldn't be that way, and maybe it doesn't make much difference, but it's so.) On the average, they have less education than the people in Editorial; I'd guess that quite a few of them never finished college, or never even attended. Some of them haven't been with *The Call* very long, and there are even a fair number of temps

working there at any one time. Without meaning to disparage the work they do, I think I can safely say that it isn't as interesting or varied or creative as that done by writers and editors. Finally, these people are probably a whole lot saner, on the whole, than Editorial types. You have to be a bit odd in the head to be a newspaperman—newspaperperson, I should say.

What conclusions do I draw from all this? Simply that it isn't likely that a single approach will work equally well in both departments. I am not an industrial psychologist—whatever that is—and we have a good Human Resources unit that can no doubt figure out the best way of involving all kinds of different people in this design process. So I am content to leave the details to them. I am just pointing out that we probably need two different models here. As Blake said, "One law for the lion and the ox is oppression"!

2. Now, to get back to my own department and the people I know best. I think that just about everyone would acknowledge that these folk are highly creative and independent types. They take pride in their professionalism, but they are largely self-motivating; they don't like to be told what to do or when to do it, beyond a certain point. They like to take the initiative, they like to organize their own work, and they like to pace themselves.

For our present purposes, these traits have both positive and negative implications. First of all, some of the people in my department won't like the idea of installing *any* automated system, no matter who designs it or how it's designed. They like the way of working that they've evolved for themselves over the years, and won't be happy about having to change it at the command of some computer whiz who never met a copy deadline. I suppose this attitude is found everywhere, and I don't think the newsroom will prove to be any exception. Some of these people would still rather be pounding a big black upright Remington Rand manual, with a limp cigarette in the corner of their mouth, à la *The Front Page*.

It should also be recognized that reporters, because of their temperaments and the special nature of their work, may have some unique problems with information systems. Reporters are, as you know, very jealous about such matters as privacy and confidentiality. They have to be. They often deal with sensitive material to which they have privileged access, and they must protect both the confidentiality of their sources and their own exclusive right to use their material as they see fit. Now, some people simply don't consider computer systems secure—password or no password. They've heard about too many breaches of computer security, and they may be reluctant to trust their notes and files to *any* information system. A few people don't even like the idea that management will always have access to the system—though it's hard to see how this could be avoided.

Such people, however, will definitely be in the minority. Most of the younger staff members are well aware of how badly out-of-date our present technology is. (God knows they complain about it enough!) They're equally aware of the constant pressure to cut costs and increase productivity while maintaining the high level of quality on which we pride ourselves. And they understand very well the potential benefits of computer-based systems in just those areas. They've read about them (maybe even written articles about them), seen them installed at other papers—quite a few of them have home systems, I'd guess. I'm sure that nearly every one of these people would just love to design some sort of high-tech system for the newsroom. But that's just the problem. Highly educated, independent, self-motivated people aren't always very skilled or experienced in group decision-making; they're too used to doing things their own way. These people are each likely to have their own very definite ideas about the system—but will they be able to work together to come up with a set of recommendations, or will we end up with fifty different proposals? I should mention, too, that there will inevitably be people who have no objection to the idea per se, but won't want to participate because they're afraid they don't know enough and don't want to expose their ignorance. This is more likely to be a problem in the Circulation Department, I should think. Finally, some people may feel that to participate in the design process is somehow to allow themselves to be "used" by management. I don't understand this attitude myself, but I can't claim that it doesn't exist. There are always a few people who are determined to be discontent, no matter what is offered them.

3. I have been dwelling on the employees, but I do not think that employee attitudes will be the key factor in determining whether this plan succeeds. Ultimately, I think that the attitude of management will be decisive. The role of management is to lead, and when a novel and controversial initiative such as this is at stake, weak or divided leadership is likely to prove fatal. A great deal depends on the degree of unity and commitment that management brings to the project. If there is opposition to this plan from the Information Systems Department, we are unlikely to accomplish anything. The same is true if middle management is ambivalent.

What we need, then, is a clear, total, and unambiguous commitment on the part of management at all levels—especially at the top, for that is the true fountainhead of authority. If top management is unwilling to commit its prestige and its resources (including, of course, financial resources) to this endeavor, it would be best not to undertake it at all. The necessary authority—and ultimate responsibility—will have to be clearly delegated, whether to Human Resources, an interdepartmental task force of some sort, or an outside consultant. Who is in charge is less important than the fact that someone truly *is* in charge. Finally, persistence will be needed. This will not be a simple project, and a fast-and-dirty approach won't work; it would simply alienate everybody, not least the end users whose cooperation we are trying to secure. If management enters the race it must be prepared to stay the course. The rewards for doing so, however, are potentially very great.

Discussion Questions

1. What problem does Jackson Olcott face? Given the issues raised by Millspaugh, Sobieski, and Aaronson, what factors above and beyond employee participation must he consider?

2. Evaluate the conflicting positions of Millspaugh and Sobieski. As Olcott, what is your reaction to each of these views? What actions would you take in response?

3. Aaronson has pointed out some differences between the circulation and the editorial departments. What additional differences in tasks and personal characteristics might you expect to find in this organization? Do these factors suggest that different types of supervision or degrees of employee participation would be effective in the two departments? Should the degree of employee participation for either of these groups be any different in managing technical change as opposed to conducting other administrative duties?

4. Assume that Olcott and the department heads agree that employee participation is desired on this project. What alternative forms of involvement might they consider?

5. Aaronson states that top managment leadership is critical to the successful implementation of the information system. What role should Olcott seek to play, and what role, if any, should he seek to avoid?

CASE 9

Creating a Computerized Medical Surveillance System

Improving Health but Risking Lawsuits

Advanced Chemicals manufactures a wide range of chemical and pharmaceutical products. One of its divisions, Bio-Life, is involved in the bioengineering of new agriculture and medical substances, some of which will be produced in space as part of the satellite-launched Space Lab programs.

Advanced Chemicals experienced a dramatic episode two years ago, when an incidence of eye cancer was found to have developed among workers using a new chemical called poly-benzene-acetate (PBA). Although company health and safety staff had been tracking studies of PBA's effects, early reports were conflicting and basically inconclusive. The labor union at Advanced Chemicals — the Chemical and Pharmaceutical Workers International — had complained from the start that workers using PBA experienced eye pain when they went home. The company's health director investigated and recommended a program of "company-paid eyewash" for all employees using PBA, but concluded that "no permanent ill effects have been demonstrated." The union then called on the National Institute for Occuational Safety and Health (NIOSH) to do a hazard evaluation study. This study had two major findings:

1. There was a 21 percent higher incidence of eye cancer among the company's workers using PBA than in a control population with matched characteristics not using PBA. However, evidence of a direct link between PBA and the incidence of eye cancer was not absolute.

2. Advanced Chemicals, like all other companies in the industry, did not maintain the kind of detailed worker health records, based on regular medical exams and reported problems, that (according to NIOSH) could help immensely in the early identification of adverse reactions to new chemicals and drugs. This lack would become even more crucial, NIOSH warned, as new biological agents were brought into manufacturing. NIOSH noted, however,

that employers were under no legal obligation from current federal or state statutes to collect such detailed worker health records, apart from accident reports and the chemical identifications required by Occupational Safety and Health Administration regulations and any applicable state or local right-to-know laws. (OSHA regulations provide that *if* a company maintains medical records, its employees must be allowed access to them; moreover, the union has certain rights of inspection as bargaining representative. Creating such records, however, is not now required by law.)

In response to the NIOSH study, Walter Prentice, President of Advanced Chemicals, set up a task force to examine ways in which the company could more effectively address future worker health problems in its operations, with special attention to the advisability of instituting a comprehensive, automated health information system. The task force has just reported its findings with majority and minority positions as presented in the pages that follow. What should Prentice decide?

Key Persons Involved

Walter Prentice, *President*
For the Majority Report:
F. Lionel, M.D., *Director, Medical Services*
Jason Prince, *Director, Corporate Information Systems*
Charles C. Hobby, *Director, Public Affairs*
Jennifer Atkins, *Vice President for Human Resources*
For the Minority Report:
Andrew Parsons, *Director, Labor Relations*
Ford C. Haskell, *Director, Manufacturing Operations*
Timothy Anderson, *General Counsel*

Findings of the Interdepartmental Task Force for the Study of Health Information Systems

MAJORITY REPORT

Submitted By:
F. Lionel, M.D., *Director, Medical Services*
Jason Prince, *Director, Corporate Information Systems*
Charles C. Hobby, *Director, Public Affairs*
Jennifer Atkins, *Vice President for Human Resources*

Executive Summary It is proposed that Advanced Chemicals immediately begin design work on a comprehensive, automated health information system, to be operational in two years. The primary objective of this system should include:

Creating a Computerized Medical Surveillance System

- Maintenance of consistent, up-to-date information on all chemicals used or produced in our operations, as well as radiation or other emissions to which employees may at times be exposed
- Identification of employees exposed to all potentially hazardous substances or conditions, and quantification of the extent of their exposure
- Provision of appropriate medical surveillance for all employees, based on the nature and extent of their exposure to specific hazards
- Identification of employees with medical conditions that might put them at increased risk from such exposure
- Continuous incorporation of epidemiological and other research data to provide advance warning of potential health risks associated with our operations

Such a system would provide improved protection for the health and safety of our employees, reduce worker disability and absenteeism, lower health benefit and insurance costs, diminish our legal exposure in the liability area, and assist and systematize our compliance with legal requirements.

Introduction The following report represents the views of four of the seven members appointed by President Walter Prentice to the Interdepartmental Task Force for the Study of Health Information Systems. This Task Force, created in March 1987, was charged with evaluating the feasibility and advisability of creating a health information system (HIS) for Advanced Chemicals, and analyzing the potential benefits, possible disadvantages, and problems that might result from the implementation of such a system. It is our conclusion that the advantages to be derived from an HIS greatly outweigh the drawbacks, that such a system is currently feasible, and that work on the system should begin as soon as possible. (A different conclusion is presented in the Minority Report that is being submitted by three members of the Task Force.)

This report is in two parts. Part I is a brief description of the proposed system. Part II discusses the benefits that are anticipated from the operation of such a system.

I.

The "state of the art" in the construction of complex, interactive databases is currently such that a comprehensive, automated HIS can be created at cost levels that are acceptable for a company the size of Advanced Chemicals. While many design details remain to be addressed, the basic configuration of the system is likely to be as follows. It would comprise four interactive modules (databases). Each module would be continuously and automatically updated, in part by input from other modules.

- Module I would contain information about every chemical substance used by Advanced Chemicals in our workplaces, as well as every substance produced or created in the course of our operations to which employees might be exposed (including radiation from our equipment or other potentially dangerous emissions). The information collected would include all available data about the potential health hazards associated with these substances and/or radiations. Specifically included would be such items as acute toxic effects, chronic toxic effects, medical conditions that should preclude exposure to the particular substance, and exposure levels or contact time (either immediate or cumulative) that would indicate the need for a medical examination. Many of the data in Module I would be derived from the Material Safety Data Sheets (MSDS) currently used in all our operations. The MSDS information is now largely in machine-readable form; where it is not, conversion to machine-readable form could be effected at relatively modest cost. Other input to Module I would be directly from Module IV, as described below.

- Module II would contain a continuously updated record of each employee's exposure to every substance or other hazard listed in Module I. Each employee would be assigned an identification number and provided with a plastic card, with which he or she would be required to log in and out when reporting to or leaving a particular location or workstation at any of our workplaces. This could be done very quickly and easily by means of card-reading machines. (Here again, the necessary technological capability is available; however, installation of such equipment would represent one of the major costs in the implementation of the system.) Information about the substances present at each location would also be incorporated into the Module II database through terminals associated with these machines. Thus, whenever an employee completed a tour of duty at a particular location, his or her file would automatically be updated to include a record of the latest exposures.

- Module III would contain the medical and health records of each employee. The information in this module would include such things as the procedures, findings, and test results from all the employee's company-sponsored medical examinations; family history; hospitalization records; and medications prescribed. These data would be compiled from a number of sources. To supplement those presently available to the company, employees would be required to submit periodically (probably annually or semiannually) a comprehensive, machine-readable health questionnaire, to be devised by Advanced Chemicals. This form would have two parts: one to be completed by the employee, and one to be completed by the employee's physician or health care provider. The first part would include questions designed to elicit information about the employee's life-style and personal habits that might be relevant to the assessment of medical risk associated with exposure to substances in the workplace. The second part would be customized by the system to reflect the employee's exposure to specific substances, as quantified in Module II. In additon to general health questions, it would include requests for information, test results, and so forth, needed to monitor symptoms that might be associated with substances to which the employee has been exposed. (Each employee would have to sign a release giving his or her physician permission to release this information.)

On the basis of the information compiled in Module II, the system would schedule regular medical examinations for each employee. It would also automatically schedule a special examination whenever an employee's exposure to a specific substance (as recorded in Module II) exceeded a designated threshold level, or when a symptom that might be associated with substance or radiation toxicity appeared in the periodic health questionnaire. The system would therefore provide for the protection of employee health an "early warning" capability that we do not now possess.

- Module IV would be a comprehensive database of epidemiological and other information from the international medical research literature on the health effects of chemical substances. This material would be updated regularly, in part by the use of existing database search systems. The information in this module would in turn be used to update Module I on an automatic and continuous basis, as well as alerting our medical staff to significant trends anywhere in the world (for example, to a rising incidence of a particular health problem thought to be associated with a substance that we use).

It can be seen from this description that there will be continuous interaction and communication among the four modules, without the need for specific instructions from the users of the system. Thus, as we have indicated above, the data compiled in Module IV will be automatically input to Module I. Similarly, Module III will draw on the worker exposure data in Module II to schedule medical examinations and flag potential health risks. But it also will draw on the medical information in Module I to determine "safe" exposure levels, and will automatically take account of any change in that information. Suppose, for example, that research were to reveal that a particular substance is actually toxic at lower exposure levels than was previously believed. This information would be compiled by Module IV and immediately used to update the files of Module I. Module III would then *immediately and automatically* reschedule medical examinations for all workers whose exposure to the substance in question (as recorded in Module II) exceeded the *new* safety levels.

Clearly the requirements for computer memory and processing speed to make possible such a comprehensive and highly interactive system are great. The technological capability, however, is now available, as is the requisite programming sophistication. Although the price of such a system will be high, it is our judgment that the anticipated benefits make such a system cost-effective for a company like Advanced Chemicals, as we shall attempt to demonstrate in the following section.

II.   The advantages that would accrue to Advanced Chemicals from an HIS such as we have described above are of several kinds. Some lend themselves relatively well to a purely quantitative, cost-benefit analysis (though the actual figures can, of course, only be estimated until the system is in operation). Others are more intangible. It is easy to attach a dollar value to X worker-days lost to illness, but how do we judge the value of such things as public trust? In this section the benefits of an HIS are discussed in general terms.

We have organized the discussion that follows under three broad headings:

    A. Practical benefits to Advanced Chemicals
    B. Ethical considerations and social responsibility
    C. Legal considerations

A. Practical Benefits to Advanced Chemicals   The main objective of an HIS is improved protection of our workforce from illness or injury caused by exposure to substances or conditions encountered in the workplace. Quite aside from our ethical obligations to maximize such protection (a consideration discussed below), affording a higher level of protection to our employees has many practical benefits for the company. Health-related problems are enormously expensive to Advanced Chemicals, as they are to most employers today. It is self-evident, therefore, that any measures for improving and safeguarding the health of its employees will translate into large savings for Advanced Chemicals.

More specifically, the system described above would enable us to detect at a much earlier point than is now possible a substance or process that poses unacceptable health risks to our employees. Such early detection would allow us to take prudent, timely precautionary measures for the protection of our work force. (We could, for example, provide better protective gear to employees, modify the process to reduce exposure levels, or even replace the dangerous substance with a safer one—as has been done with asbestos insulation, lead additives in gasoline, PCB in transformers, many pesticides, and so on.) To take an immediate example, a system such as we have described would almost certainly have alerted us to the dangers of PBA long before NIOSH discovered the high incidence of eye cancer among our workers using this chemical.

The system would also enable us to identify individual employees who, because of their cumulative level of exposure to a potentially hazardous substance, or because of some physical predispositon, are at special risk. It would then be possible to limit further exposure before serious health problems occurred—in the most severe cases, by providing comparable work in some other phase of our operations. In fact, such a system might even give us the capability of screening new employees to discover those who should not be exposed to specific substances. Job assignments could then be made accordingly.

These capabilities would result in savings in a number of areas:

- Absenteeism due to illness would be decreased, with consequent gains in productivity.

- The need to replace employees who were disabled by chemical exposure on the job would also be reduced. Such turnover, with its associated disruption of operations and expenses for recruitment and training of new employees, is a major financial drain on companies like ours.
- Health benefit costs—especially those associated with long-term disability—would be sharply curtailed.
- We could anticipate fewer liability actions by workers alleging harm from exposure to dangerous substances in the workplace, because (a) fewer workers would actually experience health problems, and (b) the availability of precise exposure and medical records would discourage the filing of unwarranted claims. The consequent savings in legal expenses and insurance premiums could be substantial.
- Improvements in employee morale would result from the perception that Advanced Chemicals is providing the safest possible work environment. This would translate into higher productivity and, again, lower turnover (that is, fewer employees leaving to take safer jobs).
- Similarly, our relations with the Chemical and Pharmaceutical Workers International union would benefit from our evident commitment to employee safety.

B. Ethical Considerations and Social Responsibility   It need hardly be said that, even if the aforementioned benefits to Advanced Chemicals did not exist, we would still be ethically obligated to provide our employees with the safest possible workplace. We at Advanced Chemicals are acutely aware of our responsibilities—to our work force, to our long-standing tradition as a leader in the areas of health and safety, and to society at large.

In this last connection, two points deserving of considerable emphasis are often overlooked. First, the efforts that we make in behalf of our employees—such as the implementation of an HIS—are likely to be enormously beneficial to society as a whole. This is because workers in a company such as ours, which leads the way in the development of new drugs and chemicals, are the nation's "canaries" (as the union has often put it). Health and safety problems that may not surface for decades in the general population will quickly become evident among chemical workers, who typically are the first people to be exposed to new substances and usually are exposed to far greater concentrations and/or for longer periods of time than ordinary citizens. Thus, the hazards of asbestos, for example, were detected not among people living in asbestos-insulated buildings but among asbestos workers in the World War II shipyards and factories.

This point takes on special significance when radically new processes and products, such as those being explored by our Bio-Life division, are concerned. The pressure of competition makes it essential that we move ahead with maximum speed in this vitally important and rapidly developing field. Such haste inevitably creates some uncertainty. It is impossible to say with absolute assurance what public health questions may ultimately be raised by bioengineering of the kind that Bio-Life is now pioneering. No one really expects an "Andromeda

Strain" scenario, but the possibility of subtle, longer-term risks cannot be ruled out. It seems clear, though, that if problems turn up, they will turn up first among our employees. We are therefore under a double obligation—to our work force and to our society—to maintain the greatest possible vigilance with regard to the health and well-being of our employees.

The second point is that such vigilance, even if it is motivated chiefly by considerations of ethics and social responsibility, is not without its potential rewards. Clearly it is better for *us* as well as for society to discover that a new product has deleterious health effects by monitoring the health records of a few workers than by finding that 10,000 customers have become ill a year after the product has come out on the market. In the words of one specialist in this area:

> Employee exposure can provide a firm that measures the effects of such exposure over time with valuable early leads to problems that new products might produce once they are on the market. . . . Given the great deal of money needed to evaluate the effects of each new product, an extensive database [derived from employee exposure and health records] could begin to generate data about toxic effects, particularly long-term effects, that would otherwise be very expensive to generate through ad hoc research studies.

Needless to say, our potential liability is also much lower if problems are discovered early; put bluntly, it is better to be sued by five workers than by 10,000 customers.

More important, we live in a time when anxiety over environment and public health issues is rife and distrust of large corporations is widespread. Advanced Chemicals has an enviable and hard-won reputation for responsibility and responsiveness to environmental, health, and safety concerns. The trust that we enjoy is an intangible asset, yet its value in the conduct of many aspects of our business operations is very real. It is a factor each time we apply for a license to open a plant, introduce a new product, sit down to negotiate a union contract, are a party in litigation, or testify before a congressional committee. Such trust is particularly crucial for the continued success and expansion of Bio-Life, which must function in the glare of publicity, under the anxious scrutiny of federal, state, and local legislators, government regulatory agencies, special interest organizations, and a jittery public. In instances like these, it is unquestionably true that a strong ethical commitment and a strong sense of social responsibility are also good business.

C. Legal Considerations Insofar as an HIS would greatly increase our ability to monitor the substances that we use, their location, their concentrations, employee exposure, employee health records, and the like, the system would facilitate our compliance with all existing occupational health and safety regulations, such as local right-to-know laws and OSHA provisions. Perhaps more important in the long run, the records that the system would compile, though not legally required at present, may be legally mandated at some future time. Certainly the current trend is toward ever more comprehensive legislation for the protection of worker health and safety. We, then, would already have in place a system that other companies would have to develop from scratch.

There is, however, one legal problem in connection with an HIS that will have to be addressed. Under current law, no company is required to collect the kinds of information that the proposed system would amass; however, if such information *is* collected, employees cannot legally be denied access to it. Thus, in any liability action brought against Advanced Chemicals by an employee, the plaintiff's counsel would be able to consult and make use of our records of plaintiff's exposure to chemicals, his or her medical examinations and test results, and even our own epidemiological data. The system itself would thus be a potential "witness for the plaintiff" in any litigation against us.

Rather than abandon the entire concept of an HIS because of this factor, we feel that the correct course of action is to press for a change in tort law so as to protect the possessor of such a system. Surely it is not the intent of any legislature to penalize a responsible employer for attempting to protect the health of its employees. Since this could indeed be the effect of present law, it stands to reason that legislators could be persuaded of the wisdom of changing that law. Our legal staff and our lobbyists should at least explore the prospects for succeeding with such an effort.

In conclusion, we feel that the chemical and pharmaceutical industries — especially vanguard companies like Advanced Chemicals — can no longer operate under the old assumption that "what we don't know can't hurt us." We can no longer bury our heads in the sand and start to deal with problems only when our employees become ill or sue us. It is our business — and good business — to know everything possible about the potential health hazards of our operations, and to know it as early as possible. Other companies have clearly come to this realization in recent years. Gusher Oil has already implemented an HIS along lines similar to what we have described above, and it has been reported that several other companies (among them at least two of our chief competitors, Superior Pharmaceuticals and Finest Drug Corporation) are either actively developing similar systems or seriously considering them. We have always taken the lead in our industry — why should we be left at the starting post now?

It may seem prudent to take a more cautious approach and wait until health information systems are as common as pension plans. It is important to keep in mind, however, that an HIS is not a "quick fix" — its value manifests itself over the long run. The sooner we implement such a system, the sooner we start to compile a database that will serve us well five, ten, or even twenty years from now. As one expert has said, "The value of an HIS database in the next five or ten years will depend heavily on whether the right individual data are collected, in sufficient detail, today." For this reason, it is best that we act as expeditiously as possible.

Findings of the Interdepartmental Task Force for the Study of Health Information Systems

MINORITY REPORT

Submitted By:
Andrew Parsons, *Director, Labor Relations*
Ford C. Haskell, *Director, Manufacturing Operations*
Timothy Anderson, *General Counsel*

Executive Summary Advanced Chemicals should continue to take a proactive posture with respect to potential health hazards in its operations, particularly those associated with innovative ventures such as bioengineering, and should do everything in its power to address the legitimate concerns of its work force and the general public. Among the steps that might be considered are expanding our medical staff; enlarging our health insurance program to pay for broader diagnostic testing of employees, as well as medical, psychological, or genetic counseling if indicated; and increasing the scope of our medical research programs and/or supporting epidemiological and toxicological research at appropriate hospital or university centers.

However, installing a comprehensive HIS is neither technically feasible nor (if it could be done) likely to be cost-effective at the present time. Such a system could, moreover, greatly increase our legal exposure as a "witness for the plaintiff" in liability actions; divert large amounts of valuable supervisor and manager time and energy from pressing operational concerns; complicate our relations with our employees; and present our unions with a weapon they could use in collective bargaining or as an organizing tool. At this time Advanced Chemicals should confine itself to monitoring changes in information technology, the regulatory climate, and the practices of other companies so as to be prepared should an HIS become a more attractive option in the future.

Introduction The following report represents the views of three members of President Prentice's Task Force for the Study of Health Information Systems. As the system under discussion has been adequately described in the Majority Report, it is not necessary to repeat that description here, except to comment on the feasibility of specific aspects. We will instead concentrate on what we consider to be the problems involved in the attempt to construct such a system and the deleterious consequences that might result if the system were to be implemented. As in the Majority Report, our arguments will be made in essentially qualitative terms.

Our essential points have been organized under four general headings:
    A. Problems in system design, implementation, and operation
    B. The problem of liability

C. Other legal and regulatory problems
   D. Labor and employee relations problems

A. Problems in System Design, Implementation, and Operation Some of the difficulties that the proposed HIS is likely to encounter are technical in nature, while others involve the human factor. For an example of a purely technical limitation, it is necessary only to consider the problem of data reliability. Experience with systems of this degree of complexity suggests that problems in input accuracy invariably arise. Such problems, which would be particularly difficult to avoid in a system that takes so much information in so many different modes (a variety of machine-readable forms, data from other modules, input from card-reading machines, and so on), could render the entire system nearly useless—or, what is perhaps worse, suspect. Who would want their health to hang upon a misplaced decimal point?

As this example suggests, human and technical factors are not always easy to separate. For instance, the effectiveness of the system will depend heavily on the cooperation of individual employees, who must not only conscientiously register their comings and goings by means of their identity cards but also entrust the system with highly sensitive personal information. Clearly, they will not be willing to do this unless they believe that the system is completely safe and secure and that their privacy is totally protected. Is it realistic to suppose that such a degree of security for huge volumes of personal information can be attained? To judge by experience with comparably elaborate systems, such a conclusion is highly problematical. Many employees will know about the security problem, and they are likely to act accordingly—that is, withhold information lest it be disclosed to the wrong people or used for the wrong purposes. (This point is discussed further below, in the context of labor relations.)

The issue of cooperation on the part of employees arises in many contexts. Some employees may resent, as a "Big Brother" device, the use of identity cards to track their movements. Others may simply be too busy or too careless to record all their changes of location during the course of a working day. Still others may think it unnecessary and pointless. In this they may be correct. Is it really possible to measure a worker's exposure to specific substances simply by recording the place to which he or she is assigned? The concentration of different chemicals varies from day to day and hour to hour, and it may also vary greatly from one side of a lab (or a mixing vat) to the other. Can the proposed system make such fine discriminations? It is our judgment that the system of "flagging" excessive exposures for the purpose of scheduling medical exams will generate large numbers of "false negatives" and "false positives"—that is, burdensome, time-consuming, and unnecessary tests and examinations will be scheduled for some employees, while others will fail to get the surveillance they really need. This will not only undermine the efficient functioning of the system and increase our liability exposure (see below) but also discredit the entire system in the minds of its users and so discourage user cooperation.

It should also be kept in mind that a system such as this cannot run itself, no matter how fully automated it is. This system will make great demands on the time and energy of computer programmers, medical staff members, and (perhaps most important) supervisors and managers, who will face the unenviable choice of either allowing themselves to be distracted from their operational responsibilities or giving the HIS less attention than it needs to function effectively. It will surely be necessary to hire additional medical and computer personnel. To this expense must be added the expense of the extra medical examinations and diagnostic tests that the system calls for. If these costs are borne by the employees, they will hardly be inclined to cooperate with this part of the program. The result may be scheduled appointments not kept, recommended tests not taken, and mandated procedures not followed. But if all such costs are assumed by the company, the financial burden may be enormous.

B. The Problem of Liability   As the Majority Report has pointed out, there is currently no legal requirement that employers collect the kinds of data that the proposed system would compile. If such information *is* collected, however, the law stipulates that it be fully accessible to individual employees, and also (to a more limited extent) to their union as bargaining representative. Thus, in any liability action against Advanced Chemicals, the plaintiff would be able to use our own records against us, to document exactly how, when, and where he or she was exposed to an injurious substance, and what we did (or didn't do) about it. Such an electronic "witness for the plaintiff" could have devastating consequences for the defense. In the words of one participant in a recent workshop on such systems:

> So what you are essentially doing—you, the health system developers—is systematically looking for disease and its causes, systematically accumulating and arranging evidence of your company's liability, systematically committing "hara-kiri" for the private corporation in a private marketplace. Because if you are any good at it, you are discovering liability. That's your ultimate aim—the goal of your activity.

The current state of case law bearing on workplace health hazards is fairly clear. Companies do not have a positive legal duty to know about all possible occupational diseases, or to conduct sophisticated employee health monitoring and research. What they *do* know, however, they are legally obligated to *disclose*. Thus, what we have here is clearly a situation in which "what you go out of your way to know" may indeed hurt you. This may be unfortunate, but it is the law right now, and we would be foolish to ignore it.

It is also essential to bear in mind the liability implications of greatly increasing the number and scope of company medical examinations of employees. As an authority in this field recently pointed out:

> An employer that voluntarily conducts a medical examination of an employee may be liable for failure to disclose a dangerous condition or disease.

Knowledge of a company doctor will be imputed to the employer. In addition, an organization may be held responsible for a company physician's negligence just as it would be for the negligence of any other corporate employee.

Thus, by instituting a system that relies so heavily on medical examinations, we may be further enlarging our liability exposure.

Other Legal and Regulatory Problems It seems virtually certain that once a rich database of information on health, exposure, and research has been amassed, the company will face demands for access from many different quarters. For a company with an HIS, this fact raises a whole series of questions, some of them very vexing. Will NIOSH, the Environmental Protection Agency, and similar government regulatory agencies have unlimited access to HIS data? Will the company be able to oppose what it regards as "fishing expeditions" by these or other agencies? What about labor unions—both those which currently represent employees and those which are seeking to do so? How will the confidentiality of employee medical information be protected if many departments within the company and external organizations have access to HIS data? Similarly, how can trade secrets be protected if detailed information about chemical exposure is available to employees, unions, government agencies, and perhaps even the general public?

Additional questions concerning access to and disposition of such data have recently been raised by a specialist in this field, who has written:

> [How] much firm, unimpeachable laboratory evidence should a company have before it reveals a possible health hazard to its employees and the public? Who should get this information? In what form should it be presented? How should the information be reported so that the message is understandable yet not unnecessarily alarming? How much effort should be expended to locate *former* employees who may have worked with hazardous substances at one time?

Clearly, many of these questions have no easy answers. Here again, we see that the more we go out of our way to know, the more we are confronted with problems concerning just how that knowledge is to be handled.

D. Labor and Employee Relations Problems We have already touched, in passing, on the problem of maintaining the confidentiality of employee medical records. For employees, however, privacy is not just a technical issue of "data security." Even if the files are properly safeguarded, employees are likely to balk at providing the kind of detailed personal information called for by the periodic questionnaire that is a central component of the proposed system. In order to monitor the reactions of individuals to potential hazards, this form must elicit information about each employee's life-style and personal experiences: smoking, drug and alcohol use, sexual practices, and so on. It is quite likely that such a questionnaire will encounter resistance, resentment, and perhaps deliberate falsification on the part of individual employees. It must be remembered that the law does not mandate an HIS; thus, there is no sanction for getting employees to cooperate with the program. The burden of securing their cooperation falls entirely on the company.

Given the unpopularity of such "prying" by the company, it is equally likely that the questionnaire will be opposed by the union. Indeed, a number of unions have already taken a stance against company-sponsored health information systems. Some base their position solely on the currently very sensitive issue of invasion of privacy; others, on the grounds that the information gathered could be used as a basis for denying an employee's rightful benefits in future worker compensation claims. A related concern, expressed by some unions and individual employees, is that medical surveillance data could be used to justify the transfer or even termination of workers on the basis of their health status or supposed vulnerability to specific chemical or other hazards in the workplace. In this view, a company could reassign or fire an employee simply because he or she had a genetic predisposition that made him or her unusually susceptible to some workplace risk factor—or even because he or she had already reached the limit of safe exposure to a particular substance. This last argument raises the highly emotional issue of management treating workers as "disposable" components of the manufacturing process, to be used to their capacity and then discarded. It is clear that such issues are powerful enough to be used not only in collective bargaining but as organizing tools.

Finally, if such a system were in operation, union representatives would surely demand access to its data. They would want not only individual data for use in liability or medical discrimination cases, but also our epidemiological data for use in identifying real or imagined "trends." They could certainly publicize alarming things that they saw (or believed that they saw, health statistics being so notoriously open to varying interpretations), and so jeopardize the reputation of Advanced Chemicals as a safe and conscientious employer. And they could use these "findings" and interpretations in collective bargaining and in organizing drives. Do we really want to put such a powerful weapon into the hands of our adversaries, when there is no need to do so?

In conclusion, we wish to make it clear that we fully share the belief that there are increasing health hazards associated with high-technology manufacturing, and that it is incumbent upon us to address these hazards in a constructive manner. This could be done in a number of ways:

- By increasing the size of our medical staff and enlarging the scope of our epidemiological and toxicological research programs
- By supporting similar research being carried out at hospitals, universities, and other institutions
- By studying possible expansion and/or improvement of our health insurance programs. We could, for example, broaden the range and increase the frequency of diagnostic tests for workers who are worried about possible

adverse effects of substances to which they are exposed. We might even consider sponsoring psychological and genetic counseling for such employees—either within the company or through outside providers.

An HIS, however, is not an attractive or even a feasible option at the present time. For one thing, we are probably some five to ten years away from having the technical capability to build and operate a system that will do the job. For another thing, although some companies have dipped their toes into the water, very few have actually committed themselves to these systems, and their experiences so far have not been very encouraging. Why spend such enormous sums to buy trouble for ourselves, when neither law nor common industry practice require us to do so? The CEO of another firm recently summed up the situation (and our attitude) very well. Asked by a company epidemiologist for funds to create an HIS, he responded incredulously,

> You want me to give you three million dollars to develop evidence that I am liable for five hundred million dollars?

Discussion Questions

1. Scientific consensus rarely exists on the likelihood of success for many advanced technologies; in fact, deep divisions in opinion are often present. What can business do when views within the scientific community differ widely on the probable benefits of new technologies like the HIS? What method might you use to assess the technical and economic feasibility of a comprehensive medical information system for Advanced Chemicals?

2. What priority would you place on the company being on the leading edge of developing experimental projects of this kind vs. waiting until the technological, economic, and legal issues are more resolved? Is it ever justifiable, in your view, to *not* gain knowledge about a health hazard when it is technologically feasible to do so?

3. Society appears to be subject to accelerating patterns of litigation and increasing readiness of employees and attorneys to sue corporations. Does this litigious system create a disincentive for companies like Advanced Chemicals, so that they do not develop advanced systems like the HIS even when to do so would be in the interests of individual employees and of the community as a whole? How do you evaluate the recommendation of the Majority Report to seek legislation to protect the company from liability in developing an HIS?

4. The Minority Report points out that the success of an HIS will depend on employee cooperation. In this sense, a system is a social and organizational, as well as a technical, innovation. What suggestions would you make to Walter Prentice for the successful implementation of this socio-technical system? Would your answer be different if the company were non-union?

5. Assume that the company proceeds to develop an HIS as described. Information from this system will naturally be intended to influence personnel decisions. What, specifically, does the company need to know about its jobs and its people in order to help managers make personnel decisions that actually

protect health? What safeguards (such as specific supervisory practices or company policies) might help ensure that information is used *only* to protect employee health, and not for extraneous reasons such as avoiding medical claims?

6. From a societal standpoint, do private employers have optimal incentives to develop advanced systems for enhancing and improving employee health? Are any of the following groups potentially more logical and efficient settings than an industrial firm for the development and implementation of comprehensive health information systems?

    a. Private health insurers
    b. An association of chemical manufacturers
    c. A public health agency

7. Assuming the role of Walter Prentice, what directions would you give to your managers in attempting to address the legitimate concerns of the Minority Report while still seeking to advance the cause of protecting employee health?

CASE 10

Choosing Corporate Strategies in State Legislative Campaigns

Responding to a Sweeping "Right-to-Know" Proposal

International Manufacturing Corporation is a 75-year-old firm that now specializes in producing plastics, mainframe computers, and consumer electronical equipment. It has eight manufacturing plants in the United States and four overseas facilities, in Formosa, Singapore, Hong Kong, and South Korea. Its corporate headquarters are in New Caledonia.

International has regularly been on the *Fortune* list of the ten "most admired" American companies and ranks among the top five "most desirable" firms in surveys of graduating college seniors entering careers in business. In addition to a well-known job-security policy and a "people first" human resources philosophy, the company has been an internationally recognized innovator over the past four decades in developing and implementing advanced manufacturing technologies and computerized information systems.

International also follows a policy of active involvement in industry associations, business educational and research groups, and local community affairs. It has testified often in federal and state legislative hearings on legislation both directly and indirectly affecting business.

You are Roy Lear, President of International, and have called on senior executives to advise you on whether the company should take a public stand on a far-reaching right-to-know bill currently under consideration in the New Caledonia State Legislature.

Key Persons Involved

Roy Lear, *President*
Phillip LaFolia, *Assistant to the President*
Fred Regan, *Vice President, Labor Relations*
Armand Goneril, *Vice President, Manufacturing*
Louis Cordelia, *Vice President, Public Affairs*

Memorandum

Date: April 2
To: Fred Regan
Vice President, Labor Relations

Louis Cordelia
Vice President, Public Affairs

Armand Goneril
Vice President, Manufacturing
From: Philip LaFolia
Assistant to the President
Re: Proposed State Right-to-Know Legislation

As you know, President Lear will be attending the New Caledonia Manufacturers Association meeting in two weeks. At that time he would like to be able to set forth International's position of the right-to-know (RTK) bill recently introduced in the New Caledonia State Senate and now being debated in that body. It is therefore essential that we have your views on this matter within a week's time.

In case you have not yet had a chance to study the proposed legislation, let me summarize its key provisions.

- New Caledonia employers who make, store, use, or process "potentially toxic or hazardous substances" must prepare or obtain a separate Material Safety Data Sheet (MSDS) for each such substance. For the purposes of this law, a potentially toxic or hazardous substance is one that appears in the National Institute for Occupational Safety and Health (NIOSH) Registry of Toxic Effects of Chemical Substances—a list that currently comprises some 65,000 chemicals and is continually being expanded. (It should be noted that inclusion on this list does not mean that a substance has actually been found to be dangerous, merely that it has been *tested* for toxicity. Many of the substances on the list are quite harmless.) The MSDS must include, among other things, the substance's chemical name, common name, and Chemical Abstract Service (CAS) number; the hazards or risks posed by the substance, including acute and chronic health effects and symptoms of overexposure; safety precautions for use of or exposure to the substance; and emergency procedures for spills, fire, disposal, and first aid.
- In addition, labels containing most or all of this information must be placed on containers holding toxic or hazardous substances (as defined above), as well as on storage tanks, mixing vats, reaction vessels, and pipelines or other conduits carrying such substances.
- Employers must keep copies of the MSDSs on file at a central location in the workplace. Any employee who believes that he or she may have been exposed to a toxic substance—or any representative designated by such an employee—can request to see the MSDS for that substance. If the employer

does not comply with such a request within 48 hours, the worker(s) has the right to refuse to work with the substance in question. The same is true in the case of unlabeled substances or mixtures of substances.
- In addition to the MSDSs, employers will have to prepare, for use by the state departments of Public Health and Environmental Protection, detailed surveys of the chemicals they have on hand. This information will be made available to emergency response units (for example, local fire departments). It will also be a matter of public record and easily accessible to anyone requesting it.
- Any employer wishing to withhold information from the surveys or the MSDSs in order to protect a trade secret must make formal, written application to the state Department of Public Health. Trade-secret protection will not be granted to any substance known or suspected to be a mutagen, carcinogen, or other health hazard, and can be denied to any other substance at the discretion of the Commissioner of Public Health.
- Employers must implement a comprehensive training program for employees using substances known or suspected of being toxic or hazardous.

You can see from this summary that the bill under consideration is a very sweeping one—perhaps the strongest piece of RTK legislation thus far proposed in any state. However, our legal counsel has pointed out that some portions of this bill, should it be enacted into law, may not survive a challenge in the courts. In the fall of 1985 and the spring of 1986, the long-awaited Hazard Communication Standards of the federal Occupational Safety and Health Administration were phased in. These regulations, in OSHA's words,

> establish uniform requirements for hazard communication in manufacturing. . . . Each employee who is exposed to hazardous chemicals will receive information about them through a comprehensive hazard communication program. . . . All covered employers will be required to provide the information to their employees by means of labels on containers, material safety data sheets, and training.

In a New Jersey case, a federal judge ruled that insofar as the state's RTK statute dealt with hazard disclosure to workers in manufacturing, it was preempted by the OSHA standards and would have to be redrawn to eliminate conflict with federal regulations. A state law mandating *public* access and disclosure, however—a so-called community RTK law—would be valid, since it would cover an area to which the OSHA standards do not apply. Thus, we can expect that the community RTK provisions of the legislation now being debated would most likely survive any legal challenge. The same is true of disclosure provisions relating to workers in jobs other than manufacturing—true, at least, until such time as the OSHA regulations are extended to cover those areas.

Regardless of its legal status, the pending legislation has the enthusiastic backing of a considerable number of unions—including the Chemical and Utility Workers of America (CUWA), which represents the majority of our own work force—and of a broad coalition of consumer, environmental, and other liberal groups such as the Public Interest Protection Alliance and the New Caledonia Coalition of Concerned Citizens. All of these pressure groups are lobbying energetically for the bill and are prepared to press actively to ensure its passage. The New Caledonia Chamber of Commerce and the Allied Industries of New Caledonia (AINC) have already stated their opposition to the bill and are preparing a campaign to defeat it.

It is clear that we cannot support the bill now pending. That being the case, it would appear that we have basically three options:

1. Take a firm stance in opposition to the legislation.
2. Remain silent on the whole issue.
3. Oppose the bill in its present form, but come forward with our own specific proposals for a more moderate and balanced bill.

Clearly each of these positions has its drawbacks. Opposing the bill could make us unpopular with civic groups and the public at large, not to mention unions; supporting an alternative bill, no matter how moderate, would probably alienate us from other firms, many of which are flatly against any such legislation whatsoever. But if we take no position and remain silent, we forfeit the opportunity to exert any influence in behalf of our own interests.

President Lear is eager to receive your input on this difficult question as soon as possible.

Memorandum

Date: April 4
To: Roy Lear
   President
From: Fred Regan
   Vice President, Labor Relations
Re: International Manufacturing's Position on Proposed RTK Legislation

In response to Phil's request for an opinion about where we should stand on the pending RTK bill, I have a simple answer: we should take no stand at all. In fact, we should keep as low a profile as possible on the whole matter. I know that International doesn't usually duck tough issues, but there are times when discretion is the better part of valor, and I think this is one of them. If we get involved in this squabble, we have a lot to lose and almost certainly nothing to gain.

First of all, there's the union to consider. As I'm sure you know, our relations with the CUWA have been uneven. Sometimes they've been remarkably amicable, thank heaven. But when they get a bee in their collective bonnet, they can be very tough (remember the confrontation over the mandatory drug-testing issue?). Lately things have been going smoothly. Do we really want to rock the boat by trying to scuttle their pet bill? You may not be aware of this, but Senator Gloucester, who introduced S 3311 (the pending bill), is a big fan of the CUWA, and the feeling is mutual: he is one of the most vociferous union spokesmen in the state legislature, and the unions—the CUWA in particular—back him heavily at election time. The union leadership helped him draft this bill, and has made an all-out commitment to getting it through. RTK is a highly emotional issue with the union membership—and we have a new contract to negotiate this coming November. Why rile them up now?

We should also have some thought for our image with the public at large. We have the reputation of being an enlightened, forward-looking company that takes its social responsibilities seriously. Standing up in opposition to a piece of legislation that many civic groups have endorsed and that large segments of the public see as progressive can only undermine popular trust in International—trust that we have taken a long time to build, and that our advertising people spend millions of dollars each year to reinforce.

Of course, we could try to cozy up to the CUWA and the environmental activists by endorsing RTK in principle and then coming in with an alternative bill (or a lot of amendments to this one) that would be easier for us to live with. But I don't think this strategy would get us anywhere. The unions (and the environmental Nervous Nellies) are very attached to S 3311 in its present form. They see it as pioneering legislation, a model for the rest of the country. They *like* the idea that it's the toughest bill of its kind ever introduced, and that's the way they want to keep it. If we draft weaker legislation—even if it's better, fairer, more balanced, and so on—the unions and their sympathizers will scream bloody murder about how we've gutted their precious bill, how we're trying to co-opt the movement for worker safety and environmental protection. . . . in short, business up to its old tricks.

At the same time, backing *any* other bill would, as Phil suggested, open our right flank to attack from some of the more hard-line business organizations, who will see us as a traitor to the cause. We don't really want to be perceived in the business community as a bunch of sellouts, do we? Remember, some of the firms with whom we do the most business are still in the dark ages as far as environmental and workers' rights issues are concerned. Can't you just see Charley Smythe of Amalgamated getting red in the face when he hears that International is backing an RTK bill? He won't care what's in the bill—just that we even considered supporting it. Smythe hasn't come to terms with workers' compensation or the corporate income tax yet, let alone RTK. But we had nearly $3 million of sales to Amalgamated last year.

I could see risking the wrath of the unions or the business community or even both if we had a fighting chance of accomplishing something worthwhile in terms of the legislation itself, but we don't. As far as our ability to defend our own interests in the state legislature is concerned, this is strictly a no-win situation. It's not likely that we can prevent some sort of RTK bill from being passed; the policital momentum is too great. If we dig in our heels for a last-ditch stand against any legislation at all, we'll just get steamrolled. (Look what happened in New Jersey.) But if we bring in a more moderate bill, our position will become ammunition for the enemy. For them, half the battle is to force us to admit that there's a problem in the first place. If we say, "Well, we know there's a problem, and here's the legislation we think is needed to deal with it," they'll be delighted. They'll just reply: "See, even the big corporations admit there's a need for such legislation. They've even drafted a watered-down version of our bill to try to deflect us from our purpose. Very sneaky—but we won't be fooled! This just proves that we really need our original strong bill."

All in all, if we get involved in this we're just going to get our fannies caught in a wringer. And it's not necessary! This isn't our responsibility alone—the bill will affect practically everybody who does business in this state. So why should we stick our necks out for everyone else? Why should we put ourselves in an exposed position and let other people take potshots at us from every direction? Let the Manufacturers Association carry the ball on this issue, or the C of C, or AINC. That's what we pay dues for. It's all very well to be a leader, but let's try to be a leader in popular causes. We're not going to make many friends on this issue, no matter what we do. And since there's no law that says we have to take a position on RTK, let's not. Let's just duck this one publicly. Privately, we can support whatever efforts the Executive Committee of the Manufacturers Association sees fit to make. Their lobbyists should have a good sense of where things stand right now in the legislature and what we might hope to accomplish behind the scenes.

Memorandum

Date: April 7
To: Roy Lear
President
From: Armand Goneril
Vice President, Manufacturing
Re: Need to Oppose S 3311

With regard to the RTK bill now pending in the state senate, I don't see how there can be any ambiguity in our position. We've got to oppose it as strenuously as possible—it's poison. I can't think of a worse bill to come along in the past twenty years—not just for us, but for any business trying to get along in New Caledonia. Of course, I am not opposed to the idea that employees should know what harmful substances they're working with and be protected from them; no one at International will quarrel with that. And I believe that members of the public have certain rights to be informed of anything that is truly dangerous to their communities. But this bill typifies the way in which a good idea can be ruined by irresponsible, overzealous proponents who don't take the time to think through the consequences of what they're proposing. (Some of the people behind this measure are probably motivated more by simple hostility to corporations per se than anything else, but let's assume these are in the minority.)

Just listing the flaws in the bill could take all day. Here are a few of the more obvious ones:

- It's completely idiotic to require MSDSs and labeling for every substance in the NIOSH registry. This list contains tens of thousands of chemicals, including things like water, salt, and sugar. Many of them are known to be completely harmless; they're only on the list because they've been tested for toxicity. What is the necessity for labeling harmless substances?
- Most firms (including ours) have, because of federal law, already instituted safety labeling, hazard communication, and worker training programs. We have spent a great deal of money already in complying with OSHA's Hazard Communication Standards, which covers the "tell the employees" side of RTK. And with threats of lawsuits from persons claiming to have been injured by manufacturer negligence (the "toxic tort" cases) and with insurance costs what they are today, we would be crazy not to have active programs of this kind. Naturally, these programs have been tailored to the needs of individual firms. They therefore vary from company to company, depending on a company's size, the nature of the materials it handles, and so on. The proposed state legislation would only result in duplication of effort, and would shoehorn each company into the same mold, whether appropriate or not. A printer employing five people would have to meet the same requirements as a research and development lab employing five thousand.
- In addition, there are already federal laws and regulations on the books that deal with chemical safety and hazard disclosure: the Federal Hazard Communication Standards of OSHA, the Clean Air Act, the Clean Water Act, the Toxic Substance Control Act—I counted fifteen federal and five state statutes that address various aspects of the "problems" that S 3311 is designed to "solve." What is the point of further and redundant legislation?
- The trade-secret protection provisions of the bill are woefully inadequate. Many substances that manufacturers need to safeguard as proprietary will be unprotectable merely because they are *suspected* of being harmful. And the machinery for obtaining exemption from disclosure, even for harmless substances, is so burdensome to manufacturers as to be almost useless.
- The costs of complying with this law will be crushing to many companies. Can you imagine the amount of paperwork it will entail? That is, assuming that compliance is even possible. Many employers probably won't be able to find out what chemicals are in every product they buy for their own use. How will they label pipelines, which carry different things at different times? The same is true for mixing vats and storage vessels. And what about the cost of enforcement? That, too, is likely to be enormous. Where is New Caledonia going to get the money? From another tax increase? We're trying to attract business to the state, not drive it away! Taxes are already too high—and if the taxes don't drive companies out of the state, the trouble and expense of complying with ill-conceived laws such as this one surely will.

- All of these drawbacks might still be acceptable *if* the law provided any real protection—but I doubt if it will. For one thing, workers don't need to know the chemical name and CAS number of everything they handle. They need to know its common name and, in brief and clear form (preferably by some sort of easily visible color code), what kind of hazard it poses. Labeling *everything*, whether harmful or not, with unpronounceable and incomprehensible chemical names will just confuse people and dilute the warning value of the labels—it will just be ineffective! And for another thing, much of the bill will ultimately be irrelevant anyway. OSHA's Hazard Communication Standards have just been published, and as Phil pointed out in his memo, these preempt many of the labeling and worker protection aspects of S 3311. The company has complied with the OSHA standards. These new provisions in S 3311 will be struck down in court as soon as someone takes the trouble to challenge them, and the only people who will benefit will be a handful of lawyers.
- S 3311 would be a boon, however, for two groups: (a) union organizers and (b) those who wish to harass or discredit businesses. Since virtually anyone would have access to chemical inventory information, could demand on-site inspections of workplaces, and could act on behalf of workers in demanding to see MSDSs, the door would be open for anyone who wanted to stir up trouble—either with a view to convincing employees that they need union representation (or stronger union representation) to "protect" them from "hazardous" substances, or with a view to convincing the public that communities are in mortal danger from insidious chemical poisons used by irresponsible corporations. It would be child's play to convince residents of any community that they have another Bhopal right in their backyard.
- In fact, however, it is the legislation itself that poses the biggest threat to public safety. Since anyone could use this law to discover the nature, quantity, and precise location of drugs, explosives, or other sensitive materials, it could be exploited by criminals or terrorists. Security for dangerous substances would be almost impossible to achieve. That is why some people are calling this the right-to-steal bill.
- The penalties proposed for violations are unduly harsh, including as they do prison sentences for many infractions. Nor does S 3311 distinguish between deliberate and accidental violations. If this bill were enacted, no sane person would want to be in a position of responsibility in this industry!
- Finally (and I could go on, but I guess that's enough for starters), the law is virtually unenforceable, for reasons that I'm sure must be abundantly clear by now.

RTK is much more than an unwise, costly, and overcomplicated piece of specific legislation. Viewed properly, it is an attempt by U.S. unions to take RTK and parlay it into the kind of radical "union right to information" that European unions have won through even broader "access" legislation and codetermination laws. This is the foot-in-the-door toward the European "democratization"-of-the-workplace ideology, and we should fight it here.

As you can probably see, I think that S3311 has a few flaws. This legislation is sure to impose unbearable hardships on many businesses. It will make more jobs for bureaucrats in government (and, for that matter, for paper-pushers and record-keepers in our own firms) while costing the jobs of countless workers as some businesses close, others move to less inhospitable states, and still others are forced into layoffs and plant closings. How can we not, then, take a firm stand against such an outrageous piece of legislation? If it passed simply because no one took the responsibility for trying to stop it, we would have only ourselves to blame. But even if the fight proves futile, what will others think of us if we don't at least try to stand up for our own rights?

Memorandum

Date: April 8
To: Roy Lear
　　　President
From: Louis Cordelia
　　　Vice President, Public Affairs
Re: Our Position on State RTK Legislation

For some time I have been keeping abreast of developments in RTK legislation nationwide. Thus, even before your inquiry regarding the bill now pending in the New Caledonia State Senate, I had come to the conclusion that we would sooner or later have to face up to this issue. Having given a good deal of thought to the matter, I feel that it would be a big mistake either to avoid taking a position or to oppose any RTK legislation. Instead, we should carefully consider our own needs as well as those of our work force and the public, and try to present a compromise bill that both we and the RTK proponents can live with.

I realize that this may not be a popular suggestion. Nobody loves a compromise, and in this case we can't even be certain that the compromise bill will pass; we may simply have to absorb a defeat. So many people, I imagine, will say, "Why not at least try for victory" and opt for an Alamo-style resistance to RTK in any form. Others may say, "If it looks as though we can't win, why not just stay out of the fight?" Both of these positions seem reasonable; however, I don't feel that in the long run they are in our best interests.

First of all, we must consider this problem from a broad perspective. These days, people are nervous about chemicals, and they are nervous about large corporations. When both chemicals *and* large corporations are involved, they can get downright paranoid. It's no use just saying to them, "Well, don't be nervous—you can trust us." Heaven knows, our PR people spend a good deal of time and money to get that message across to the public, and I don't question the value of their efforts, but there's a limit to how much can be accomplished that way. Everyone has heard about Bhopal; many people remember Love Canal. Just recently New York banned the eating or sale of striped bass because of the presence of PCBs in their tissues. People know how those PCBs got there.

My point is that the public is better informed, more sophisticated, and more anxious than ever before. We must show them that we are making a good-faith effort to address their concerns, or we will just be buried under an avalanche of protective legislation. Much of it will not be at all to our liking—unfair, ill conceived, impractical, ruinously expensive, you name it—but it will be passed. Legislatures ultimately respond to popular pressure. The longer we try to keep the lid on, the more violent will be the explosion, and we will have only ourselves to blame. What we must do instead is to channel these energies (which derive from concerns that are at bottom legitimate, even if somewhat exaggerated) in such a way that *our* legitimate needs are also addressed.

It is also important that the public not be allowed to view industry (and, of course, International Manufacturing in particular) as an adversary. We are proud of our reputation and have worked hard to gain and hold the public's respect. We have never been concerned with the bottom line to the exclusion of all other concerns. We have always taken seriously our responsibilities to our employees and to the community. If we take a stance against RTK legislation (and no one will care that it was because the bill was a bad one—all they'll remember is that we were against RTK), we will jeopardize a public image that took years to create. All our responsible, voluntary actions—the affirmative action program, the funding of environmental research, our very liberal policies on employee privacy issues, our pioneering of quality circles, and the rest—could be quickly forgotten. These days, one black mark is often enough to brand a company as an enemy of the people.

At the same time, we should understand the intensity of feelings about RTK on the part of the unions. Although we may tend to impute organizational self-interest and liberal-ideological motives to union advocacy, behind these are genuine problems with earlier industry conduct, and we all know that some "outlaw" conduct continues today. I have attached two short articles, one covering the valid emotional concerns that unions express and the other showing the determination that the RTK coalition in nearby Michigan displayed right before they passed their RTK law.

For these reasons, I propose that we develop a bill of our own that attempts to protect our workers, the communities in which we do business, *and* International Manufacturing, and have this new bill introduced by a leading moderate legislator. (Alternatively, we might be able to develop and support amendments to S 3311 that would make it acceptable; that is a determination for our legal staff to make.) I do not believe that these three aims are incompatible, if we are willing to compromise. Nor do I think that the labor and environmental lobbies would be averse to working with us to hammer out such a compromise. After all, they, too, have to worry about ending up with nothing because they asked for too much. They don't want to provoke die-hard opposition, nor do they want to have to contend with legal challenges to an overly ambitious bill that won't stand up in court. They, too, I'm sure, would rather have a law that everyone can live with. In Massachusetts, industry representatives and RTK advocates wound up producing a compromise bill that all sides ultimately supported, even though they were still bitterly apart on matters of principle.

I'm certainly not the person to spell out just what form such a compromise should take or what provisions would make such a bill satisfactory to us. These are matters for our Legal, Health and Safety, Manufacturing, Labor, and Community Relations people to hash out. But I think that even I can see the broad outlines. If S 3311 calls for preparation of MSDSs for 65,000 substances, perhaps the list can be limited to a more manageable number—say, only those known to pose serious health or safety hazards. If the paperwork involved in preparing elaborate MSDSs is too burdensome, perhaps some simpler way of providing the same information can be found. If S 3311 calls for labeling all containers that contain *any* measurable amount of a listed substance, perhaps a concentration threshold can be introduced. If S 3311's requirements for labeling and public disclosure would jeopardize trade secrets, perhaps an easier mechanism for claiming an exemption when trade secrets are involved could be built into the law. The provisions for work-site inspections by employees and/or community residents could be modified—for example, by requiring that a petition be filed or a warrant issued before inspection was permitted. Access of community residents to MSDSs could be similarly restricted to prevent frivolous or harassing inquiries. Penalties for unintentional violations of the law could be made less harsh than for willful violations. And so on.

I would like to address briefly two counterarguments to the position outlined above. First, some might object that we should not enter fights unless we are fairly confident of winning them. Since we cannot be sure that a compromise bill will prevail over the original "hard-line" S 3311, why take the risk of entering the fray? My answer is that we have to be concerned with the future as well as the present. State legislatures are extraordinarily volatile; a bill passed today may be heavily amended or even repealed in the next session. It is true that the unsatisfactory S 3311 may win this time around. Nevertheless, it is important that we stake out a position now and that the mass media report our position as part of the debate. Otherwise, the fact that some companies favor a reasonable and practical regulatory system will *not* appear in the mass media, and the public will assume that *all* companies accept the Neanderthal positions that will be offered.

Speaking out will not only help our public image but also give us a great advantage in credibility when the legislature recognizes the shortcomings of its first efforts and tries to formulate a better bill in some later session. At that point we will be in a position to say, "We told you so." I can guaranteee that the legislature, anxious to correct its past miskates, will then treat our serious, well-thought-out proposals with the respect they deserve. In short, we may lose this battle but ultimately win the war.

Second, some might also object that we should not take a position before other businesses in New Caledonia have done so, for fear of damaging our standing with our peers. On this point I feel very strongly. We are leaders and always have been. We don't have to run with the herd or wait for the stragglers. We are proud to be trailblazers—let others follow *our* example, not vice versa. This is particularly true where issues of social responsibility are concerned. Our record on health, safety, and environmental issues is an enviable one. Just this year, you yourself served on the Conference Board's special panel on the Future of Worker Health in a High-Tech Society. Why should we back down to appease some troglodyte? Let's stay where we belong—in the forefront.

Attachment   **RIGHT TO KNOW**
by Michael Kenny

"What you don't know won't hurt you," the old saying goes.

Baloney.

What you *don't* know about the substances and the hazards around your workplace can be dangerous. NOT knowing these things can endanger your health—even your life. Remember, you have a right to know.

Toxic effects have been reported for more than 50,000 chemicals that are thought to be in the workplaces of America. New ones are being added all the time. More than 2,000 of these are suspected of causing cancer in humans, based on laboratory animal studies.

There is valid scientific evidence of increased health risks for about 20 hazardous substances. Investigators say future studies may lengthen the list.

Approximately 21 million Americans are exposed to substances regulated by OSHA, the federal Occupational Safety and Health Administration.

The National Institute for Occupational Safety and Health (NIOSH) now includes neurotoxic (nerve poison) disorders as one of the 10 leading work-related diseases and injuries. NIOSH says it added nerve disorders because of their potential severity and because of the large number of workers potentially at risk.

A conservative estimate of the workers exposed full time to one or more neurotoxic agents is 7.7 million, and the number of potentially neurotoxic chemicals found in the workplace exceeds 850, according to the federal agency.

While precise figures are hard to come by, it is roughly estimated that about 100,000 Americans die each year from occupational diseases, and some 400,000 new cases of occupational diseases are recognized annually.

In addition to the toll of illness, employer reports show that about 5,000 workers—nearly 100 a week—die each year in job-related accidents, and 2 million more are disabled.

You have a right to know what causes all this sickness and dying.

No one has the right to expose you to a machine, a chemical, or a work practice on your job that may harm or kill you.

Workers' "right-to-know" is a logical extension of the watershed legislation which created OSHA in 1970—one of the most important gains achieved for all workers under the strong leadership of the trade union movement. That law, among other things, guarantees workers the right to a safe and healthful workplace.

A major victory in the Right-to-Know Campaign was a 1982 ruling by the National Labor Relations Board which gave unions access to lists of chemicals used in the workplace.

OSHA rules provide workers access to a broad range of information, including hazard exposure data and medical records.

Right-to-know laws have been passed in many states and cities. Many union contracts have incorporated right-to-know provisions. Similarly, the right to refuse work under dangerous conditions is recognized in many union contracts.

These are hard-won rights, meant to be understood and used by you or your representative for your protection. But such rights are like muscles — unless you exercise them, they become flabby and useless.

Source: *Utility Workers of America, LIGHT*, October 1986. Used with permission of *LIGHT*.

RIGHT-TO-KNOW BATTLE CONTINUES: LEGISLATORS ON HOT SEAT

Attachment

LANSING, MICHIGAN — For the past two years, a coalition of labor, environmental, and community organizations has been battling to enact a state law that would give Michigan workers, emergency personnel, and community residents access to information on toxic workplace chemicals. The battle will resume this September when the Michigan legislature takes up H.B. 4111, the Right-to-Know bill.

The bill, which was introduced by state representative Juanita Watkins (D-Detroit), is similar to legislation which passed the House but died in the Senate Labor Committee in 1984. Again in 1985, industry lobbyists have mounted a campaign to defeat strong Right-to-Know protections for workers and citizens in Michigan.

Because of wide public support for Right-to-Know legislation and because common sense dictates that workers should know the nature of the hazards they are exposed to, industry lobbyists have been unable to oppose the concept of Right-to-Know. Instead, the State Chamber of Commerce, the Michigan Manufacturers Association, and the Michigan Chemical Council, to name a few, are supporting a substitute bill which would gut the major provisions of H.B. 4111. In this way they can claim to favor Right-to-Know, while they oppose specific provisions which give Right-to-Know some teeth.

In September your state representative and state senator will be deciding the fate of your right to information on chemicals which cause cancer, birth defects, lung disease, and other illnesses. They will have to choose between a strong bill which truly gives people access to information or a watered-down substitute which would severely restrict public and worker access. We think it's time Michigan joined the 23 other states which have already adopted effective Right-to-Know legislation.

In August the Right-to-Know Task Force will be sending a survey to every state representative and senator asking for their position on the crucial provisions of H.B. 4111. We are hoping the survey will inform the legislature on the issue, guide our state and local lobbying efforts, and flush out those legislators who claim to support Right-to-Know, but who have opposed the most important sections of the bill. We will be announcing the results of the survey in September when the legislature convenes. We believe voters have a right-to-know where their legislators stand on this critical public health issue.

Source: Right to Know Task Force Action Sheet, September/October 1985.

Discussion Questions

1. Weigh the merits of the differing recommendations of Regan, Goneril, and Cordelia. What are the benefits and the risks to the firm of each of these strategies? Which person's position best captures your sympathies?

2. What are the key differences (for business and in terms of likely operational systems) between the *employee* and the *community* RTK provisions?

3. Companies having favorable responses to the principle of employee/community RTK legislation often have specific concerns about (a) maintaining trade-secret protection and (b) avoiding harassment and interference with business operations. What procedures involving these issues might provide an effective balance between the rights of employees and community residents and the interests of industry?

4. Opinion is divided on the extent to which companies should be held liable for personal injury claims due to exposure to hazardous substances. For example:

   a. Should a corporation that complies with all warning and disclosure ordinances and regulations be exempt from personal liability claims?

   b. If not, should the company be liable only for failing to warn persons about *foreseeable* risks in the intended use of its product? Is there any basis on which a company should be liable for injuries resulting from failure to warn about *undiscoverable* or *unknowable* risks at the time of use?

5. This case leads to an evaluation of public policy outcomes when immediate, conflicting objectives (e.g., economic efficiency vs. safety) are advocated by different interest groups (e.g., business vs. community members). Would not an effective political process be capable of developing a workable plan, that is, a plan capable of being supported by all interested parties? What factors might make a difference in whether a "workable" plan is attained or not?

CASE 11

Selling High Technology To Anti-Democratic Regimes

Are There Ethics in the International Marketplace?

Omnibus Information Systems (OIS) manufactures mainframe computers, peripherals, and communications systems. The company had sales of $750 million in the last fiscal year and employs 28,000 persons worldwide, 18,000 of them in the United States. OIS specializes in several application areas, one of them consisting of integrated information systems (hardware plus software plus telecommunications) for law enforcement and intelligence agencies. Some 60 percent of its market for these systems is in the United States, where OIS is the leader in the field; the remaining 40 percent is divided among twenty-six other countries in Europe, Asia, Latin America, and Africa.

The company has recently introduced the "Ordered Society Information System" (OSIS), a sophisticated hardware and software system that is based on a series of modules and attached-machine capabilities. Its components include:

- Advanced optical scanning of various licensing, taxation, residence, voting, and health data files maintained by the government. File-matching capabilities permit the compilation of "comprehensive dossiers" on targeted individuals. The system can also sort randomly through large files for "funcos"—"funny coincidences" of interest to the regime.
- An advanced voiceprint-based system that can pass wiretapped or microphoned conversations through a voice scanner (either while conversations are in progress or after they have been recorded on tape) and identify the voices of specific persons.
- An advanced "typewriter and word-processor duplicating system" that uses parabolic intercept techniques to "tap" the electrical impulses being generated by the keyboard of a typewriter or word processor. The material being typed can be reproduced simultaneously on the screen of an intercept monitor and printed out on command.

- A "psychological stress evaluator" module that enables the operator to analyze the frequency modulations of a live or recorded voice and ascertain, with a claimed 93 percent accuracy, whether the speaker is consciously lying.

The Friends of Freedom, an activist group similar to Amnesty International, has announced to the press that it will protest at OIS's upcoming annual shareholders meeting (four months away) the continuing sale of OIS police computer systems such as OSIS to the internal security forces of three countries:

1. Boersland, a white-ruled nation in southern Africa that maintains an apartheid-style policy toward its black majority

2. San Miguel, a Central American nation ruled by a military junta that has imposed strict limits on political opposition and civil liberties

3. East Onan, a Moscow-like repressive regime in the Middle East with alleged links to the Abu Mustapha terrorist group

The Friends of Freedom contends that OIS is wrong in selling advanced intelligence systems to these regimes when it:

> . . . knows or certainly should know that its computers will be used to assist in the harassment of political enemies, surveillance and arrest of dissidents, and ultimately torture of prisoners. American corporations should not provide the tools by which enemies of democracy can violate human rights and tighten their hold on their peoples.

The Friends of Freedom has said that if OIS does not voluntarily end such sales, the group will organize a national boycott of OIS products. It will also put pressure on local, state, and federal governmental agencies not to buy OIS computer systems—not only those used in law enforcement, but also those used in the areas of health and welfare, taxation, and driver registration.

Company President Max Interface is the decision maker who must propose to the OIS board of directors a management response to the Friends of Freedom. He will read the following memos and use them to make his decision.

Key Persons Involved

Max Interface, *President*
Bernie Closefast, *Vice President, Sales*
Priscilla Flack, *Vice President, Public Affairs*
Sue N. Prevail, *Legal Counsel*

Memorandum

Date: July 9
To: Max Interface
            President
From: Bernice Closefast
            Vice President, Sales
Re: Demands of the Friends of Freedom

As you requested, I have been studying the press releases of this Friends of Freedom outfit and thinking about how we might deal with them. The experience hasn't improved my opinion of these people any. They preach at you, and then if you don't agree with them, they threaten you. As it happens, I don't like being preached at and I don't like being threatened. More to the point, though, I don't think their arguments hold water. We are a business concern, and righting all the world's wrongs isn't part of our business. We don't have the duty, or the right, or (frankly) the expertise to make fine political distinctions or evaluate the ideology of each of our customers. How can any business function if it has to worry about whether the client is virtuous? If you were in the clothing business, would you refuse to sell a guy a pair of pants because he might wear them to a Ku Klux Klan rally? Would you worry about selling someone a car because he might use it in a bank robbery? Business is tough enough without having to think about such things.

Even if we wanted to impose some sort of ideological litmus test on our customers, how would we set it up? Who is to decide who are the good guys and who are the bad guys? Governments don't come to us and say, "We want to buy a computer to oppress our citizens"; they say that they want to control terrorists, criminals, and foreign-backed revolutionaries. The Friends of Freedom seems to feel that it has the exclusive right to decide who is on the side of the angels. But it isn't so simple. The regime that some people call "fascist" is seen by other people as "committed to the defense of free enterprise and traditional values." The same is true on the left—one man's "people's democracy" or "progressive socialist state" is another's "communist tyranny" or "Soviet puppet." I think the simplest solution is to stay out of politics and do what we do best—sell computers.

In fact, if we must get into the issue of morality, I think that selling computers to undemocratic regimes is probably beneficial in the long run. The more trade we have with these societies, the greater will be their exposure to our values. The more advanced technology they import, the faster will be the development of an educated class that can spearhead social and political progress. I certainly don't see how *not* selling computers to undemocratic regimes is going to make them more democratic or make them treat their people any better. And it won't keep them from getting the computers in any case. If we don't sell police systems to these countries, there are plenty of other people who will. The Japanese and the French are not so inhibited about who they sell to—they don't demand character references, only cash. Even one of our domestic competitors may be tempted to brave the wrath of the Friends of Freedom and grab a share of the market. (Can't you see Pineapple Computer just jumping at the chance?) And as far as San Miguel and Boersland are concerned, the Russians would love to get a toehold in those markets with their new KGB 111 system.

Actually, opting out of these markets would have only one consequence that I can see, and it wouldn't be for our customers, it would be for us. As you know, the OSIS systems sell for between $3 million and $8 million, depending on the capabilities selected. So if we adopt the policy that the Friends of Freedom is proposing, we stand to lose somewhere in the neighborhood of $35 million to $75 million in sales next year. How do we explain that to Wall Street?

The Friends of Freedom's statements contain a lot of hysterical rhetoric about OSIS and similar systems; you'd think we were selling medieval instruments of torture. People should be made to understand what we're really talking about here. It's not as if we're providing governments with electrical shock devices, or drugs for chemical brainwashing. There's nothing in these systems that's beyond the pale, nothing that violates basic decency. The systems just represent a technical step forward, an augmentation of existing technology for widely accepted practices—record keeping, authorized wiretapping, polygraph testing, and the like. Law enforcement, security, and intelligence organizations need these capabilities, and they're going to get them, one way or another. We don't have any control over how they will use these capabilities, any more than the maker of a battery can determine whether it will be used to power a smoke detector or detonate a bomb. The technology itself is neutral; it may be abused, but there's not much we can do to prevent that.

I guess it's pretty clear where I stand on this. If we let one pressure group push us around, we'll only be encouraging everybody and his brother to jump in with their favorite causes. The Arabs will demand that we not sell to Israel, the IRA will demand that we not deal with Britain, the right-wingers will demand that we blacklist leftist states in the Third World—where will it end? The next thing you know, you'll get up one morning and find headquarters surrounded by a picket line of turbaned Sikhs, carrying swords, protesting our sales to India! So to heck with the Friends of Freedom and their moralizing. If they don't like what's going on in San Miguel, they should be pressuring the government there, or the government in Washington—not us. Our job is to sell computers. I think that most people, when they understand the issues involved, will back us up.

Memorandum

Date: July 14
To: Max Interface
President
From: Priscilla Flack
Vice President, Public Affairs
Re: Friends of Freedom Protests

Your inquiry about the position we should take in response to the demands of the Friends of Freedom has led me to take a fresh look at the question of sales to foreign governments. I think that the whole issue should be considered with great care and circumspection; it needs close attention at the highest levels of management, not just a superficial response for the Board of Directors meeting. The stakes are high, and the decisions aren't easy ones to make.

First of all, we must understand that these latest demands are just the culmination of something that has been brewing for quite a while. Protests by the Friends of Freedom have been gathering momentum for several years, and have already picked up the backing of many church and student groups, as well as the major civil liberties organizations. Their strategy of targeting a variety of unpopular regimes, on the left and on the right, has helped them to rally support from a number of quarters. Civil rights, black power, and some labor organizations back their demand for an end to computer sales to Boersland; Jewish groups join in because they oppose selling any high-tech systems to East Onan; liberals and other foreign policy critics, who don't like our involvement with military regimes in Central America, jump on the bandwagon because San Miguel is also on the Friends of Freedom's list. They have chosen their targets well; Boersland and San Miguel are not very well loved in this country, and virtually no one likes East Onan.

More important, the current demands are unlikely to mark the end of the matter. I anticipate that this issue will only loom larger as time goes on. And it isn't only the threatened boycotts that we have to worry about. You will recall that a few years ago the federal government banned the export of computer technology to South Africa; it seems likely that supporters of the Friends of Freedom will press for federal legislation outlawing the sale of computer systems to these regimes and perhaps others like them. Moreover, Friends of Freedom sympathizers already have considerable clout in several states, and are said to be drafting bills for introduction in three state legislatures.

Thus, if we continue to do business with unsavory regimes we are faced in all probability with a long, uphill, and perhaps ultimately futile battle. And the longer we carry on the struggle, the more we expose ourselves to unfavorable publicity, denunciations, boycotts, and the like. The long controversy over Nestle's marketing tactics used to promote infant formula in developing nations serves to remind us of the furor and hostility that can result from these emotionally charged situations.

OIS has a well-earned reputation as an enlightened and progressive company; do we want to jeopardize that reputation over an issue on which we will have few supporters and incur the enmity of many segments of the general population, as well as important pressure groups? It is easy to put a dollar value on lost sales, and hard to measure the worth of goodwill and public trust. Nevertheless, these are among our greatest assets, and we should be prepared to make sacrifices to preserve them.

Bernie Closefast has argued that it is very difficult to make political and moral judgments—to pick and choose among the people we do business with. This is true, but in reality, don't we always have to make such judgments? On the one hand, when the Council of Concerned Black Clergy petitioned us for an affirmative action program to increase minority hiring, we acquiesced—even though we weren't required to by law, and didn't feel that we had been discriminatory in our hiring practices. We were careful to set up a program that didn't violate EEO guidelines (and so didn't leave us open to any "reverse discrimination" complaints). The results have been successful, and highly satisfactory to all concerned. On the other hand, when the Jewish Vigilance Committee demanded that we pull our office out of Riyadh and stop doing business with *all* Arab countries, we refused—and rightly so, I believe. My point is that many of the tough business decisions we must face in today's world have moral and political dimensions; there's just no way to escape them.

I also think that the moral issues here are not as problematical as Bernie suggests. Technology may be morally neutral, but technology designed for specific ends, put into the hands of specific people, is not necessarily neutral. For example, nuclear technology may be neutral, in that it can be used in medicine, or to generate electricity, or to make weapons. Even nuclear weapons can be thought of as neutral, in that they can be used for aggression or for self-defense. Nevertheless, our government strictly limits the export of nuclear technology and fissionable materials. It decides who it trusts and who it doesn't; it takes pains to make sure that this technology doesn't get into the wrong hands, and even in the right hands, that it is used only for peaceful purposes. Similarly, a gun is neutral in that it can be used in law enforcement, or for self-defense, or for crime. Nevertheless, many states have laws that limit the right to carry handguns and attempt to keep criminals from acquiring them. And certain devices, such as machine guns and dumdum bullets, are outlawed for sale to the general public because they are considered to have no legitimate civilian use commensurate with their potential for misuse.

Now, we are not just selling computers that can be used for any task, depending on the software. I can see that such computers could truly be considered neutral. But we have created sophisticated systems for very specific purposes: wiretapping, record scanning and compilation of dossiers, voiceprint recognition, and so on. The design of such advanced systems is beyond the capabilities of most of the countries to which they are sold, and indeed beyond the capabilities of most other suppliers. Don't we have some moral obligation to take precautions against these systems being used to oppress people? Aren't we responsible for the results if we let these tools fall into the hand of tyrants—just as we would be if we sold a gun to an assassin?

Therefore, much as I dislike some of the tactics of the Friends of Freedom and similar groups, I think that in this instance it would be wise to yield and to accept the principle that we should not sell police systems to "unsavory" regimes. This does not necessarily mean that we must accept the Friends of Freedom's judgment as to who we should blacklist. It would be far better to have some sort of objective criterion, so that we can't easily be pressured or second-guessed, as Bernie fears. For example, the Sullivan Principles sought for a number of years to provide a morally defensible basis upon which U.S. companies could do business in South Africa. In effect, these principles created an external reference point that helped to reconcile opposing prescriptions for companies' investments amidst the complex issues facing that troubled country.

Moreover, each year the State Department conducts a review of human rights progress in all countries receiving U.S. aid, and defines a standard of "acceptable human rights progress." We could establish a policy of selling only to those countries which meet this standard. In this way we would appear to be shaping our policy to conform to national aims and values, rather than merely caving in to a pressure group. What we'd lose in sales we would more than make up for in increased prestige—and a lot of us might sleep better at night, too.

Memorandum

Date: July 18
To: Max Interface
President
From: Sue N. Prevail
Legal Counsel
Re: Computer Sales to "Repressive" Regimes

Herewith my response to your recent request for an analysis of our position on the sale of police computer systems to certain unpopular regimes.

One point in particular must be emphasized from the outset: *U.S. law does not currently forbid our selling these systems to the countries in question (Boersland, East Onan, and San Miguel)*. As you know, national policy in this area is guided by the Export Administration Act of 1969, which states that U.S. policy is "to encourage trade with all countries . . . except those countries with which such trade has been determined by the President to be against the national interest." Specifically, the law calls for export restrictions when trade is detrimental to the national security of the United States, and none of the three countries in question have been argued by anyone in this debate to be risks to our national security. In this respect, the situation is very different from that involving Soviet bloc countries. The export of computer technology to Eastern bloc states was banned under this law some years ago, but the restriction does not apply to the countries you have inquired about. (East Onan is generally regarded as a Soviet ally, but it is not officially considered part of the Communist bloc.)

Now, if it should become official U.S. policy to outlaw the sale of computer systems to these three countries — that is, should these countries be added to the ban list — we would of course obey the law and discontinue selling to them. But in the absence of such action by the government, why should we take the initiative and stop selling our products to these states? As a corporation, our obligation is to stay within the law, not to conduct our own foreign policy. Such an action on our part might even be seen as embarrassing to the U.S. government, which after all maintains ties with all three countries (though with varying degrees of cordiality).

Moreover, we must not forget that we have an obligation to our stockholders. If we relinquish the profits to be made by selling computer systems to these nations, we are in effect forcing the stockholders to make financial sacrifices in support of moral judgments and political values that they may not necessarily share. Conceivably, we might even be providing the basis for a suit by stockholders against management, on the grounds that it is the duty of management to pursue all opportunities to maximize profits.

Therefore, I would strongly advise that we continue our present policy of selling to any buyer who can afford our systems, so long as such transactions do not violate any state or federal statutes. If and when the laws change, we will change our practices, but we have no obligation — to our stockholders, to the government, or to special interest groups such as the Friends of Freedom — to adopt the role of moral trailblazer in such matters.

Discussion Questions

1. What actual situations come to mind in which U.S. companies in recent years have been challenged on political or ideological grounds for doing business with specific foreign countries? How were these issues resolved?

2. Bernie Closefast presents three arguments against halting company sales to the countries in question:
   a. That OIS equipment does not differ from many other products that can be and are use for suppression and terrorism
   b. That it is impossible to draw the line between addressing moral concerns and yielding to the claims of all groups
   c. That the long-run effect of the sales may be beneficial to democratic values

   Respond to each of these points. What level of validity do you place in them?

3. Flack appears to hold different assumptions from those of Closefast and Prevail about shareholder interests in this situation. State what you believe shareholder interests are in this case. How should they be determined for a particular company?

4. Flack and Prevail propose different decision-making bases to help the company resolve the issue of the contested sales. Flack advocates relying on an external evaluation of foreign countries' internal political regimes, such as the U.S. government's assessment of a country's progress on human rights. Prevail suggests using established state and federal laws and administrative rulings as the only guides. Evaluate the advantages and disadvantages of each of these proposals.

5. Considering the various issues and recommendations of persons in this case, what policy would you, as Max Interface, propose to your board?

CASE 12

Business and Political Action Committees

How to Define the Corporate Interest

A diversified manufacturer with more than $3 billion in sales, Technidyne Corporation is currently faced with a problem of corporate strategy. A major defense contractor, Technidyne has used profits over the past decade to acquire companies primarily in the automotive supply industry in an effort to level corporate earnings. Its diversification efforts, however, are currently not as profitable as its defense division.

For example, Rubberguard, Inc., a large Technidyne subsidiary producing retreaded tires, is suffering from recent labor troubles and is experiencing intense competition from new low-cost imported tires. The Ferrous Steel Corporation, another wholly owned Technidyne subsidiary, has been troubled with the forces of foreign competition and is lagging the industry in new product development. Profits from both of these subsidiaries, among several others, have been waning; Technidyne's corporate growth and profitability have largely been attributable to the more robust $1.8 billion Technidyne Defense Division.

Beginning with the production of electromechanical controls for automobiles, Technidyne has grown rapidly over the last quarter of a century. As a result of an aggressive acquisition effort and successful internal development, leading-edge technologies now being developed by Technidyne Defense Laboratories include sophisticated electronic sensors, diagnostic software for rapid interpretation of massive electronic inputs, and integrated communications interface equipment. It has also committed substantial resources to the development of proven mil-spec, micro-miniature circuitry and assemblies suitable for use in military systems.

Student materials for this case have drawn on material prepared for the authors by Frederick S. Freer III of Cleveland, Ohio.

The company is well positioned to participate in the development of a new national defense system called the Electronic Guidance Advanced Defense System (EGADS), sponsored by the Air Force. As a key participant in early stage feasibility studies of this system, Technidyne stands to receive major development contracts, if the EGADS program is funded. However, the program is large and costly, and has become a public issue. A variety of antinuclear groups and some members of Congress oppose the program, calling it an unnecessary and unworkable expenditure of public funds. Others fear that evidence of a U.S. intention to develop and install this system would constitute a direct challenge to the Soviet Union and would represent another step in what critics consider a senseless and destabilizing arms race. At the very least, the EGADS program promises to be a hot political issue in the next Congress. If funded, it offers Technidyne the potential to triple annual revenues for its defense division within ten years.

The new Chairman and Chief Executive Officer at Technidyne, Herb Dillings, is impressed with the view that a public role is part of the job of a corporate executive, especially an executive in the defense contracting industry. Are responsibilities to employees, shareholders, and Technidyne's communities served by letting others alone decide the course of national policy? Then too, virtually all other defense contractors of Technidyne's size have active PACs; although they appear to follow no single philosophy or style of operation, there seems to be no reluctance on the part of Technidyne's competitors to participate actively in electoral and governmental politics.

The Technidyne PAC has languished in recent years, with little attention or cultivation of new members. With less than $8,000 in its account, the PAC has had few funds available for distribution to federal officeholders or hopefuls. Hal "Bucky" Buckfiler, a new manager, age 38, and recently appointed Assistant to the CEO, has expressed strong interest in seeing the Technidyne PAC play an active role in the current elections.

Buckfiler's capabilities were tested earlier when he served as a United States Marine. Earning both a Purple Heart and the Distinguished Service Medal in Vietnam, Buckfiler went on to complete his undergraduate education at the University of Southern California. Buckfiler enjoyed a range of assignments before joining Technidyne; at other firms his responsibilities included product management and sales management, and he was a divisional marketing manager.

At the recommendation of Buckfiler's supervisor, a division chief at the Bureau of the Budget, CEO Dillings arranged for Buckfiler to join Technidyne at the close of his third year in Washington.

Hal Buckfiler now reports directly to Dillings. Technidyne senior staffers have privately speculated that Dillings may be evaluating Buckfiler as heir apparent to Robert Sterling, Vice President of Public Relations. Dillings has recently talked about changing the focus of Sterling's office to "governmental affairs."

Business and Political Action Committees

At 63, Sterling is nearing retirement. He has led a respectable but conservative career, with involvements predominantly in advertising, media relations, and community relations. Sterling has long been concerned about governmental abuse of power and has often spoken about the need for an ethical overhaul in government.

Meanwhile, Herb Dillings has been giving thought to the importance to Technidyne's future of a number of current political campaigns. For example, Dillings is aware that Senator Robert Winfield, the current chairman of the Senate Armed Services Committee, favors the EGADS program. Winfield has had a long legislative career. Now, however, at age 75, he is being seriously challenged for his Senate seat in this year's election. While Winfield is a staunch advocate of the EGADS program, his challenger has been a frequent critic of government waste and has built her campaign platform partly on opposition to the EGADS program.

With an expected confrontation on EGADS funding next spring, the impending November elections may prove pivotal for Technidyne. A few days ago, Dillings raised the issue of political action at a meeting of corporate officers and requested written comments about what role, if any, Technidyne should play in the coming elections. In response, he has received the memos that follow. After reading these memos, what course of action do you think best serves the interests of the firm, its employees, and its shareholders?

Key Persons Involved

Herb Dillings, *Chairman and Chief Executive Officer*
Hal Buckfiler, *Assistant to the CEO*
Malcolm G. Weber, *Executive Vice President*
Robert V. Sterling, *Vice President of Public Relations*
Lawrence Workmann, *Director, Department of Legal Affairs*

Memorandum

Date: May 10
To: Herb Dillings, Chairman and Chief Executive Officer
From: Hal Buckfiler, Assistant to the CEO
Re: Political Action Strategies

At Tuesday's officers' meeting you opened the question of political action for Technidyne with your senior officers, and we can now move ahead to place this corporation exactly where it needs to be today, as a political force in the mainstream of governmental influence. Technidyne has stood on the sidelines long enough. I joined the firm because of your rightful concern about the company's dependence on governmental policy, and I look forward to using my knowledge of the people and the inner workings of Washington to make Technidyne a major corporate player in the game.

I fully expect that this initial phase of discussion among your staff members will involve considerable hemming and hawing and gnashing of teeth about the proper role of the firm in the political process. But I don't want there to be any misunderstanding between you and me. Politics is where the big decisions are made, and either you are in the game or out of it. You may want to put some window dressing on the front entrance, but I came here to put Technidyne in the big leagues, and I assume that's what you brought me here for.

Few executives truly understand the way decisions are made in this country. High-minded idealism may make them feel good, but it doesn't get anything done for them. Politicians and bureaucrats have interests, just like everyone else. And, like everybody else, they act in terms of their interests. Just as in your economic markets, everything that gets done works on the basis of an exchange. Both parties have to find value in the arrangement based on their interests. Any other way of approaching government policy is misguided, wasteful, and naive. Why shouldn't we expect markets to work for political decisions as well as for economic decisions?

In case you personally have any reservations about approaching your relations with government in this way, let met put your mind at rest. There is nothing immoral or evil, and certainly there is nothing illegal, about trading in the influence game. Practically everybody in Washington understands this game, and everyone expects you to play it if you want to get something done. Moreover, the system is enormously successful. The proof is in the pudding; the system gets results.

In some respects I feel sorry for those who stand on the sidelines, wringing their hands at "special interest politics." They ought to get off that number if they can't show us a better system. What other political system allows so many interests to have access at so many points in governmental decisions? You may have to play hardball to win, but you really understand what democracy is all about when you climb into the political dugouts. In most cases, all you can say about a bleeding heart is that he isn't a very good player.

Also, I don't know if you appreciate the level of sophistication needed today to get your way with the government. Blatant approaches to buying influence went out in the 1800s. Today, you have to spend time and money establishing and maintaining your contacts. People who demand, threaten, or make explicit appeals to self-interest don't get past the first receptionist. Politics *is* professionalism. It is practiced by professionals with other professionals. Crudeness doesn't sell. You have to be a good communicator. And most of all, you have to be an excellent listener.

Within two weeks I will have prepared detailed strategies for (a) our congressional campaign contributions and (b) our independent expenditure campaign. In order to impact on next fall's elections, we are going to have to get these programs in gear immediately. In general, I see our strategy shaping up something like the following. First, with your approval I will prepare a detailed report of the voting record on defense of every congressional candidate up for election or reelection. Second, I will identify the two dozen or so "key" races

Business and Political Action Committees

from a defense policy point of view. A "key" race is a close contest in which defense funding is a major campaign issue, namely, one in which there is either a strong "pro" or a strong "anti" candidate. Third, I will identify races in which one or the other or both candidates have made the EGADS program a campaign issue.

Using this information, I suggest we put priority for our campaign-financing contributions on the "key" races. Working within the Federal Election Commission's limit of a $5,000 contribution to any one candidate in an election, we will determine the amount of contribution to a specific candidate on the "probability of winning" principle. Remember that our goal is to gain maximum accessibility to governmental influence *after* the election. Therefore, the last thing we want to do is close our options with any single candidate. A close reading of the political press gives a good indication of each candidate's position in the race. What's more, we already know that incumbents consistently win more than 60 percent of the elections. Determining a candidate's probability of winning is relatively easy.

I suggest we initially allocate $5,000 to each race, divided among the candidates according to each candidate's probability of winning. Take, for example, the House race between Representative Marian Tosser and her challenger, Frank Quell, in our own state of New Caledonia. Any candid estimate of Tosser's probability of winning would have to give her close to a 70 percent likelihood. Consequently, we would initially target $3,500 to Tosser's campaign and $1,500 to Quell's.

Then, because this is a "key" race on defense issues, a bonus contribution can be given to the pro-defense candidate, depending on the strength of his or her position. In this case, Representative Tosser has taken public positions strongly favorable to the defense program, and she has a voting record to back it up. Consequently, Tosser would receive another $1,500, bringing her up to the $5,000 limit for this election. Similarly, we will have a "take away" policy for candidates having an anti-defense posture. If Quell comes out with negative views on defense, we will take away amounts from his $1,500. The amount of the "take away" will depend on how strong his views are.

This system achieves several purposes. First, it allocates a greater portion of our resources to persons most likely to hold elected positions in Washington. Believe me, this is simply smart politics, and it is common practice for PACs to allocate funds to candidates based on the "probability of winning" principle. Otherwise, why do you think incumbent candidates received fully 80 percent of PAC funding in the 1983–84 election cycle? Second, this method keeps a door open with less favored candidates. A cardinal rule of politics is never, ever to burn your bridges with anyone. The real world is too full of surprises and unexpected twists and turns in political affairs. We always want to leave a door open. Third, this system places a premium on support for pro-defense candidates. There should be no debate about this at Technidyne. In addition, the system provides some encouragement or incentive to candidates to change or at least moderate negative views on defense. In my experience, no other campaign funding strategy offers comparable benefits.

A second major thrust to our political action strategy will focus on winning races critical to the EGADS program. As you know, this program is somewhat of a political lightning rod, since opposing EGADS is serving as a convenient issue for some candidates to make names for themselves. There is at least one "must" race involving a key committee position that may determine whether EGADS will survive the next defense budget: Senator Winfield's reelection. In this case, campaign contributions are too indirect and passive. Where the stakes are this high, we have to be more aggressive and action oriented.

Here we will organize two publicity campaigns. One will be in favor of Winfield. The other will be in opposition to his challenger, Mary Richman. This will involve creating and airing paid TV spots in the last two months of the election. These will be hard-hitting 15- and 30-second commercials designed to influence votes and, frankly, win the election.

As long as we don't coordinate our activities with Winfield's campaign committee, federal election rules allow us to make unlimited expenditures in his behalf. This is a constitutionally recognized freedom of speech right of corporations as well as individuals. We understand the issues and the candidates, and will be able to do the good senator justice. These ads are going to be very powerful. Believe me, you are going to love them.

Herb, this is the game plan for the elections. You need only do three things to ensure its success. First, put as much money as possible in the PAC account. This program is going to cost a lot of money. We won't even be able to get off the ground for under half a million dollars. Malcom Weber has got some good ideas for raising this money. Give him a free rein and we'll be in business.

Second, you need to give some thought as to how to organize our Washington lobbying effort after the new Congress gets seated next year. All of this work on the election will go for naught unless we have a strong presence in Washington. Undoubtedly, you will want me involved in this piece of the action as well, and at your request I will draw up a game plan for working the legislature and executive agencies after the election.

Third, keep your other managers off my back as we implement this strategy. Needless to say, I don't have the time or patience to bring Sterling along. Do what you want with him, but realize that we can't afford to compromise our goals or our time schedule. The stakes are just too high.

I'll have the detailed strategies on your desk within two weeks.

Business and Political Action Committees

Memorandum

Date: May 10
To: Herb Dillings, Chief Executive Officer
From: Malcom G. Weber, Executive Vice President
Re: Technidyne Political Involvement

Your presentation Tuesday on corporate political action went very well. Knowing that you were concerned about how the officers would react to a higher profile for Technidyne in the current campaign, I observed their reactions closely. Your comments were well received. There is little doubt that we will be involved in the electoral process, although *how* we go about it remains to be determined.

As you know, I have long supported an organization named Citizens for a Free America. CFA is a public interest organization engaged in research, writing, and education to promote traditional American values of free enterprise and to reduce the role of the government in the lives of American citizens. From my viewpoint, these are values we must be concerned about in this country, both as private and as corporate citizens.

The country may be facing an important crossroads in this election. Many people are concerned that the forces of uninhibited government spending and thoughtless business regulation are about to reassert themselves with a vengeance. I sense that our Congress is again tempted to endorse the attitude that government will do for you whatever you do not want to do for yourself. I call this the "negative responsibility" society. People also fail to realize the importance of national defense to our cherished freedoms. Many seem to think our national security can be maintained without continual new investments in technology.

In addition to widespread misperceptions about America's need to modernize its defenses, there is a popular notion that all corporations are run by crooks. Yes, we are only too painfully aware that business in general and our industry in particular have had a number of unfortunate ethical and legal problems in the past few years. This doesn't mean that society needs to throw out the free enterprise system. Yet that seems to be the implication of a lot of careless talk one reads in the papers.

There is in this country a shortage of politicians who have the fortitude to stand up to unreasonable pressure group demands and irresponsible journalism. Even though in private many legislators convey deep personal doubts about the coming pressures of the country's "social agenda," few have the courage to stand up in public for what they believe. To speak out for private enterprise today is not very popular, and those who do take public positions in favor of our economic system need our support. In my view, the Technidyne PAC should be unabashedly partisan. We have an important stake in the system, so let's defend it.

It amazes me to read, as I recently did, that 80 percent of all corporate PAC contributions went to incumbents in the last election cycle. I consider this extremely unfortunate. Corporations should be shaping the political landscape of the future, not acting to reduce their political risks. U.S. business has the need and the duty to claim the political process no matter what the risk. Our future and indeed the future of the entire country depend on it. I cannot stress too strongly that Technidyne PAC contributions should go only to candidates—Bobby Winfield is a good example—who have declared themselves advocates and spokespersons of the private enterprise system. I suggest you direct Bucky Buckfiler and Bob Sterling, who seem to be assuming responsibility for administering our PAC, to compile a list of congressional candidates on the basis of their avowed ideologies. Contributions should be alloted to pro–private enterprise candidates only.

I also feel strongly about a second point. The fortunes of all Technidyne employees will rise or fall with the corporation, and I believe each person should contribute to our PAC. Recently I ran across a news clipping in my files from *The New York Times* in which Bob Wright, President of NBC, articulated the importance of corporate and employee involvement in the political process. In his usual forthright way, Bob said that employees who refuse to contribute should question their own dedication to the company. I feel the same way about Technidyne. Don't you? I've enclosed a copy of the clipping for your own review and interest.

Fortunately, Technidyne can solicit employees of our union, the Amalgamated Factory Workers of America, just as the union can solicit management employees. Although neither company nor union has taken advantage of the opportunity to request political contributions from the members of the other group, I feel that we now should. Nonmanagement employees as well as managers should realize that we are all in this together.

How much money can we expect to raise from these efforts? I have heard that executives of the Dart Corporation on the averge contributed more than $1,000 each in the last election. That kind of support would be absolutely terrific. Realistically, across our nearly 5,000 exempt employees I would hope that in our first serious effort we can average $50 per manager, or about $250,000. Our exempt employees are more of a question for me. I would expect an average contribution of about $3 apiece from our 70,000 nonexempt employees. Conservatively, then, we should be able to participate this year with close to half a million dollars.

The issue is not how much money we can raise for this campaign. You can see that plenty of money will be available. The more important issue is to make a public statement about where our hearts and our values lie. By asking for money from employees and shareholders, and by making campaign donations to those who care about the future of our system, we will be fulfilling our moral duty.

I hope these comments are useful. I will be pleased to assume whatever role you wish in the coming campaign.

Business and Political Action Committees

Attachment

NBC HEAD PROPOSES STAFF POLITICAL CONTRIBUTIONS
By Peter J. Boyer

The president of NBC has urged that the network start a political action committee and that NBC employees who refuse to contribute to it "question their own dedication to the company," according to sources at NBC.

The executive, Robert D. Wright, addressed a memorandum on the proposal last month to Corydon B. Dunham, NBC's executive vice president and general counsel, and distributed copies to several other NBC officials.

Neither CBS nor ABC has a political action committee. However, the General Electric Company, NBC's parent corporation, has one.

The NBC memorandum has raised questions of propriety at the network, especially within NBC News. Lawrence K. Grossman, the president of NBC News, said that there was "no ambiguity" about his opposition to news division participation in the proposed committee, and that he had so informed Mr. Wright.

"It was very quickly stated," Mr. Grossman said yesterday. "The news division's policies would preclude anybody from NBC News from participating in anything like that. So we are not part of anything that may be going on."

Mr. Wright said in a telephone interview yesterday that NBC News employees would not be expected to participate in the political action committee.

A former executive at G.E., Mr. Wright became president of NBC on Sept. 1. His memorandum notes that NBC's business and the political process are intertwined, and adds:

"Employees that earn their living and support their families from the profits of our business must recognize a need to invest some portion of their earnings to ensure that the company is well represented in Washington, and that its important issues are clearly placed before Congress."

"Employees who elect not to participate in a giving program of this type," the memorandum went on, "should question their own dedication to the company and their expectations."

Several NBC employees, who asked that their names not be used, said they saw a hint of coercion in that passage.

Told of that concern, Mr. Wright said in a telephone interview yesterday: "I think it's fair for me to say that people should be active in the political process. When I hear people say that they don't care one way or the other who's elected, or who's in office, I guess I don't have a lot of tolerance for that. I don't care who you support, but you should do something."

Sharon Snyder, a spokesman in Washington for the Federal Election Commission, said yesterday that election laws prohibited companies from contributing political action committee money secured from employees "by physical force, job discrimination, financial reprisal or the threat of physical force, job discrimination or financial reprisal."

Other Networks' Positions

George F. Schweitzer, vice president of communications for the CBS Broadcast Group, said yesterday that CBS had no political action committee because "we operate a large, worldwide news organization and we feel that would be in conflict with the basic operation of that news organization."

However, ABC, which recently merged into Capital Cities Communications Inc., is considering "an appropriate way to participate more actively" in the political process, according to Patricia Matson, a network spokesman. She said that forming a political action committee was "one of the things we're looking at."

Direct corporate contributions to candidates are illegal. NBC executives for years have been privately encouraged by management to contribute to political candidates sympathetic to network television's position on various issues. Forming a political action committee would be a method of simplifying that process, Mr. Wright said.

The issue of increasing political involvement by the networks is partly a response to the networks' abiding tension over Hollywood's ownership of television programs. For several years, network ownership of the programs they broadcast has been limited, giving rise to a lucrative syndication market that principally benefits the owners of the programs—the Hollywood producers.

The networks have hoped to enlist the aid of Washington in freeing themselves from such restrictions, but have found the lobbying efforts of the Hollywood producers difficult to overcome.

The broadcasting industry already has several political action committees, including the National Association of Broadcasters Television and Radio P.A.C., which contributed $49,040 to various Congressional campaigns this year, according to Broadcasting magazine. Several network executives contribute to that P.A.C. as individuals, according to Ms. Matson of ABC.

Cable Industry's Contribution

According to Broadcasting, the national cable television industry's Cable P.A.C. contributed $64,050 to candidates this year. In comparison, according to Broadcasting, the Hollywood production community contributed $70,319 to political action committees.

Several station groups as well as cable television companies have political action committees, a point that Mr. Wright, who once headed Cox Cable Communications, noted yesterday.

"I was in the cable industry and they, like the motion-picture business, are very politically active," said Mr. Wright. "It's a little unfair. If you're going to be involved in the political process, you should at least support the process of electing people. There's nothing to be embarrassed about."

In his memorandum, Mr. Wright said the purpose of the proposed NBC political action committee would not be to favor any particular party or ideology.

"The point is simply to reinforce our support for the men and women of Congress who must continually face the continuing and time-consuming task of re-election," the memorandum said. "By showing our tangible support for the proces, the office and the officeholders, we are in a better position to have our views intelligently and fairly viewed by Congress."

Source: Copyright © 1986 by The New York Times Company. Reprinted by permission.

Memorandum

Date: May 15
To: Herb Dillings, Chairman and Chief Executive Officer
From: Robert V. Sterling, Vice President of Public Relations
Re: Technidyne Political Involvement

In response to our meeting last week, I have taken the opportunity to organize my thoughts about Technidyne's role in this year's congressional campaigns. First, let me reinforce my view that political involvement by Technidyne is vitally important. Corporations in this country are presented by law and tradition with an unusual chance to participate in the political process. This is a franchise that we *must* take seriously. We cannot walk away from the opportunity to help determine the direction of the country. You are to be commended for recognizing this fact and encouraging our involvement.

While few would disagree with such noble principles, various immediate objectives for political involvement are present, and there are many ways for the firm to become involved. As you know, I feel strongly that our activities should follow several principles in particular. Politics can be an exciting and heady activity, but it can also be a hot stove. Most important, I don't want Technidyne to act naively or foolishly and come away burned.

First and foremost, we have to decide what our goals are in the political arena. Certainly, I want the EGADS program to be fully funded in Congress. This program is important to the future of the company, and I deeply believe it is a program essential to the security of the United States. Having said this, however, I still think it would be a mistake to establish the funding of this defense program as our political goal. This is much too narrow and self-serving. In my view, Technidyne's true interest is broader and longer term—namely, improvement in the general quality of leadership in Washington. Let me explain more fully.

I have long said that integrity is the biggest problem this country faces in government and in business. I have no doubt that recent events in Washington and throughout our industry are proving me correct. The secret and possibly illegal operations that came to light as part of the "Junketscam" revelations last year are only illustrative of the mediocre quality of our elected officials. What levels of arrogance and disrespect are public officials capable of? This incident was an insult and an embarrassment to every well-meaning and honest citizen.

And the problem of integrity in public service is by no means limited to elected officials in Washington—it seems to permeate public bureaucracy at the state level as well. Witness the fact that the Director of Procurement of our own state of New Caledonia was recently found to be exercising his authority from a variety of motels in the state capital.

In some respects, we have grown immune to incompetent and often unethical conduct by public officials. Americans have come to expect little from their representatives—and that's exactly what we get. I believe it's time that corporations such as ours begin drawing the line. We need to use whatever political muscle we have to improve the quality of people in public office. No other goal may be as important to our future as a company.

Second, you are well aware that our industry has been badly damaged in the public eye from a continuing series of revelations about abuse and fraud in government contracting. A recent opinion poll reported in *Business News* placed defense contractors in the lowest 20 percent of fifty U.S. industries in terms of honesty in doing business. Fortunately, Technidyne has not been implicated in the investigations into overcharging on government contracts and using deceptive accounting practices. I need not remind you, however, that our public image is extremely vulnerable to suspicion and guilt by association.

What does this mean for us in terms of PAC activities? First, Bucky Buckfiler may be right that greater PAC activity is warranted. We have been on a learning curve since we initiated the PAC four years ago, and as PAC chairman, I have urged that we go slowly with small steps before becoming more visible in the campaign-financing game. We now know a lot more about how the system works than we did then.

I want to stress, though, my strong belief that the integrity of our political system *and* the credibility of business should be our foremost priorities. Our greatest threat is the loss of confidence of the American public in the country's institutions. Technidyne PAC contributions should be dictated equally by our assessment of the best overall candidates and by the need to enhance the image of our company and our industry.

I know that I face great pressure to use the PAC to maximize the likelihood of funding for the EGADS program. To do so would not be at all unusual. Washington is literally crawling with special interest groups, each trying to get its share of the national largesse, and I know that many companies view their PAC activities as simply one more way to obtain influence. This country has created a system that allows, even encourages, politicians to pander to narrow interests. In 1986 more than $300 million was spent on national congressional campaigns, and more than $100 million was spent by PACs alone. Senate campaigns today are literally costing several millions of dollars, and it seems legislators spend more time raising money than they do familiarizing themselves with the public issues before them. And what is the result of this incredibly expensive, chaotic system? Cynicism and corruption in government. It seems that the more money there is in the system, the poorer is the quality of people drawn into politics.

Let's not drag Technidyne down into the muck and grovel for big government funding on EGADS. We should stand apart from this disgusting display of trough feeding. Under its current circumstances, our industry can't afford to be viewed as buying influence in Washington. Working for honest, above-board representatives in Washington will be better for the company, and for the country, in the long run. Good public officials — those who vote with an eye to principle, not to the next election — are easily identified. They deserve our support, and we will help ourselves by being identified with them.

I should also add that public leadership transcends party affiliation. There are honest and dedicated persons in both parties who deserve to be elected. Restricting our contributions to advocates of one or another political philosophy unwisely eliminates some superb public servants from our consideration and most likely directs funds to lesser candidates. Similarly, we should pay close attention to challengers and open races in order to identify and support genuine leaders.

Finally, I have been somewhat disturbed that the whole agenda about Technidyne's political involvement has focused on campaign-financing — PAC — activity. While I agree that the PAC is an important way for us to be involved, it is by no means the only, or the best, avenue. Personally, I feel Technidyne has a message that the American public should hear, and I think we should consider "public service" advertising in the major media. Technidyne could focus on the importance of ethical behavior and prudent management at all levels of government. We could add our voice to those seeking to awaken the public to the despicable performance among all too many elected officials. Our efforts in this light would be a true public service, and would position the firm as an ethical and patriotic company worthy of additional government contracts, should the EGADS program be funded.

Herb, I applaud your sensitivity to the firm's political role. Moving the firm more in that direction has been long overdue here. Nothing would please me more in the twilight of my career at Technidyne than to assume leadership for stronger and well-focused political participation by the firm. I know that other persons in the company endorse different views of our political role and call for brasher and more aggressive programs with respect to EGADS. Whatever course we take, be sure that our PAC is not used as a tactical tool. Focusing on short-term results is a trap. Given the politicians' insatiable appetite for campaign funds, a dollar contributed this year will have to be reinforced with two dollars next year. We need to employ more *strategic* techniques — techniques that create indelible impressions and improve the quality of decisions throughout government.

Memorandum

Date: May 20
To: Herbert Dillings, Chairman and Chief Executive Officer
From: Lawrence Workmann, Director, Department of Legal Affairs
Re: Issues of Corporate Political Action

In response to your recent questions, I am happy to clarify some points of confusion about legal issues involving corporate political action. In addition, however, I would like to take this opportunity to communicate my own perspective on the issues of political action being discussed among our corporate officers.

First, you are quite right in thinking that an individual or a corporation can make unlimited expenditures in behalf of a candidate for political office. The courts have steadfastly refused to place limitations on the right of political free speech, and there is no reason to think that they will waiver in support of this principle now. Of course, such expenditures must be independent from the candidate's campaign in actuality and in appearance. They key issue here is that corporate activities must be entirely *un*coordinated with any campaign officials or committees. *Any* form of communication would place our expenditures under the strict limitation of a direct campaign contribution, that is, $5,000 per election per candidate. In short, the firm may campaign without monetary restriction so long as it acts as an independent, autonomous entity.

Second, Technidyne is permitted to solicit members of our central bargaining unit, District 13, of the Amalgamated Factory Workers of America. Thus, the Technidyne PAC can legally request contributions from all employees. However, I would also remind you that our master contract is due for renegotiation a year from now. Technidyne's relations with the union have been sensitive over the past three years, since in our last contract the company obtained a below-industry average wage increase as well as a three-year deferment of the Cost of Living Adjustment provision.

We have done a credible job of maintaining a decent working relationship through a stressful period. Needless to say, our current contract is distasteful to the union leaders as well as its members. I am concerned that soliciting nonmanagement employees may strain this relationship. Asking money from union members who have already agreed to a financial sacrifice may not be in our best interest as groundwork for opening contract negotiations next year. So while employee solicitation for the corporate PAC is perfectly legal, is it completely wise? In addition, our action in soliciting union employees creates the opportunity, by law, for the union to solicit management employees. Do we want to open ourselves to this, even if it doesn't gain the union much money?

Another word of caution pertains to solicitation of management *and* nonmanagement. Whomever you do decide to solicit, be very careful to make contributions completely voluntary. It is illegal for companies to use any sort of coercion, intimidation, or actual or threatened reprisal in raising money that is contributed to a campaign through a PAC. I know that several of your officers and advisers have a full head of steam for involvement in the coming election. Be extremely careful not to let them ride roughshod over any of our employees. To do so will surely backfire on the reputation of the firm as well as reflect negatively on your political goals. Today's stimulating experience of political action may look inconsequential in light of tomorrow's lawsuit for coercive fund-raising.

Third, it is legal for the firm to solicit its shareholders. Although the precedents are few, they do exist and the firms involved have had no difficulty with the Federal Election Commission. Mention of shareholders, however, raises what for me is the most perplexing aspect of corporate political action. You may recall that the major legal precedent for corporate freedom of speech occurred in 1978, when the Supreme Court struck down a Massachusetts law prohibiting corporate expenditures on matters not materially affecting the assets of the firm. In voiding this law, the court affirmed first amendment rights for corporations (*Bellotti v. First National Bank of Boston*).

However, I believe this decision opened an issue that is still less than completely resolved and that may have implications for Technidyne. In *Bellotti*, the state of Massachusetts argued that prohibiting the use of corporate resources to influence a referendum was necessary in order to protect shareholders whose views differed from the "corporation's" view. Rejecting this argument, the court stated that shareholders are competent to protect their own interests. Through "procedures of corporate democracy," said the court, shareholders possess adequate mechanisms to influence company policy pertaining to campaign expenditures.

My question becomes, What procedures can ensure that Technidyne's PAC expenditures are consistent with the perceived interests of shareholders? You and I both know of the great variety of political views in our company. Discussions of Technidyne PAC activity within your own management group in the past several weeks illustrate how views can differ within a relatively small group. Technidyne has nearly 75,000 employees and more than 120,000 shareholders. Political preferences are bound to conflict in groups this large. What expenditures could begin to represent shareholder and employee interests? Do you have confidence that a small group of senior managers, or even the board as a whole, can safely speak for the "corporation"?

A related question is whether we should report our PAC expenditures to shareholders and to contributing employees. Certainly our actions will be a matter of public record in our filings with the Federal Election Commission, but few persons have access to these materials. Ought not the firm report directly to interested parties? Such reporting is not required by law and, to my knowledge, is rarely performed. Expenditure disclosure, however, might go a long way toward fulfilling requirements for corporate democracy.

Herb, you asked several important questions about the legalities of corporate involvement in electoral politics. You can see that there are few absolute bars to active and full participation. Implications of PAC activity on employee, union, and shareholder relations, however, may not be entirely visible. In light of these uncertainties, I believe caution is warranted. Specifically, my recommendations on the matters you raised are as follows:

1. If they plan to engage in independent expenditures, ensure that your PAC administrators engage in *no* form of communication with the candidates they support or those candidates' campaign organizations.

2. Reevaluate the question of soliciting nonmanagement employees. At the very least, you should ask Ernie Franco, Director of Industrial Relations, to evaluate the pros and cons of this proposal.

3. Instruct Buckfiler and Sterling to avoid the chain of command in soliciting employees.

4. Consider the possibility of disclosing our PAC expenditures to shareholders.

Finally, I would like to take the liberty of presenting a different perspective on campaign finance than has yet surfaced in our discussions at Technidyne. Serious concerns about the escalating cost of congressional campaigns and about PAC dominance in campaign contributions have emerged over the past several national election cycles. In fact, a number of legislators have proposed modifications to current campaign-financing rules in an effort to deal with these issues. As I see it, most of the discussion about reform is focusing on a proposal by Senators David Boren and Robert Byrd, who want to reduce the current limit of PAC contributions from $5,000 per candidate per election to $3,500, and to increase the limit for individuals from $1,000 per candidate per election to $1,500. Another provision of their bill is to limit the total dollar amount of PAC contributions that can go to a single candidate for national office. Depending on the population of a state, Senate candidates would be restricted to accepting between $175,000 and $750,000. A companion proposal in the House would limit PAC contributions to all House candidates to $100,000.

Another aspect of the Boren-Byrd proposal is to offer partial public financing to all national congressional candidates. This arrangement would parallel our system of public financing for Presidential candidates from the two major political parties. Public funds are provided, up to a limit, as a match for campaign money candidates raise privately, and aggregate limits are placed on total campaign expenditures. For example, Senate candidates in our most populous states would be able to spend roughly $4 million, of which $3 million would be public funds. Much lower limits would apply to Senate candidates from less populated states. Public financing seeks to put all candidates on a level playing field.

I mention this bill and these specific provisions because they make a lot of sense to me, and I think they should be present in our deliberations. I personally think the whole system of campaign financing needs restructuring. These proposals seem to me to work in the right direction, and I believe that a viable strategy for Technidyne is to put our support behind the efforts of persons like Boren and Byrd to make the whole system fairer and less costly for all.

Business and Political Action Committees

Discussion Questions

1. As a citizen, do you think it makes any difference to society as a whole if Technidyne Corporation is politically active? Does it make any difference if it is not?

2. Several Technidyne managers express strong interest in PAC activity but hold different assumptions about how PAC funds should be allocated. What are the views of Sterling, Buckfiler, and Weber concerning PAC contributions? Taking the role of Dillings, which of their philosophies do you endorse, and why?

3. Do you agree with Weber's rationale for soliciting nonmanagement employees for corporate PAC contributions? As a lower-level, non-union employee, how would you respond to a request from your corporate PAC? Would your response be different if you were a union member? What specific guidelines might Dillings insist upon to avoid threats or reprisals connected with employee solicitations?

4. Legal counsel Lawrence Workmann cautions Dillings against assuming that corporate interests, as viewed by a small group of officers, would adequately represent the interests of shareholders and contributing employees. Do you agree with his proposal for expenditure disclosure? As a shareholder, what would you feel is necessary for the firm to comply with the principles of the *Bellotti* decision?

5. Respond to Buckfiler's interest in campaign advertising. What are the advantages and disadvantages of unlimited independent expenditures on political campaigns? Is there any merit in creating limitations on political advertising?

6. Respond to Workmann's support for reforming the rules of campaign finance. Specifically, evaluate (a) changing the allowed contributions for both individuals and PACs and (b) limiting total PAC contributions. Also, first as a manager and second as a citizen, how do you feel about the Boren-Byrd proposal for partial public financing for Senate candidates?

Student Response Form (anonymous)

(Do *not* put your name on this form.)

1. Generally, how do you like using case materials in your classes?

    Very much _____   Somewhat _____   Not at all _____

2. Taking all the cases that you read and treating them as a whole, please indicate your responses to the following questions:

   ITEM (Check *one* answer below.)

	ITEM	Very valuable / Definitely	Fairly valuable / Generally	Not too valuable / Not sufficiently / Only sometimes	Not valuable at all / Not at all
a.	How valuable were these cases in stimulating your interest and involvement in this course?	Very valuable	Fairly valuable	Not too valuable	Not valuable at all
b.	Did the memos provide enough facts to get you into the key issues of the case?	Definitely	Generally	Not sufficiently	Not at all
c.	Did the competing positions presented in the memos ring true to you as "real" managerial expressions?	Definitely	Generally	Not sufficiently	Not at all
d.	Do you feel that the Discussion Questions posed for you after the memos captured the essential factual and value issues managers should be considering?	Definitely	Generally	Only sometimes	Not at all
e.	How much do you feel that the reading and discussion of these cases improved your skills in decision making?	Definitely	Generally	Only sometimes	Not at all

3. For each of the cases that you read, please indicate your evaluation of its overall quality:

CASE	Poor	Fair	Good	Excellent
1. Medical Ethics (American Pharmaceutical Company)				
2. Sexual Discrimination (Computer Inventory Systems)				
3. Reproductive Risk (Longlife Rubber Company)				
4. Employee Protest (Great Rainbow Furniture Company)				
5. Testing Employees (United Companies)				
6. Due Process (All State Bank)				
7. "Big Brother" (Acme Office Products)				
8. New Information System (*The Call*)				
9. Medical Surveillance System (Advanced Chemicals)				
10. State Legislative Strategy (International Manufacturing Corporation)				
11. Selling High Technology Abroad (Omnibus Information Systems)				
12. Business and PACs (Technidyne Corporation)				

Return to: Editor, Ballinger Publishing Company, 54 Church Street, Cambridge, MA 02138.